# BIOFILM ERADICATION AND PREVENTION

# BIOFILM ERADICATION AND PREVENTION

## A Pharmaceutical Approach to Medical Device Infections

**TAMILVANAN SHUNMUGAPERUMAL**
Department of Pharmaceutical Technology
International Medical University
Kuala Lumpur, Malaysia

WILEY

**A JOHN WILEY & SONS, INC., PUBLICATION**

Published by John Wiley & Sons, Inc., Hoboken, New Jersey
Published simultaneously in Canada



***Library of Congress Cataloging-in-Publication Data:***

Shunmugaperumal, Tamilvanan.
Biofilm eradication and prevention : a pharmaceutical approach to medical device infections / Tamilvanan Shunmugaperumal.
p. ; cm.
Includes bibliographical references and index.
ISBN 978-0-470-47996-4 (cloth)
1. Medical instruments and apparatus–Microbiology. 2. Biofilms. 3. Medical instruments and apparatus–Sterilization. I. Title.
[DNLM: 1. Biofilms. 2. Anti-Infective Agents–therapeutic use. 3. Drug Carriers–therapeutic use. 4. Prostheses and Implants–microbiology. 5. Prosthesis-Related Infections–prevention & control. QW 90 S562 2010]
RA762.S58 2010
610.28′4—dc22

2010003433

10 9 8 7 6 5 4 3 2 1

To my beloved wife, Suriya Prapha, and to my son, Arunachalam.
The perseverance and tolerance of my spouse over the years when my eyes were glued on the computer screen, as well as the play-time sacrifice of my son, are highly appreciated.

—TAMILVANAN SHUNMUGAPERUMAL

# CONTENTS

# PREFACE

Microbial biofilms are microcosms attaching irreversibly to abiotic or biotic surfaces and are promulgated as congregates of single or multiple populations. Since there is an increased use of implanted medical devices, the incidence of these biofilm-associated diseases is increasing. Moreover, the nonshedding surfaces of these devices provide ideal substrata for colonization by biofilm-forming microbes. The consequences of this mode of growth are far-reaching. Microbes in biofilms exhibit increased tolerance toward antimicrobial agents and decreased susceptibility to host defense systems. Hence, biofilm-associated diseases are becoming increasingly difficult to treat. Not surprisingly, therefore, interest in biofilms has increased dramatically in recent years. The application of new microscopic and molecular techniques has revolutionized our understanding of biofilm structure, composition, organization, and activities, which result in important advances in the prevention and treatment of biofilm-related diseases.

This book can conveniently be divided into three parts depending on the biofilms' importance in the medical field and the necessity of eradicating them from forming over medical devices. Part I deals with the development and characterization of a biofilm onto the surfaces of implanted or inserted medical devices. Some of the specific answers concerning the reasons why biofilms form over medical device surfaces and what triggers biofilm formation are discussed. Part I consists of five chapters. Chapter 1 is an introduction to the subject matter of this book. A comprehensive overview on the subject matter is provided to the readers so that anyone with little knowledge of medical biofilms can acquire and understand the seriousness of the biofilm formed over the implanted or inserted medical devices. The rationale for biofilm eradication from medical devices is fully explored in Chapter 2. A range of medical devices used in modern day medical practices are categorically shown in order to see the discussed subject matter clearly. Here calculations are also given in terms of expenses to support biofilm prevention and eradication to patients who have been given medical devices to salvage the already lost normal function of organs present in their body. A need-based reason to look for an approach to prevent biofilm formation over medical devices is thus explained *per se*. Chapter 3 contains the stepwise development of biofilms onto the implanted or inserted medical devices. Hence, pathogenesis of device-related nosocomial infections starts with justifications from environmental, biochemical, physiological, and biomechanical points of view, and so on. Some details concerning the consequences of medical biofilms to cause a particular infection are also discussed in this chapter along with case studies. Chapter 4 focuses on how biofilm microbes build up resistance–tolerance against conventional antimicrobial agents when they are treating device-related nosocomial infections. This partially explains the need for alternate strategies to administer antimicrobial agents in order to prevent the development of tolerance by biofilm microbes, but at the same time to demonstrate the ramification of pharmaceutical knowledge on this subject matter. Chapter 5 briefly covers the important topic of studying and investigating medical biofilms using various analytical techniques. Starting from conventional plate counting and continuing through modern day electron microscopic and molecular methods, these are arranged by year of application in analyzing biofilms as they appeared in the reports published by different research groups across the world.

Although medically relevant biofilms develop commonly on inert surfaces, such as medical devices or on dead tissue (sequestra of dead bone), they can also form on living tissues, as in the case of endocarditis. Tissue samples taken from patients with dental caries, periodontitis, otitis media, biliary tract infections, and bacterial prostatitis also show the presence of bacterial microcolonies surrounded by an exopolymeric matrix (i.e., somehow biofilm-related). Therefore, these established infections could be termed as nondevice-related chronic infections. Part II elaborates on these types of biofilm-mediated chronic infections that occurred in various organs. Biofilm-related infections developed in ocular tissues, oral cavity, topical skin regions and lung with cystic fibrosis (CF) are selected to illustrate cases of potential interest. Chapter 6 studies biofilm-related infections occurring in both intra- and extraocular tissues. The usual way of treating these ocular diseases (i.e., topical application of aqueous-based and collyre-type eyedrops containing antimicrobial agents) are not so effective. Thus, it becomes necessary to discuss the potential of oil- and polymer-based nanocarriers (e.g., nanosized emulsions and nanoparticles) for eradicating the ocular infections found in this chapter. Chapter 7 incorporates biofilm-related infections occurring in the oral cavity. This always moist site should provide impetus to the development of chronic infections specifically due to the presence of biofilm-forming microbes in the oral cavity. Here too, treating with conventional antimicrobial agents would not produce a desired effect. Hence, shifting toward modern pharmaceutical approaches to treat infections of the oral cavity becomes more precocious. Some of these pharmaceutical approaches, although at the research level, are not far from becoming commercialized. Most promising antimicrobial agent-laden novel drug delivery systems are explored as case studies in this chapter. In addition, the formation of a biofilm over dental chair units, as well as dental waterliners in conjunction with currently developed prevention strategies including a centralized, automated dental hospital water quality and biofilm management system, are discussed.

The epidemic increase in obese humans worldwide is followed by a similar increase in diabetes and cardiovascular diseases. Such patients are particularly prone to the development of chronic wounds, which become colonized by a number of bacterial species. Chapter 8 presents a hypothesis aimed at explaining why venous leg ulcers, pressure ulcers, and diabetic foot ulcers develop into a chronic state. The lack of proper wound healing is at least in part caused by inefficient eradication of infecting, opportunistic pathogens, a situation reminiscent of chronic *Pseudomonas aeruginosa* infections found in patients suffering from CF. Implications of biofilm formation in chronic wounds and CF are shown here.

The introduction of novel drug delivery carriers in pharmaceutical sciences helps the physician to achieve the required therapeutic concentration of the drug at the diseased region of the body while minimizing drug exposure to nondiseased normal organs. To extend the pharmaceutical knowledge gained over the decades on novel drug delivery carriers to medical biofilm prevention and eradication, it has become necessary first to explore the already developed strategies for prevention of device-related nosocomial infections. Recommended technological and nontechnological strategies in conjunction with electrical, ultrasound, and photodynamic stimulation to disrupt biofilms by enhancing the efficiency of certain antimicrobial agents are shown. Thus, Chapter 9 begins with already available strategies followed by pharmaceutical approaches like the potential of lipid- and polymer-based drug delivery carriers for eradicating biofilm consortia on device-related nosocomial infections.

Liposomes loaded with antimicrobial agents could effectively be applied as an antibiofilm coating to reduce microbial adhesion–colonization onto medical devices, and as drug delivery carriers to biofilm interfaces in intracellular infection. All these applications are discussed in detail in Chapter 10.

Many polymer-based carrier systems also have been proposed, including those based on biodegradable polymers [e.g., poly(lactide-*co*-glycolide)], as well as fibrous scaffolds and thermoreversible hydrogels and surface (properties) modified polymeric catheter materials (e.g., antimicrobial, antiseptic, or metallic substances-coated polymeric materials). Their contribution to the prevention–resolution of infection is reviewed in Chapter 11. Additionally, the Chapter 12 explores an interesting topic of novel small molecules (e.g., iron and its complexes) to control medical biofilm formation.

Through these three parts of the book we intend to cover recent advances in pharmaceutical approachs to prevent medical device- and nondevice-related infections caused by biofilm-forming microorganisms. Many other approaches either within the pharmaceutical sciences or other allied disciplines are still in their rudimentary research stages. On the other hand, biofilm structural elucidation observed through different advanced analytical techniques is constantly progressing as usual. An intriguingly combined research based on knowledge derived both from the pharmaceutical approach and biofilm structural elucidation should contribute to a more efficacious way for biofilm prevention and eradication in the future. Nevertheless, further studies are warranted to translate knowledge on the mechanisms of biofilm formation into applicable therapeutic and preventive strategies.

# ACKNOWLEDGMENTS

My appreciation goes to bachelor and master degree students from Addis Ababa University, Ethiopia, Arulmigu Kalasalingam College of Pharmacy, Krishnankoil, Srivilliputtur, Tamil Nadu State, India, Sankaralingam Bhuvaneswari College of Pharmacy, Sivakasi, Tamil Nadu State, India, and International Medical University (IMU) SDN BHD, Kuala Lumpur, Malaysia, for their support and encouragement given during the preparation of this manuscript. Similarly, short-term financial support provided to me by the University of Antwerp, Belgium, to conduct preliminary research works on biofilm-related infections over medical devices is acknowledged.

TAMILVANAN SHUNMUGAPERUMAL

# PART I

# DEVELOPMENT AND CHARACTERIZATION OF BIOFILMS

CHAPTER 1

# INTRODUCTION AND OVERVIEW OF BIOFILM

## 1.1. INTRODUCTION

Any surface, whether synthetic or biomaterials, is primarily coated initially with local environmental constituents (e.g., water, electrolytes, and then organic substances). This conditioning film often exists before the arrival of any microorganisms onto the material surfaces. Indeed, the presence of water, electrolytes, and organic substances could give impetus for microbial growth and its further colonization onto the material surfaces *in vitro*. Subsequently, the presence of surface-bound microorganisms can provide a profound effect on the materials performance. If the material is meant for assisting in any course of medical treatment, then, it is essential that the biomaterial should be free from harmful microorganisms (e.g., bacteria, fungi, and protozoa).

The idea that bacteria grow preferentially on surfaces has come to the fore at regular intervals, for >150 years [1]. Steadily, throughout the history of microbiology, a very small proportion of microbiologists, by performing direct microscopic examinations, have found, however, that these free-floating or "planktonic" bacteria grow differently after they adhere to a surface and initiate biofilm formation. Moreover, in microbiology, knowledge has traditionally been gained from studies of suspensions of cells grown from a single cell in laboratory culture plates. These planktonic cells, for example, have been used in studies of how well antibiotics can kill bacteria. However, microbes can also aggregate themselves termed as biofilms (i.e., organized layers of cells attached

*Biofilm Eradication and Prevention: A Pharmaceutical Approach to Medical Device Infections*, By Tamilvanan Shunmugaperumal

to a surface). Therefore, over the past century, the study of microbial adhesion has generated a language all its own. Since many of these terms are still in use, a brief discussion of their meaning with reference to biofilm development would be apropos. The terms sessile and planktonic have evolved to describe surface-bound and free-floating microorganisms, respectively. The surface of interest to which sessile organisms are attached can be either abiotic (inert materials) or biotic (living tissue or cells) [2].

The definition of a biofilm has evolved over the last 35 years. In 1976, Marshall [3] noted the involvement of "very fine extracellular polymer fibrils" that anchored bacteria to surfaces. Costerton et al. [4] observed that communities of attached bacteria in aquatic systems were found to be encased in a "glycocalyx" matrix that was found to be polysaccharide in nature, and this matrix material was shown to mediate adhesion. In 1987, Costerton et al. [5] stated that a biofilm consists of single cells and micro colonies, all embedded in a highly hydrated, predominantly anionic exo-polymer matrix. Characklis and Marshall in 1990 [6] went on to describe other defining aspects of biofilms (e.g., the characteristics of spatial and temporal heterogeneity and involvement of inorganic or abiotic substances held together in the biofilm matrix). Again Costerton et al. in 1995 [7] emphasized that biofilms could adhere to surfaces and interfaces and to each other, including in the definition microbial aggregates and floccules and adherent populations within pore spaces of porous media. Costerton and Lappin-Scott [8] at the same time stated that adhesion triggered expression of genes controlling production of bacterial components necessary for adhesion and biofilm formation, emphasizing that the process of biofilm formation was regulated by specific genes transcribed during initial cell attachment. For example, in studies of *Pseudomonas aeruginosa*, Davies and Geesey [9] have shown that the gene (*algC*) controlling phosphomannomutase, involved in alginate (exopolysaccharide) synthesis, is upregulated within minutes of adhesion to a solid surface. Recent studies have shown that *algD*, *algU*, *rpoS*, and the genes controlling polyphosphokinase synthesis are all upregulated in biofilm formation and that as many as 45 genes differ in expression between sessile cells and their planktonic counterparts. Costerton et al. [10] defined a biofilm as "a structured community of bacterial cells enclosed in a self-produced polymeric matrix and adherent to an inert or living surface", whereas Carpentier and Cerf [11] simplify the concept as "a community of microbes embedded in an organic polymer matrix, adhering to a surface". Tamilvanan et al. [12] defined microbial biofilms as microcosm attaching irreversibly to abiotic or biotic surfaces and promulgated as congregates of single or multiple populations. Underlying each of these definitions are the three basic ingredients of a biofilm: microbes, glycocalyx, and surface. If one of these components is removed from the mix, a biofilm does not develop. A glycocalyx is the glue that holds the biofilm fast to the colonized surface and is a complex of exopolysaccharides of bacterial origin and trapped exogenous substances found in the local environment, including nucleic acids, proteins, minerals, nutrients, cell wall material, and so on [5]. Slime was a term used in 1940 [13] to describe a bacterial biofilm layer and resurrected in 1982 [14] to designate the glycocalyx produced by highly adherent strains of *Staphylococcus epidermidis* recovered from infected biomedical implants.

A current new definition for a biofilm must therefore take into consideration not only readily observable characteristics [i.e., cells irreversibly attached to a surface or interface, embedded in a matrix of extracellular polymeric substances (EPS) that these cells have produced, and including the noncellular or abiotic components], but also other physiological attributes of these organisms, including such characteristics as altered growth rate and the fact that biofilm organisms transcribe genes that planktonic organisms do not. The new definition of a biofilm is a microbially derived sessile community characterized by cells that are irreversibly attached to a substratum or interface or to each other, are embedded in a matrix of extracellular polymeric substances that they have produced, and exhibit an altered phenotype with respect to growth rate and gene transcription. This definition will be useful, because some bacterial populations that fulfilled the earlier criteria of a biofilm, which involved matrix formation and growth at a surface, did not actually assume the biofilm phenotype. These "nonbiofilm" populations, which include colonies of bacteria growing on the surface of agar, behave like planktonic cells "stranded" on a surface and exhibit none of the inherent resistance–tolerance characteristics of true biofilms. We can now speak of biofilm cells within matrix enclosed fragments that have broken off from a biofilm on a colonized medical device and now circulate in body fluids with all the resistance–tolerance characteristics of the parent community.

In nature, probably 99% or more of all bacteria exist in biofilms. For example, in an alpine stream there is typically only 10 bacteria $mL^{-1}$, whereas bacteria living in slimy biofilms on nearby rocks can occur in numbers like $5 \times 10^8$ $cm^{-2}$. Biofilms can form on various surfaces, including biotic surfaces (e.g., mucosal membranes, teeth), medical devices, and household surfaces. When a bacterium attaches to a hard surface in a moist environment, gene expression is adapted to the new environment. Some genes are upregulated, whereas others are depressed or turned off. Consequently, microbial biofilm systems are studied by many scientific disciplines (microbiology, ecology, immunology, biotechnology, engineering, medical microbiology) and across diverse research fields (environment, industry, medicine). Biofilms can be beneficial when they break down contaminants in soil and water as used in wastewater treatment, but can also cause severe problems in industrial settings, corroding everything from pipes in heating systems to computer chips or causing problems on the hulls of ships.

In the human host, biofilms exist as a community of sessile bacteria embedded in a matrix of EPS they have produced, which adhere to a foreign body or a mucosal surface with impaired host defense [15,16] or ample roughness [17]. Biofilm formation represents a protected mode of growth that allows microbes to survive in hostile environments and also disperse to colonize new areas. In addition, biofilm formation is an ancient and integral component of the prokaryotic life cycle and is a key factor for survival in diverse environments [18]. With the use of scanning electron microscopy (SEM) and confocal laser scanning microscope (CLSM), it became clear that biofilms are not unstructured, homogeneous deposits of cells and accumulated slime, but complex communities of surface-associated cells enclosed in a polymer matrix containing open water channels [19].

## 1.2. MICROBIAL ATTACHMENT SURFACES IN MODERN MEDICAL DEVICES

The materials meant for assisting in any course of medical treatment are simply called medical devices, and they are routinely being made up of many different types of surfaces. The medical device surfaces may act as a substrate for microbial attachment. In general, the substrate materials can be made with a variety of polymeric, ceramic, and metallic materials, as well as combinations of two or more of the same (e.g., hybrid materials). A "polymeric material" is a material that contains one or more types of polymers (e.g., from 50 or less to 75 to 90 to 95 to 97.5 to 99 wt%). Polymers are molecules containing multiple copies of one or more constitutional units, commonly referred to as monomers. Polymers may take on a number of configurations including linear, cyclic, and branched configurations. Homopolymers are polymers that contain multiple copies of a single constitutional unit, whereas copolymers are polymers that contain multiple copies of at least two dissimilar constitutional units, examples of which include random, statistical, gradient, periodic (e.g., alternating), and block copolymers. The term "monomers" may refer to free monomers and to those that are incorporated into polymers, with the distinction being clear from the context in which the term is used.

Examples of polymeric substrate materials include those that contain one or more suitable biostable or biodegradable polymers. Polymers of potential interest to develop medical devices are polycarboxylic acid polymers and copolymers including polyacrylic acids; acetal polymers and copolymers; acrylate and methacrylate polymers and copolymers (e.g., *n*-butyl methacrylate); cellulosic polymers and copolymers, including cellulose acetates, cellulose nitrates, cellulose propionates, cellulose acetate butyrates, cellophanes, rayons, rayon triacetates, and cellulose ethers (e.g., carboxymethyl celluloses and hydroxyalkyl celluloses); polyoxymethylene polymers and copolymers; polyimide polymers and copolymers (e.g., polyether block imides, polyamidimides, polyesterimides, and polyetherimides); polysulfone polymers and copolymers including polyarylsulfones and polyethersulfones; polyamide polymers and copolymers including nylon 6,6, nylon 12, polyether-block *co*-polyamide polymers (e.g., Pebax® resins), polycaprolactams and polyacrylamides; resins including alkyd resins, phenolic resins, urea resins, melamine resins, epoxy resins, allyl resins, and epoxide resins; polycarbonates; polyacrylonitriles; polyvinylpyrrolidones (cross-linked and otherwise); polymers and copolymers of vinyl monomers including polyvinyl alcohols (PVA), polyvinyl halides (e.g., poly(vinyl chloride (PVC)), ethylene-vinylacetate (EVA) copolymers, polyvinylidene chlorides, polyvinyl ethers (e.g., polyvinyl methyl ethers), vinyl aromatic polymers, and copolymers (e.g., polystyrenes), styrene-maleic anhydride copolymers, vinyl aromatic-hydrocarbon copolymers including styrene-butadiene copolymers, styrene-ethylene-butylene copolymers (e.g., a polystyrene-polyethylene–butylene-polystyrene (SEBS) copolymer, available as Kraton® G series polymers), styrene-isoprene copolymers (e.g., polystyrene-polyisoprene-polystyrene), acrylonitrile-styrene copolymers, acrylonitrile-butadiene-styrene copolymers, styrene-butadiene copolymers and styrene-isobutylene copolymers (e.g., polyisobutylene-polystyrene block copolymers (e.g., styrene-isobutylene poly isobutylene-polystyrene, SIBS), polyvinyl ketones, polyvinylcarbazoles, and polyvinyl esters (e.g., polyvinyl acetates); polybenzimidazoles; ionomers; polyalkyl oxide polymers and copolymers including polyethylene oxides (PEO); polyesters including polyethylene terephthalates, polybutylene terephthalates, and aliphatic polyesters (e.g., polymers and copolymers of lactide, which includes lactic acid, as well as D-,L-, and *meso*-lactide), ε-caprolactone, glycolide (including glycolic acid), hydroxybutyrate, hydroxyvalerate, *p*-dioxanone, trimethylene carbonate (and its alkyl derivatives), 1,4-dioxepan-2-one, 1,5-dioxepan-2-one, and 6,6-dimethyl-1,4-dioxan-2-one [a copolymer of polylactic acid (PLA) and polycaprolactone is one specific example]; polyether polymers and copolymers including polyarylethers (e.g., polyphenylene ethers, polyether ketones, polyether ether ketones); polyphenylene sulfides; polyisocyanates; polyolefin polymers and copolymers, including polyalkylenes (e.g., polypropylenes), polyethylenes (low and high density, low and high molecular weight), polybutylenes (e.g., polybut-1-ene and polyisobutylene), polyolefin elastomers (e.g., santoprene), ethylene propylene diene monomer (EPDM) rubbers, poly-4-methyl-pen-1-enes, ethylene-α-olefin copolymers, ethylene-methyl methacrylate copolymers and ethylene-vinyl acetate copolymers; fluorinated polymers and copolymers, including polytetrafluoroethylenes (PTFE), poly(tetrafluoroethylene-co-hexafluoropropene) (FEP), modified ethylene-tetrafluoroethylene (ETFE) copolymers, and polyvinylidene fluorides (PVDF); silicone polymers and copolymers; polyurethanes; *p*-xylylene polymers; polyiminocarbonates; copoly(ether-esters) (e.g., polyethylene oxide-PLA copolymers); polyphosphazines; polyalkylene oxalates; polyoxaamides and polyoxaesters (including those containing amines and/or amido groups); polyorthoesters; biopolymers [e.g., polypeptides, proteins, polysaccharides, and fatty acids (and esters thereof)], including fibrin, fibrinogen, collagen, elastin, chitosan, gelatin, starch, and glycosaminoglycans (e.g., hyaluronic acid).

A "ceramic material" is a material that contains one or more ceramic species (e.g., from 50 or less to 75 to 90 to 95 to 97.5 to 99 wt%). Specific examples of ceramic substrate materials are composed of the following: metal oxides, including aluminum oxides and transition metal oxides (e.g., oxides of Ti, Zr, Hf, Ta, Mo, W, Rh, and Ir); Si; Si-based ceramics [e.g., those containing silicon nitrides, silicon carbides, and silicon oxides (sometimes referred to as glass ceramics)]; calcium phosphate ceramics (e.g., hydroxyapatite); C and C-based, ceramic-like materials (e.g., carbon nitrides, etc.).

A "metallic material" is a material that contains one or more metals (e.g., from 50 or less to 75 to 90 to 95 to 97.5 to 99 wt%). Substrate examples that fall into this category include substantially pure metals (biostable metals, e.g., Au, Pt, Pd, Ir, Os, Rh, Ti, Ta, W, and Rh, and bioresorbable metals, e.g., Mg and Fe), metal alloys comprising Fe and Cr (e.g., stainless steels, including Pt-enriched radiopaque stainless steel), alloys comprising Ni and Ti (e.g., Nitinol), alloys comprising Co and Cr, including alloys that comprise Co, Cr, and Fe (e.g., elgiloy alloys), alloys comprising Ni, Co, and Cr (e.g., MP 35N) and alloys comprising Co, Cr, W, and Ni (e.g., L605), alloys comprising Ni and Cr (e.g., inconel alloys), and bioabsorbable metal alloys (e.g., Mg and Fe alloys, including their combinations with Ce, Ca, Zn, Zr, Li, etc.).

Depending on the substrate characteristics (hydrophilicity, hydrophobicity, surface charge, etc.), the material surfaces may vary in their ability to support microbial attachment and then biofilm formation [20–22]. Considering this fact, most of the research work was focused on comparative *in vitro* evaluation of microbial attachment onto different substrate materials. Table 1.1 shows a nonexhaustive list of substrate materials taken for this type of comparative evaluation studies. Rogers et al. [23] found that counts of attached *Legionella pneumophila* cells were highest on latex surfaces ($2.2 \times 10^5$ colony-forming

**TABLE 1.1. Nonexhaustive List of Substrate Materials Used for Comparative In Vitro Evaluation of Microbial Attachment**

| Biostable and biodegradable polymers |
|---|
| Glass |
| Ceramics |
| Mild steel |
| Stainless steel |
| Cu |
| Ag |
| Natural rubber latex |
| Si |
| Plastics |
| Polybutylene |
| Polyethylene |
| Polypropylene |
| Polyurethane |
| Ethylene–propylene |
| PVC |
| Chlorinated PVC |
| Unplasticized PVC |

units $cm^{-2}$) followed, in decreasing order, by ethylene-propylene, chlorinated PVC, polypropylene, mild steel, stainless steel, unplasticized PVC, polyethylene, and glass surfaces. These authors suggested that latex and other plastic surfaces leached nutrients into the surrounding growth medium, which was thought to encourage biofilm development [23]. Similarly, *Aeromonas hydrophila* preferentially colonized polybutylene followed by stainless steel surfaces, but only a few cells attached to Cu surfaces [22]. These authors proposed that *A. hydrophila* cells attached preferentially to hydrophobic surfaces (e.g., polybutylene), compared with more hydrophilic surfaces (e.g., stainless steel and Cu) [22]. In addition, Bakker et al. [24] reported that the attachment of three bacterial strains isolated from colonized medical devices (*S. epidermidis GB 9/6, Acinetobacter baumannii 2* and *P. aeruginosa*) decreased on various polyurethane surfaces with increasing surface free energy of the substrata. Surface topography may also play a role in bacterial biofilm formation. Reportedly, surfaces that are porous with rough-surface microtopography entrap more bacteria compared with smoother surfaces. For example, Gough and Dodd [25] reported that wooden chopping boards, especially those scored through use, retained more *Salmonella typhimurium* cells than smoother plastic chopping boards.

A high-glucose medium promotes the formation of biofilms [26], particularly of *Candida parapsilosis*, reflecting its potential to cause device-related infections in patients receiving parenteral nutrition [27]. Cell surface hydrophobicity correlates positively with *Candida* biofilm formation [28], and gentle shaking [29] also enhances biofilm formation. Note that all of these conditions are encountered *in vivo* (e.g., in the circulation and urinary system) as well, favoring biofilm formation when devices are inserted. An *in vitro* study showed that *C. parapsilosis*, *C. pseudotropicalis*, and *C. glabrata* produced significantly less biofilm on PVC disks than did the more pathogenic *C. albicans*, as determined by dry-weight, colorimetric, or radioisotope assays [26].

## 1.3. MICROBIAL ATTACHMENT SURFACES IN CHRONIC INFECTIONS *IN VIVO*

A typical example of a microbial biofilm is dental plaque, consisting of a well-defined surface (dental enamel), a matrix of polysaccharides (mainly dextran), and microbial cells (including *Streptococcus mutans*) [19]. However, not all biofilms fit this standard definition that easily. For example, in the case of mucosal biofilms in cystic fibrosis (CF) lungs [30] and otitis media [31], and for "amniotic fluid sludge" biofilms [32], the term "surface" has to be interpreted in a broader sense. In these biofilms, the thick mucous layer, which is essentially abiotic (i.e., nonliving material), provides anchorage for the microbial cells and acts as a surface for biofilm formation. The crystalline lens of the eye is the structure in charge of focusing objects at different distances from the eye. In order to focus, it must change its shape. With time, the lens loses its transparency and becomes opaque enough to impair vision. At this moment it is called a cataract. Cataracts are the main cause of blindness worldwide. Four out of 10 people >60-years old develop cataracts and >2 million patients are operated each year in the United States. Cataract surgery is a simple, relatively painless procedure carried out to regain vision. This surgery is the most frequently performed ophthalmologic procedure in the world, and also one of the most successful. For sight to be restored, the cloudy natural lens is replaced by a transparent artificial lens, called an implant or an intraocular lens (IOLs) (Fig. 1.1). Different biomaterials are used for IOLs. The first one was rigid poly(methyl methacrylate) (PMMA). Since the late 1970s, when the implantation of IOL became widespread due to improved technology, millions of PMMA IOLs were implanted worldwide. The PMMA can be native or treated on its surface in the case of heparinized or fluorine PMMA. With the development of phacoemulsification and the search for new IOL materials that would preserve the small cornea incision, foldable IOL appeared. The first one, made of a Si elastomer, was implanted in 1984 [33]. The biomaterials commonly used for IOL are Si, hydrophilic or hydrophobic acrylic, and hydrogel (Table 1.2).

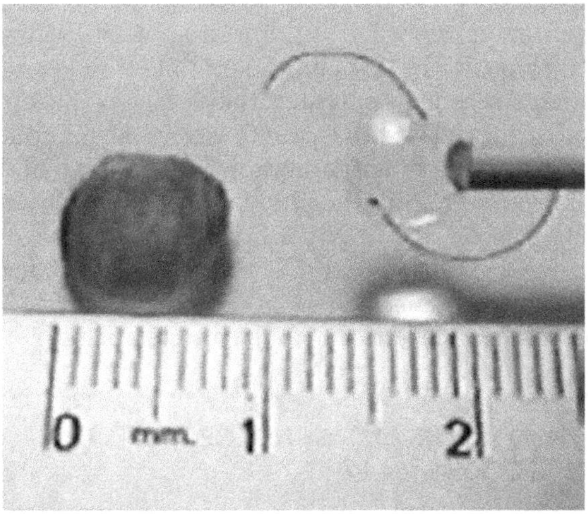


**Figure 1.1.** A cloudy natural lens (a cataract) and an artificial lens (an implant or an intraocular lens).

**TABLE 1.2. The Sterile Intraocular Lens (Inserts) Manufactured by Various Firms in France[a]**

| Material | Manufacturer |
|---|---|
| PMMA | Alcon, Paris |
| Si | AMO, Paris |
| Hydrophobic acrylic | Alcon, Paris |
| Hydrophilic acrylic | Bausch & Lomb, Paris |

[a]Taken from Baillif et al. [34].

However, the presence of a foreign body creates a risk of infection. Indeed, postoperative endophthalmitis following IOL implantation is one of the most dreaded complications of cataract surgery, giving rise to poor vision and sometimes blindness [35,36].

## 1.4. MICROBIAL BIOFILM FORMATION PROCESS

Now that we concede that bacteria form biofilms in essentially the same manner in whatever ecosystem they inhabit. Therefore, it is important that we take full advantage of the elegant studies of this process that fill the environmental and industrial microbiology literature. The scientific and engineering community has already examined biofilm formation in some detail and has published a couple of books [37,38] on this subject. Many aspects of biofilm formation are counterintuitive, and it may be useful to summarize these issues, so that the medical community does not repeat this work.

Perhaps the first surprise, for the medical community, is that bacteria form biofilms preferentially in very high shear environments (i.e., rapidly flowing milieus). Planktonic bacteria can adhere to surfaces and initiate biofilm formation in the presence of shear forces that dwarf those of heart valves and exceed Reynolds numbers of 5000 [37]. The Reynolds number is a dimensionless number describing the turbulent flow of a liquid; if this number is high, turbulent flow exists; if it is low, laminar flow conditions prevail. Engineers speculate that turbulent flow enhances bacterial adhesion and biofilm formation by impinging the planktonic cells on the surface, but whatever the mechanism, biofilms form preferentially at high-shear locations in natural and industrial systems.

Studies of bacterial adhesion with laboratory strains of bacteria, many of which had been transferred thousands of times and lost their ability to adhere, first indicated that very smooth surfaces might escape bacterial colonization. Subsequent studies with "wild" and fully adherent bacterial strains showed that smooth surfaces are colonized as easily as rough surfaces and that the physical characteristics of a surface influence bacterial adhesion to only a minor extent [7]. Once a biofilm has formed and the exopolysaccharide matrix has been secreted by the sessile cells, the resultant structure is highly viscoelastic and behaves in a rubbery manner [39]. When biofilms are formed in low-shear environments, they have a low tensile strength and break easily, but biofilms formed at high shear are remarkably strong and resistant to mechanical breakage.

The conversion of bacterial life, as free-floating planktonic forms, to complex sessile communities has been extensively investigated. The process is one that has emerged from billions of years of evolution and is likely to have multiple redundant pathways for its development. The local, low-concentration signal production and reception in cell–cell signaling systems is called quorum-sensing (QS) [40–42]. Quorum-sensing is a density-dependent cell-signaling mechanism and is one way by which bacteria "talk" to one another (i.e., a

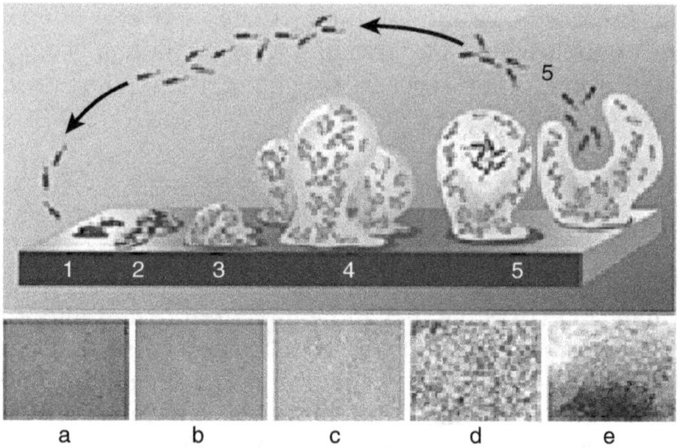


**Figure 1.2.** The biofilm life cycle: attachment, adhesion, aggregation, growth and maturation, and detachment. (Reprinted with permission from *Ann. Rev. Microbiol.*, 56, 187–209, 2002 [44].)

density-dependent mode of interbacterial signaling. Quorum-sensing is commonly associated with adverse health effects (e.g., biofilm formation, bacteria pathogenicity, and virulence).

From proteomic studies of *Pseudomonas*, five main steps of development have been established [43–45] (see Figs. 1.2 and 1.3). Hence, the formation of these microbial accretions is a dynamic five-step process as shown below in a flow chart [47].

**Flow chart illustrating the sequential phases involved during the formation of biofilm over material surfaces**

Surface conditioning → Reversible attachment → Irreversible attachment → Colonization → Detachment

## 1.5. SURFACE CONDITIONING

The first substances associated with the surface area of colonization may actually not be bacteria, but trace organics. These organics are thought to form a layer, which neutralizes excessive surface charge and surface free energy, which may prevent the initial bacterial approach, as it has been acknowledged that microorganisms attach more rapidly to hydrophobic, nonpolar surfaces [48–50]. Surfaces of attachment are thus conditioned by adsorption of organic and inorganic nutrients that influence subsequent bacterial attachment [51,52]. For example, Landry et al. showed that *P. aeruginosa* biofilms developed large cellular aggregates and had increased tolerance to the antibiotic tobramycin

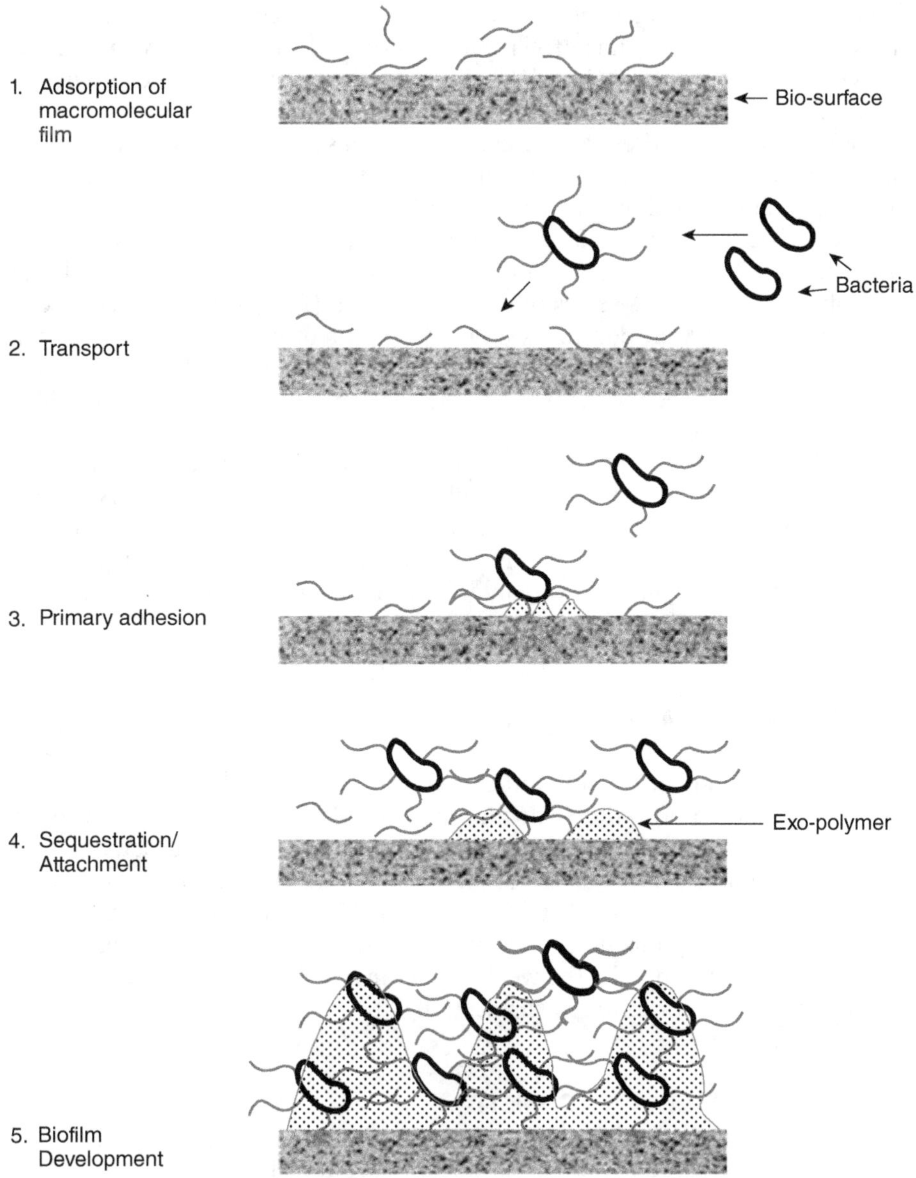


**Figure 1.3.** Schematic diagram of biofilm formation on microbiota or any biosurface with sequential steps. (Modified from Doyle [46].)

when grown on surfaces conditioned with the glycoprotein mucin, compared with corresponding biofilms grown on glass or surfaces coated with actin or deoxyribonucleic acid (DNA) [53]. Furthermore, these organic molecules often serve as nutrients for the attached bacteria. The rate of bacterial settling and association with the area of colonization also depends on the velocity characteristics of the surrounding liquid medium because individual cells in a liquid environment behave as particles [51]. The attachment of bacteria onto a surface initiates a cascade of changes. In fact, it has been shown that a whole different set of genes is triggered by cell attachment, which is responsible for the biofilm phenotype. A series of ribonucleic acid (RNA)-polymerase associated sigma factors that derepress a large number of genes have been implicated in this process [7,54]. In *P. aeruginosa* biofilms grown for 6 days, only 40% of the expressed proteins were identical to the planktonic form [45]. Moreover, *algD*, *algU*, *rpoS*, and genes controlling polyphosphokinase synthesis were found to be upregulated [55]. However, detailed studies of differential gene expression in *P. aeruginosa* biofilms using sophisticated DNA microarray technology showed that, as a percentage, genes that are differentially expressed in planktonic and biofilm cells are relatively few (1%) [56]. The phenotypic change is guided by an interbacterial communicating system called "QS" [57].

## 1.6. REVERSIBLE ATTACHMENT

Initial transport and reversible attachment of bacterial cells to a surface can occur by sedimentation and Brownian motion of microbial cells, convection currents within a bulk liquid transporting bacteria to the surface, active movement by motile bacteria, or electrostatic and physical interactions between the bacterial cell surface and substratum [58,59]. The reversible adhesion state may result in an equilibrium distribution between adhering and suspended cells, and is considered to be the weakest link in the chain of events connecting bacterial cells to a conditioned surface [58,60].

## 1.7. IRREVERSIBLE ATTACHMENT

Bacterial cells attached reversibly to surfaces produce EPS due to stimulation of membrane-bound sensory proteins of the bacterial cell [61], which allows for the development of cell–cell bridges that, in turn, cement the cells to the surface [18,51]. For example, alginate, the EPS produced by biofilms of *P. aeruginosa* in CF infections, is reportedly integral in biofilm development and "cementing" *P. aeruginosa* cells to surfaces [9,18].

Quorum-sensing employs the use of small, diffusible molecules, members of the class of N-acylated homoserine lactones, which are released by biofilm bacteria into their local environment, where they can interact with neighboring cells [62]. Furthermore, the QS systems rely on self-generated signaling molecules to coordinate gene expression in response to population density. The majority of signalling molecules identified thus far can be classified into three main groups: acylhomoserine lactones (AHLs), oligopeptides, and the LuxS–autoinducer 2 [63]. Nevertheless, the types of chemicals associated with cell–cell signaling represent an ever-expanding collection of molecules that are structurally quite diverse. The QS is crucial in determining the density of the bacterial population, and it increases locally as more bacteria attach. Regulation of this type coordinates bacterial behavior at the population level [62]. At this stage, attachment is reversible because it is based on electrostatic attraction rather than chemical bonds. However, some of the cells form structures for firmer anchoring, thus advancing in the second step of biofilm formation, the irreversible adhesion. This step requires the mediation of bacterial surface proteins, the cardinal of which is similar to *S. aureus* autolysin and is denominated AtlE [15].

The Gram-negative bacterium *Pseudomonas aeruginosa* has become a model organism for independently studying these two social phenomena, namely, QS and biofilm formation. In a seminal investigation in 1998, Davies et al. [57] discovered a link between them. Quorum-sensing was reported to be required for elaboration of mature, differentiated *P. aeruginosa* biofilms. Since that time exhaustive effort has been directed toward uncovering the mechanism(s) by which QS regulates biofilm production. One of the end goals being that elucidating the pathways of biofilm development will make it possible to control their formation. Over the past decade, significant strides have been made toward understanding biofilm development in *P. aeruginosa*. We now have a much clearer picture of the mechanisms involved (see the very recent review by de Kievit [64] for complete details). An underlying message that is emerging is that development of these sessile communities may proceed by many different pathways. A model of *P. aeruginosa* biofilm development is depicted in Fig. 1.4, with connections to QS indicated [64].

## 1.8. COLONIZATION

The final stage in biofilm establishment is surface colonization [52]. Attached bacteria grow and divide, forming microcolonies that are considered to be the basic organizational units of a biofilm [65]. Entrapment of other planktonic cells in the EPS also occurs, resulting in the formation of a biofilm [58]. The colonization of a surface by one bacterium (i.e., primary colonizers) is also often found to influence the attachment of others to the same surface (i.e., secondary colonizers). For example, the plaque bacterium, *Streptococcus cristatus*, reportedly produces a 59-kDa surface protein that specifically inhibits attachment and biofilm development in *Porphyromonas gingivalis* and *Prevotella intermedia* [66,67]. The completed biofilm has a complex architecture, consisting of biofilm bacteria in EPS enclosed microcolonies

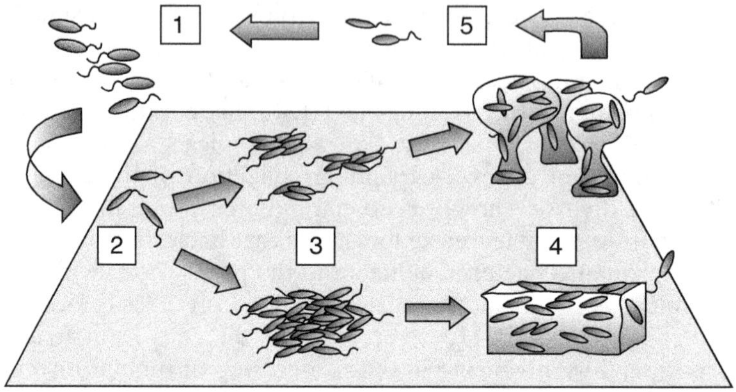


**Figure 1.4.** *Pseudomonas aeruginosa* biofilm development. Planktonic cells (stage 1) attach onto a solid surface (stage 2) and microcolonies are formed (stage 3). Under conditions that promote bacterial migration (e.g., succinate, glutamate), cells will spread over the substratum, ultimately developing into a flat, uniform mat (stage 4). Under motility-limiting conditions (e.g., glucose), the microcolonies proliferate forming stalk- and mushroom-like structures (stage 4). At various points throughout biofilm maturation, cells detach and resume the planktonic mode of growth (stage 5). The QS controlled rhamnolipid production impacts microcolony formation (stage 3), maintenance of open channels (stage 4), mushroom cap formation (stage 4), and dispersion from the biofilm (stage 5). In addition, production of Pel polysaccharide and DNA release, which are both important for the EPS matrix (stages 3 and 4), are under QS control (Reproduced with permission from de Kievit *Environ. Microbiol.*, 11, 279–288, 209, Wiley Interscience [64]).

interspersed with less dense regions of the matrix that include highly permeable water channels carrying nutrients and waste products [65,68].

The aggregation of bacteria and the production of the EPS represent the third step of biofilm formation. In staphylococci, the EPS matrix is a polymer of β-1, 6-linked *N*-acetylglucosamine, whose synthesis is mediated by the *ica* operon [15]. The chemistry of EPS, in general, is quite complex and includes polysaccharides, nucleic acids, and proteins [69,70]. The EPS polysaccharides differ between Gram-positive and -negative bacteria. In the latter, bacteria polysaccharides are neutral or polyanionic. By contrast, Gram-positive bacteria have primarily cationic polysaccharides [69]. The composition and structure of polysaccharides determine the primary EPS conformation [69].

Step 4 of the process is the maturation of the biofilm structure. The latter includes cell growth (and potential reproduction) within a given microenvironment, as determined by exopolysaccharide substances, neighboring cells, and proximity to a water channel [71]. The open water channels represent a primitive circulatory system for the preservation of homeostasis within the biofilm. In the mature biofilm, more volume is being occupied by the EPS matrix

(70–95%) than by bacterial cells (5–25%) [72]. At this stage, secondary colonizers (other bacteria or fungi) can become associated with the biofilm surface [73].

Finally, bacteria can be detached from the biofilm (step 5) either by external forces or as a part of a wavelike migrating physical movement [74] or even as a self-induced process to disseminate to the environment. Even though biofilm dispersion is an almost untouched area of research, it has been reported that the RNA binding protein *CsrA* acts as an activator of biofilm dispersal in *Escherichia coli* by way of regulation of intracellular glycogen biosynthesis and catabolism [75].

## 1.9. DETACHMENT

It was initially suggested that turbulent shear forces may be responsible for detachment of clumps of biofilm cells and subsequent transfer to other surfaces for attachment. This type of detachment mechanism only seems to be accurate for biofilms that are grown under laminar shear forces and are more likely to detach when shear forces become more turbulent [76]. However, recent studies have suggested that detachment, often termed "dispersion" or "dissolution", is an active process that is highly regulated by the attached cell populations [77]. Several strategies have been suggested regarding how biofilm bacteria disseminate into other areas for further surface colonization.

- One such proposal suggests that cells located at the periphery of the biofilm are released into the surrounding environment, return to the planktonic state, and find new surfaces for biofilm development. This is an active process that is regulated by the attached cell populations, and is often termed "dispersion" or "dissolution" [77]. This strategy is adopted by *P. aeruginosa*, which produces an alginate lyase enzyme that, in turn, dissolves the alginate matrix, releasing cells into the surrounding environment [18].
- A further mechanism of dispersion has been shown in bacterial cells that display "swarmer" motility. For example, *Proteus mirabilis*, the causative agent in many urinary and catheter associated infections, differentiates into swarmer cells once attached to surfaces in biofilms, and migrates over catheter tubing. In some cases, *P. mirabilis* transports other bacterial species by this process [78,79]. Furthermore, differentiated *P. mirabilis* swarmer cells are associated with virulence and invasion of host cells [79].
- Gliding bacteria attached to surfaces (e.g., *Myxococcus Xanthus*), have been speculated to produce a slime trail that may facilitate motion over surfaces [80].
- It has also been proposed that certain bacteria may alter various surface components (e.g., glycolipids, peptidolipids, lipopolysaccharides, and

lipoteichoic acids), which may, in turn, alter cell surface hydrophobicity, facilitating release of a surface-bound cell. Reportedly, *E. coli* and *P. aeruginosa* cells were increased from biofilms by increasing their cell surface hydrophobicity [66].

- Quorum-sensing (cell–cell signaling) has also been suggested to mediate the exodus of cells from crowded biofilms of *Serratia* spp. [81].

## 1.10. FUNGAL BIOFILM FORMATION

Fungi are organisms that lack chlorophyll, but resemble plants. These organisms are saprophytic, but can also utilize living matter. Fungi are subdivided into yeasts, which are unicellular and molds that are multicellular with filamentous hyphae.

Fungal biofilm formation is a complex and diverse phenomenon. *Candida* species are emerging as important nosocomial pathogens, and an implanted device with a detectable biofilm is frequently associated with these infections [82]. The evidence linking *Candida* biofilms to device-related infections is growing as more standardized methods for evaluating *Candida* biofilms *in vitro* emerge. *Candida albicans* biofilm formation has been studied more extensively than biofilms of other *Candida* species. *Candida albicans* biofilm formation has three developmental phases: adherence of yeast cells to the device surface (early phase), formation of a matrix with dimorphic switching from yeast-to-hyphal forms (intermediate phase), and increase in the matrix material taking on a three-dimensional (3D) architecture (maturation phase) [83,84]. Fully mature *Candida* biofilms have a mixture of morphological forms and consist of a dense network of yeasts, hyphae, and pseudohyphae in a matrix of polysaccharides [83], carbohydrate, protein, and unknown components. For a comprehensive review of fungal biofilm formation onto various medical devices, the readers are referred to Ref. [85], which deals with the formation and structure of *Candida* biofilms, and the influence of the nature of the contact surface, environmental factors, *Candida* morphogenesis, and the *Candida* species involved.

## 1.11. DEVICE-RELATED NOSOCOMIAL INFECTIONS

The term nosocomial was derived from two Greek words Noscos (disease) and Komeion (to take care of). Nosocomial infections are otherwise called Hospital acquired infections (HAI). The earliest available advice on hospital construction and hygiene is contained in the Charaka-samhita, a Sanskrit notebook of medicine, which was probably written in the fourth century BC. The HAI do have a significant impact on medical costs, hospital stay, and mortality among patients.

Though the tendency for patients with indwelling medical devices to develop infections has reportedly been known since the 14th century [86], the use of surgically implanted or nonsurgically inserted medical devices has received an escalating interest in modern medical practices. This is due to a result of their beneficial effect on quality of life and in some circumstances, on patient survival rates. Therefore, the insertion of indwelling or implanted foreign polymer bodies (e.g., prosthetic heart valves, cardiac pacemakers, total artificial hearts, and total joint replacements or other orthopedic devices), as well as intravascular catheters, renal dialysis shunts, cerebrospinal fluid (CSF) shunts, or continuous ambulatory peritoneal dialysis catheters, has become an integral and indispensable part of modern medical care. Exemplary implantable or insertable medical devices for use contain stents (including coronary vascular stents, peripheral vascular stents, cerebral, urethral, ureteral, biliary, tracheal, gastrointestinal, and esophageal stents), stent coverings, stent grafts, vascular grafts, catheters (urological or vascular catheters, e.g., balloon catheters and various central venous catheters), guide wires, balloons, filters (e.g., vena cava filters and mesh filters for distil protection devices), abdominal aortic aneurysm (AAA) devices (e.g., AAA stents, AAA grafts), vascular access ports, dialysis ports, embolization devices including cerebral aneurysm filler coils (including Guglilmi detachable coils and metal coils), embolic agents, hermetic sealants, septal defect closure devices, myocardial plugs, patches, pacemakers, lead coatings including coatings for pacemaker leads, defibrillation leads, and coils, ventricular assist devices including left ventricular assist hearts and pumps, total artificial hearts, shunts, valves including heart valves and vascular valves, anastomosis clips and rings, cochlear implants, tissue bulking devices, and tissue engineering scaffolds for cartilage, bone, skin, and other *in vivo* tissue regeneration, sutures, suture anchors, tissue staples and ligating clips at surgical sites, cannulae, metal wire ligatures, urethral slings, hernia "meshes", artificial ligaments, orthopedic prosthesis and dental implants, among others. While medical devices are increasingly used in almost all fields of medicine for diagnostic and/or therapeutic procedures (see also in Table 1.3 for a very specific overview), they are particularly necessary for managing the care of critically ill patients. Implantable and insertable medical devices are used for systemic treatment, as well as for the localized treatment of any mammalian tissue or organ. Nonlimiting and particular examples are tumors; organs including the heart, coronary and peripheral vascular system (referred to overall as "the vasculature"), the urogenital system, including kidneys, bladder, urethra, ureters, prostate, vagina, uterus and ovaries, eyes, ears, spine, nervous system, lungs, trachea, esophagus, intestines, stomach, brain, liver and pancreas, skeletal muscle, smooth muscle, breast, dermal tissue, cartilage, tooth and bone.

In general, the placement of a vascular access with increasingly sophisticated catheters is widely used. A considerable number of patients have one or more vascular catheters in place during their hospital stay [87]. In the United States, 15 million central venous catheter (CVC) days (i.e., the total number of days of exposure to CVCs by all patients in the selected population during

**TABLE 1.3. Implanted Medical Devices**

| |
|---|
| *Intravascular* |
| Peripheral catheters (venous, arterial) |
| Midline catheters |
| Central venous catheters |
| nontunneled catheters (Cook, Arrow) |
| tunneled catheters (Hickman, Broviac, Groshong) |
| Pulmonary artery catheters |
| Totally implanted ports (Port-a-Cath, MediPort, Infusaport) |
| *Cardiovascular* |
| Mechanical heart valves |
| Implantable defibrillators and related devices |
| Vascular grafts |
| Ventricular assist devices |
| Coronary stents |
| Implantable patient monitors |
| *Neurosurgical* |
| Ventricular shunts |
| Ommaya reservoirs |
| Intracranial pressure devices |
| Implantable neurological stimulators |
| *Orthopedic* |
| Joint prostheses and other reconstructive orthopedic implants |
| Spinal implants |
| Fracture-fixation devices |
| *Urological* |
| Inflatable penile implants |
| *Gynaecological* |
| Breast implants |
| *Otolaryngological* |
| Cochlear implants |
| Middle-ear implants |
| *Ophthalmological* |
| Intraocular lenses |
| Glaucoma tubes |
| *Dental* |
| Dental implants |

the selected time period) occur in intensive care units (ICUs) each year [88]. However, the use of foreign material has led to special complications associated with the presence of such material because insertion or implantation of medical devices is associated with a definitive risk of bacterial and fungal infections, that is, foreign body-related infections (FBRI). Upon implantation or insertion into the patient's body for exerting the intended purpose, like salvage of normal functions of vital organs, these medical devices are unfortunately becoming the sites of competition between host cell integration and microbial adhesion [89]. Thus, clinicians who deal with device-related and other chronic bacterial infections have gradually defined a new category of infectious disease that differs radically from the acute epidemic bacterial diseases that predominated until the middle of thc last century [90].

Foreign body-related infections comprise all entities with respect to local and bloodstream infections (BSI) associated with inserted or implanted medical devices. On the subject of catheter-related infections (CRI), the confusing terminology and varying definitions in the medical literature have historically been a barrier to effective communication [91,92]. The definitions of CRI and their microbiological criteria are given in Table 1.4 [93–96]. The CRI

**TABLE 1.4. Definition of Intravascular Catheter-Related Infections**[a]

| Type of Infection | Definition |
|---|---|
| Catheter colonization | Cultured catheter segment yields a significant number of bacteria according to the culture methods used (in the absence of any clinical signs of infection at the insertion site) |
| Localized catheter infection (exit site infection) | Infection at the insertion site: periorificial cellulitis, purulence, erythema, tenderness, induration, tunnelities, and pocket infections (for totally implanted devices) |
| Catheter-related bloodstream infection[b] | Isolation of the same microorganism[c] from a (semi) quantitative culture of the distal catheter segment and from the blood of a patient with clinical symptoms of sepsis and no other apparent source of infection.<br>Defervescence after removal of an implicated catheter from a patient with primary bloodstream infection (indirect evidence in the absence of catheter culture) |

[a]See Refs. [93–96].

[b]This term is preferred to the term "catheter-related sepsis" because "sepsis" does not imply the presence of bacteraemia and because this is used to define the systemic inflammatory response syndrome associated with a septic focus. "Catheter-related bacteraemia" is less accurate as blood cultures may grow fungal species (fungaemia) [93].

[c]That is, clonally identical isolates of the same species (ideally proven by genotyping techniques, practically at least by antibiogram).

include colonization of the device, localized catheter infections (exit site, pocket, tunnel infection), and catheter-related bloodstream infection (CRBI) [93–96]. In the absence of a standard reference technique ("gold" standard), microbiological diagnostics of FBRI are still a matter of debate [97–101]. There are two major sources of CRBI: (1) colonization of the intravenous devices (IVD), or "catheter-related infection," and (2) contamination of the fluid administered through the device, or "infusate-related infection" [102]. Contaminated infusate is the cause of most epidemic CRBI and this source has been reviewed elsewhere [103]. In contrast, catheter-related infections are responsible for most endemic CRBI and constitute the focus of this book.

For microorganisms to cause CRI, they must first gain access to the extraluminal or intraluminal surface of the device, where they can adhere and become incorporated into a biofilm that allows sustained infection and hematogenous dissemination [104]. Microorganisms gain access by one of the three following mechanisms: (1) skin organisms invade the percutaneous tract, probably facilitated by capillary action [105], at the time of insertion or in the days afterward; (2) microorganisms contaminate the catheter hub (and lumen) when the catheter is inserted over a percutaneous guidewire or when it is later manipulated [106]; or (3) organisms are carried hematogenously to the implanted IVD from remote sources of local infection (e.g., a pneumonia) (Fig. 1.5) [107,108].

With short-term IVD (i.e., those in place <10 days) like peripheral intravenous catheters, arterial catheters, and noncuffed, nontunneled CVC, most

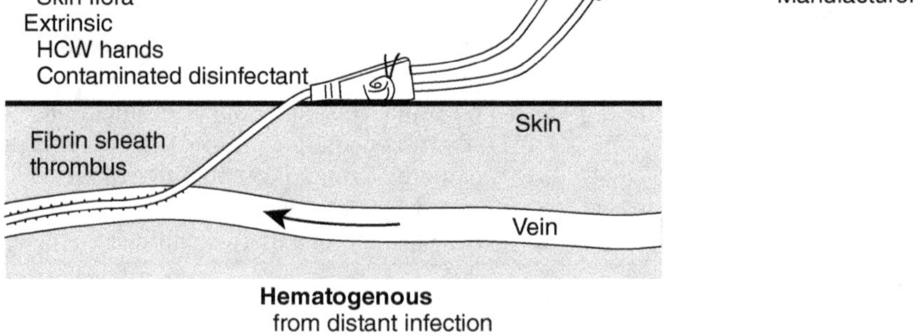


**Figure 1.5.** Potential sources of infection of a percutaneous IVD: the contiguous skin flora, contamination of the catheter hub and lumen, contamination of infusate, and hematogenous colonization of the IVD from distant, unrelated sites of infection. (HCW-healthcare worker.)

CRBI are of cutaneous origin, are from the insertion site, and gain access extraluminally [109–111] and, occasionally, intraluminally. In contrast, contamination of the catheter hub and lumen appears to be the predominant mode of BSI with long-term, permanent IVD (i.e., those in place ≥10 days), such as cuffed Hickman- and Broviac-type catheters, cuffed hemodialysis CVC, subcutaneous central venous ports, and peripherally inserted central catheters [112–115].

As stated above, microbial adherence and biofilm production proceed in two steps: first, an attachment to a surface and, second, a cell-to-cell adhesion, with pluristratification of microorganisms onto the biomaterial surfaces-like medical devices. Though development of device-related nosocomial infections begins with colonization of the medical device material, followed by a complex metamorphosis by the microorganisms with resultant biofilm formation, the microbial source for these infections may be acquired either from indigenous microbiota *in vivo* or from those micros present in exogenous sources *in vitro* [116]. The contamination of the medical device most likely occurs by inoculation with only a few microorganisms from the patient's skin or mucous membranes during implantation. Sometimes, the pathogens may also be acquired from the hands of the surgical or clinical staff. According to the underlying patient characteristics, the microorganisms that are implicated, and the type of device, morbidity and mortality of device-associated infections may vary. But FBRI, particularly CRI, significantly contribute to the increasing problem of nosocomial infections [87,117]. It also has been suggested that hospital water distribution systems are one of the most overlooked, important, and controllable sources of HAI [118].

According to Cozad and Jones [119], environmental surfaces could harbor pathogenic organisms that are able to cause infectious diseases. The Centers for Disease Control and Prevention (CDC) in their *Guidelines for Environmental Infection Control in Health-Care Facilities* pinpoints the evidence of nosocomial infections associated with fomites [120]. Brady et al. [121] emphasized the increase of the role of contaminated surfaces as a potential source for infectious diseases. Furthermore, Richards et al. [122] reported that at least half of all cases of nosocomial infections are associated with medical devices. The majority of device-related infections, thus, occur in the early postoperative period and likely are due to contamination at the time of implantation or insertion. Even though the chances of obtaining tailor-made medical devices in a sterile package are highly possible, the maintenance of the sterility of medical devices especially after opening from their package during implantation or insertion particularly in hospital premises is somehow always speculative. Because of the eventual dissemination of blood, saliva, urine, and other secretions due to routine healthcare practices, the microorganisms could be spread throughout the environment even after the adaptation of some essential precautionary–preliminary disinfection and sterilization procedures within the hospital premises. It was Hungarian obstetrician Ignaz Semmelweis in 1847 who formulated the first disinfectant by using chlorinated lime to wash hands, but he lost his job as a result of his discovery. His bosses at the Vienna General Hospital were not impressed that his findings implicated them in many (unintentional) deaths. In 1869, Dr. Joseph Lister invented a pump to spray carbolic acid into the air in operating theatres. The chemical killed the bugs and kept patients safe. Dr. Lister did not discover a new drug, but he did make the connection between lack of cleanliness in hospitals and death after operations. For this reason, he is known as "Father of Antiseptic Surgery".

The medical consequences of device-related infections can be disastrous; they include potentially life-threatening systemic infections and device malfunction that may require device removal, often complicated by tissue destruction. A further complication that may be associated with urological medical devices is encrustation (biofilm mineralization), a phenomenon that frequently results in impairment of urine patency [123]. For example, in one study, it was reported that blockage of the lumen of ureteral stents due to the presence of encrustation occurred in ~75% of stents by 24 weeks [124]. Additional complications of encrustation on urinary medical devices include pain and tissue trauma upon device removal and the harboring of bacteria within the crystalline encrustation, which may then become a focus of infection [123,125]. Hence, device-related infections are important to understand because of their association with increased morbidity, mortality, additional hospital cost to patient, blocking encrustation, and mechanical failure (fracture) [86,123]. Therefore, owing to the problems associated with the use of medical devices, there is a clinical need for the development of novel materials [126] and novel coatings including diamond-like carbon (DLC) coatings over existing materials that will offer resistance to infection and encrustation [127–130], design of biomaterials with antimicrobial surfaces through some new surface modification techniques [131], and finally the use of novel approaches to deliver antimicrobial agents for eradicating the biofilm consortia from medical devices (as described in this book). Nevertheless, biofilm-associated bacteria are large enough to defeat the immune system [132], and have been protected from complement-mediated opsonic factors and phagocytic cells. Structurally, biofilms are composed primarily of surface-adherent microbial consortia encased in a matrix of EPS. The components of EPS of biofilm can also modulate the cellular immune responses [132].

A number of reviews have been solely devoted to the subject of medical device-related infections [18,51,133–138]. In particular, various pharmaceutical strategies are reviewed to prevent ventilator-associated pneumonia (VAP), which refers to a subset of nosocomial pneumonia occurring in patients in whom the pneumonia was neither present nor incubating at intubation and who have been receiving mechanical ventilation via an endotracheal tube [139]. Furthermore, biofilms are a major concern in the field of medical, pharmaceutical, and biosciences especially in the case of a variety of biomaterial-centered infections in human [140]. Fungal, protozoal, and bacterial biofilms have been found on a variety of indwelling devices removed from patients with associated biomaterial-centered infections [141–144]. Although device-related infections are most frequently caused by bacterial pathogens (e.g., *Staphylococcus epidermidis* and *aureus*) fungal etiologies are uncommon and carry an exceedingly high mortality rate [85,145].

A comprehensive list of common causative organisms of biofilm to produce device related nosocomial infections are shown in Table 1.5. On the other hand, it is believed that >60–65% of all microbial infections in humans are caused by biofilms [18,146,147]. Infections ascribed to biofilms include common diseases (e.g., childhood middle-ear infection and gingivitis); infections of all known indwelling devices (e.g., catheters, orthopaedic prostheses, and heart valves); and biofilm infections also occur in sufferers of incurable CF. Although, the biofilm formation is thought to be the concern of industrial and environmental microbiologists who are interested in phenomena such as biofouling [148–152], the microbial biofilms are unequivocally responsible for the recalcitrance of many infections to conventional antimicrobial therapy [151,152]. In other words, clinical failure is often due not to infections with bacteria harboring mechanisms resulting in high-level antibiotic resistance–tolerance, but rather to bacteria that are phenotypically resistant *in vivo* [153,154]. In fact, the microbial biofilms are known to be involved in persistent sources of infection. Persisters are specialized cells that have evolved to survive all possible natural threats. Over one-half of a century has passed since the discovery of drug-tolerant persisters, but their study is still an emerging field. The presence of persisters in biofilms provides an important incentive to understand their nature. Recent advances in isolating persisters, determining their transcriptome, and finding candidate persister genes are hopeful indications that the pace of progress in understanding this elusive problem is picking up. Formidable obstacles remain, due to difficulty in isolating sufficient amounts of persister cells, the apparent redundancy of persister genes, and the temporary phenotype of these cells. Furthermore, the mechanism of drug tolerance appears to be mechanistically distinct from resistance and is based on shutting down antibiotic targets (see Chapter 4 for details).

The objectives of Part I are (1) to present a rationale for biofilm eradication from modern medical devices followed by a short overview on pathogenesis of device-related infections (2) to give an outline of biofilm resistance–tolerance to conventional antimicrobial agents, and (3) to explore various analytical techniques used for biofilm identification and characterization.

**TABLE 1.5. A Comprehensive, but Not Exhaustive, List of Common Causative Organisms of Biofilm Responsible to Produce Device Related Nosocomial Infections**

| Names of Pathogenic Organisms Including Bacteria, Fungi, and Yeast |
|---|
| *Staphylococcus aureus* |
| *Staphylococcus epidermidis* |
| *Staphylococcus haemolyticus* |
| *Streptococcus mutans* |
| Coagulase-negative *staphylococci* (CoNS) |
| *Enterococcus faecalis* and *E. faecium* |
| *Burkholderia cepacia* |
| *Pseudomonas aeruginosa* |
| *Corynebacterium* spp. |
| *Propionibacterium* spp. |
| *Bacillus* spp. |
| *Micrococcus* spp. |
| *Enterobacter* spp. |
| *Serratia* spp. |
| *Bacteroides fragilis* |
| *Mycobacterium fortuitum* and *M. chelonei* |
| *Escherichia coli* |
| *Proteus mirabilis* |
| *Klebsiella pneumoniae* |
| *Acinetobacter baumannii* |
| *Stenotrophomonas maltophilia* |
| *Candida albicans* |
| Non-albicans *Candida* spp. |
|   *C. parapsilosis* |
|   *C. glabrata* |
|   *C. krusei* |
|   *C. tropicalis* |
|   *C. guillermondii* |
|   *C. dubliniensis* |
|   *C. lusitaniae* |
| HACEK group of organisms |
|   *Haemophilus aphrophilus* |
|   *H. paraphrophilus* |
|   *Actinobacillus actinomycetemcomitans* |
|   *Cardiobacterium hominis* |
|   *Eikenella corrodens* |
|   *Kingella kingae* |
| *Malassezia* spp. |
| *Rhodotorula* spp. |
| *Hansenula anomala* |
| *Aspergillus* spp. |

## REFERENCES

1. Costerton, J.W. (1999), Introduction to biofilm, *Int. J. Antimicrob. Agents*, 11, 217–221; discussion 237–239.
2. Dunne, Jr., W.M. (2002), Bacterial adhesion: seen any good biofilms lately? *Clin. Microbiol. Rev.*, 15, 155–166.
3. Marshall, K.C. (1976), *Interfaces in microbial ecology*, Harvard University Press, Cambridge, MA, pp. 44–47.
4. Costerton, J.W., Geesey, G.G., and Cheng, G.K. (1978), How bacteria stick, *Sci. Am.*, 238, 86–95.
5. Costerton, J.W., Cheng, K.-J., Geesey, G.G., Ladd, T.I., Nickel, J.C., Dasgupta, M., and Marrie, T.J. (1987), Bacterial biofilms in nature and disease, *Annu. Rev. Microbiol.*, 41, 435–464.
6. Characklis, W.G. and Marshall, K.C. (1990), Biofilms: a basis for an interdisciplinary approach: in Characklis, W.G., and Marshall, K.C., Eds., *Biofilms*, John Wiley & Sons, Inc., New York, pp. 3–15.
7. Costerton, J.W., Lewandowski, Z., Caldwell, D.E., Korber, D.R., and Lappin-Scott, H.M. (1995), Microbial biofilms, *Annu. Rev. Microbiol.*, 49, 711–745.
8. Costerton, J.W. and Lappin-Scott, H.M. (1995), Introduction to microbial biofilms, in Lappin-Scott, H.M., and Costerton, J.W., Eds., *Microbial Biofilms*, Cambridge University Press, Cambridge, United Kingdom, pp. 1–11.
9. Davies, D.G. and Geesey, G.G. (1995), Regulation of the alginate biosynthesis gene *algC* in *Pseudomonas aeruginosa* during biofilm development in continuous culture, *Appl. Environ. Microbiol.*, 61, 860–867.
10. Costerton, J.W., Stewart, P.S., and Greenberg, E.P. (1999), Bacterial biofilms: a common cause of persistent infections, *Science*, 284, 1318–1322.
11. Carpentier, B. and Cerf, O. (1993), Biofilms and their consequences, with particular reference to hygiene in the food industry, *J. Appl. Bacteriol.*, 75, 499–511.
12. Tamilvanan, S., Venkateshan, N., and Ludwig, A. (2008), The potential of lipid-and polymer-based drug delivery carriers for eradicating biofilm consortia on device-related nosocomial infections, *J. Controlled Rel.*, 128, 2–22.
13. Heukelekian, H.H.A. (1940), Relation between food concentration and surface for bacterial growth, *J. Bacteriol.*, 40, 546–558.
14. Christensen, G.D., Simpson, W.A., Bisno, A.L., and Beachey, E.H. (1982), Adherence of slime-producing strains of *Staphylococcus epidermidis* to smooth surfaces, *Infect. Immun.*, 37, 318–326.
15. Costerton, J.W., Montanaro, L., and Arciola, C.R. (2004), Biofilm in implant infections: its production and regulation, *Int. J. Artif. Organs.*, 28, 1062–1068.
16. Ferguson, B.J. and Stolz, D.B. (2005), Demonstration of biofilm in human bacterial chronic rhinosinusitis, *Am. J. Rhinol.*, 19, 452–457.
17. Characklis, W.G., Mc Feters, G.A., and Marshall, K.C. (1990), Physiological ecology in biofilm systems, in: Characklis, W.G. and Marshall, K.C., Eds., *Biofilms*, John Willey & Sons, Inc., New York, pp. 341–394.
18. Hall-Stoodley, L., Costerton, J.W., and Stoodley, P. (2004), Bacterial biofilms: from the natural environment to infectious diseases, *Nat. Rev. Microbiol.*, 2, 95–108.

19. Donlan, R.M. and Costerton, J.W. (2002), Biofilms: survival mechanisms of clinically relevant microorganisms, *Clin. Microbiol. Rev.*, 15, 167–193.
20. Mafu, A.S., Roy, D., Goulet, J., and Magny, P. (1990), Attachment of *Listeria monocytogenes* to stainless steel, glass, polypropylene and rubber surfaces after short contact times, *J. Food Prot.*, 53, 742–746.
21. Rijnaarts, H.M., Norde, W., Bouwer, E.J., Lyklema, J., and Zehnder, A.J.B. (1993), Bacterial adhesion under static and dynamic conditions, *Appl. Environ. Microbiol.*, 59, 3255–3265.
22. Assanta, M.A., Roy, D., and Monpetit, D. (1998), Adhesion of *Aeromonas hydrophila* to water distribution system pipes after different contact times, *J. Food Prot.*, 61, 1321–1329.
23. Rogers, J., Dowsett, A.B., Dennis, P.J., Lee, J.V., and Keevil, C.W. (1994), Influence of plumbing materials on biofilm formation and growth of *Legionella pneumophila* in potable water systems, *Appl. Environ. Microbiol.*, 60, 1842–1851.
24. Bakker, D.P., Postmus, B.R., Busscher, H.J., and van der Mei, H.C. (2004), Bacterial strains isolated from different niches can exhibit different patterns of adhesion to substrata, *Appl. Environ. Microbiol.*, 70, 3758–3760.
25. Gough, N.L. and Dodd, C.E.R. (1998), The survival and disinfection of *Salmonella typhimurium* on chopping board surfaces of wood and plastic, *Food Control.*, 6, 363–368.
26. Hawser, S.P. and Douglas, L.J. (1994), Biofilm formation by *Candida* species on the surface of catheter materials in vitro, *Infect. Immun.*, 62, 915–921.
27. Shin, J.H., Kee, S.J., Shin, M.G., Kim, S.H., Shin, D.H., Lee, S.K., Suh, S.P., and Ryang, D.W. (2002), Biofilm production by isolates of *Candida* species recovered from non-neutropenic patients: comparison of bloodstream isolates with isolates from other sources, *J. Clin. Microbiol.*, 40, 1244–1248.
28. Li, X., Yan, Z., and Xu, J. (2003), Quantitative variation of biofilms among strains in natural populations of *Candida albicans*, *Microbiology*, 149, 353–362.
29. Hawser, S.P., Baillie, G.S., and Douglas, L.J. (1998), Production of extracellular matrix by *Candida albicans* biofilms, *J. Med. Microbiol.*, 47, 253–256.
30. Moreau-Marquis, S., Stanton, B.A., O'Toole, G.A. (2008), *Pseudomonas aeruginosa* biofilm formation in the cystic fibrosis airway, *Pulm. Pharmacol. Ther.*, 21, 595–599.
31. Post, J.C. (2001), Direct evidence of bacterial biofilms in otitis media, *Laryngoscope*, 111, 2083–2094.
32. Romero, R., Schaudinn, C., Kusanovic, J.P., Gorur, A., Gotsch, F., Webster, P., Nhan-Chang, C.L., Erez, O., Kim, C.J., Espinoza, J., Gonçalves, L.F., Vaisbuch, E., Mazaki-Tovi, S., Hassan, S.S., and Costerton, J.W. (2008), Detection of a microbial biofilm in intra-amniotic infection. *Amer. J. Obstet. Gynecology*, 198, 135. e1–5.
33. Mazzocco, T.R. (1985), Early clinical experience with elastic lens implants, *Trans. Ophthalmol. Soc. UK* 104(Pt 5), 578–579.
34. Baillif, S., Ecochard, R., Casoli, E., Freney, J., Burillon, C., and Kodjikian, L. (2008), Adherence and kinetics of biofilm formation of *Staphylococcus epidermidis* to different types of intraocular lenses under dynamic flow conditions, *J. Cataract Refract. Surg.*, 34, 153–158.

35. Alfonso, E.C. and Flynn, Jr., H.W. (1995), Controversies in endophthalmitis prevention. The risk for emerging resistance to vancomycin, *Arch. Ophthalmol.*, 113, 369–370.

36. Kodjikian, L., Burillon, C., Roques, C., Pellon, G., Renaud, F.N.R., Hartmann, D., and Freney, J. (2004), Intraocular lenses, bacterial adhesion and endophthalmitis prevention: A review, *Bio-Medical Mater. Eng.*, 14, 395–409.

37. Characklis, W.G. and Marshall, K.C., Eds., *Biofilms*, John Wiley & Sons, Inc., New York, 1990.

38. Lappin-Scott, H.M. and Costerton, J.W., Eds., *Microbial biofilms*, Cambridge University Press, Cambridge, United Kingdom, 1995.

39. Stoodley, P., Lewandowski, Z., Boyle, J.D., and Lappin-Scott, H.M. (1998), Oscillation characteristics of biofilm streamers in turbulent flowing water as related to drag and pressure drop, *Biotechnol. Bioeng.*, 57, 536–544.

40. Hooshangi, S. and Bentley, W.E. (2008), From unicellular properties to multicellular behavior: bacteria quorum sensing circuitry and applications, *Curr. Opin. Biotechnol.*, 19, 550–555.

41. Zhang, L.H. and Dong, Y.H. (2004), Quorum sensing and signal interference: diverse implications, *Mol. Microbiol.*, 53, 1563–1571.

42. Whitchurch, C.B., Alm, R.A., and Mattick, J.S. (1996), The alginate regulator AlgR and an associated sensor FimS are required for twitching motility in *Pseudomonas aeruginosa*, *Proc. Natl. Acad. Sci. USA*, 93, 9839–9843.

43. Sauer, K. (2003), The genomics and proteomics of biofilm formation, *Genome Biol.*, 4, 219.

44. Stoodley, P., Sauer, K., Davies, D.G., and Costerton, J.W. (2002), Biofilms as complex differentiated communities, *Ann. Rev. Microbiol.*, 56, 187–209.

45. Sauer, K., Camper, A.K., Ehrlich, G.D., Costerton, J.W., and Davies, D.G. (2002), *Pseudomonas aeruginosa* displays multiple phenotypes during development as a biofilm, *J. Bacteriol.*, 184, 1140–1154.

46. Doyle, R.J. (1999), Biofilms, in: Abelson, J.N. and Simon, M.I. Eds., *Methods in Enzymology*, Academic Press, New York, pp. 131–194.

47. Watnick, P. and Kolter, R. (2000), Biofilm, city of microbes, *J. Bacteriol.*, 182, 2675–2679.

48. Bendinger, B., Rijnaarts, H.H., Altendorf, K., and Zehnder, A.J.B. (1993), Physicochemical cell surface and adhesive properties of coryneform bacteria related to the presence and chain length of mycoid acids, *Appl. Environ. Microbiol.*, 59, 3973–3977.

49. Pringle, J.H. and Fletcher, M. (1983), Influence of substratum wettability on attachment of fresh water bacteria to solid surfaces, *Appl. Environ. Microbiol.*, 45, 811–817.

50. Fletcher, M. and Loeb, G.I. (1979), Influence of substratum characteristics on the attachment of a marine pseudomonad to solid surfaces, *Appl. Environ. Microbiol.*, 37, 67–72.

51. Donlan, R.M. (2002), Biofilms: microbial life on surfaces, *Emerg. Infect. Dis.*, 8, 881–890.

52. Zottola, E.A. and Sasahara, K.C. (1994), Microbial biofilms in the food processing industry-should they be of concern? *Int. J. Food Microbiol.*, 23, 125–148.

53. Landry, R.M., An D, Hupp, J.T., Singh, P.K., and Parsek, M.R. (2006), Mucin–*Pseudomonas aeruginosa* interactions promote biofilm formation and antibiotic resistance, *Mol. Microbiol.*, 59, 142–151.
54. Schurr, M.J., Martin, D.W., Mudd, M.H., and Deretic, V. (1994), Gene cluster controlling conversion to alginate-overproducing phenotype in *Pseudomonas aeruginosa*: functional analysis in a heterologous host and role in the instability of mucoidy, *J. Bacteriol.*, 176, 3375–3382.
55. Pulcini, E. (2001), The effects of initial adhesion events on the physiology of *Pseudomonas aeruginosa*, Ph.D. dissertation, Bozeman, MT, Montana State University, 2001. [*Quoted in*: Donlan, R.M. (2002), Biofilms: microbial life on surfaces, *Emerg. Infect. Dis.*, 8, 881–890.]
56. Whitely, M., Bangera, M.G., Bumgarner, R.E., Parsek, M.R., Teitzel, G.M., Lory, S., and Greenberg, E.P. (2001), Gene expression in *Pseudomonas aeruginosa* biofilms, *Nature (London)*, 413, 860–864.
57. Davies, D.G., Parsek, M.R., Pearson, J.P., Iglewski, B.H., Costerton, J.W., and Greenberg, E.P. (1998), The involvement of cell-to-cell signals in the development of a bacterial biofilm, *Science*, 280, 295–298.
58. van Loosdrecht, M.C.M., Lyklema, J., Norde, W., and Zehnder, A.J.B. (1990), Influence of interfaces on microbial activity, *Microbiol. Rev.*, 54, 75–87.
59. Flemming, C.A., Palmer Jr, R.J., Arrage, A.A., van der Mei, H.C., and White, D.C. (1998), Cell surface physiochemistry alters biofilm development of *Pseudomonas aeruginosa* lipopolysaccharide mutants, *Biofouling*, 13, 213–231.
60. Busscher, H.J., Bos, R., and van der Mei, H.C. (1995), Microbial adhesion is a determinant for the strength of biofilm adhesion, *FEMS Microbiol. Lett.*, 128, 229–234.
61. Boyd, A. and Chakrabarty, A.M. (1995), *Pseudomonas aeruginosa* biofilms: role of the alginate exopolysaccharide, *J. Indust. Microbio.*, 15, 162–168.
62. Davies, D.G. (2003), Understanding biofilm resistance to antibacterial agents, *Nature (London)*, 2, 114–122.
63. Keller, L., and Surette, M.G. (2006), Communication in bacteria: an ecological and evolutionary perspective, *Nat. Rev. Microbiol.*, 4, 249–258.
64. de Kievit, T.R. (2009), Quorum sensing in *Pseudomonas aeruginosa* biofilms, *Environ. Microbiol.*, 11, 279–288.
65. Costerton, J.W., Lewandowski, Z., De Beer, D., Caldwell, D., Korber, D., and James, G. (1994), Biofilms, the customised microniche, *J. Bacteriol.* 176, 2137–2142.
66. Neu, T.R. (1996), Significance of bacterial surface-active compounds in interactions of bacteria with interfaces, *Microbiol. Rev.*, 60, 151–166.
67. Xie, H., Cook, G.S., Costerton, J.W., Bruce, G., Rose, T.M., and Lamont, R.J. (2000), Intergeneric communication in dental plaque biofilms, *J. Bacteriol.*, 182, 7067–7069.
68. Stoodley, P., De Beer, D., and Lewandowski, Z. (1994), Liquid flow in biofilm systems, *Appl. Environ. Microbiol.*, 60, 2711–2716.
69. Sutherland, I.W. (2001), The biofilm matrix- an immobilized but dynamic microbial environment, *Trends. Microbiol.*, 9, 222–227.
70. Flemming, H.C., Wingender, J., Mayer, C., Korstgens, V., and Borchard, W. (2000), Cohesiveness in biofilm matrix polymers, in: Allison, D., Gilbert, P., Lappin-Scott,

H.M., and Wilson, M., Eds., Community Structure and Cooperation in Biofilms, *SGM Symposium Series*, 59. Cambridge University Press, Cambridge, United Kingdom, pp. 87–105.

71. Post, J.C., Stoodley, P., Hall-Stoodley, L., and Ehrlich, GD. (2004), The role of biofilms in otolaryngologic infections, *Curr. Opin. Otolaryngol. Head Neck Surg.*, 12, 185–190.
72. Geesey, G.G., Lewandowski, Z., and Flemming, H.C., Eds. (1994), *Biofouling and Biocorrosion in Industrial Water Systems*, Ann. Arbor, Lewis Publishers, MI.
73. Borenstein, S.B. (1994), *Microbiologically Influenced Corrosion Handbook*, Industrial Press, Inc., New York.
74. Stoodley, P., Lewandowski, Z., Boyle, J.D., and Lappin-Scott, H.M. (1999), The formation of migratory ripples in a mixed species bacterial biofilm growing in turbulent flow, *Environ. Microbiol*, 1, 447–455.
75. Jackson, D.W., Suzuki, K., Oakford, L., Simecka, J.W., Hart, M.E., and Romeo, T. (2002), Biofilm formation and dispersal under the influence of the global regulator CsrA of *Escherichia coli*, *J. Bacteriol.*, 184, 290–301.
76. Hall-Stoodley, L. and Stoodley, P. (2005), Biofilm formation and dispersal and the transmission of human pathogens, *Trends Microbiol.*, 13, 7–10.
77. Parsek, M.R. and Fuqua, C. (2004), *Biofilms* 2003: emerging themes and challenges in studies of surface-associated microbial life, *J. Bacteriol.*, 186, 4427–4440.
78. Sabbuba, N., Hughes, G., and Stickler, D.J. (2002), The migration of *Proteus mirabilis* and other urinary tract pathogens over Foley catheters, *BJU Int.*, 89, 55–60.
79. Rather, P.N. (2005), Swarmer cell differentiation in *Proteus mirabilis*, *Environ. Microbiol.*, 7, 1065–1073.
80. Merz, A.J. and Forest, K.T. (2002), Bacterial surface motility: slime trails, grappling hooks and nozzles, *Curr. Microbiol.*, 12, R297–R303.
81. Parsek, M.R. and Greenburg, E.P. (2005), Sociomicrobiology: the connections between quorum sensing and biofilms, *Trends Microbiol.*, 13, 27–33.
82. Douglas, L.J. (2003), *Candida* biofilms and their role in infection, *Trends Microbiol.*, 11, 30–36.
83. Chandra, J., Kuhn, D.M., Mukherjee, P.K., Hoyer, L.L., McCormick, T., and Ghannoum, M.A. (2001), Biofilm formation by the fungal pathogen *Candida albicans:* development, architecture, and drug resistance, *J. Bacteriol.*, 183, 5385–5394.
84. Hawser, S.P. and Douglas, L.J. (1994), Biofilm formation by *Candida* species on the surface of catheter materials *in vitro*, *Infect. Immun.*, 62, 915–921.
85. Kojic, E.M. and Darouiche, R.O. (2004), *Candida* infections of medical devices, *Clin. Microbiol. Rev.*, 17, 255–267.
86. Tenke, P., Kovacs, B., Jäckel, M., and Nagy, E. (2006), The role of biofilm infection in urology, *World J. Urol.*, 24, 13–20.
87. NNIS System (2003), National Nosocomial Infections Surveillance (NNIS) System Report, data summary from January 1992 through June 2003, issued August 2003, *Am. J. Infect. Control.*, 31, 481–498.
88. Mermel, L.A. (2000), Prevention of intravascular catheter-related infections. *Ann. Intern. Med.*, 132, 391–402.

89. Vinh, D.C. and Embil, J.M. (2005), Device-related infections: a review, *J. Long Term Effic. Medical Implants*, 15, 467–488.

90. Costerton, J.W. (1999), Introduction to biofilm, *Int. J. Antimicrob. Agents.*, 11, 217–221.

91. Mermel, L.A. (1997), Defining intravascular catheter-related infections: a plea for uniformity, *Nutrition 4 Suppl.*, 2S–4S.

92. Samore, M.H. and Burke, J.P. (2000), Infections of long intravenous lines: new developments and controversies, *Curr. Clin. Top Infect. Dis.*, 20, 256–270.

93. Sitges-Serra, A. and Girvent, M. (1999), Catheter-related bloodstream infections, *World J. Surg.*, 6, 589–595.

94. Garner, J.S., Jarvis, W.R., Emori, T.G. Horan, T.C., Hughes, J.M. (1988), CDC definitions for nosocomial infections, *Am. J. Infect. Control.*, 3, 128–140.

95. Eggimann, P. and Pittet, D. (2002), Overview of catheter-related infections with special emphasis on prevention based on educational programs, *Clin. Microbiol. Infect.*, 5, 295–309.

96. Pearson, M.L. (1996), Guideline for prevention of intravascular device-related infections: the Hospital Infection Control Practices Advisory Committee, *Am. J. Infect. Control.*, 4, 262–293.

97. Bouza, E., Burilllo, A., and Munoz, P. (2002), Catheter-related infections: diagnosis and intravascular treatment, *Clin. Microbiol. Infect.*, 5, 265–274.

98. Hodge, D. and Puntis, J.W. (2002), Diagnosis, prevention, and management of catheter-related bloodstream infection during long term parenteral nutrition, *Arch. Dis. Child Fetal. Neonatal Ed.*, 87, F21–F24.

99. Raad, I.I. and Hanna, H.A. (2002), Intravascular catheter-related infections: new horizons and recent advances, *Arch. Intern. Med.*, 162, 871–878.

100. Kristinsson, K.G. (1997), Diagnosis of catheter-related infections, in: Seifert, H., Jansen, B., and Farr, B.M., Eds., *Catheter-Related Infections*, Marcel Dekker Inc., New York, pp. 31–57.

101. von Eiff, C. and Peters, G. (2003), Pathogenesis and detection of biofilm formation on medical implants, in: Jass, J., Surman, S., and Walker, J., Eds., *Medical Biofilms: Detection, Prevention and Control*, John Wiley & Sons, Chichester, United Kingdom, pp. 51–72.

102. Maki, D.G., Goldman, D.A., and Rhame, F.S. (1973), Infection control in intravenous therapy, *Ann. Intern. Med.*, 79, 867–887.

103. Maki, D. and Mermel, L. (1998), Infections due to infusion therapy, in: Bennett, J.V. and Brachman, P.S., Eds., *Hospital infections*, 4th ed., Lippincott-Raven, Philadelphia, pp. 689–724.

104. Marrie, T.J. and Costerton, J.W. (1984), Scanning and transmission electron microscopy of *in situ* bacterial colonization of intravenous and intra arterial catheters, *J. Clin. Microbiol.*, 19, 687–693.

105. Cooper, G.L., Schiller, A.L., and Hopkins, C.C. (1988), Possible role of capillary action in pathogenesis of experimental catheter-associated dermal tunnel infections, *J. Clin. Microbiol.*, 26, 8–12.

106. Sitges-Serra, A., Linares, J., and Garau, J. (1985), Catheter sepsis: the clue is the hub, *Surgery*, 97, 355–357.

107. Maki, D.G., Jarrett, F., and Sarafin, H.W. (1977), A semiquantitative culture method for identification of catheter-related infection in the burn patient, *J. Surg. Res.*, 22, 513–520.

108. Maki, D.G. and Hassemer, C.A. (1981), Endemic rate of fluid contamination and related septicemia in arterial pressure monitoring, *Am. J. Med.*, 70, 733–738.

109. Bjornson, H.S., Colley, R., Bower, R.H., Duty, V.P., Schwartz-Fulton, J.T., and Fischer, J.E. (1982), Association between microorganism growth at the catheter insertion site and colonization of the catheter in patients receiving total parenteral nutrition, *Surgery*, 92, 720–727.

110. Cooper, G.L. and Hopkins, C.C. (1985), Rapid diagnosis of intravascular catheter-associated infection by direct Gram staining of catheter segments, *N. Engl. J. Med.*, 312, 1142–1147.

111. Mermel, L.A., McCormick, R.D., Springman, S.R., and Maki, D.G. (1991), The pathogenesis and epidemiology of catheter-related infection with pulmonary artery Swan-Ganz catheters: a prospective study utilizing molecular subtyping, *Am. J. Med.*, 91, 197S–205S.

112. Cheesbrough, J.S., Finch, R.G., and Burden, R.P. (1986), A prospective study of The mechanisms of infection associated with hemodialysis catheters, *J. Infect. Dis.*, 154, 579–589.

113. Flynn, P.M., Shenep, J.L., Stokes, D.C., and Barrett, F.F. (1987), *In situ* management of confirmed central venous catheter-related bacteraemia, *Pediatr. Infect. Dis. J.*, 6, 729–734.

114. Weightman, N.C., Simpson, E.M., Speller, D.C., Mott, M.G., and Oakhill, A. (1988), Bacteraemia related to indwelling central venous catheters: prevention, diagnosis and treatment, *Eur. J. Clin. Microbiol. Infect. Dis.*, 7, 125–129.

115. Maki, D.G., Narans, L.L., and Banton, J. (1998), A prospective study of the Pathogenesis of PICC-related BSI [abstract K-10], in: *Program and abstracts of the 38th Interscience Conference of Antimicrobial Agents and Chemotherapy*, San Diego, American Society for Microbiology, Washington, DC, p. 502.

116. Rutala, W.A. and Weber, D.J. (1997), Use of inorganic hypochlorite (bleach) in health-care facilities, *Clin. Microbiol. Rev.*, 10, 597–610.

117. Safdar, N., Kluger, D.M., Maki, D.G. (2002), A review of risk factors for catheter-related bloodstream infection caused by percutaneously inserted, noncuffed central venous catheters: implications for preventive strategies, *Medicine (Baltimore)*, 81 (6), 466–479.

118. Anaissie, E., Penzak, S., and Dignani, C. (2002), The hospital water supply as a source of nosocomial infections, *Arch. Int. Med.*, 162, 1483–1492.

119. Cozad, A. and Jones, R.D. (2003), Disinfection and the prevention of infectious disease, *Am. J. Infection Control*, 31, 243–254.

120. Sehulster, L. and Chinn, R.Y.W. (2003), Guidelines for environmental infection Control in health-care facilities. Recommendations of CDC and the Healthcare Infection Control Practices Advisory Committee (HICPAC). Centers for Disease Control and Prevention, *M.M.W.R. Recommendation Rep.*, 52(RR-10), 1–42.

121. Brady, M.J., Lisay, C.M., Yurkovetskiy, A.V., and Sawan, S.P. (2003), Persistent silver disinfectant for the environmental control of pathogenic bacteria, *Am. J. Infect. Control*, 31, 208–214.

122. Richards, M.J., Edwards, J.R., Culver, D.H., and Gaynes, R.P. (1999), Nosocomial infections in medical intensive care units in the United States National Nosocomial Infections Surveillance System, *Crit. Care Med.*, 27, 887–892.

123. Gorman, S.P. and Jones, D.S. (2003), Complications of urinary devices, in: Wilson, M., Ed., *Medical Implications of Biofilms*, Cambridge University Press, Cambridge, UK, pp. 136–170.

124. Keane, P.F., Bonner, M.C., Johnston, S.R., Zafar, A., and Gorman, S.P. (1994), Characterization of biofilm and encrustation on ureteral stents *in vivo*, *Br. J. Urol.*, 73, 687–691.

125. Tunney, M.M., Jones, D.S., and Gorman, S.P. (1998), Assessment of biofilm and biofilm-related problems associated with urinary tract devices, in: Doyle, R.J., Ed., *Methods in Enzymology: Biofilms*, Academic Press, FL, pp. 558–566.

126. Malcolm, R.K., McCullagh, S.D., Woolfson, A.D., Gorman, S.P., Jones, D.S., and Cuddy, J. (2004), Controlled release of a model antibacterial drug from a novel self-lubricating silicone biomaterial, *J. Control. Rel.*, 97, 313–320.

127. Jones, D.S., Garvin, C.P., Dowling, D., Donnelly, K., and Gorman, S.P. (2006), Examination of surface properties and *in vitro* biological performance of amorphous diamond-like carbon-coated polyurethane, *J. Biomed. Mater. Res. Part B: Appl. Biomater.*, 78B, 230–236.

128. Matthew, A., Myer, B., and Rushton, N. (2001), *In vitro* and *in vivo* investigations into the biocompatibility of diamond-like carbon (DLC) coatings for orthopedic applications, *J. Biomed. Mater. Res.*, 58, 319–328.

129. Jones, M.I., McColl, R., Grant, D.M., Parker, K.G., and Parker, T.L. (2000), Protein adsorption and platelet attachment and activation, on TiN, TiC, and DLC coatings on titanium for cardiovascular applications, *J. Biomed. Mater. Res.*, 52, 413–421.

130. Dowling, D.P., Donnelly, K, and O'Brien, T.P. (1996), Application of diamond like carbon films as hermetic coatings on optical fibres, *Diamond Relat. Mater.*, 5, 492–495.

131. Duran, L.W. (2000), Preventing medical device related infections, *Med. Device Technol.*, 11, 14–17.

132. Stiver, H.G., Zachidniak, Z, and Speert, D.P. (1988), Inhibition of polymorphonuclear leucocyte chemotaxis by the mucoid exopolysaccharide of *P. aeruginosa*, *Clin. Invest. Med.*, 11, 247–252.

133. Costerton, J.W. (2005), Biofilm theory can guide the treatment of device-related orthopaedic infections, *Clin. Orthop. Relat. Res.*, 437, 7–11.

134. Fitzpatrick, F., Humphreys, H., and O'Gara, J.P. (2005), The genetics of staphylococcal biofilm formation—will a greater understanding of pathogenesis lead to better management of device-related infection? *Clin. Microbiol. Infect.*, 11, 967–973.

135. Smith, A.W. (2005), Biofilms and antibiotic therapy: Is there a role for combating bacterial resistance by the use of novel drug delivery systems? *Adv. Drug Del. Rev.*, 57, 1539–1550.

136. von Eiff, C., Jansen, B., Kohnen, W., and Becker, K. (2005), Infections associated with medical devices: pathogenesis, management and prophylaxis, *Drugs*, 65, 179–214.

137. Reisnera, A., Høibyb, N., Tolker-Nielsena, T., and Molina, S. (2005), Microbial pathogenesis and biofilm development, in: Russell, W. and Herwald, H., Eds., *Concepts in Bacterial Virulence, Contrib. Microbiol.* Vol. 12, Karger, Basel, pp. 114–131.

138. Sihorkar, V. and Vyas, S.P. (2001), Biofilm Consortia on biomedical and biological surfaces: delivery and targeting strategies, *Pharm. Res.*, 18, 1247–1254.

139. McCrory, R., Jones, D.S., Adair, C.G., and Gorman, S.P. (2003), Pharmaceutical strategies to prevent ventilator-associated pneumonia, *J. Pharm. Pharmacol.*, 55, 411–428.

140. Gristina, A.G. (1987), Biomaterial-centered infection: microbial adhesion versus tissue integration, *Science*, 237, 1588–1595.

141. Kuhn, D.M., George, T, Chandra, J, Mukherjee, P.K., and Ghannoum, M.A. (2002), Antifungal susceptibility of *candida* biofilms: unique efficacy of amphotericin B lipid formulations and echinocandins, *Antimicrob. Agents Chemother.*, 46, 1773–1780.

142. Doggett, M.S. (2000), Characterization of fungal biofilms within a municipal water distribution system, *Appl. Environ. Microbiol.*, 66, 1249–1251.

143. Brown, M.R, and Barker, J. (1999), Unexplored reservoirs of pathogenic bacteria: protozoa and biofilms, *Trends Microbiol.*, 7, 46–50.

144. Khardori, N. and Yassien, M. (1995), Biofilms in device related infections, *J. Ind. Microbiol.*, 15, 141–147.

145. Hindupur, S. and Muslin, A.J. (2005), Septic shock induced from an implantable cardioverter-defibrillator lead-associated *Candida albicans* vegetation, *J. Interv. Card. Electrophysiol.*, 14, 55–59.

146. Lewis, K. (2001), The riddle of biofilm resistance, *Antimicrob. Agents Chemother.*, 45, 999–1007.

147. Mah, T.F. and O'Toole, G.A. (2001), Mechanisms of biofilm resistance to antimicrobial agents, *Trends Microbiol.*, 9, 34–39.

148. Dobretsov, S., Dahms, H.U., and Qian, P.Y. (2006), Inhibition of biofouling by marine microorganisms and their metabolites, *Biofouling*, 22, 43–54.

149. Coetser, S.E. and Cloete, T.E. (2005), Biofouling and biocorrosion in industrial water systems, *Crit. Rev. Microbiol.*, 31, 213–232.

150. Yan, T. and Yan, W.X. (2003), Fouling of offshore structures in China-a review, *Biofouling, Suppl.*, 1, 133–138.

151. Fux, C.A., Stoodley, P., Hall-Stoodley, L., and Costerton, J.W. (2003), Bacterial biofilms: a diagnostic and therapeutic challenge, *Expert Rev. Anti-Infect. Ther.*, 1, 667–683.

152. Stewart, P.S. and Costerton, J.W. (2001), Antibiotic resistance of bacteria in biofilms, *Lancet*, 358, 135–138.

153. Lindsay, D. and von Holy, A. (2006), Bacterial biofilms within the clinical setting: what healthcare professionals should know, *J. Hosp. Infect.*, 64, 313–325.

154. Mehta, A. (2007), Say no to nosocomial infections, *Express Pharma.*, March 16–31, 76–78.

# CHAPTER 2

# RATIONALE FOR BIOFILM ERADICATION FROM MODERN MEDICAL DEVICES

## 2.1. INTRODUCTION

Healthcare institutions purchase millions of intravascular catheters each year since they are integral and indispensable parts in modern-day medical practice, particularly in intensive care units (ICU). In fact, >5 million central venous catheters (CVC) are inserted annually in the United States, accounting for 0.28–0.8 central-line days patient$^{-1}$ day$^{-1}$ [1,2]. It is estimated that 2–12% of CVCs result in sepsis [3]. Analysis of the National Nosocomial Infections Surveillance (NNIS) data shows that 87% of primary blood stream infections occurred in patients with a central line. Approximately 80,000 catheter-related bloodstream infections occur each year in the ICU in US hospitals, and result in up to 20,000 deaths [4]. The impact and cost of such infections is thus enormous. Before 1992, there was a steady increase in the incidence of candidemia in combined medical–surgical ICU [2], but the contribution of *Candida* to bloodstream infections stabilized between 1992 and 1998 at ~11.5% [5]. A total of 72–87% of bloodstream infections, including candidemia, are considered to be catheter related in ICU patients [5,6]. The role of catheters in neutropenic patients is less clear than that in ICU patients because gastrointestinal mucositis is a probable source of candidemia in these patients. Candidemia is an independent risk determinant for predicting death in patients with nosocomial bloodstream infections [7]. The crude mortality due to candidemia has been estimated to be as high as 57%, but the attributable mortality is reported to

*Biofilm Eradication and Prevention: A Pharmaceutical Approach to Medical Device Infections*, By Tamilvanan Shunmugaperumal

be 38% [8]. The attributable mortality of candidemia was correlated, in a multivariate logistic analysis, with the APACHE II score, the duration of candidemia, and rapidly fatal underlying illnesses (29% of study patients were neutropenic) [9]. Other predictors of adverse outcome included evidence of neutropenia and visceral dissemination [10]. In general, the likelihood of developing CVC related infections depends on the type of catheter, the hospital service, the site of insertion, and the duration of catheter placement [4]. Risk factors for CVC related infections include neutropenia for >8 days, hematologic malignancy, total parenteral nutrition, duration of site use, frequent manipulation of the catheter, improper insertion and maintenance of the catheter, and high APACHE II score [3,11].

Although such catheters provide necessary vascular access, their use puts patients at risk for local and systemic infectious complications, including local site infection, catheter-related bloodstream infections (CRBSI), septic thrombophlebitis, endocarditis, and other metastatic infections (e.g., lung abscess, brain abscess, osteomyelitis, and endophthalmitis). A fairly comprehensive, but by no means exhaustive, list of catheters used for venous and arterial access in modern scientific medicine is shown in Table 2.1 [12] in conjunction with their indications for possible device-related infections.

Peripheral venous catheters are the devices most frequently used for vascular access. Although the incidence of local or BSI associated with peripheral venous catheters is usually low, serious infectious complications produce considerable annual morbidity because of the frequency with which such catheters are used. However, the majority of serious catheter-related infections are associated with CVC, especially those that are placed in patients in ICU. In the ICU setting, the incidence of infection is often higher than in the less acute in-patient or ambulatory setting. In the ICU, central venous access might be needed for extended periods of time; patients can be colonized with hospital-acquired organisms; and the catheter can be manipulated multiple times per day for the administration of fluids, drugs, and blood products. Moreover, some catheters can be inserted in urgent situations, during which optimal attention to aseptic technique might not be feasible. Certain catheters (e.g., pulmonary artery catheters and peripheral arterial catheters) can be accessed multiple times per day for hemodynamic measurements or to obtain samples for laboratory analysis, augmenting the potential for contamination and subsequent clinical infection.

The magnitude of the potential for CVC to cause morbidity and mortality resulting from infectious complications has been estimated in several studies [13]. In the United States, 15 million CVC days (i.e., the total number of days of exposure to CVC by all patients in the selected population during the selected time period) occur in ICUs each year [13]. If the average rate of CVC associated bloodstream infections (BSI) is 5.3 per 1000 catheter days in the ICU [14], ~80,000 CVC associated BSIs occur in ICUs each year in the United States. The attributable mortality for these BSI has ranged from no increase in mortality in studies that controlled for severity of illness [15–17], to 35%

**TABLE 2.1. Venous and Arterial Access Catheters[a]**

| Catheter Type | Insertion/ Implantation (entry) Site | Length | Indications |
|---|---|---|---|
| Peripheral venous catheters (short) | veins of forearm or hand | <3 in. | Phlebitis on chronic use |
| Peripheral arterial catheters | radial artery, femoral, axillary, brachial, posterior tibial arteries | <3 in. | Low infection risk |
| Midline catheters | Proximal basilic or cephalic veins; does not enter central veins | 3–8 in. | Anaphylactoid reactions with elastomeric hydrogel |
| Nontunneled central venous catheters | Percutaneously inserted into central veins (subclavian, internal jugular, or femoral) | >8 cm depending on patient size | Account for majority of CRBSI |
| Pulmonary artery catheters | With Teflon® introducer in a central veins | >30 cm depending on patient size | Usually heparin bonded; CRBSI similar to CVCs |
| Peripherally inserted central venous catheters (PICC) | Basilic, cephalic, or brachial veins and enter the superior vena cava | >20 cm depending on patient size | Lower rate of infection than nontunneled CVCs |
| Tunneled central venous catheters | Subclavian, internal jugular, or femoral veins | >8 cm depending on patient size | Cuff inhibits migration of organisms into catheter tract; lower infection rate than nontunneled CVC |
| Totally implantable | Tunneled beneath skin and have subcutaneous port accessed with a needle; implanted in subclavian or internal jugular vein | >8 cm depending on patient size | Lowest risk for CRBSI; surgery required for catheter removal |
| Umbilical catheters | Umbilical vein or artery | <6 cm depending on patient size | Risk for CRBSI similar with catheters placed in umbilical vein versus artery |

[a]Modified from O'Grady et al. [12].

increase in mortality in prospective studies that did not use this control [18,19]. Thus, the attributable mortality remains unclear. The attributable cost per infection is an estimated $34,508–56,000 [15,20], and the annual cost of caring for patients with CVC associated BSI ranges from $296 to $2.3 billion [21]. Advanced surveys also indicate that catheter-associated bacteremia following catheter-related infection, is by far the leading cause of nosocomial BSI in ICU [22]. Therefore, by several analyses, the cost of CVC associated BSI is substantial, both in terms of morbidity and in terms of financial resources expended.

Another well-known case of biofilm-mediated nosocomial infection is associated with artificial joints. Even with the use of perioperative antimicrobial prophylaxis and a laminar air-flow surgical environment, the risk of intraoperative infection is still ~1% for hip and shoulder replacement and 2% after knee replacement [23]. With >200,000 hip replacements and 200,000 knee replacements each year in the United States alone, the healthcare costs are high. Nevertheless, the magnitude of this infectious complication is quite remarkable, considering that in 1995 ~216,000 total knee replacements were performed. This number is expected to more than double by the year 2030 [24]. This increase in the number of implanted joint prostheses is stimulated by the growing size of the patient population, particularly of older persons, who are most likely to require the implantation and revision of joint prostheses [24]. *Candida* infections of prosthetic joints mostly involves hip and knee prostheses, with only a case report involving other joint prostheses [25]. Implantation of knee and hip prostheses carries a higher risk for infection than smaller joint prostheses due to the longer duration of these operations, the inherently low blood flow to cortical bone, and the formation of a hematoma in a larger dead space around such larger devices. These hematomas can devascularize the surrounding tissue and prevent the entry of antibiotics [24]. In general, the mean cost of management of an episode of joint infection is estimated to exceed $50,000. The cost is probably even higher for *Candida* infection because of frequent delays in diagnosis and more prolonged treatment of this fungal infection compared with bacterial infection. The mortality due to prosthetic joint infections is low [1], but mortality due to *Candida* infections is not known in this setting. Risk factors for infection of prosthetic joints include prior surgery at the site of the prosthesis, rheumatoid arthritis, immunocompromised state, diabetes mellitus, poor nutritional status, obesity, psoriasis, and advanced age [26].

Urinary tract infections are the second-most common type of bacterial infection, after those of the respiratory tract, and the bacterial biofilms formation on urinary tract catheters represents one segment of the urinary tract infections market (http://www.leaddiscovery.co.uk/reports/Urinary%20Tract%20Infections%20-%20Ciprofloxacin%20Leads%20the%20Way.html, accessed and collected on June 13, 2008). Because the range of medical devices employed now within the urinary tract is truly vast, although urinary catheters and stents still account for the overwhelming majority [22]. The process of encrustation (i.e., biofilm mineralization) is a concern in the urinary devices especially in long-term indwelling catheters and stents. Encrustation with or without an infectious origin is primarily composed of mineralized deposits containing magnesium ammonium phosphate (struvite) and calcium phosphate (hydroxyapatite). In both cases, the composition of the concretion is similar to urinary stones [23]. Studies suggest that ~50% of patients undergoing catheterization for >28 days develop recurrent encrustation or blockage [27], whereas as many as 75% of stents are encrusted after 24 weeks [28]. These mineralized deposits frequently obstruct the lumen of devices, leading to urinary retention, painful distension of the bladder, or more severe complications (e.g., urolithiasis, pyelonephritis, septicemia, and shock) [29]. Finally, the abrasive nature of encrusted urinary devices may lead to permanent damage to the uroepithelium. In addition, foley catheter infections lead to ~900,000 nosocomial urinary track infections. Overall, the biofilms thus formed have a high tolerance to various anti-infective strategies, and biofilm colonies on urinary catheters can have >1000 times more tolerance to antibiotics than their planktonic counterparts [30–34]. In addition to the financial burden, biofilm formation frequently leads to the infection of surrounding tissue and often requires removal of the catheter subjecting the patient to discomfort.

## 2.2. AN IMPLANT-ASSOCIATED INFECTION

An implant-associated infection is defined as a host immune response to one or more microbial pathogens on an indwelling implant. The host response may be local (e.g., septic prosthetic knee) or systemic, and are usually suspected on the basis of clinical manifestations, microbiogical investigations, and/or radiological imaging.

Definitive diagnosis, however, requires surgical exploration and culture of the implant [35]. Surgically implanted device infections (SIDIs) are associated with increased morbidity and mortality, particularly because traditional principles of management include surgical intervention and prolonged course of antimicrobial therapy. Thus, they are also associated with increased healthcare costs [36]. Furthermore, issues in management are complicated by an inability (1) to determine accurately the presence of a SIDI without surgical exploration; (2) the reluctance to operate on a patient who may not have an infection; (3) unclear guidelines on optimal surgical intervention; and (4) timing of reimplantation, and unclear optimal duration of antibiotics.

## 2.3. SURGICALLY IMPLANTED DEVICE INFECTIONS

This section reviews the major classes of implanted devices and their infectious complications. Thus, the rationale behind eradicating biofilm-associated and device-related infections are being explored in detail in each cases.

### 2.3.1. Neurosurgical Devices

Currently, the most frequently implanted neurosurgical devices used are CSF shunts, intraspinal pumps, and neurostimulators.

### 2.3.2. Cerebrospinal Fluid Shunts

Cerebrospinal fluid (CSF) shunts are inserted in both children and adults to relieve elevated intracranial pressure from a variety of causes. The most commonly employed device is a ventriculo-peritoneal (VP) shunt, in which a catheter is inserted into the frontal horn of the lateral ventricle of the nondominant cerebral hemisphere, with the drainage tubing passing via a subcutaneous tunnel and inserting into the peritoneal space. Variants of this device include external ventricular drains (EVD), in which the catheter is connected to a closed external drainage system, and the ventriculo-atrial (VA) shunt, in which the drainage tubing is inserted into the right atrium.

Placement of a CSF shunt involves implantation of foreign devices or material is thus considered a "clean-contaminated" producer [37,38]. A clean-contaminated surgery is associated with a 10.1% risk of infection in the absence of preoperative antimicrobial prophylaxis [39]. As such, perioperative antibiotic prophylaxis is used routinely to minimize the risk of postoperative surgical site infection, including SIDIs, although few randomized, placebo-controlled trials addressing the efficacy of such a practice have been conducted [38].

The frequency of CSF shunt infections is variable, depending on the patient population studied and the experience of the neurosurgery team involved, but it typically ranges from <1 to 30%, with an overall incidence of ~10% [40]. Most of these infections are caused by bacteria originating from the patient's skin that are introduced into the wound or into the device at the time of surgery [38,41].

Most CSF shunt infections occur within the first month after the procedure [42]. Identifiable risk factors for CSF shunt infections from pediatrics have implicated the etiology of hydrocephalus [43,44], patient age [45,46], previous revisions [38,45], and duration of shunt surgery [47]. Studies have, however, been conflicting, and it is difficult to apply such findings to widely differing pediatric and adult populations [48].

The clinical manifestations of CSF shunt infections depend on which part of the shunt is implicated and thus can be divided into proximal and distal. Proximal central nervous system (CNS) shunt infection involves the catheter placed within the ventricles and thus manifests itself as a ventriculitis (without meningitis). Meningismus rarely occurs, unless there is a communication between the ventricles and the leptomeninges, allowing for infected CSF to cause meningitis [41]. Most often, ventriculitis presents with nonspecific symptoms (e.g., lethargy and malaise) [41].

Distal CNS shunt infections present with symptoms related to the site of shunt drainage [38,41,49]. Distal VP shunt infections manifest as infected intraperitoneal fluid cysts or frank peritonitis. The VA shunts, because they are directly inserted into the right atrium, frequently cause fever and bacteramia. Both proximal and distal CNS shunt infections frequently present with fever, being present in up to 100% of patients in various studies [50,51]. However, the absence of fever is not useful in excluding such an infection [49]. Another common symptom is focal pain, although this is present in only up to 40% of patients [49]. The pain may be localized either to the wound (proximal infection) or to the peritoneum (distal infection). Some patients will have easily recognizable evidence of wound infection, including signs of inflammation and purulent drainage.

The most common organisms causing CNS shunt infections are coagulase-negative staphylococci (CoNS), typically *Staphylococcus epidermidis* (40–50%) followed by *Staphylococcus aureus* (25%) [38,41,48]. The remaining causes include Gram-negative bacilli (*Enterobacteriaceae*, e.g., *Klebsiella spp.* and *Enterobacter spp.*, and non-lactose fermenting organisms, e.g., *Pseudomonas aeruginosa* and *Acinetobacter baumanii*) [52–54], skin flora (*Corynebacterium spp.*, *Propionibacterium spp.*) [55], and *streptococci/enterococci* [41].

Diagnosis of CNS shunt infections require CSF analysis for cell count, Gram stain culture, and biochemistry (glucose and protein). The CSF profile is similar to that for community-acquired bacterial meningitis. A CSF pleocytosis of >100 cells $mm^{-3}$ has a positive predictive value of 89% [50], but the absence of pleocytosis does not rule out infection [41]. Blood cultures typically are not of value for VP shunts, although they have a sensitivity of ~90% when multiple blood cultures are taken in suspected VA shunts [56]. Abdominal sonography for evidence of distal shunt abnormalities in the peritoneum can be useful as well [38].

Management of CSF shunt infections is best accomplished with a multidisciplinary team approach. The selected antimicrobial agent(s) should be effective against the above-mentioned microorganisms. For the *staphylococci*, vancomycin 1-g intravenously (iv) every 6–12 h and rifampin 300 mg orally twice a day in adults is recommended [38]. For the Gram-negative bacilli, a third-generation cephalosporin (e.g., cefotaxime 2-g iv every 4–6 h can be used) [38]. Antimicrocrobial therapy must, however, be modified to optimally target the recovered pathogens once results of cultures become available. Ideally, the shunt should be removed because it represents the source of infection and because an infected shunt typically is nonfunctional, resulting in obstruction and pyo/hydrocephalus [38,41]. Attempts to clear the infection with antibiotics alone have been associated with a high failure rate because of the inability of the antimicrobial agent to penetrate the biofilm of the avascular device [41]. If possible, temporary drainage may need to be inserted and maintained for 8–10 weeks before implanting a new replacement shunt [38]. In patients with refractory CNS shunt infections or for whom surgical removal of the infected shunt is not possible, intraventricular antibiotics may be considered as an option [38]. The British Neurosurgery Working Party recommends intraventricular antibiotics as the first-line approach [57]. However, well-controlled studies supporting this or any other approach are lacking. For distal VP shunt infections, the intraperitoneal process must be treated first, along with a temporary external drainage.

### 2.3.3. Intraspinal Pumps

Implanted intraspinal catheters increasingly are being used to treat patients with refractory pain, especially those with advanced malignancies. There are three major systems that are currently used: an externalized system (similar to an intravascular catheter tunneled subcutaneously to an exit site on the chest or abdominal wall); a subcutaneous reservoir (similar to an implanted port intravascular catheter); and an implanted catheter-pump system with a subcutaneous medication chamber [58]. These catheters are placed either in the intrathecal (spinal subarachnoid) space or in the epidural space.

Infections of tunneled intraspinal catheter systems can be divided anatomically into those involving the CNS (i.e., intrathecal infection/meningitis, epidural abscessed) and those not involving the CNS (i.e., tunnel, subcutaneous pocket, or exit site infection). These two categories are not mutually exclusive. The rate of infectious complications of these catheters is not well defined, but it is believed to be similar to that of VP shunts [38]. Although such infections rarely lead to any significant morbidity or death they can be associated with local or systemic symptoms and loss of pain control. In a retrospective cohort study of cancer patients with chronic tunneled intraspinal catheters, it was noted that the only factor indentified that was independently associated with increased risk for infection was a prolonged duration of catheter placement surgery (>100 min) [58].

The majority of infections occurred in the first 2 weeks after insertion. The main pathogens identified were skin flora (*Streptococci*, CoNS, *Corynebacterium*). However, there was one pocket infection caused by *P. aeurginosa.* Although the clinical manifestations of the patients that led to a suspicion of infection were not provided, the authors noted that fever and leukocytosis were not always present, even in patients with meningitis. Management of infected intraspinal catheters was individualized to the patient. In those in whom the catheter could be removed, the infection was cured. However, this study also demonstrated that catheter removal is not mandatory in all cases and that it may be possible to treat such infections by leaving the catheter *in situ* and using oral antimicrobial therapy for suppression of symptoms. Such an approach may be appropriate given the demographics of the patients that typically have long-term intraspinal catheters (i.e., those with terminal illnesses, short-life expectancy, or poor performance status). Nevertheless, if there is CNS involvement (meningitis, epidural abscess) or a subcutaneous pocket infection, antibiotic therapy alone is unlikely to be sufficient, and the intraspinal catheter should be removed.

### 2.3.4. Neurostimulators

Neurostimulation therapy involves applying low-voltage electricity to different regions of the central or peripheral nervous system. This electrical impulse triggers a neurological response that interferes with the transmission of unwanted pain or motor dysfunction signals to the brain. Currently, the primary uses of neurostimulation therapy are management of intractable neuropathic pain (usually via spinal cord or peripheral nerve stimulators) and treatment of movement disorders (e.g., Parkinson's disease, via deep brain stimulation). Neurostimulator devices consist of leads, which are thin, coiled wires with electrodes at their tips that are surgically placed in the specific area of the nervous system targeted, and a neurostimulator (or power supply) that is implanted under the skin (e.g., near the collarbone for deep brain stimulators, in the wall of the lower abdomen for spinal cord stimulators). The number of leads placed, the number of electrodes placed, and the placement of the electrodes depends on the type of stimulator and the condition being treated.

Data on infections associated with neurostimulator devices are scarce. The largest series addressing this issue has been in patients with spinal cord stimulators (SCS) [59]. The typical risk factors for surgical site infections (e.g., diabetes mellitus, debilitation, malnutrition, corticosteroid use) were not identified as risk factors for neurostimulator-related infections. Furthermore, the majority (87%) of patients with SCS infections received perioperative antibiotics, suggesting that this practice does not affect risk. The pocket was the most common site of infection, followed by the lead tracks. Microbiological data were limited. *Staphylococcal species* (not further identified) was the most common isolate. However, no growth was seen in almost 20% of cases. The next most commonly identified organism was *P. aeruginosa* (3%) and fungi were not isolated. The majority of infected SCS devices were explanted: 82% underwent total explantation, 12% partial, 4% remained *in situ*, and 4% were not reported. Most device removals (56%) were performed within 2 months of implantation. A clinical outcome of cure was achieved in 91% of patients and no deaths were reported.

In the absence of such limited data, there are no definite guidelines for the management of neurostimulator-related infections. Diagnosis may be difficult because of subtle signs and symptoms. Clinical manifestations may include new onset confusion, neurologic deterioration of the underlying disorder, device malfunction, or signs of tunnel or pocket infection. Appropriate cultures, including a lumbar puncture (LP) and aspirate of pocket fluid if present, should be performed. Diagnostic imaging may be helpful. Until evidence from studies suggests otherwise, the principle of device removal and systemic antibiotics seems safe. Explantation ideally should be total, or at least partial, considering that the majority of patients who underwent removal were cured [59]. Surgical specimens should be sent for culture. Because *staphylococcal species* predominate, a reasonable empiric choice would be vancomycin. Therapy should be tailored based on Gram stain and culture.

### 2.3.5. Neurological Prostheses

Infection of CNS (hydrocephalus) shunts is a major problem in patients with ventricular drainage. Therefore, efforts have also been made to develop infection-resistant hydrocephalus shunts or other neurological prostheses. Bridgett et al. [60] reported on the reduced staphylococcal adherence to Hydromer®—coated and, thus, hydrophilic CSF shunts. However, there were technical difficulties in achieving a uniform Hydromer layer on the Si rubber [60]. Bayston et al. [61–64] published a considerable amount of experimental work on impregnation of Si shunt catheters with various antimicrobials. In particular, a combination of rifampicin and clindamycin proved to be clearly superior to other agents tested. In a newer study, it was shown *in vitro* that the rifampicin-clindamycin impregnated catheters are able to kill adhered staphylococci completely within 48–52 h [65,66].

Schierholz et al. [67] developed an incorporation method for rifampicin and other hydrophobic antibacterials into Si ventricular catheters. In an animal model using New Zealand white rabbits, rifampicin loaded catheters were implanted into the ventricular space and infection was induced by inoculation of certain dosages of *S. epidermidis* or *S. aureus* [68]. None of the animals that received the rifampicin-loaded catheter showed clinical signs of infection, nor could the infecting strain be recovered from the catheter, brain tissue, or CSF. In contrast, all animals with the uncoated catheters showed signs of severe meningitis or ventriculitis, and the infecting strains were cultivated in each case from the catheter and from surrounding tissue. As an improvement of the catheter, especially to prevent development of resistance of staphylococci to rifampicin, a combination of rifampicin and trimethoprim was used for the impregnation process [69]. Furthermore, two cases were reported in which the rifampicin catheter was successfully used for the treatment of patients with a complicated course of shunt infection [70].

In addition, a Si catheter with a combination of three antimicrobials (rifampicin, fusidic acid, and mupirocin) has been described with a long-lasting drug release of up to ~100 days; however, no animal or clinical data are available so far for this type of catheter [71]. Zabramski et al. [72] performed a prospective, randomized clinical trial with an external ventricular drain catheter coated with minocycline and rifampicin. The antibacterial-impregnated catheters were one-half as likely to become colonized as the control catheter (17.9 vs 36.7%, $p < 0.0012$), and CSF cultures were seven times less frequently positive in patients with the modified catheters than in the control group (1.3 vs 9.4%, $p = 0.002$).

### 2.3.6. Cochlear Implants

Cochlear implants are electronic devices that provide a high-quality sense of hearing to severely and profoundly deaf children and adults. The U.S. Food and Drug Administration (FDA) regulates manufacturers of cochlear implants.

For manufacturers to sell cochlear implants in the United States, they must first show the FDA that their implants are safe and effective. As a matter of policy, FDA does not rate or recommend brands of cochlear implants or medical facilities that implant them.

The basic model of a cochlear implant consists of an external and an internal component [73]. The external component consists of a microphone, which detects speech signals, connected to a signal processor that transforms the speech signals into digital impulses. The internal receiver, which is surgically implanted beneath the skin, connected via an internal receiver-stimulator that traverses the temporal bone, enters the middle ear, and connects to an electrode array that has been placed surgically in the lumen of the cochlea. The digital impulses generated by the external component are transmitted percutaneously to the internal receiver, and then conducted to the electrode array, where they stimulate the sensory fibers of the auditory nerve, allowing for the perception of sound in the auditory cortex (Fig. 2.1).

The rate of infection complications following cochlear implantation is quite low, ranging from 1.7 to 3.3%. The majority of these complications involve the surgical incision site, manifesting as skin flap necrosis, wound dehiscence, or wound infection, which is typically due to contamination at the time of implantation. These complications tend to manifest in the early postoperative state, although the definition of "early" has varied from either <30 days after surgery to <3 months [74,75].

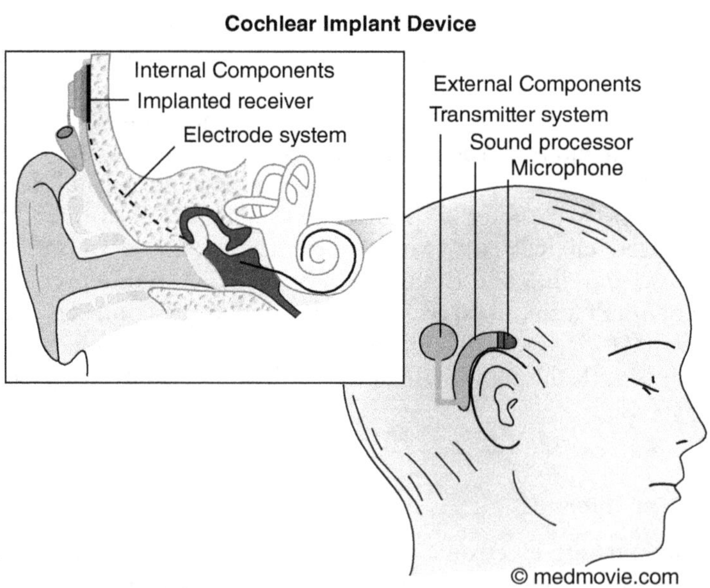


**Figure 2.1.** External and internal components of cochlear implants. (See color insert.)

There are few reports concerning the microbiology or the management of these infectious complications, although one study identified *S. aureus* as the most common pathogen [74]. Traditionally, it was felt that removal of the implant was required to allow for proper wound healing [75,76]. However, if the wound infection is detected early enough, advances in surgical techniques (e.g., distancing the incision in site from the implant bed and ensuring that there is no intersection of the two [75]), minimizes the likelihood that a superficial infection will spread to involve the implant. More recent studies suggest that postoperative wound infections that are not clinically severe may be managed conservatively (specifically, with incision and drainages, prolonged intravenous and /or oral antibiotics, and local wound care) thus allowing for retention of the implant [74,76]. Such an approach was sufficient in 25% of causes in one study [74]. The limitation of this suggestion, however, is that "infection severity" was not well defined and was actually based on the treatment modalities needed to cure infection. Such definitions prevent a firm conclusion as to the optimal management of these infections.

For wound infections with extrusion of the device, as well as for patients who develop implant-related sepsis, device explantation is recommended [74,77]. When this approach is necessary, the electrode array may be left in the cochlea [74,75]. This practice allows the intracochlear space to remain open and thus increase the likelihood of success for future reimplantation.

Recently, another major problem that has emerged is the association of cochlear implants with the development of subsequent bacterial meningitis among children in the United States [78]. The most frequent pathogen was *Streptococcus pneumoniae*, but *Haemophilus influenzae* was also identified. The origin of infection was felt to be otogenic (middle and inner ear). Onset of symptoms varied from days to months after surgery. Epidemiologic investigations linked this increased risk of bacterial meningitis to the use of a specific cochlear implant model that incorporated a positioner (a Silastic wedge placed next to the electrode array to position it against the medial wall of the cochlea to facilitate transmission of the electrical signal). As a result of these cases, the devices were recalled in July, 2002. Moreover, public health notifications in 2007 and 2006, and advices for patients in 2007 and 2006 from FDA repeatedly warn about the increased, life-threatening risk of bacterial meningitis in cochlear implant recipients and the importance of fully immunizing them.

The increased risk of bacterial meningitis, however, was not fully explained by the presence of this additional component, because the rate of *S. pneumoniae* meningitis among children with an implant that had no positioner remained 16 times higher than the background rate (age-matched control in the general population). This finding may be explained by the fact that children who require cochlear implants are more likely to have other risk factors for meningitis, such as inner-ear malformations and CSF leak. It is interesting that the association of cochlear implants with increased rate of bacterial meningitis has not been identified in the United Kingdom (the Manchester Cochlear

Implant Programme) [77]. Nonetheless, international committees recommend that all patients considered cochlear implant candidates, as well as patients with established implants, be vaccinated for *S. pneumoniae* and *H. influenzae* type b according to their age-appropriate schedules at least 2 weeks before surgery [78,79].

### 2.3.7. Intracardiac Devices

The major surgically implanted devices relevant to the heart are prosthetic cardiac valves, permanent transvenous pacemakers (TVPM)–implanted cardioverter defibrillators (ICDs), and ventricular assist devices (VAD).

### 2.3.8. Prosthetic Value Endocarditis

Treatment of native valvular heart disease has resulted in an increasing number of patients with a prosthetic valve. Infection of prosthetic cardiac valves, with or without surrounding cardiac structures, defines prosthetic valve endocarditis (PVE). These valves can be either mechanical prostheses or bioprostheses. Mechanical prostheses include ball-and-cage valves (e.g., the Starr–Edwards valve), single tilting disk valves (e.g., the Bjork–Shiley, Lillehei–Kaster, and Medtronic Hall models), and bileaflet tilting disk valves (e.g., the St. Jude Medical and CarboMedics products). The bileaflet valves are the most commonly used mechanical prostheses today [80]. Bioprosthetic valves include porcine heterografts, which are porcine valves fixed with glutaraldehyde, and bovine pericardial valves, which are constructed from bovine pericardium formed into three cusps mounted in a stent [80]. Prosthetic valves are most commonly inserted in the aortic or the mitral position.

Long-term follow-up studies indicate that 3–5.7% of valve recipients develop PVE [81]. However, the actual frequency of this infection is not uniform and varies with time after cardiac surgery. The rate peaks in the initial 3 months after surgery, remains elevated through the sixth month, and then declines to 0.3–0.6% by 12 months, where it remains constant thereafter [82]. Following a similar trend, the mortality rate is highest for PVE that develops within the first few months after surgery, ranging from 33–45% in some studies to as high as 77% in others [83,84]. Although it had been previously felt that valve position (aortic versus mitral) and prosthesis type (mechanical versus biological) influenced the propensity for PVE, recent studies indicate that the risk of PVE is similar for mitral and aortic valves, irrespective of prosthesis type. Furthermore, beyond the first postoperative year, the incidence of PVE of mechanical valves is similar to that of bioprosthetic valves [82,84].

Prosthetic valve endocarditis can be categorized into three periods, based on likely microbiology and method of infection: early (within the first valve surgery), delayed (between the second and twelfth months after surgery), and late (>12 months after surgery). The early period is predominantly caused by staphylococci, CoNS, and specifically methicillin-resistant *S. epidermidis*, accounts for a large majority of early PVE, followed by *S. aureus* [81,82,84]. Late PVE in otherwise healthy patients reflects community acquisition and is caused by organisms that typically cause native valve endocarditis, (i.e., viridans group streptococci, *S. aureus*, and the enterococci—although there is an increasing frequency of CoNS) [82]. The HACEK group of organisms (*Haemophilus aphrophilus*, *H. paraphrophilus*, *Actinobacillus actinomycetemcomitans, Cardiobacterium hominis*, *Eikenella corrodens*, *Kingella Kingae*) is also possible. The pathogens for delayed PVE consist of those that cause both early and late PVE [82].

Early PVE, the most common form, typically results from contamination of the prosthetic valve either directly at the time of implantation stemming from a break in sterile technique, or via transient bacteremia introduced from elsewhere, (e.g., from infected intravascular catheters or skin wounds). The surface of the prosthesis resists microbial adhesion. The sewing ring, however, and the valve annulus are not yet endothelialized and are therefore a site for platelet–fibrin thrombi formation. These platelet–fibrin aggregates are potential sites for the adherence of microorganisms. Adherent microorganisms that survive can then proliferate and form biofilm, which favors growth along the surface of the valve. With further thrombus formation, vegetation develops, which may eventually lead to functional obstruction or incompetence of the valve. If the infection that is initially in the annulus–prosthesis junction infiltrates the surrounding perivalvular tissue, the stability of the valve weakens, leading to a paravalvular leak. As the infection progresses, abscesses, fistulae, and progressive tissue destruction (annulus, cardiac conduction system) ensues. If not adequately controlled, the infection may erode to involve the aortic root, causing a mycotic aortic aneurysm, or the pericardium, causing an empyemic pericaditis. Early PVE, even with relatively "benign"skin flora (e.g., CoNS) may actually produce a very destructive infection associated with a high mortality.

The clinical manifestations of early PVE depend on the stage and severity of the infection. Fever is almost universally present [84], and its presence should prompt a search for PVE. Other evidence of early PVE includes new or changing regurgitant murmurs, congestive heart failure, shock, or electrocardiographic evidence of cardiac conduction disturbances. Systemic emboli, especially to the CNS, can occur in up to 40% of patients [82].

The diagnosis of PVE can be made using the modified Duke criteria. In the absence of antimicrobial therapy, blood cultures will be positive in ≥90% of patients [81]. Persistent bacteremia, defined as multiple positive blood cultures obtained independently over time, is usually present and can help confirm the pathogenic role of questionable isolates (e.g., the CoNS and *diphtheroids*). The role of cardiac echocardiography is clearly helpful. Although transthoracic echocardiography (TTE) has a sensitivity ranging from 17 to 36%, that of transesophageal echocardiography (TEE) ranges from 82 to 86% and is especially useful to assess valves in the mitral position [82,84]. The positive predictive value of a TEE for PVE is almost 90% [82]. However, a negative TEE does not necessarily exclude the diagnosis of PVE if there is other strong clinical evidence.

The management of PVE depends on the severity of the illness and may consist of medical therapy alone or combined medical and surgical interventions. In patients with a subacute course and hemodynamic stability, delaying antibiotic therapy while awaiting results of blood cultures has been suggested [82]. However, patients who present with evidence of significant valve dysfunction (e.g., septic shock, congestive heart failure) should be treated empirically with antibiotics after blood cultures have been obtained. Given the predominance of methicillin-resistant CoNS and *S. aureus* in early PVE, empiric therapy with vancomycin at 15 mg $kg^{-1}$ iv every 12 h should be initiated. Gentamicin at 1 mg $kg^{-1}$ iv every 8 h is recommended for synergy because, in the absence of resistance, it is an effective anti-staphylococcal antimicrobial [81]. A third agent, rifampin at 300 mg $kg^{-1}$ iv every 8 h is also recommended if there are no contraindications because of its unique bactericidal activity against *staphylococci* that have adhered to foreign material [82]. The *staphylococci*, however, have a high intrinsic mutation rate that confers resistance to rifampin, such that this antimicrobial should never be used alone to treat PVE. Once the susceptibility results of the isolate are available, the antibiotic regimen may be modified. Particularly if the *staphylococcal species* is identified as susceptible to β-lactams (e.g., cloxacillin), this class of antimicrobials is preferred because they are associated with more effective sterilization and better clinical outcomes in infective endocarditis [85].

Surgical management is indicated if certain prognostic markers associated with poor outcome are present. These include new or increasing murmur, moderate-to-severe congestive heart failure resulting from valve dysfunction, new cardiac conduction disturbances, persistent fever despite 10 or more days of appropriate antibiotics, or selected echocardiogram findings [86]. These features suggest complicated disease with significant valve dysfunction or paravalvular tissue invasion and are not likely to respond to antibiotic therapy alone [82,84]. In these situations, antibiotic therapy with valve replacement is necessary, because the combined modality treatment results in higher survival rates (44–64%) [84] and decreased risk of relapse.

*Staphylococcus aureus* PVE is also an alarming scenario. Medical therapy alone for this condition is associated with mortality rates ranging from 42 to 100%, but when antimicrobial therapy is combined with surgery, the mortality rate decreases 20-fold [81]. Thus, *S. aureus* PVE should strongly be considered for early surgical intervention. Replacing an infected valve with a new prosthesis in the context of a bacteremia raises the concern of recurrence of the infection onto the new valve. Recurrence of PVE, however, develops in only 6–15% of patients with active bacterial infection who are operated upon [84]. Following surgery, antibiotics should be continued for at least 6 weeks [82,84].

Fungi are responsible for 2–10% of all cases of PVE, and *Candida* accounts for up to 90% of these fungal infections [87,88]. Patients with prosthetic heart valves who develop nosocomial candidemia have a notable risk of developing

*Candida* PVE, often months or years later (up to 690 days later). In a review of 44 cases of candidemia in patients with prosthetic heart valves, *Candida* PVE developed in 25% of such patients [89]. Fungal PVE is associated with a higher mortality rate than bacterial PVE. A review [90] of fungal endocarditis concluded that untreated fungal PVE was uniformly fatal, antifungal agents or surgery alone reduced the mortality rate to 82%, whereas combined surgery and antifungal therapy decreased the mortality rate to 33%. Others [88] found a 33% overall mortality rate due to fungal PVE with a mean follow-up of 4.5 years. Patients with complicated PVE (congestive heart failure or persistent fungemia) had higher mortality rates regardless of the mode of therapy [91]. Surprisingly, one study reported that patients with *Candida* PVE had a better survival than those with candidemia alone (mortality rates of 25 and 83%, respectively, at 1 year) [89]. Combined medical–surgical therapy is the current standard of therapy of fungal PVE [91,92]. Although one study reported that the mortality in patients with uncomplicated PVE was rather similar in those who received medical therapy alone and in those who underwent combined medical–surgical therapy (40 and 33%, respectively) [91], the authors of that study concluded that medical therapy alone should be considered only for patients for whom surgery is regarded as unduly hazardous. However, fungal PVE should be suspected in the presence of negative bacterial blood cultures, bulky vegetations, metastatic infection, perivalvular invasion, embolization to large blood vessels, and disseminated fungal infection [88,91,93].

### 2.3.9. Transvenous Permanent Pacemakers/Automatic Implantable Cardioverter Defibrillators

Cardiac pacemakers and automatic implantable cardioverter defibrillators (AICDs) have revolutionized the management of arrhythmias, decreasing mortality in patients with known ventricular arrhythmias, as well as in those with ischemic cardiomyopathy who are at risk for ventricular arrhythmias [94]. It is therefore, anticipated that the use of these devices will continue to increase.

Both TVPMs and AICDs consist of a generator or defibrillator inserted subcutaneously into the chest (or abdominal ) wall, as well as lead wires that pass through the soft tissue, enter the subclavian vein, pass through the right side of the heart, and terminate with electrodes implanted in the right atrial and/or right ventricular endocardium. Infection rates of these devices have not been well quantified, but range from 1 to 7% [82,95]. The TVPM–ICD infections can be divided into four major groups, depending on which component of the system is involved. Infection can involve the generator–defibrillator pocket, the lead wires as they pass through the soft tissue, thrombi that form on the lead wires as they pass through the venous system, or endocardial structures that communicate with the lead wires (e.g., the tricuspid valve or the insertion sites of the electrodes into the right atrium or right ventricle). The TVPM–ICD infections can also be classified by time of onset following implantation. Early infection is defined as that developing within the first month, while late infection develops after the first month. These classifications help determine pathogenesis and likely microbial etiologies and can thus guide management.

Early infection most often occurs as a result of intraoperative contamination of the TVPM–ICD or the pocket at the time of implantation. Late implantation usually develops as a result of erosion of the generator–defibrillator through the skin, with subsequent contamination of the pocket, followed by spread of the infection to the electrodes and eventually to the endocardial surfaces. An alternative method, albeit rate, in which infection of these devices may occur, is with bacteremia from a distance site of infection, with hematogenous spread to the TVPM–CID. This situation has been most frequently described with *S. aureus* [96].

The majority of TVPM–AICD infections are caused by staphylococci. Early infection tends to be predominated by *S. aureus*, whereas late infections, probably resulting from exposure of the device after skin erosion, are more commonly caused by CoNS [82]. Other skin flora have been implicated, however, including *propionibacterium acnes* and *Micrococcus* spp., as have a variety of Gram-negative bacilli (e.g., *E.coli*, *Klebsiella* spp., *Enterobacter* spp., *Serratia* spp.), mycobacteria, and fungi. In one series, 13% of cases were polymicrobial infections [95].

The clinical manifestations of TVPM–ACID infections are variable and depend on which component is affected. Infections involving the pocket, either isolated or in association with infections of deeper components, manifest as "pocket cellulites" with erythema, pain, and warmth. A draining sinus tract or erosion through the overlying skin may be present as well. In addition to these local symptoms, systemic symptoms (e.g., fever, chills, malaise, nausea) may also be present. Isolated systemic complaints in the absence of any local findings may occur. Note that despite the potential endovascular infection associated with these devices, a history of fever is present in less than one-third of patients, highlighting the need to consider this infection in patients who have a TVPM–AICD and who are systemically unwell, even in the absence of fever [95,97]. Furthermore, the absence of fever does not exclude bacteremia. In one report, 40% of patients with bacteremia related to TVPM–AICD infections did not have fever or systemic symptoms [95]. Hence, infections of these devices must always be considered when a patient presents with local or systemic complaints, and bacteremia must always be pursued with serial blood cultures, even in the absence of systemic symptoms. If infection of the endocardial surfaces is suspected (i.e., tricuspid valve or lead vegetations), echocardiography can be helpful in establishing the diagnosis [95]. As with other types of endocarditides, pacemaker endocarditis is best evaluated with a TEE rather than a TTE (respective sensitivities of 91–96 vs 23–30%) [82].

The optimal management of TVPM–AICD infections involves a combined medical–surgical approach consisting of complete explantation of all hardware with prolonged antibiotic therapy. In retrospective studies [95,98–100], this approach has been associated consistently with lower mortality rates when compared to medical therapy alone. Although the older reports noted that electrode removal, either via traction or via cardiotomy, was associated with significant morbidity and mortality [98], the availability of newer devices to facilitate wire removal (e.g., excimer laser sheath technology) may allow for safer complete extractions [101]. Previous attempts at partial device removal, specifically, the removal of only the generator and adjacent lead wires with concomitant antibiotic therapy, typically have been associated with low success rates [95,98,102]. This is presumably because the infection had extended to involve the deeper wires and electrodes.

Based upon the microorganisms responsible for TVPM–AICD infections, specifically the microflora from the skin, empiric antimicrobial therapy with the regimen recommended for PVE (i.e., vancomycin, gentamicin, and rifampin) is prudent pending the results of blood and device cultures. The duration of antibiotic therapy is not well defined for this entity. However, extrapolation from the reports about management of infective endocarditis suggest at least 4 weeks of iv antibiotic therapy after complete removal of the device, if there is evidence of pacemaker endocarditis (i.e., persistent positive blood cultures, vegetations on lead wires or electrodes).

For infections that are limited to the device pocket, some authors favor device removal and parenteral antibiotics initially, with or without subsequent oral therapy [82]. The duration of therapy depends on whether bacteremia was present, as well as the clinical response. For patients without fever, bacteremia, or detectable vegetations on echocardiography, but who have positive electrode tip cultures, the optimal duration of therapy remains unclear. However, 45 weeks of parenteral therapy seems prudent, given the potential for infection involving the endocardium of the right atrium and/or right ventricle [82]. For patients in whom removal of the device is not feasible, prolonged (or even indefinite) chronic suppressive oral antibiotics are given, with selection of agents based on culture results, bioavailability of the agent from the gastrointestinal tract, frequency of dosing, and tolerability.

The optimal timing for reimplantation of a pacemaker or AICD is unknown. Some authors have recommended reimplantation of the device as early as 36-h explantation in patients with only local pocket infection [103]. Other authors, however, suggest waiting at least several days before considering reimplantation, because most blood cultures become positive within 48 h [95], during which time the need for a TVPM or AICD can be re-evaluated.

### 2.3.10. Left Ventricular Assist Devices

Left ventricular assist devices (LVADs) are temporary mechanical circulatory support devices that act as a bridge for patients awaiting cardiac transplantation. An LVAD is an implanted device that replaces the pump function of the damaged left ventricle. Many models of LVADs are available and are reviewed elsewhere [104]. A representative model (e.g., the HeartMate®) (Thoratec), consists of a Ti encased blood pump, an inflow cannula from the left ventricle, and an outflow cannula inserted into the ascending aorta [105]. In this model, two porcine valves maintain unidirectional flow. The blood pump is placed into the abdominal cavity or a preperitoneal pocket. A driveline connects the device to an external power source via an exit site in the abdominal wall. There are, however, only a limited number of donor organs, and thus some centers have focused on the use of LVADs as long-term devices to entirely avoid heart transplants [106]. Whether temporary or permanent, infection of LVAD remains a major complication, occurring in 25–50% of patients. It is unclear whether an LVAD infection is associated with increased risk for mortality in this patient population [107].

Although the epidemiology of LVAD related infections is not well studied because of the relative novelty of this technology, one study suggests the majority of infections occurred between weeks 2 and 6 postimplantation, suggesting this period as a high-risk time for infection [107]. A longer-term study following patients for up to 2.5 years after implantation found that one-third of LVADs eventually became infected within 3 months [82]. Currently, there is no division of the timing of LVAD related infections into discrete periods postinsertion, as there is with the other intracardiac devices. The early onset of infections (i.e., <6 weeks) is likely related to biofilm formation by particular microorganisms [107]. The later onset (i.e., <3 months) may be related to technical aspects, considering that device malfunction may precede almost 25% of infections [107].

The microbiology of these infections is also different, because the majority of patients who require LVAD have end-stage cardiomyopathy, are critically ill, and infections of these devices tend to be with microorganisms acquired from the healthcare environment. Staphylococcal species, especially methicillin-resistant ones, predominate. Other common etiologies are the Gram-negative rods (*Pseudomonas* spp., *Klebsiella* spp., *E. coli*, *Enterobacter* spp., *Proteus* spp., *Serratia* spp.), *Candida* spp., and *Enterococcus* spp. [84,107,108]. As such, empiric therapy with vancomycin and an antipseudomonal agent (e.g., ceftazidime or ciprofloxacin) would be appropriate. Empiric antifungal therapy should also be considered [107,109] although the choice of ideal agent (e.g., azoles, amphotericin B formulations, or echinocandins) is not supported by good evidence.

The LVAD related infections can be categorized anatomically into driveline infection, pump pocket infection, pump endocarditis (i.e., involving the pump's interior or inflow–outflow tracts), and infection of the sternum–substernal space (mediastinitis). This categorization correlates with clinical manifestations and helps to provide a rational for management, although there is little definitive evidence to guide investigation or therapy.

A driveline infection is characterized by the presence of poor healing and localized evidence of inflammation at the exit site. Systemic manifestations (e.g., fever, leukocytosis) may not occur. Thus, in the evaluation of a patient with an LVAD, it is essential to distinguish between a driveline and a deeper wound infection that would require more aggressive treatment. Management of a driveline infection combines two approaches: local wound care, which entails debridement and local cleansing with a bactericidial detergent to remove or destroy microbial biofilm and then directed systemic antimicrobial therapy [110,111].

A pump pocket infection can present in various ways. It may occur in an indolent fashion with subtle symptoms, usually indicating the presence of less virulent organisms (e.g., *S. epidermidis*). Alternatively, it can present as a pocket abscess, or with frank sepsis. It may also manifest as a new and persistent drainage from the driveline exit site. It has been suggested that multiple blood cultures in patients with LVADs who present with even low-grade fevers and constitutional symptoms (e.g., unwellness, malaise) will help to establish the diagnosis [110,111]. If the blood cultures yield *Staphylococcus* spp., all indwelling prosthetic vascular channels are cultured and considered for removal. If the bacteremia persists despite an adequate trial of systemic antibiotics, an ultrasound of the pump pocket is obtained. Most pump pockets, especially in the initial weeks after surgery, will contain fluid and thus the presence of fluid is not diagnostically helpful. However, absence of fluid should prompt a search for another source of infection. Abundant or complicated fluid collections, especially in the context of bacteremia, may warrant pocket exploration, during which time fluid is sampled for culture. Anecdotal experience suggests that the simultaneous insertion of PMMA antibiotic beads containing vancomycin and gentamicin in the pocket may be helpful. However, systemic antibiotics would also be required, but the duration of therapy is uncertain [110–112].

Pump endocarditis is defined as infection of any surface of the system (e.g., the pump's interior surface), the inflow tract, or the outflow that is in contact with blood, typically leading to bacteremia. As with other endocarditides, persistent bacteremia in the absence of another source in a patient with an LVAD is suggestive of pump endocarditis. Other clues that may indicate the presence of this type of infection include septic embolization without evidence of vegetations on the native heart valves and dysfunction or insufficiency of the inflow or outflow valves with accompanying bacteremia [111]. The best methods of definitively identifying the pump as the source of endocarditis are unknown. Echocardiography and magnetic resonance imaging (MRI) are not technically possible. Computed tomography and nuclear imaging with isotope labeled leukocytes have been used, with unknown sensitivities or specificities. Pump endocarditis is also challenging with regard to management, because removal of the LVAD to eliminate the source of bacteremia is not usually feasible without concomitant cardiac transplantation and subsequent immunosuppression [105]. Replacement of the infected LVAD with new devices is associated with substantial risk for death and complications [111]. Prolonged iv antibiotics with a minimum of 4–6-weeks has been suggested [84,113] if transplantation particularly is in the near future. One study suggests continuing adequate antibiotic therapy until transplantation, particularly if *S. aureus* is the offending agent [114]. Failure to control pump endocarditis (i.e., persistent bacteremia despite appropriate antibiotics, relapsing bacteremia after completion of antibiotics, or fungaemia) may necessitate cardiac transplantation as the only treatment option available that can successfully manage the infection if the patient is stable enough to undergo the procedure [115–118].

Infection of the sternal–substernal space is the least common manifestation of LVAD related infections. Treatment consists of prolonged systemic antibiotics with aggressive wound drainage and revision, possibly with muscle flap coverage [119].

### 2.3.11. Breast Implants

The use of breast implants for cosmetic and reconstructive surgeries is increasingly common. Although the more common complications related to breast implants involve tissue disfigurement resulting from abnormal healing, the more medically serious adverse events relate to implant infection. The rate of such infections after augmentation mammaplasty ranges from 1.7 to 2.5% [120]. However, the rate of infection may be higher in patients who received implants after mastectomy for cancer or for cancer prophylaxis [121].

The manifestations of breast-implant-related infections are much less dramatic than with other device-related infections, because frank sepsis almost never occurs [122]. Infections arising from breast implants typically present with unilateral local symptoms in an acute or subacute–chronic manner. Acute infections typically occur as postoperative surgical site infections in the few weeks subsequent to the surgery, with erythema, edema, and poor healing, which may be accompanied by purulent discharge or systemic symptoms. Subacute–chronic infections manifest more subtly. In one of the larger studies on breast-implant infections, 139 infected breast implants from 72 patients were evaluated [123]. The most common inflammatory symptoms identified were breast pain (93%), axillary pain (39%), and upper extremity paresthesia (29%) [123]. Poor aesthetic appearance, for which there is mounting evidence that subacute–chronic infections may play a role [124,125], were also very common; specifically, capsular contracture occurred in 91% of patients, and breast shape change was noted in 80% [123].

The microbiology of breast-implant infections has not been studied as extensively as other SIDIs. The most frequency reported microorganisms causing acute infections are *S. aureus*, the peptostreptococci, and *C. perfringens* [123]. In the previously mentioned study, 47% of the 139 removed implants were culture positive. The most commonly identified organisms of these subacute–chronic infections were *P. acnes* (57.5%), *S. epidermidis* (41%), and *E. coli* (1.5%). Fungal infections were not detected. Consistent with the indolent nature of most of these bacteria, culture positivity was not significantly associated with systemic symptoms.

Augmentation mammaplasty procedures may also be complicated by infection with rapidly growing atypical mycobacteria, specifically *M. fortuitum* and

*M. chelonei* complex [126]. The onset of clinical signs of infection varies between 1 week to 2 years after the procedure. The presenting clinical features are predominantly local and include swelling, tenderness, erythema, and discharge. Systemic symptoms (e.g., fever, chills, and other signs of sepsis) may be absent. The discharge is usually odorless and serous, but may be purulent. Routine wound cultures may be negative or suggestive of diphtheroid organisms. Gram-staining may reveal white blood cells with no organisms. Although experience in management is limited, removal of the prosthesis and oral antimycobacterial therapy for at least 6 months is suggested [126].

The optimal management of bacterial breast-implant infections is not well established and is at best derived from case studies [127]. For suspected infections, wound or fluid culture is obtained and empiric antibiotics directed at the previously mentioned organisms started. Response to antibiotics would determine whether surgical intervention were required. For persistent mild infections, local debridement (e.g., capsulectomy with closed-suction drainage), in addition to antibiotic therapy, may be successful.

### 2.3.12. Penile Implants

Penile prosthetic implants to restore erectile function are being used increasingly. Although many devices exist, the inflatable penile prosthesis is the most frequency implanted prosthesis in the United States [128]. Infection rates of penile implants range from 2 to 8% [129], although the types of infections included range from superficial wound infections to penile gangrene, and the rates vary with primary implantation versus reoperation for revision or reimplantation [130].

Penile prosthetic infections (PPIs) are divided into clinically apparent and subclinical infections [129]. The former group is characterized by an acute presentation involving erythema, induration, and tenderness overlying the portion of the prosthesis, and it may also be accompanied by fever, discharge, or even device extrusion. The majority of these infections occurs in the early postoperative period, and is likely due to contamination at the time of implantation. These infections previously have been reported to be most commonly due to *S. epidermidis* [131,132], although quantitative culture analysis has demonstrated that low colony counts of *S. epidermidis* may be found in the absence of prosthetic infection as a result of either asymptomatic colonization or culture contamination, and that the reported incidence of prosthetic infections attributed to this organism may be overestimated [130].

In infected, prostheses, surgical intervention and antibiotics are required. Depending on the severity of the infection and on particular host characteristics, different surgical approaches are considered. If there is exposure of the device, presence of purulent discharge, immune compromise, or sever diabetes, the removal of all foreign body penile prostheses components is required, along with drainage of any fluid collections [129,133]. The wound is allowed to heal by secondary intention, although skin closure may be attempted after

3–5 days. A conservative estimate for optimal time before prosthesis replacement is 90–120 days [129]. However, in patients without evidence of tissue necrosis, significant purulence, immune compromise, diabetes, or rapidly progressing infection or cylinder erosion, salvage procedures may be attempted. This surgical approach consists of the following steps: prosthesis removal, wound irrigation with serial antibiotic and cleansing solutions, and subsequent reimplantation in the same operation [134]. The reported success rate after following this approach is 85% [135].

Subclinical prosthetic infections are the most frequent presentation of PPIs, typically manifesting a chronic prosthesis-associated pain or even device migration [136]. A proposed management strategy for this type of infection consists of empiric antibiotic therapy with ciprofloxacin or with a cephalosporin [129,137]. If antibiotics are successful in suppressing pain, with an expected response rate of 60% for ciprofloxacin and 25–30% for cephalexin [129], it should be continued for 10–12 weeks [136]. If pain persists or recurs after antibiotics are discontinued, surgical intervention is recommended. Intraoperatively, irrigation with vancomycin and protamine was used to emulsify bacterial biofilm, followed by prosthesis removal and substitution of a new device [137]. Vancomycin was used systemically for the following 24–48 h to target *S. epidermidis*. This approach was reported to be successful in 90% of patients, although the sample size was quite small (14 painful penile prostheses evaluated; 13 were culture positive, 10 of the culture-positive devices were exchanged, resulting in 9 of 10 patients with resolution of pain) [136,137].

The most serious PPI is genitourinary gangrene, in which management is similar to that of Fournier's gangrene. An immediate surgical intervention for device removal is obligatory, along with thorough irrigation and broad-spectrum antibiotics (including coverage for anaerobes, e.g., *Bacteroides fragilis*) [136]. There may be a potential role for hyperbaric oxygen therapy, but only adjunctive to the above-mentioned therapy. Unfortunately, these infections are usually so severe that salvage of the penis is impossible.

In summary, the implantation of inflatable penile prostheses for the treatment of erectile dysfunction continues to be widely practiced in the United States and internationally. As third-line therapy for erectile dysfunction, the numbers of implants continue to rise as the population of men treated for erectile dysfunction increases. Complications of penile prosthesis implantation continued to decline as mechanical malfunctions have decreased as a result of re-engineering inflatable penile prostheses. Inflatable penile prostheses from both available vendors continue to be reliable, effective methods for restoring erectile function with high satisfaction rates. The most troublesome complication of these prostheses, however, is not mechanical, but rather that of prosthesis infection. Prosthesis infections may result in further surgery, loss of penile tissue, and even the inability to replace penile prosthesis. While standard sterile technique perioperative antibiotics and careful surgical procedures continue to be the cornerstone of penile prosthesis infection avoidance, newer designs of penile prostheses for antibiotic coating have resulted in an improvement in the prevalence and incidents of penile prosthesis infection. For those patients in whom penile prostheses become infected despite adequate prophylaxis, newer techniques of salvage have demonstrated increasing success. Once and still the most dreaded complication of penile prosthesis implantation, prothesis infections can now be avoided by perioperative preparation and antibiotics, as well as antibiotic-coated penile prostheses. Treatment of penile prosthesis infections once associated with severe loss of function can often be successful with modern salvage techniques. Implanting urologists must be familiar with prophylaxis, avoidance, and treatment of penile prosthesis infections.

## 2.4. ORTHOPEDIC-DEVICE-RELATED INFECTIONS

Orthopedic implants are used for bone fixation or joint replacement. Although the actual current incidence of infections of these devices is very low when the procedures are performed by experienced surgeons (<1–2%) [21], the absolute number of these infections is anticipated to rise as the number of elderly patients and victims of trauma requiring fixation or joint replacements increase. These infections, when they do develop, are associated with significant morbidity with attributable mortality rates of 2.7–18% [138] and additional expenditures to the healthcare system [139].

Traditional principles in the management of orthopedic-device-related infections (ODRIs) consisted of complete removal of the arthroplasty or fixation device, followed by prolonged antimicrobial therapy and possible future reimplantation. This approach, however, has been empiric, hindered by the lack of consensus definitions or classification of such infections, as well as the paucity of statistically well-powered studies to address the optimal management of this condition.

Progress has been made in terms of identifying patients at highest risk for ODRIs. A case-control study has clearly demonstrated that patients with the following risk factors were at high risk for prosthetic joint infections: postoperative surgical site infection, an elevated NNIS score, systemic malignancy, and previous joint arthroplasty [138]. The joint replaced is also an important consideration for risk of infection. Knee arthroplasties are associated with a higher risk of infection than are hip arthroplasties [140].

An additional advancement was in the establishment of definitions and a classification of ODRIs. The initial classification scheme proposed by Coventry [141] divided the periods after implantation into three stages. Based on subsequent clinical evidence of successful management, a revised classification system has been suggested by Widmer [140].

- Infection in the early postoperative period, a "type 1" infection, typically has its onset <2–4 weeks after surgery. It manifests as an acute surgical site infection, with local and systemic features. The predominant local findings consist of persistent pain, accompanied by evidence of wound infection. Fever is present and may be accompanied by chills, rigors, and diaphoresis. One of the major clinical objectives at this point is to distinguish a superficial infection from a deep surgical site infection in which there is continuous spread of the infection to the fascia, muscles, and tissue around the implant [140].

- The late chronic stage infection, or "type 2" ODRI, is characterized by an onset at least 1 month after prosthesis insertion [140]. The onset is insidious and typically occurs between 16 months and 2 years [142]. The signs and symptoms are also very subtle, manifesting most often as persistent pain or early, loosening of the prosthesis. However, loosening may also be related to an aseptic process (e.g., repetitive wear with secondary osteolysis) [143]. Distinction between loosening from a chronic infection and aseptic loosening remains difficult. Occasionally, these infections manifest as sinus tract formation with chronic discharge; the presence of this finding is highly suggestive of a type 2 infection [144].
- A "type 3" ODRI consists of hematogenous infection with subsequent seeding of the orthopedic implant, typically occurring with joint prostheses. Observations suggest that this type of infection usually occurs >2 years after surgery [21,140]. However, peri- or-postoperative bacteremia may also infect hardware sooner. Its presentation is acute, with sudden, rapid deterioration in the function of an implant that previously had been functioning well, along with systemic features of infection.

Based on these definitions, the spectrum of causative organisms can be classified. Type 1 infections usually occur as the result of intraoperative contamination and are most commonly caused by *S. aureus* and CoNS [138,145]. Type 2 ODRIs are also speculated to originate at the time of surgery with inoculation of skin flora present in low inoculum (i.e., *S. aureus*, CoNS) or organisms of low virulence (e.g., *Propionibacterium* species and other anaerobes) [146,147]. Type 3 infections are due to hematogenous seeding following transient or catheter-related bacteremia and most commonly are caused by the *streptococci* and Gram-negative bacilli [140,146].

Diagnostic investigations to be pursued in the face of possible ODRIs are not standardized. A frequent approach consists of a complete blood count to determine the presence of leukocytosis accompanied by a left shift, an erythrocyte sedimentation rate (ESR), and a C-reactive protein (CRP). Although the last two measurements can be nonspecifically elevated from a variety of inflammatory processes, including the orthopedic surgery, they should not remain elevated beyond 2–3 weeks if the postoperative course was uncomplicated [140]. A normal ESR and CRP essentially eliminate the possibility of an ODRI [140]. There have been no studies assessing the diagnostic potential of procalcitonin in ODRI, although it may prove useful in identifying patients with suspected bacterial infections [148]. Imaging is also frequently used to help establish whether a prosthesis is infected. Standard radiography may be helpful, although in a recent multicenter review, its performance characteristics fared worse than the use of ESR or CRP [149]. Nuclear imaging with various radioisotopes is not significantly better than plain radiography, with the major limitation being the inability to distinguish acute bacterial infection from noninfectious inflammation, and occult infection from aseptic loosening [149].

Preoperative joint aspiration appears to be the most useful test, as an adjunct to the ESR and CRP, in diagnosing an ODRI [140,149]. It should, however, be performed under sterile technique, and given the possibility of the infectious organisms adhering to the prosthesis in a biofilm, ultrasound-guided needle placement is suggested to obtain an ideal sample. Multiple independent specimens, ideally at least three, should be obtained at the time of the procedure to assist in the interpretation of microbiological results. The presence of the same organisms in all three specimens, in the context of a clinically suspected ODRI, supports the etiologic role of the identified bacteria [150].

Once the decision for surgical intervention is made, multiple intraoperative aspirates should be obtained for culture, in conjunction with biopsy specimens for histopathology. The combination of multiple positive microbiological analysis with evidence of inflammation on histopathology (≥5 neutrophils–high-power field) is felt to be the current reference standard for the diagnosis of ODRIs [140,147].

Operative interventions for the management of ODRIs have expanded. Current options include salvage procedures, consisting of surgical drainage and debridement with retention of prosthesis and long-term treatment with antibiotics; one or two stage replacement (with or without use of antimicrobial cement and long-term treatment with antibiotics); girdlestone arthroplasty (resection of the proximal femur for persistent deep-seated hip infections); arthrodesis; and amputation [140,147]. Chronic–life-long suppressive antimicrobial therapy may be indicated in patients who are not surgical candidates [151].

Optimal management appears to be based on the type of ODRI present [140,147,152]. For type 1 infections, the preferred method is surgical exploration and debridement as soon as possible after onset of symptoms, prior to the use of empiric antibiotics, so that multiple biopsy samples of periprosthetic tissue can be obtained for microbiological analysis. Antibiotics should be given as soon as the joint capsule has been incised and appropriate specimens are obtained for microbiological and histopathological analysis. Criteria have been proposed to determine which patients could be considered eligible for salvage procedure, thus minimizing their need for further surgical intervention and delayed recuperation [140]. These factors include (1) an acute infection manifesting either within 2–4 weeks after surgery or as an acute-onset hematogenous infection; (2) a stable prosthesis with no evidence of loosening; (3) microbiological evidence of infection, as determined by the presence of a single or predominant organism from multiple specimens obtained intraoperatively by aspiration; (4) histopathologic evidence of infection , determined intraoperatively, preferably by frozen section; (5) the availability of an antibiotic regimen with preferably bactericidal activity (based on susceptibilities determined from the preoperative joint aspirate); and (6) a patient willing to be compliant with long-term antimicrobial therapy [140].

In a single, randomized, placebo-controlled study of patients with ODRIs who fulfilled the above criteria, those treated by salvage procedure, combined with ciprofloxacin 750 mg po bid and rifampin 450 mg po bid, had cure rates for staphylococcal infections of 100% [153]. The key to successful therapy appeared to be a short duration of infection-related symptoms prior to therapy. These results are far superior-to-salvage procedures combined with prolonged β-lactam antibiotics, which is associated with failure rates of 32–86% [147].

However, if the above criteria are not fulfilled, a salvage procedure is discouraged. Instead, a one- or two-stage revision should be considered. Antibiotic therapy is recommended initially to be given intravenously [140]. Although the duration of antibiotics is controversial, a minimum of 2 weeks is required. A change to oral therapy to complete at least 3 months for internal fixation devices and hip prostheses and at 6 months for total knee prostheses is recommended [140]. If quinolones are not tolerated, or the organism is resistant, combined fusidic acid and rifampicin appears equally affective [154]. Duration of therapy may be more prolonged, depending on clinical signs and symptoms, as well as the trend of serologic parameters (ESR, CRP).

For type 2, or late chronic infection, a salvage procedure is not an option, particularly if there is loosening of the implant or a fistulous communication to the prosthesis [140]. A preoperative joint aspirate may allow for the identification of organisms and their susceptibilities, so that antimicrobial therapy may be given in the weeks preceding surgical intervention. A one-stage revision is recommended for patients with simple ODRIs. Patients with sinus tracts, deep abscesses, or infections with microorganisms for which an oral regimen with good bioavailability is not available (specifically, methicillin-resistant *S. aureus*, the *enterococci*, and quinolone-resistant *P. aeruginosa*) will have unacceptable failure rates with a one-stage revision and thus should undergo a two-stage revision [147]. In a two-stage revision, the initial prosthesis is removed, and surrounding tissue is sampled multiple times for culture (aerobic and anaerobic) and for histopathology. Subsequently, a space-holder or external fixation devices are installed.

Antimicrobial therapy is given based on available culture results. The duration of antibiotics before reimplantation depends on the causative organisms identified. If antibiotics with efficacy against surface-adhering bacteria can be employed (e.g., rifampin), then 2–4 weeks appears adequate; otherwise, 6 weeks is recommended [147]. All antibiotics should be stopped 2–4 days before reimplantation. This practice allows for the subsequent tissue biopsies to be properly assessed for bacterial culture. If the specimens have no growth of bacteria, then a further 6 weeks of antimicrobial therapy is given [147]. However, if organisms are identified, then directed antibiotic therapy is given for another 3 months for hip replacement and for 6 months in knee replacement. If *staphylococci* are identified, an antibiotics combination containing rifampin is highly recommended, provided that the organism is susceptible.

To improve patient outcome and reduce healthcare costs, strategies should be implemented to reduce the incidence of implant-associated nosocomial infections. Before proceeding with different strategies, pathogenesis of device-related nosocomial infections, biofilm resistance to conventional antimicrobial agents, and analytical techniques for identifying and characterizing biofilm are being shortly outlined one by one in Chapters 3–5.

## REFERENCES


1. Darouiche, R.O. (2001), Device-associated infections: a macroproblem that starts with microadherence, *Clin. Infect. Dis.*, 33, 1567–1572.
2. National Nosocomial Infections Surveillance System. (2001), NNIS System report. Data summary from January 1992–June 2001, issue August 2001, *J. Infect. Control*, 29, 404–420.
3. Tacconelli, E., Tumbarello, M., Pittiruti, M., Leone, F., Lucia, M.B., Cauda, R., and Ortona, L. (1997), Central venous catheter-related sepsis in a cohort of 366 hospitalised patients, *Eur. J. Clin. Microbiol. Infect. Dis.*, 16, 203–209.
4. Mermel, L.A., Farr, B.M., Sherertz, R.J., Raad, I.I., O'Grady, N., Harris, J.S., and Craven, D.E., Infectious Diseases Society of America; American College of Critical Care Medicine; Society for Healthcare Epidemiology of America. (2001), Guidelines for the management of intravascular catheter-related infections, *Infect. Control Hosp. Epidemiol.*, 22, 222–242.
5. Richards, M.J., Edwards, J.R., Culver, D.H., and Gaynes, R.P. (2000), Nosocomial infections in combined medical-surgical intensive care units in the United States, *Infect. Control Hosp. Epidemiol.*, 21, 510–515.
6. Rex, J.H., Bennett, J.E., Sugar, A.M., Pappas, P.G., van der Horst, C.M., Edwards, J.E., Washburn, R.G., Scheld, W.M., Karchmer, A.W., Dine, A.P., Levenstein, M.J., and Webb, C.D., for The Candidemia Study Group and the National Institute of Allergy and Infectious Diseases Mycoses Study Group. (1994), A randomized trial comparing fluconazole with amphotericin B for the treatment of candidemia in patients without neutropenia: Candidemia Study Group and the National Institute, *N. Engl. J. Med.*, 331, 1325–1330.
7. Miller, P.J. and Wenzel, R.P. (1987), Etiologic organisms as independent predictors of death and morbidity associated with bloodstream infections, *J. Infect. Dis.*, 156, 471–477.
8. Wenzel, R.P. (1995), Nosocomial candidemia: risk factors and attributable mortality, *Clin. Infect. Dis.*, 20, 1531–1534.
9. Fraser, V.J., Jones, M., Dunkel, J., Storfer, S., Medoff, G., and Dunagan, W.C. (1992), Candidemia in a tertiary care hospital: epidemiology, risk factors, and predictors of mortality, *Clin. Infect. Dis.*, 15, 414–421.
10. Anaissie, E.J., Rex, J.H., Uzun, Ö., and Vartivarian, S. (1998), Predictors of adverse outcome in cancer patients with candidemia, *Am. J. Med.*, 104, 238–245.

11. Raad, I. and Bodey, G.P. (1992), Infectious complications of indwelling vascular catheters, *Clin. Infect. Dis.*, 15, 197–210.
12. O'Grady, N.P., Alexander, M., Dellinger, E.P., Gerberding, J.L., Heard, S.O., Maki, D.G., Masur, H., McCormick, R.D., Mermel, L.A., Pearson, M.L., Raad, I.I., Randolph, A., and Weinstein, R.A. (2002), Guidelines for the prevention of intravascular catheter-related infections, Centers for Disease Control and Prevention, *MMWR Recomm. Rep.*, 51(RR-10), 1–29.
13. Rello, J., Ochagavia, A., Sabanes, E., Roque, M., Mariscal, D., Reynaga, E., and Valles, J. (2000), Evaluation of outcome of intravenous catheter-related infections in critically ill patients, *Am. J. Respir. Crit. Care Med.*, 162, 1027–1030.
14. Digiovine, B., Chenoweth, C., Watts, C., and Higgins, M. (1999), The attributable mortality and costs of primary nosocomial bloodstream infections in the intensive care unit, *Am. J. Respir. Crit. Care Med.*, 160, 976–981.
15. Soufir, L., Timsit, J.F., Mahe, C., Carlet, J., Regnier, B., and Chevret, S. (1999), Attributable morbidity and mortality of catheter-related septicemia in critically ill patients: a matched, risk-adjusted, cohort study, *Infect. Control. Hosp. Epidemiol.*, 20, 396–401.
16. Collignon, P.J. (1994), Intravascular catheter associated sepsis: a common problem. The Australian study on intravascular catheter associated sepsis, *Med. J. Aust.*, 161, 374–378.
17. Pittet, D., Tarara, D., and Wenzel, R.P. (1994), Nosocomial bloodstream infection in critically ill patients, excess length of stay, extra costs, and attributable mortality, *JAMA*, 271, 1598–1601.
18. Dimick, J.B., Pelz, R.K., Consunji, R., Swoboda, S.M., Hendrix, C.W., and Lipsett, P.A. (2001), Increased resource use associated with catheter-related blood-stream infection in the surgical intensive care unit, *Arch. Surg.*, 136, 229–234.
19. Mermel, L.A. (2000), Correction: catheter related bloodstream-infections, *Ann. Intern. Med.*, 133, 395.
20. Brub-Buisson, C. (2001), New technologies and infection control practices to prevent intravascular catheter-related infections, *Am. J. Respir. Crit. Care Med.*, 164, 1557–1558.
21. Zimmerli, W., Trampuz, A., and Ochsner, P.E. (2004), Prosthetic-joint infections, *New Engl. J. Med.*, 351, 1645–1654.
22. Hamill, T.M., Gilmore, B.F. Jones, D.S., and Gorman, S.P. (2007), Strategies for the development of the urinary catheter, *Expert Rev. Med. Devices*, 4, 215–225.
23. Rouprêt, M., Daudon, M., Hupertan, V., Gattegno, B., Thibault, P., and Traxer, O. (2005), Can urinary stent encrustation analysis predict urinary stone composition? *Urology*, 66, 246–251.
24. Stocks, G. and Janssen, H.F. (2000), Infection in patients after implantation of an orthopedic device, *Asaio J.*, 46, S41–S46.
25. Lim, E.V. and Stern, P.J. (1986), *Candida* infection after implant arthroplasty. A case report, *J. Bone Joint Surg. Am.*, 68, 143–145.
26. Brause, B.D. (1989), Prosthetic joint infections, *Curr. Opin. Rheumatol.*, 1, 194–198.
27. Stickler, D.J., Morris, E.A., and Hughes, G. (2002), Strategies for the control of catheter encrustation, *Int. J. Antimicrob. Agents*, 19, 499–506.

28. Bonner, M.C., Keane, P.F., and Gorman, S.P. (1993), Characterization and antibiotic sensitivities of isolates from urethral stent biofilm, *J. Pharm. Pharmacol.*, 45, 1445.
29. Warren, J.W. (2001), Catheter-associated urinary tract infections, *Int. J. Antimicrob. Agents*, 17, 299–303.
30. Harrison, J.J., Turner, R.J., and Ceri, H. (2005), Persister cells, the biofilm matrix and tolerance to metal cations in biofilm and planktonic *Pseudomonas aeruginosa*, *Environ. Microbiol.*, 7, 981–994.
31. Harrison, J.J., Turner, R.J., and Ceri, H. (2005), Metal tolerance in bacterial biofilms, *Recent Res. Dev. Microbiol.*, 9, 33–55.
32. Harrison, J.J., Ceri, H., Roper, N.J., Badry, E.A., Sproule, K.M., and Turner, R.J. (2005), Persister cells mediate tolerance to metal oxyanions in *Escherichia coli*, *Microbiology*, 151, 3181–3195.
33. Harrison, J.J., Ceri, H., Stremick, C.A., and Turner, R.J. (2004), Biofilm susceptibility to metal toxicity, *Environ. Microbiol.*, 6, 1220–1227.
34. Harrison, J.J., Ceri, H., Stremick, C.A., and Turner, R.J. (2004), *Differences* in biofilm and planktonic cell mediated reduction of metalloid oxyanions, *FEMS Microbiol. Lett.*, 235, 357–362.
35. Darouiche, R.O. (2003), Antimicrobial approaches for preventing infections associated with surgical implants, *Clin. Infect. Dis.*, 36, 1284–1289.
36. Whitehouse, J.D., Friedman, N.D., Kirland, K.B., Richardson, and W.J., Sexton, D.J. (2002), The impact of surgical-site infections following orthopedic surgery at a community hospital and a university hospital: adverse quality of life, excess length of stay, and extra cost, *Infect. Control. Hosp. Epidemiol.*, 23, 183–189.
37. Voth, D. (1985), Perioperative prevention of infection in neurosurgery, *Antibiotic Chemother.*, 33, 165–183.
38. Cardona-Bonet, L.L. and Cortes, A. (2004), Management of perioperative infectious complications in the neurologic patient, *Neurol. Clin.*, 22, 329–345.
39. Kernodle, D.S. and Kaiser, A.B. (2000), Chapter 308: Postoperative infections and antimicrobial prophylaxis, in: Mandell, G.L., Bennett, J.E., Dolin, R., Eds., *Principles and practice of infectious disease*, Vol 2. Churchill Livingstone, Philadelphia.
40. Brook, I. (2002), Meningitis and shunt infection caused by anaerobic bacteria in children, *Pediatr. Neurol.*, 26, 99–105.
41. Morris, A. and Low, D.E. (1999), Nosocomial bacteria meningitis, including central nervous system shunt infections, *Infect. Dis. Clin. North. Am.*, 13, 735–750.
42. Geroge, R., Leibrock, L., and Epstein, M. (1979), Long term analysis of cerebrospinal fluid shunt infections. A 25 year experience, *J. Neurosurg.*, 51, 804–811.
43. Ammirati, M. and Raidmondi, A.J. (1987), Cerebrospinal fluid infections in children: a study on the relationship between the etiology of hydrocephalus, age at the time of shunt placement, and infection rate, *Childs Nerv. Syst.*, 3, 106–109.
44. Serlo, W, Fernell, E., Heikkinen, E., Anderson, H., and vonWendt, L. (1990), Functions and complications of shunts in different etiologies of childhood hydrocephalus, *Childs Nerv. Syst.*, 6, 92–94.
45. Renier, D., Lacombe, J., Pierre-Kahn, A., Sainte-Rose, C., and Hrisch, J.F. (1984), Factors causing acute shunt infection; computer5 analysis of 1174 operations, *J. Neurosurge.*, 61, 1072–1078.

46. Mancao, M., Miller, C., Cochrane, B., Hoff, C., Sauter, K., Weber, E. (1998), Cerebrospinal fluid shunt infections in infants and children in Mobile, Alabama, *Acta Paediatr.*, 87, 667–670.
47. Kontny, U., Hofling, B., Gutjahr, P., Voth, D., Schwartz, M., and Schmitt, H.J. (1993), CSF shunt infections in children, *Infection*, 21, 89–92.
48. McGrit, M.J., Zaas, A., Fuchs, H.E., George, T.M., Kaye, K., and Sexton, D.J. (2003), Risk factors for pediatric ventriculoperitoneal shunt infection and predictors of infection pathogens. *Clin. Infect. Dis.*, 36, 858–862.
49. Tunkel, A.R. and Scheld, W. (2000), Acute Meningitis, in: Mandell, G.L., Bennett, J.E., and Dolin, R., Eds., *Principles and practice of infections diseases*, Vol 2. Churchill Livingstone, Philadelphila.
50. Schoenbaum, S.C., Gardner, P., and Shilliton, J. (1975), Infections of cerebrospinal fluid shunts: epidemiology, clinical manifestations, and therapy, *J. Infect. Dis.*, 131, 543–552.
51. Bell, W.O. (1992), Management of infected cerebrospinal fluid shunts in children, *Contemp. Nerurosurg.*, 14, 6–16.
52. Lyke, K.E., Obasanjo, O., Williams, M.A., O'Brien, M., Chotani, R., and Perl, T.M. (2001), Ventriculities complicating use of intraventricular catheters in adult neurosurgical patients, *Clin. Infect. Dis.*, 33(12), 2028–2033.
53. Rodriguez, W.J., Khan, W., Cocchetto, D.M., Feris, J., Puig, J.R., and Akram, S. (1990), Treatment of Pseudomonas meningitis with ceftazidime with or without concurrent therapy, *Ped. Infect. Dis. J.*, 9, 83–87.
54. Nguyen, M.H., Harris, S.P., Muder, R.R., and Pasculle, A.W. (1994), Antibiotic-resistant *Acinetobacter meningitis* in neurosurgical patients, *Neurosurgery*, 35, 851–855.
55. Strand, C.L. and Dubois, R. (1981), *Propionibacterium acnes* central nervous system shunt infection. Commercial blood culture medium-dependent isolation of the bacterium, *Am. J. Clin. Pathol.*, 75, 743–746.
56. Forward, K.R., Fewer, H.D., and Striver, H.G. (1983), Cerebrospinal fluid shunt infection. A review of 35 infections in 32 patients, *J. Neurosurg.*, 59, 389–394.
57. Brown, E.M. (2000), For the Infection in Neurosurgery Working Party of the British Society for Antimicrobial Chemotherapy. The management of neurosurgical patients with postoperative bacterial or aseptic meningitis or external ventricular drain-associated ventriculitis, *Br. J. Neurosurg.*, 14, 7–12.
58. Byers, K., Axelrod, P., Michael, S., and Rosen, S. (1995), Infections complicating tunneled intraspinal catheter systems used to treat chronic pain, *Clin. Infect. Dis.*, 21, 403–408.
59. Follett, K.A., Boortz-Marx, R., Drake, J.M., DuPen, S., Schneider, S.J., Turner, M.S., and Coffey, R.J. (2004), Prevention and management of intrathecal drug delivery and spinal cord stimulation system infections, *Anesthesiology*, 100, 1582–1594.
60. Bridgett, M.J., Davies, M.C., Denyer, S.P., and Eldridge, P.R. (1993), *In vitro* assessment of bacterial adhesion to Hydromer-coated cerebrospinal fluid shunts, *Biomaterials*, 14, 184–188.
61. Bayston, R., Zdroyewski, V., and Barsham, S. (1988), Use of an *in vitro* model for studying the eradication of catheter colonisation by *Staphylococcus epidermidis*, *J. Infect.*, 16, 141–146.

62. Bayston, R. and Barsham, S. (1988), Catheter colonisation: a laboratory model suitable for aetiological, therapeutic and preventive studies, *Med. Lab. Sci.*, 45, 235–239.

63. Bayston, R., Grove, N., Siegel, J., Lawellin, D., and Barsham, S. (1989), Prevention of hydrocephalus shunt catheter colonisation *in vitro* by impregnation with antimicrobials, *J. Neurol. Neurosurg. Psychiatry*, 52, 605–609.

64. Bayston, R. and Milner, R.D. (1981), Antimicrobial activity of silicone rubber used in hydrocephalus shunts, after impregnation with antimicrobial substances, *J. Clin. Pathol.*, 34, 1057–1062.

65. Bayston, R. and Lambert, E. (1997), Duration of protective activity of cerebrospinal fluid-shunt catheters impregnated with antimicrobial agents to prevent bacterial catheter-related infection, *J. Neurosurg.*, 87, 247–251.

66. Bayston, R., Ashraf, W., and Bhundia, C. (2004), Mode of action of an antimicrobial biomaterial for use in hydrocephalus shuns, *J. Antimicrob. Chemother.*, 53, 778–782.

67. Schierholz, J., Jansen, B., Jaenicke, L., and Pulverer, G. (1994), *In-vitro* efficacy of an antibiotic releasing silicone ventricle catheter to prevent shunt infection, *Biomaterials*, 15, 996–1000.

68. Hampl, J., Schierholz, J., Jansen, B., and Aschoff, A. (1995), *In vitro* and *in vivo* efficacy of a rifampin-loaded silicone catheter for the prevention of CSF shunt infections, *Acta. Neurochir. (Wien)*, 133, 147–152.

69. Kohnen, W., Schaper, J., Klein, O., Tieke, B., and Jansen, B. (1998), A silicone ventricular catheter coated with a combination of rifampin and trimethoprim for the prevention of catheter-related infections, *Zentralbl. Bakteriol.*, 287, 147–156.

70. Hampl, J.A., Weitzel, A., Bonk, C., Kohnen, W., Roesner, D., and Jansen, B. (2003), Rifampin-impregnated silicone catheters: a potential tool prevention and treatment of CSF shunt infections, *Infection*, 31, 109–111.

71. Schierholz, J.M. and Pulverer, G. (1998), Investigation of a rifampin, fusidic acid and mupirocin releasing silicone catheter, *Biomaterials*, 19, 2065–2074.

72. Zabramski, J.M., Whiting, D., Darouiche, R.O., Horner, T.G., Olson, J., Robertson, C., and Hamilton, A.J. (2003), Efficacy of antimicrobial-impregnated external ventricular drain catheters: a prospective, randomized, controlled trial, *J. Neurosurg.*, 98, 725–730.

73. Gates, G.A. and Miyamoto, R. (2003), Cochlear implants, *N. Engl. J. Med.*, 349, 421–423.

74. Cunningham, D.C., Slattery, W.H., and Luzford, W.M. (2004), Postoperative infection in cochlear implant patients, *Otolaryngol. Head Neck Surg.*, 131, 109–114.

75. Kempf, H.G., Johann, K., and Lenarz, T. (1999), Complications in pediatric cochlear implant surgery, *Eur. Arch. Otorhinolaryngol.*, 256, 128–132.

76. Yu, K.C.Y., Hegarty, J., Gantz, B.J., and Lalwani, A.K. (2001), Conservative management of infections in cochlear implant recipients, *Otolaryngol. Head Neck Surg.*, 125, 66–70.

77. Green, K.M., Bhatt, Y., Saeed, S.R., and Ramsden, R.T. (2004), Complications following adult cochlear implantation: experience in Manchester, *J. Laryngol. Otol.*, 118, 417–420.

78. Reefhuis, J., Honein, M.A., Whitney, C.G., Chamany, S., Mann, E.A., Biernath, K.R., Broder, K., Manning, S., Avashia, S., Victor, M., Costa, P., Devine, O., Graham, A., and Boyle, C. (2003), Risk of bacterial meningitis in children with cochlear implants, *N. Engl. J. Med.*, 349, 435–445.
79. Callanan, V. and Poje, C. (2004), Cochlear implantation and meningitis, *Int. J. Pediatr. Otorhinolaryngol.*, 68, 545–550.
80. Thamilarasan, M. and Griffin, B. (2002), Choosing the most appropriate valve operation and prosthesis, *CCJM*, 69, 688–703.
81. Karchmer, A.W. (2000), Infections of prosthetic valves and intravascular devices, in: Mandell, G.L., Bennett, J.E. and Dolin, R. Eds., *Principles and Practice of Infections Diseases*, 5th ed., Vol 1, Churchill Livingstone, Philadelphia. pp. 903–911.
82. Karchmer, A.W. and Longworth, D. (2003), Infections of intracardiac devices, *Cardiol. Clin.*, 21, 253–271.
83. Karchmer, A.W. (2001), Infective endocarditis, in: Braunwald, E., Zipes, D.P., Libby, P., Eds., *Heart disease: a textbook of cardiovascular medicine*, 6th ed., WB Saunders Company, Philadelphia. pp. 1723–1745.
84. Giamarellou, H. (2002), Nosocomial cardiac infections, *J. Hosp. Infect.*, 50, 91–105.
85. Sexton, D.J. and Spelman, D. (2003), Current best practices and guidelines: assessment and management of complications of infective endocarditis, *Cardiol. Clin.*, 21, 273–282.
86. Calderwood, S.B., Swinski, L., Karchmer, A.W., Waternaux, C.M., Buckley, M.J. (1986), Prosthetic valve endocarditis. Analysis of factors affecting outcome of therapy, *J. Thorac. Cardiovasc. Surg.*, 92, 776–783.
87. Ivert, T.S., Dismukes, W.E., Cobbs, C.G., Blackstone, E.H., Kirklin, J.W., and Bergdahl, L.A. (1984), Prosthetic valve endocarditis, *Circulation*, 69, 223–232.
88. Melgar, G.R., Nasser, R.M., Gordon, S.M., Lytle, B.W., Keys, T.F., and Longworth, D.L. (1997), Fungal prosthetic valve endocarditis in 16 patients. An 11-year experience in a tertiary care hospital, *Medicine (Baltimore)*, 76, 94–103.
89. Nasser, R.M., Melgar, G.R., Longworth, D.L., and Gordon S.M. (1997), Incidence and risk of developing fungal prosthetic valve endocarditis after nosocomial candidemia, *Am. J. Med.*, 103, 25–32.
90. Turnier, E., Kay, J.H., Bernstein, S., Mendez, A.M., and Zubiate, P. (1975), Surgical treatment of *Candida endocarditis*, *Chest*, 67, 262–268.
91. Nguyen, M.H., Nguyen, M.L., Yu, V.L., McMahon, D., Keys, T.F., and Amidi, M. (1996), *Candida* prosthetic valve endocarditis: prospective study of six cases and review of the literature, *Clin. Infect. Dis.*, 22, 262–267.
92. Yu, V.L., Fang, G.D., Keys, T.F., Harris, A.A., Gentry, L.O., Fuchs, P.C., Wagener, M.M., and Wong, E.S. (1994), Prosthetic valve endocarditis: superiority of surgical valve replacement versus medical therapy only, *Ann. Thorac. Surg.*, 58, 1073–1077.
93. Seelig, M.S., Speth, C.P., Kozinn, P.J., Taschdjian, C.L., Toni, E.F., and Goldberg, P. (1974), Patterns of *Candida* endocarditis following cardiac surgery: importance of early diagnosis and therapy (an analysis of 91 cases), *Prog. Cardiovasc. Dis.*, 17, 125–160.

94. Gollob, M.H. and Seger, J.J. (2001), Current status of the implantable cardioverter-Defibrillator, *Chest*, 119 (4), 1210–1221.
95. Chua, M.H., Wilkoff, W.L., Lee, I., Juratli, N., Longworth, D.L., and Gordon, S.M. (2000), Diagnosis and management of infections involving implantable electrophysiologic cardiac devices, *Ann. Int. Med.*, 133 (8), 604–608.
96. Chamis, A.L., Peterson, G., Cabell, C.H., Corey, G.R., Sorrentino, R.A., Greenfield, R.A., Ryan, T., Reller, L.B., and Fowler, V.G. Jr. (2001), *Staphylococcus aureus* bacteremia in patients with permanent pacemakers or implantable cardioverter-defibrillators, *Circulation*, 104 (9), 1029–1033.
97. Spratt, K.A., Blumberg, E.A., Wood, C.A., Kutalek, S.P., and Reboli, A.C. (1993), Infections of implantable cardioverter defibrillators: approach to management, *Clin. Infect. Dis.*, 17, 679–685.
98. Cacoub, P., Leprince, P., Nataf, P., Hausfater, P., Dorent, R., Wechsler, B., Bors, V., Pavie, A., Piette, J.C., and Gandjbakhch, I. (1998), Pacemaker infective endocarditis, *Am. J. Cardiol.*, 82, 480–484.
99. Klug, D., Lacroix, D., Savoye, C., Goullard, L., Grandmougin, D., Hennequin, J.L., Kacet, S., and Lekieffre, J. (1997), Systemic infection related to endocarditis on pacemaker leads: clinical presentation and management, *Circulation*, 95, 2098–2107.
100. Arber, N., Pras, E., Copperman, Y., Schapiro, J.M., Meiner, V., Lossos, I.S., Militianu, A., Hassin, D., Pras, E., and Shai, A. (1994), Pacemaker endocarditis. Report of 44 cases and review of the literature, *Medicine*, 73, 299–305.
101. Wilkoff, B.L. (1999), Pacemaker lead extraction with the laser sheath: results of the pacing lead extraction with the excimer sheath (PLEXES) trial, *J. Am. Coll. Cardiol.*, 33, 1671–1676.
102. Lai, K.K. and Fontecchio, S.A. (1998), Infections associated with implantable cardioverter defibrillators placed transvenously and via throacotomies, *Clin. Infect. Dis.*, 27, 265–269.
103. Byrd, C.L. (2000), Management of implant complications, in: Ellenbogen, K.A., Kay, G.N., and Wilkoff, B.L. Eds., *Clinical Cardiac Pacing and Defibrillation*, 2nd ed, WB Saunders, Philadelphia. pp. 669–694.
104. Hirsch, D.J. and Cooper, J. Jr. (2003), Cardiac failure and left ventricular assist devices, *Anesthesiol. Clin. North. Am.*, 21, 625–638.
105. Malani, P.N., Dyke, D.B.S., Pagani, F.D., and Chenowetch, C.E. (2002), Nosocomial infections in left ventricular assist device recipients, *Clin. Infect. Dis.*, 34, 1295–1300.
106. Rose, E.A., Gelijins, A.C., Moskowitz, A.J., Arons, R., Gupta, L., Weinberg, A., Faries, P.L., Nowygrod, R., and Kent, K.C. (2001), Long-term use of a left ventricular assist device for end-stage heart failure, *N. Engl. J. Med.*, 345, 1435–1443.
107. Sivaratnam, K. and Duggan, J. (2002), Left ventricular assist device infections: three case reports and a review of the literature, *ASAIO J.*, 48, 2–7.
108. Fischer, S.A., Trenholme, G.M., Costanzo, M.R., and Piccione, W. (1997), Infections: complications in left ventricular assist device recipients, *Clin. Infect. Dis.*, 24, 18–23.
109. Goldstein, D.J., el-Amir, N.G., Ashton, R.C. Jr., Catanese, K., Rose, R.A., Levin, H.R., and Oz, M.C. (1995), Fungal infections in left ventricular assist device recipients. Incidence, prophylaxis, and treatment, *ASAIO J.*, 41, 873–875.

110. Holman, W.L., Fix, R.J., Foley, B.A., McGiffin, D.C., Rayburn, B.K., and Kirklin, J.K. (1999), Management of wound and left ventricular assist device pocket infection, *Ann. Thorac. Surg.*, 68, 1080–1082.
111. Holman, W.L., Rayburn, B.K., McGiffin, D.C., Foley, B.A., Benza, R.L., Bourge, R.C., Pinderski, L.J., and Kirklin, J.K. (2003), Infections in ventricular assist devices: prevention and treatment, *Ann. Thorac. Surg.*, 75, S48–S57.
112. McKellar, S.H., Allred, B.D., Marks, J.D., Cowley, C.G., Classen, D.C., Gardner, S.C., and Long, J.W. (1999), LVAD pocket infection controlled with antibiotic-impregnated polymethylmethacrylate beads, *Ann. Thorac. Surg.*, 67, 554–555.
113. Lutwick, L.I., Vaghjimal, A., and Connolly, M.W. (1998), Postcardiac surgery infections, *Critical Care Clin.*, 14, 221–250.
114. McCarthy, P.M., Schmitt, S.K., Vargo, R.L., Gorden, S., Keys, T.F., and Hobbs, R.E. (1996), Implantable LVAD infections: implications for permanent use of the device, *Ann. Thorac. Surg.*, 61, 359–365.
115. Vilchez, R.A., McEllistrem, M.C., Harrison, L.H., McCurry, K.R., Kormos, R.L., and Kusne, S. (2001), Relapsing bacteremia in patients with ventricular assist device: an emergent complication of extended circulatory support, *Ann. Thorac. Surg.*, 72, 96–101.
116. Nurozler, F., Argenziano, M., Oz, M.C., and Naka, Y. (2001), Fungal left ventricular assist device endocarditis, *Ann. Thorac. Surg.*, 71, 641–648.
117. Prendergast, T.W., Todd, B.A., Beyer, A.J. 3rd, Furukawa, S., Eisen, H.J., Addonizio, V.P., Browne, B.J., and Jeevanandam, V. (1997), Management of left ventricular assist device infection with heart transplantation, *Ann. Thorac. Surg.*, 64, 142–147.
118. Morgan, J.A., Park, Y., Oz, M.C., and Naka, Y. (2003), Device related infections while on left ventricular assist device support do not adversely impact bridging to transplant or posttransplant survival, *ASAIO J.*, 49, 748–750.
119. Hutchinson, O.Z., Oz, M.C., and Ascherman, J.A. (2001), The use of muscle flaps to treat ventricular assist device infections, *Plast. Reconstr. Surg.*, 107 (2), 364–373.
120. De Cholnoky, T. (1970), Augmentation mammaplasty: survey of complications in 10 941 patients by 265 surgeons, *Plast. Reconstr. Surg.*, 45, 573–577.
121. Gabriel, S.E., Woods, J.E., O'Fallon, W.M., Beard, C.M., Kurland, L.T., and Melton, L.J. 3rd. (1997), Complications leading to surgery after breast implantation, *N. Engl. J. Med.*, 336, 677–682.
122. Freedman, A. and Jackson, I. (1989), Infections in breast implants, *Infect. Dis. Clin. North Am.*, 3, 275–287.
123. Ahn, C., Ko, C., Wagar, E., Wong, R.S., and Shaw, W.W. (1996), Microbial evaluation; 139 implants removed from symptomatic patients, *Plast. Reconstr. Surg.*, 98, 1225–1229.
124. Pajkos, A., Deva, A.K., Vickery, K., Cope, C., Chang, L., and Cossart, Y.E. (2003), Detection of subclinical infection in significant breast implant capsules, *Plast. Reconstr. Surg.*, 111, 1605–1611.
125. Darouiche, R.O., Meade, R., Mansouri, M.D., and Netscher, D.T. (2002), *In vivo* efficacy of antimicrobe-impregnated saline-filled silicone implants, *Plast. Reconstr. Surg.*, 109, 1352–1357.

126. Haiavy, J. and Tobin, H. (2002), *Mycobacterium fortuitum* infection in prosthetic breast implants, *Plast. Reconstr. Surg.*, 109, 2124–2128.
127. Spear, S.L., Howard, M.A., Boehmler, J.H., Ducic, I., Low, M., and Abbruzzesse, M.R. (2004), The infected or exposed breast implant: management and treatment strategies, *Plast. Reconstr. Surg.*, 113 (6), 1634–1644.
128. Carson, C.C. (1999), Reconstructive surgery using urologic prostheses, *Curr. Opin. Urol.*, 9, 233–239.
129. Carson, C.C. (2001), Penile prosthesis implantation and infection for Sexual Society of North America, *Int. J. import. Res.*, 13 (Suppl. 3), S35–S38.
130. Licht, M.R., Montague, D.R., Angermier, K.W., and Lakin, M.M. (1995), Cultures from genitourinary prostheses at reoperation; questioning the role of *Staphylococcus epidermidis* in periprosthetic infection, *J. Urol.*, 154, 387–390.
131. Blum, M.D. (1989), Infections of genitourinary prostheses. *Inf. Dis. Clin. North Am.*, 3, 259–247.
132. Carson, C.C. (1989), Infections in genitourinary prostheses, *Urol. Clin. North Am.*, 16, 139–147.
133. Mulcahy, J.J. (1999), Management of the infected penile implant-concepts on salvage techniques, *Int. J. Impot. Res.*, 11 (Suppl. 1), S58–S59.
134. Mulcahy, J.J., Brant, M., and Ludlow, J.K. (1995), Management of infected penile implants, *Tech. Urol.*, 1, 115–119.
135. Mulcahy, J.J. (2003), Treatment alternatives for the infected penile implant, *Int. J. Impot. Res.*, 15 (Suppl. 5), S147–S149.
136. Carson, C.C. 3rd. (1999), Management of prosthesis infections in urologic surgery, *Urol. Clin. North Am.*, 26, 829–839.
137. Parsons, C.L. (1993), Diagnosis and therapy of subclincally infected prostheses, *Surg. Gynecol. Obstet.*, 177, 504–506.
138. Berbari, E.F., Hanssen, A.D., Duffy, M.C., Steckelberg, J.M., Ilstrup, D.M., Harmsen, W.S., and Osmon, D.R. (1998), Risk factors for prosthetic joint infection: case-control study, *Clin. Infect. Dis.*, 27, 1247–1254.
139. Gillespie, W.J. (1997), Prevention and management of infection after total joint replacement, *Clin. Infect. Dis.*, 25, 1310–1317.
140. Widmer, A.F. (2001), New development in diagnosis and treatment of infection on orthopedic implants, *Clin. Infect. Dis.*, 33 (Supp. 12), S94–S106.
141. Coventry, M.B. (1975), Treatment of infections occurring in total hip surgery, *Orthop. Clin. North Am.*, 6, 991–1003.
142. Spangehl, M.J., Younger, A.S., Masri, B.A., and Duncan, C.P. (1998), Diagnosis of infection following total hip arthroplasty, *Inst. Course Lect.*, 47, 285–295.
143. Duan, K., Fan, Y., and Wang, R. (2004), Electrolytic deposition of calcium etidronate drug coating on titanium substrate, *J. Biomed. Mater. Res. B. Appl. Biomater.*, 72, 43–51.
144. Spangehl, M.J., Masri, B.A., O'Connell, J.X., and Duncan, C.P. (1999), Prospective analysis of preoperative and intraoperative investigations for the diagnosis of infection at the sites of two hundred and revision total hip arthroplasties, *J. Bone Joint Surg. Am.*, 81, 672–683.
145. Fitzgerald, R.J. (1989), Infections of hip prosthesis and artificial joints, *Inf. Dis. Clin. North Am.*, 3, 329–338.

146. Sanderson, P.J. (1991), Infection in orthopaedic implants, *J. Hosp. Infect.*, 18 (Suppl. A), 367–375.

147. Zimmerli, W. and Ochsner, R.E. (2002), Management of infection associated with prosthetic joints, *Infection*, 30, 99–108.

148. Simon, L., Gauvin, F., Amre, D.K., Saint-Louis, P., and Lacroix, J. (2004), Serum procalcitonin and C-reactive protein levels as markers of bacterial infection; a systemic review and meta-analysis, *Clin. Infect. Dis.*, 39, 206–217.

149. Bernard, L., Lubbeke, A., Stern, R., Bru, J.P., Feron, J.M., Peyramond, D., Denormandie, P., Arvieux, C., Chirouze, C., Perronne, C., Hoffmeyer, P., and Groupe D'Etude Sur L'Ostéite, (2004), Value of preoperative investigations in diagnosing prosthetic joint infection: retrospective cohort study and literature review, *Scand. J. Infect. Dis.*, 36, 410–416.

150. Atkins, B.L., Athanasou, N., Deeks, J.J., Crook, D.W., Simpson, H., Peto, T.E., McLardy-Smith, P., and Berendt, A.R. (1998), Prospective evaluation of criteria for microbiogical diagnosis of prosthetic-joint infection at revision arthroplasty, *J. Clin. Microbiol.*, 36, 2932–2939.

151. Segreti, J., Nelson, J.A., and Trenholme, G.M. (1998), prolonged suppressive antibiotic therapy for infected orthopedic prostheses, *Clin. Infect. Dis.*, 27, 711–713.

152. Darouiche, R.O. (2004), Treatment of infections associated with surgical implants, *N. Engl. Med.*, 350, 1422–1429.

153. Zimmerli, W., Widmer, A.F., Blatter, M., Frei, R., and Ochsner, P.E. (1998), Role of rifampin for treatment of orthopedic implant-related staphylococcal infection, *JAMA.*, 279, 1537–1541.

154. Drancourt, M., Stein, A., Argenson, J.N., Roiron, R., Groulier, P., and Raoult, D. (1997), Oral treatment of *Staphylococcus* spp. infected orthopaedic implants with fusidic acid or ofloxacin in combination with rifampicin, *J. Antimicrob. Chemother.*, 39, 235–240.

# CHAPTER 3

# PATHOGENESIS OF DEVICE-RELATED NOSOCOMIAL INFECTIONS

## 3.1. INTRODUCTION

A fundamental concept in the pathogenesis of device-related infections is the formation of biofilms by the infecting organisms. While a variety of Gram-positive and -negative bacteria, as well as fungi, have been involved as causative organisms in foreign body-related infections (FBRI), staphylococci, particularly *Staphylococcus epidermidis* and other CoNS account for the majority of infections, both of temporarily inserted and of permanently implanted materials [1,2]. Coagulase-negative staphylococci (CoNS) are so named for their inability to clot plasma due to the lack of production of the secreted enzyme coagulase [3]. The CoNS, which often occur as skin commensals, were historically considered innocuous or, rarely, opportunistic pathogens of low virulence [4]. However, the important role of CoNS as pathogens, with particular regard to infections associated with indwelling medical devices, is becoming increasingly appreciated [5–7]. The CoNS are now a leading cause of bacteremia in the United States, Canada, Latin America, and Europe [6,7], and many of these CoNS are resistant to multiple classes of antimicrobial agents [4]. Despite this, and in contrast to the case for *S. aureus*, infections caused by CoNS typically manifest as less severe and subacute diseases that are infrequently associated with mortality. In 1988, Freney et al. [8] described two new coagulase-negative species, *Staphylococcus schleiferi* and *Staphylococcus lugdunensis*, isolated from human clinical specimens. Since

*Biofilm Eradication and Prevention: A Pharmaceutical Approach to Medical Device Infections*, By Tamilvanan Shunmugaperumal

that time, *S. lugdunensis*, named after Lyon, France, the city where the organism was initially isolated [8], has emerged as an important human pathogen with notable clinical and microbiological characteristics that stand out among those of other CoNS. Described previously as "surreptitious" [9] and a "wolf in sheep's clothing", *S. lugdunensis* behaves more like *S. aureus* than other CoNS in many respects, including exhibiting an elevated degree of virulence. *Staphylococcus lugdunensis* is both a skin commensal and a pathogen responsible for nosocomial and community-acquired infections that may proceed aggressively and with a level of severity reminiscent of that of *S. aureus* infections. Unlike *S. aureus*, *S. lugdunensis* does not possess secreted coagulase. However, some isolates produce a membrane bound form of the enzyme (clumping factor) that yields a positive result in slide coagulase and/or rapid latex agglutination tests, potentially leading to misidentification of the organism as *S. aureus* in the clinical laboratory. *Staphylococcus lugdunensis* has a propensity to cause native valve endocarditis with a fulminant and highly destructive clinical course that is quite remarkable for a coagulase-negative species, which are otherwise more frequently the etiologic agents of prosthetic valve endocarditis [4]. Equally surprising, compared to CoNS, most *S. lugdunensis* isolates remain susceptible to a large number of antimicrobial agents [10]. Frank et al. [11] provided a comprehensive overview of the body of literature on *S. lugdunensis* since its description 20 years ago, with particular emphasis on the aspects of clinical infection (including particularly virulent cases of endocarditis), clinical microbiology, antimicrobial susceptibility, and virulence that are unique to this organism.

Note that an understanding of biofilm properties provides not only a rationale for aggressive management, but also explains the often ineffective course of conservative medical management. Considering bacteria and fungus as the main infecting organisms able to form the biofilms over modern medical devices, the following discussions are also confined initially to bacteria and then to fungus.

Bacteria exist in two forms: free-floating (planktonic) and surface-associated (biofilm) [12,13]. The later stage is characterized by the ability to irreversibly attach and grow on living and inanimate surfaces, facilitated initially by the production of adhesion molecules and subsequently by the production of extracellular polymers and matrix formation [12]. Bacteria residing in this biofilm have an altered phenotype in comparison to their planktonic form, particularly with respect to growth rate and gene expression [13]. The fact that staphylococci represent the major organisms associated with infections of medical devices has greatly spurred research on pathogenic mechanisms, resulting in important advances in our understanding of biofilm formation. Thus, a battery of staphylococcal virulence factors have been identified and characterized in the past two decades leading to important insights, particularly with respect to the interaction of the bacteria with the surface of the implanted or inserted device. Normally, CoNS live in balanced harmony on our skin, forming the major component of the cutaneous microflora. Outside

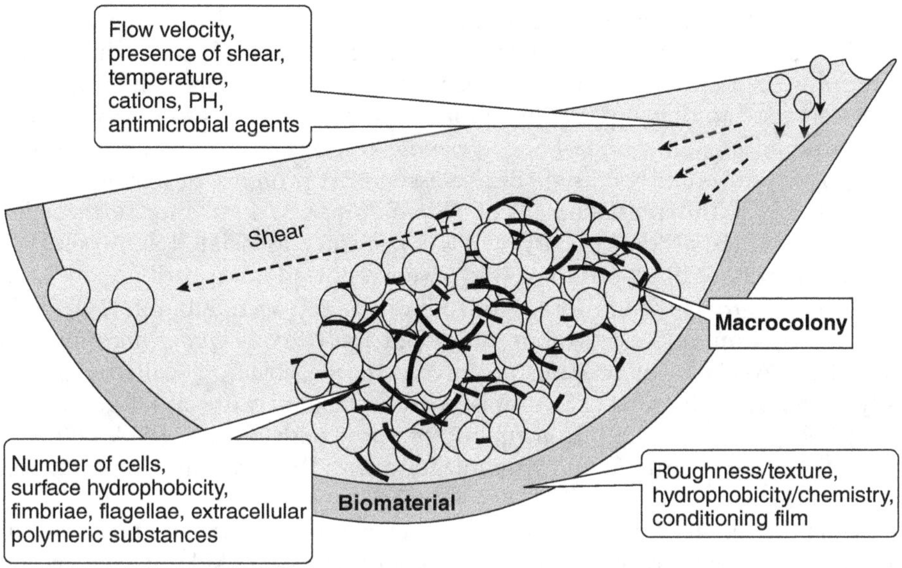


**Figure 3.1.** Pathogenesis of FBR infection: steps of biofilm formation on the surface of an intravascular catheter with rapid initial adhesion and attachment of microorganisms to the polymer surface followed by a more prolonged accumulation phase that involves cell proliferation and intercellular adhesion. Finally, microorganisms may disaggregate from the macrocolony and drift into the bloodstream resulting in metastatic and embolic complications.

the setting of a medical device, these organisms rarely cause infections. However, in relation to an inserted or implanted foreign body, these bacteria are able to colonize the surface of a foreign body by the formation of a thick, multilayered biofilm [14,15].

Figure 3.1 depicts the various steps involved in the pathogenesis of FBRI. Biofilm formation progresses in an orderly fashion. A rapid attachment of the bacteria to the surface of the implanted device is followed by a more prolonged accumulation phase that involves cell proliferation and intercellular adhesion (Fig. 3.1). Whenever devices are implanted, successful biointegration (healing) requires that host cells colonize the highly reactive surface of the foreign body, with subsequent formation of granulation tissue (or in the case of orthopedic implants, bone callus). Bacteria, however, will compete with the host cells for colonization of these surfaces. The ability, rate, and extent of various bacteria to adhere onto implanted device surfaces are dependent on properties of the material (i.e., composition) and on the cell surface characteristics of the bacteria. With respect to bacterial characteristics, virulence factors implicated in adhesion include flagella, pili, fimbriae, and glycocalyx formation. These appendages allow the bacteria once attached to the substratum, to overcome the electrostatic repulsive forces inherent to all biomaterials (i.e., indwelling implants) [12,13].

Factors involved during biofilm development over medical devices include physicochemical forces (e.g., polarity, London–van der Waal's forces, and hydrophobic interactions) [16]. In general, the rougher and more hydrophobic materials tend to develop biofilms more rapidly [13]. Cell surface hydrophobicity and initial adherence of *S. epidermidis* to polystyrene have been attributed to two different bacterial surface-associated proteins, designated SSP-1 and SSP-2 [17]. Initial attachment of *S. epidermidis* to a polymer surface also may be mediated at least in part by AtIE, a surface-associated autolysin [18]. The biofilm-associated protein (Bap) was reported to contribute to both phases of *S. aureus* biofilm formation, adhesion, and accumulation [19], while Bhp, a Bap-homologous protein, may contribute to *S. epidermidis* biofilm formation. Aside from proteins, a polysaccharide structure called capsular polysaccharide–adhesion (PS/A) has been associated with initial adherence and slime production [20]. In a rabbit model of endocarditis, PS/A deficient mutants were less virulent and immunization with PS/A resulted in protection against infection [21].

While the direct interaction between bacteria on one side and the unmodified and naked surface of the foreign body on the other side plays a crucial role in the early stages of the adherence process *in vitro* and probably also *in vivo*, additional factors may be important in later stages of adherence *in vivo*. With respect to properties of the material, smooth surfaces of the implanted or inserted medical devices do not allow for cell in-growth. Furthermore, smooth surfaces commonly exhibit inferior cell adhesion and growth relative to textured surfaces. For example, implanted devices rapidly become coated with plasma and connective tissue proteins, [e.g., fibronectin, fibrinogen, vitronectin, thrombospondin, laminin, collagen, and von Willebrand factor (vWf)], which subsequently may serve as specific receptors for colonizing microorganisms [22–24]. In addition, textured surfaces having feature sizes <100 nm are also believed to promote adhesion of all of the above-said proteins to the surface, and to provide a conformation for these proteins that better exposes amino acid sequences (e.g., RGD and YGSIR), which enhance endothelial cell binding. Moreover, small surface features are associated with an increase in surface energy, which is believed to increases cell adhesion [25]. In this regard, submicron topography, including pores, fibers, and elevations in the sub-100-nm range, has been observed for the basement membrane of the aortic valve endothelium, as well as for other basement membrane materials [26,27]. Goodman et al. [28] employed polymer casting to replicate the topographical features of the subendothelial extracellular matrix surface of denuded and distended blood vessels. They found that endothelial cells grown on such materials spread faster and appeared more like cells in their native arteries than did cells grown on untextured surfaces (see also Yap and Zhang [29]).

In the vascular system at sites of increased flow, vWf may also play an important role in adhesion of staphylococcal cells to polymer surfaces because under high shear rates platelets do not appreciably bind to extracellular matrix proteins other than vWf [30] (see Fig. 3.2). Several host factor-binding proteins

| **Dispersion** ← | **Aggregation and colonization** ← | **Initial adhesion and attachment** |
|---|---|---|
| Metastatic seeding | Exoplysaccharide production | van der Waals forces |
| (e.g.,osteomyelitis, abscesses) | Chemical interactions | Hydrophobic interactions |
| Septic thrombophlebitis | Quorum sensing | Electrostatic interactions |
| Embolic complications | Generation of resistant bacteria | Protein adhesion |

**Figure 3.2.** Flow chart showing the steps and factors associated with biofilm formation over implanted/inserted medical devices.

from *S. aureus* (e.g., the fibrinogen receptor ClfA and the fibronectin-binding proteins FbpA and FbpB) and from CoNS (e.g., the fibrinogen-binding protein Fbe and the fibronectin-binding autosin Aas) have been identified and characterized. The *S. epidermidis* autolysin AtlE that mediates primary attachment to a polymer surface (as said previously) was also found to exhibit vitronectin-binding activity, suggesting not only a function in the early stages of adherence, but also a contribution to later stages of adherence involving specific interactions with plasma proteins deposited on the polymer surface [18]. Aside from proteins, teichoic acid was suggested to function as a bridging molecule between the bacteria- and fibronectin-coated polymer [30].

## 3.2. PATHOGENESIS OF STAPHYLOCOCCAL DEVICE-RELATED INFECTION

### 3.2.1. The Intercellular Adhesin Operon, Biofilm, and Pathogenesis

Production of the extracellular polysaccharide, termed poly-*N*-acetylglucosamine (PNAG) in *S. aureus* and polysaccharide intercellular adhesin (PIA) in *S. epidermidis*, is currently the best understood mechanism of biofilm development. The PNAG–PIA is synthesized by enzymes encoded by the intercellular adhesion (*ica*) operon [31–33]. The *ica* operon is associated more commonly with invasive isolates of *S. epidermidis* than with carriage strains [34,35]. In animal models of foreign-body infection, inactivation of *ica* has been reported to be associated with decreased virulence in *S. epidermidis* [36,37]. Given that *S. epidermidis* is frequently a contaminant of sterile sites, the development of a discriminatory test to distinguish pathogenic from nonpathogenic isolates would assist greatly in the diagnosis of significant infections. The polymerase chain reaction (PCR) based detection of the *ica* locus has been proposed as such a test [34,38]. However, high rates (50%) of *S. epidermidis ica* positivity have been found in ICU isolates representative of specimen contamination [38], suggesting that detection of *ica* alone should not guide clinical decision making [35].

In contrast to the situation in *S. epidermidis*, the *ica* operon is found in up to 100% of clinical isolates of *S. aureus* [33,39–41]. However, in animal models, the relationship between PNAG–PIA production and *S. aureus* virulence is uncertain. The PNAG–PIA has been shown to be a virulence factor in a rat model of endocarditis [42]; however, deletion of *ica* and absence of PNAG are responsible in guinea pig tissue cage infection. Production of PIA had no effect on virulence [43]. Deletion of the *ica* locus in *S. aureus* isolate UAMS-1 did not impact on biofilm formation *in vitro* or in an *in vivo* murine model of catheter-related infection [44]. Consistent with this finding, four clinical isolates of methicillin-resistant *S. aureus* in which glucose-mediated biofilm development was independent of the *ica* operon have been reported [45]. In addition, the *rbf* (regulator of biofilm formation) gene in *S. aureus* encodes a putative AraC type transcription factor required for biofilm development in media supplemented with glucose or NaCl, but does not regulate *ica* operon expression [46]. Clearly, *ica*-independent mechanisms of biofilm formation exist in *S. aureus*, and detection of the *ica* operon in clinical isolates of *S. aureus* has no useful role in diagnosing isolates associated with biofilm-mediated device-related infection.

Congo red agar (CRA) has been used in the past to detect biofilm production by *S. epidermidis* [47]. This approach correlates well with a biofilm-positive phenotype observed *in vitro* [34]. Among clinical isolates, a correlation appears to exist between the phenotype on CRA and the presence of the *ica* locus [41,48]. However, there is a poor correlation between the phenotype on CRA and biofilm formation among hospital isolates of *S. epidermidis* [38]. Similarly, the phenotype on CRA was found to be an unreliable indicator of biofilm-forming capacity among clinical isolates of *S. aureus* [40]. Therefore, while screening on CRA may be easier to perform than a molecular analysis of the genes implicated in biofilm production, and could be performed easily in a diagnostic laboratory, it may be a poor method for determining the biofilm-forming capacity of clinical isolates in the diagnostic laboratory.

### 3.2.2. Cell Proliferation and Intercellular Adhesion

Once adhered to the surface of the foreign body, microorganisms multiply and accumulate in multilayered cell clusters, which require intercellular adhesion. A specific polysaccharide antigen termed PIA, which is involved in intercellular adhesion and biofilm accumulation and is chemically related to PS/A, has been detected and analyed in staphylococci [49]. The Tn*917* mutants lacking PIA were not able to accumulate in multilayered cell clusters. The *icaADBC* operon that mediates cell clustering and the intercellular adhesion synthesis in *S. epidermidis* has been cloned and sequenced [31,50]. Later, three other gene loci were identified, which have a direct or indirect regulatory influence on expression of the synthetic genes for PIA and biofilm formation [51]. In a mouse model of subcutaneous foreign body infection as well as in a rat model of CVC associated infection, a PIA negative mutant was shown to be significantly less virulent than the isogenic wild-type strain [36,37]. A PIA–hemagglutinin-positive *S. epidermidis* strain was significantly more likely to cause a subcutaneous abscess than its isogenic PIA–hemagglutinin-negative mutant and was significantly less likely to be eradicated from the inoculation site by host defence. Furthermore, the wild-type strain was found to adhere to the implanted catheters more abundantly than the PIA–hemagglutinin-negative mutant [36]. In an investigation designed to study the pathogenic properties of *S. epidermidis* strains obtained from the blood of patients with FBRI, a strong association was detected between pathogenesis and both biofilm formation and the presence of the *ica* gene cluster [34]. Recently, it was shown that induction of biofilm formation could be completely inhibited by choramphenicol, which (given at a later stage of biofilm accumulation) also inhibited further development of preformed biofilm. This indicates that continuous translation of an additional, *ica*ADBC independent factor is required for the expression of a biofilm-positive phenotype [52].

Other factors [e.g., the 140-kDa extracellular protein AAP (accumulation-associated protein)], also seem to be necessary for accumulation and biofilm formation [53]. The AAP, which is lacking in an accumulation-negative mutant and detectable only in extracellular products from bacteria grown under sessile conditions, was shown to be essential for accumulative growth in certain *S. epidermidis* strains on polymer surfaces. Of 58 CoNS studied, 55% were 140-kDa antigen-positive and produced significantly larger amounts of biofilm than the other strains that were 140-kDa antigen-negative. An antiserum specific for AAP inhibited accumulation by up to 98% of the wild-type strain [53].

Taken together, the factors described here lead to the consequence that bacteria, particularly staphylococci, are able to adhere rapidly to the surface of a foreign body. During the following accumulation phase, the bacteria proliferate to form multilayered cell clusters on the surface of a medical device. The presence of such large adherent biofilms on the surfaces of foreign bodies, particularly on explanted intravascular catheters, has been demonstrated by scanning electron microscopy (SEM) [15,54].

Intercellular signaling, often referred to as quorum-sensing (QS), has been shown to be involved in biofilm development by several Gram-positive and -negative bacteria (e.g., *Streptococcus mutants*, *Burkholderia cepacia*, and *Pseudomonas aeruginosa*). For example, under certain conditions, a QS defective mutant of *P. aeruginosa* is (in contrast to its parent strain) unable to form a highly differentiated biofilm structure. The *S. aureus* QS system is encoded by the accessory gene regulator (agr) locus that contributes to virulence in model biofilm-associated infections. It was also shown that, under some conditions, disruption of *agr* expression had no discernible influence on biofilm formation, while under others it either inhibited or enhanced biofilm formation. Under those conditions where *agr* expression enhanced biofilm formation (tested in a rotating-disk reactor), biofilms of an *agr* signaling mutant were particularly sensitive to rifampicin, but not to oxacillin [55].

Biofilms provide protection by blunting the host immune response. The major components of the host response to infected implanted material is the innate immune system, particularly the shear forces of body fluids (e.g., blood, saliva, urine) and phagocytosis by inflammatory cells (neutrophils and macrophages). Organisms that reside within a biofilm are able to avoid this washing effect of body fluids [12]. Extracellular slime has also been shown to decrease the phagocytic and intracellular killing capacity of neutrophils and to inhibit their proliferation [56]. Furthermore, surgically implanted devices are avascular. If the initial steps of wound healing are successful and allow granulation tissue formation, angiogenesis in the tissue surrounding the implant can occur; this vascularization step favors wound healing and biointegration [57]. The regulation of angiogenesis is highly complex and requires a coordinated effort from multiple cell types. In the face of an infected implant, formation of granulation tissue is impaired (as with any surgical site infection), and angiogenesis is also impaired, leaving the implant site relatively hypovascular [58]. This hypovascularity means that the implant site is surrounded by a relatively nonviable soft-tissue envelop, which favors the survival of bacteria. It also reduces delivery of immune-related proteins and antimicrobial drugs.

In summary, after implantation of a surgical (or medical) device, bacteria and host cells compete for colonization. If bacteria are able to successfully colonize the device and undergo biofilm formation, they will be able to evade the host immune response, as well as become tolerant to antimicrobials, hence explaining the difficulty in eradicating device-related infections once they are established. Given the ability of biofilms to favor bacterial survival, the rational for removal of infected devices is provided.

### 3.2.3. Clinical Factors Related to the Pathogenesis of Candida Infections

*Candida* organisms are commensals, and to act as pathogens, interruption of normal host defenses is necessary. Therefore, general risk factors for *Candida* infections include immunocompromised states, diabetes mellitus, and iatrogenic factors (e.g., antibiotic use, indwelling devices, intravenous drug use, and hyperalimentation fluids). There are several specific risk factors for particular non-*albicans* species: *C. parapsilosis* is related to foreign-body insertion, neonates, and hyperalimentation; *C. krusei* is related to azole prophylaxis and, along with *C. tropicalis*, to neutropenia and bone marrow transplantation; *C. glabrata* is related to azole prophylaxis, surgery, and urinary or vascular catheters; and *C. lusitaniae* is related to previous polyene use [59].

The source of *Candida* infections has been the subject of considerable debate. An endogenous source has been shown by using DNA typing of paired *Candida* samples from colonized sites and subsequent bloodstream infection in patients with hematological abnormalities [60] and in nonneutropenic patients [61]. A review of published studies on potential sources of candidemia found support for a gastrointestinal origin of candidemia based on experimental, clinical, and molecular similarity studies [62]. However, a cutaneous origin is suggested for *C. parapsilosis* since it is frequently recovered from skin samples and since it occurs more frequently in patients with venous catheters in place. Furthermore, the skin may be the origin of candidemia in patients with burns.

The authors note that the data used to support skin as the source of candidemia are rather incomplete. The gastrointestinal tract is the primary source of candidemia in neutropenic patients, probably due to the gastrointestinal mucositis and subsequent gut invasion by *Candida* in this patient population.

The immune system of a healthy person will swiftly and effectively deal with opportunistic pathogens. Not so for those with a compromised immune system, (e.g., after severe trauma or invasive surgery). In these patients, opportunistic infections pose a severe health risk and can be fatal. Two pathogens causing such opportunistic infections, the Gram-negative bacterium *Acinetobacter baumannii* and the yeast *C. albicans*, have gained notoriety for causing increased outbreaks in intensive care units (ICUs) of hospitals. Both can survive the harsh environment of an ICU, both thrive on plastic tubing used for catheters, and both colonize the respiratory tract of their victims. To replicate/verify what is often seen in a hospital setting, a research highlight reported by eminent scientists in *Nature Methods* journals [63] brings out an unexpected discovery. It is shown that the pathogens were indeed antagonists when they coinfected worms with *A. baumannii* and *C. albicans*. In a monomicrobial infection model, the yeast transforms into a filamentous form after entering the worm and kills it by destroying its organs. In the coinfection model, the scientists observed prolonged survival of the worms. The bacteria were keeping the yeast in check, partly by killing it outright and partly by preventing its transformation into filamentous form. How exactly *A. baumannii* interferes with the yeast is not known, but a screen of 600 *A. baumannii* transposon mutants yielded an exciting hit, a gene that is an important virulence regulator in other Gram-negative bacteria and is involved in secretion. In turn, *C. albicans* is not a helpless victim of the bacteria, given an opportunity to develop into filamentous form, it retaliates. *Candida* in a biofilm environment produces more of the QS molecule called farnesol that inhibits the bacteria. With this coinfection in worm, the scientists modeled the beginning of pathogen interactions in a host. By shedding light on the pathways that *C. albicans* uses to protect itself against *A. baumannii*, and vice versa, therapeutic targets may come to light. Although the worms live longer as a result of the pathogens battling each other, the scientists warn of a simplistic extrapolation of these results to higher organisms. Having two pathogens in human's airways is not necessarily better than having only one, as there is a delay in the development of respiratory infection and then a more severe infection. Through the course of evolution, the two pathogens have learned to deal with each other, and these are lesions we need to learn to help combat problematic infections.

## REFERENCES

1. Donlan, R.M. (2001), Biofilms and device-associated infections, *Emerg. Infect. Dis.*, 7, 277–281.

2. Hugonnet, S., Sax, H., Eggimann, P., Chevrolet, J.C., Pittet, D. (2004), Nosocomial blood-stream infection and clinical sepsis, *Emerg. Infect. Dis.*, 10, 76–81.
3. Bannerman, T.L. and Peacock, S.J. (2007), *Staphylococcus*, *Micrococcus*, and other catalase-positive cocci, in: Murray, P.R., Baron, E.J., Jorgensen, J.H., Landry, M.L., and Pfaller, M.A., Eds., *Manual of Clinical Microbiology*, 9th ed., Vol. 1. ASM Press, Washington, DC, pp. 390–411.
4. Huebner, J. and Goldmann, D.A. (1999), Coagulase-negative staphylococci: role as pathogens, *Annu. Rev. Med.*, 50, 223–236.
5. Diekema, D.J., Pfaller, M.A., Schmitz, F.J., Smayevsky, J., Bell, J., Jones, R.N., and Beach, M. (2001), Survey of infections due to *Staphylococcus* species: frequency of occurrence and antimicrobial susceptibility of isolates collected in the United States, Canada, Latin America, Europe, and the Western Pacific region for the SENTRY Antimicrobial Surveillance Program, 1997–1999, *Clin. Infect. Dis.*, 32, S114–S132.
6. Fluit, A.C., Verhoef, J., and Schmitz, F.J. (2001), Frequency of isolation and antimicrobial resistance of gram-negative and gram-positive bacteria from patients in intensive care units of 25 European university hospitals participating in the European arm of the SENTRY Antimicrobial Surveillance Program 1997–1998, *Eur. J. Clin. Microbiol. Infect. Dis.*, 20, 617–625.
7. Pfaller, M.A., Jones, R.N., Doern, G.V., Sader, H.S., Kugler, K.C., Beach, M.L. (1999), Survey of blood stream infections attributable to gram-positive cocci: frequency of occurrence and antimicrobial susceptibility of isolates collected in 1997 in the United States, Canada, and Latin America from the SENTRY Antimicrobial Surveillance Program, SENTRY Participants Group, *Diagn. Microbiol. Infect. Dis.*, 33, 283–297.
8. Freney, J., Brun, Y., Bes, M., Meugnier, H., Grimont, F., Grimont, P.A.D., Nervi, C., and Fleurette, J. (1988), *Staphylococcus lugdunensis* sp. nov. and *Staphylococcus schleiferi* sp. nov., two species from human clinical specimens, *Int. J. Syst. Bacteriol.*, 38, 168–172.
9. Sotutu, V., Carapetis, J., Wilkinson, J., Davis, A., and Curtis, N. (2002), The "surreptitious *Staphylococcus*": *Staphylococcus lugdunensis* endocarditis in a child, *Pediatr. Infect. Dis. J.*, 21, 984–986.
10. Frank, K.L., Reichert, E.J., Piper, K.E., and Patel, R. (2007), *In vitro* effects of antimicrobial agents on planktonic and biofilm forms of *Staphylococcus lugdunensis* clinical isolates, *Antimicrob. Agents Chemother.*, 51, 888–895.
11. Frank, K.L., del Pozo, J.L., and Patel, R. (2008), From clinical microbiology to infection pathogenesis: how daring to be different works for *Staphylococcus lugdunensis*, *Clin. Microbiol. Rev.*, 21, 111–133.
12. Jefferson, K.K. (2004), What drives bacteria to produce a biofilm? *FEMS Microbiol. Lett.*, 236, 163–173.
13. Donlan, R.M. (2001), Biofilm formation: a clinically relevant microbiological process, *Clin. Infect. Dis.*, 33, 1387–1392.
14. Von Eiff, C., Peters, G., and Heilmann, C. (2002), Pathogenesis of infections due to coagulase-negative staphylococci, *Lancet Infect. Dis.*, 2, 677–685.
15. Peters, G., Locci, R., and Pulverer, G. (1982), Adherence and growth of coagulase-negative staphylococci on surfaces of intravenous catheters, *J. Infect. Dis.*, 146, 479–482.

16. Dickinson, G.M. and Bisno, A.L. (1989), Infections associated with indwelling devices: infections related to extravascular devices, *Antimicrob. Agents Chemother.*, 33, 602–607.

17. Veenstra, G.J., Cremers, F.F., van Dijk, H., and Fleer, A. (1996), Ultrastructural organization and regulation of a biomaterial adhesion of *Staphylococcus epidermidis*, *J. Bacteriol.*, 178, 537–541.

18. Heilmann, C., Hussain, M., Peters, G., and Götz, F. (1997), Evidence for autolysin mediated primary attachment of *Staphylococcus epidermidis* to a polystyrene surface, *Mol. Microbiol.*, 24, 1013–1024.

19. Cucarella, C., Solano, C., Valle, J., Amorena, B., Lasa, I., and Penadés, J.R. (2001), Bap, a *Staphylococcus aureus* surface protein involved in biofilm formation, *J. Bacteriol.*, 183, 2888–2896.

20. Muller, E., Hübner, J., Gutierrez, N., Takeda, S., Goldmann, D.A., and Pier, G.B. (1993), Isolation and characterization of transposon mutants of *Staphylococcus epidermidis* deficient in capsular polysaccharide/adhesin and slime, *Infect. Immun.*, 61, 551–558.

21. Shiro, H., Muller, E., Gutierrez, N., Boisot, S., Grout, M., Tosteson, T.D., Goldmann, D., and Pier, G.B. (1994), Transposon mutants of *Staphylococcus epidermidis* deficient in elaboration of capsular polysaccharide/adhesin and slime are avirulent in a rabbit model of endocarditis, *J. Infect. Dis.*, 169, 1042–1049.

22. Herrmann, M., Vaudaux, P.E., Pittet, D., Auckenthaler, R., Lew, P.D., Schumacher-Perdreau, F., Peters, G., and Waldvogel, F.A. (1988), Fibronectin, fibrinogen, and laminin act as mediators of adherence of clinical staphylococcal isolates to foreign material, *J. Infect. Dis.*, 158, 693–701.

23. Herrmann, M., Hartleib, J., Kehrel, B., Montgomery, R.R., Sixma, J.J., and Peters, G. (1997), *Interaction* of von Willebrand factor with *Staphylococcus aureus*, *J. Infect. Dis.*, 176, 984–991.

24. Dickinson, G.M. and Bisno, A.L. (1989), Infections associated with indwelling devices: concepts of pathogenesis; infection associated with intravascular devices, *Antimicrob. Agents Chemother.*, 33 (5), 597–601.

25. Lim, J.Y., Liu, X., Vogler, E.A., and Donahue, H.J. (2004), Systematic variation in osteoblast adhesion and phenotype with substratum surface characteristics, *J. Biomed. Mater. Res. A*, 68, 504–512.

26. Brody, S., Anilkumar, T., Liliensiek, S., Last, J.A., Murphy, C.J., and Pandit A. (2006), Characterizing nanoscale topography of the aortic heart valve basement membrane for tissue engineering heart valve scaffold design, *Tissue Eng.*, 12, 413–421.

27. Flemming, R.G., Murphy, C.J., Abrams, G.A., Goodman, S.L., and Nealey, P.F. (1999), Effects of synthetic micro-and nano-structured surfaces on cell behavior, *Biomaterials*, 20, 573–588.

28. Goodman, S.L., Sims, P.A., and Albrecht, R.M. (1996), Three-dimensional extracellular matrix textured biomaterials, *Biomaterials*, 17, 2087–2095.

29. Yap, F.L. and Zhang Y. (2007), Protein and cell micropatterning and its integration with micro/nanoparticles assembly, *Biosensors Bioelectronics*, 22, 775–778.

30. Hussain, M., Heilmann, C., Peters, G., and Herrmann, M. (2001), Teichoic acid enhances adhesion of *Staphylococcus epidermidis* to immobilized fibronectin, *Microb. Pathog.*, 31, 261–270.

31. Heilmann, C., Schweitzer, O., Gerke, C., Vanittanakom, N., Mack, D., and Gotz, F. (1996), Molecular basis of intercellular adhesion in the biofilm-forming *Staphylococcus epidermidis*, *Mol. Microbiol.*, 20, 1083–1091.
32. McKenney, D., Pouliot, K.L., Wang, Y., Murthy, V., Ulrich, M., Döring, G., Lee, J.C., Goldmann, D.A., and Pier, G.B. (1999), Broadly protective vaccine for *Staphylococcus aureus* based on an in vivo expressed antigen, *Science*, 284, 1523–1527.
33. Cramton, S.E., Gerke, C., Schnell, N.F., Nichols, W.W., and Gotz, F. (1999), The intercellular adhesion (*ica*) locus is present in *Staphylococcus aureus* and is required for biofilm formation, *Infect. Immun.*, 67, 5427–5433.
34. Ziebuhr, W., Heilmann, C., Gotz, F., Meyer, P., Wilms, K., Straube, E., Hacker, J. (1997), Detection of the intercellular adhesion gene cluster (*ica*) and phase variation in *Staphylococcus epidermidis* blood culture strains and mucosal isolates, *Infect. Immun.*, 65, 890–896.
35. Frebourg, N.B., Lefebvre, S., Baert, S., and Lemeland, J.F. (2000), PCR-Based assay for discrimination between invasive and contaminating *Staphylococcus epidermidis* strains, *J. Clin. Microbiol.*, 38, 877–880.
36. Rupp, M.E., Ulphani, J.S., Fey, P.D., Bartscht, K., and Mack, D. (1999), Characterization of the importance of polysaccharide intercellular adhesin/hemagglutinin of *Staphylococcus epidermidis* in the pathogenesis of biomaterial-based infection in a mouse foreign body infection model, *Infect. Immun.*, 67, 2627–2632.
37. Rupp, M.E., Ulphani, J.S., Fey, P.D., and Mack, D. (1999), Characterization of *Staphylococcus epidermidis* polysaccharide intercellular adhesin/hemagglutinin in the pathogenesis of intravascular catheter-associated infection in a rat model, *Infect. Immun.*, 67, 2656–2659.
38. Fitzpatrick, F., Humphreys, H., Smyth, E.G., Kennedy, C.A., and O'Gara, J.P. (2002), Environmental regulation of biofilm formation in intensive care unit isolates of *Staphylococcus epidermidis*, *J. Hosp. Infect.*, 52, 212–218.
39. Fowler, V.G., Fey, P.D., Reller, L.B., Chamis, A.L., Corey, G.R., and Rupp, M.E. (2001), The intercellular adhesin locus ica is present in clinical isolates of *Staphylococcus aureus* from bacteremic patients with infected and uninfected prosthetic joints, *Med. Microbiol. Immunol. (Berlin).*, 189, 127–131.
40. Knobloch, J.K., Horstkotte, M.A., Rohde, H., and Mack, D. (2002), Evaluation of different detection methods of biofilm formation in *Staphylococcus aureus*, *Med. Microbiol. Immunol.*, 191, 101–106.
41. Arciola, C.R., Baldassarri, L., and Montanaro, L. (2001), Presence of *icaA* and *icaD* genes and slime production in a collection of staphylococcal strains from catheter-associated infections, *J. Clin. Microbiol.*, 39, 2151–2156.
42. Maira-Litran, T., Kropec, A., Abeygunawardana, C., Joyce, J., Mark, G. 3rd., Goldmann, D.A., and Pier, G.B. (2002), Immunochemical properties of the staphylococcal poly-*N*-acetylglucosamine surface polysaccharide, *Infect. Immun.*, 70, 4433–4440.
43. Francois, P., Tu Quoc, P.H., Bisognano, C., Kelley, W.L., Lew, D.P., Schrenzel, J., Cramton, S.E., Götz, F., and Vaudaux, P. (2003), Lack of biofilm contribution to bacterial colonisation in an experimental model of foreign body infection by *Staphylococcus aureus* and *Staphylococcus epidermidis*, *FEMS Immunol. Med. Microbiol.*, 35, 135–140.

44. Beenken, K.E., Dunman, P.M., McAleese, F., Macapagal, D., Murphy, E., Projan, S.J., Blevins, J.S., and Smeltzer, M.S. (2004), Global gene expression in *Staphylococcus aureus* biofilms, *J. Bacteriol.*, 186, 4665–4684.

45. Rohde, H., Burdelski, C., Bartscht, K., Hussain, M., Buck, F., Horstkotte, M.A., Knobloch, J.K., Heilmann, C., Herrmann, M., and Mack, D. (2005), Induction of *Staphylococcus epidermidis* biofilm formation via proteolytic processing of the accumulation-associated protein by staphylococcal and host proteases, *Mol. Microbiol.*, 55, 1883–1895.

46. Lim, Y., Jana, M., Luong, T.T., and Lee, C.Y. (2004), Control of glucose-and NaCl-induced biofilm formation by rbf in *Staphylococcus aureus*, *J. Bacteriol.*, 186, 722–729.

47. Freeman, D.J., Falkiner, F.R., and Keane, C.T. (1989), New method for detecting slime production by coagulase negative staphylococci, *J. Clin. Pathol.*, 89, 872–874.

48. Arciola, C.R., Campoccia, D., Gamberini, S., Cervellati, M., Donati, E., and Montanaro, L. (2002), Detection of slime production by means of an optimized Congo red agar plate test based on a colourimetric scale in *Staphylococcus epidermidis* clinical isolates genotyped for *ica* locus, *Biomaterials*, 23, 4233–4239.

49. Mack, D., Fischer, W., Krokotsch, A., Leopold, K., Hartmann, R., Egge, H., and Laufs, R. (1996), The intercellular adhesin involved in biofilm accumulation of *Staphylococcus epidermidis* is a linear beta-1,6-linked glucosaminoglycan: purification and structural analysis, *J. Bacteriol.*, 178, 175–183.

50. Gerke, C., Kraft, A., Süβmuth, R., Schweitzer, O., and Götz, F. (1998), Characterization of the *N*-acetylglucosaminyltransferase activity involved in the biosynthesis of the *Staphylococcus epidermidis* polysaccharide intercellular adhesin, *J. Biol. Chem.*, 273, 18586–18593.

51. Mack, D., Rohde, H., Dobinsky, S., Riedewald, J., Nedelmann, M., Knobloch, J.K., Elsner, H.A., and Feucht, H.H. (2000), Identification of three essential regulatory gene loci governing expression of *Staphylococcus epidermidis* polysaccharide intercellular adhesin and biofilm formation, *Infect. Immun.*, 68, 3799–3807.

52. Dobinsky, S., Kiel, K., Rohde, H., Bartscht, K., Knobloch, J.K., Horstkotte, M.A., and Mack, D. (2003), Glucose-related dissociation between icaADBC transcription and biofilm expression by *Staphylococcus epidermidis*: evidence for an additional factor required for polysaccharide intercellular adhesin synthesis, *J. Bacteriol.*, 185, 2879–2886.

53. Hussain, M., Herrmann, M., von Eiff, C., Perdreau-Remington, F., and Peters, G. (1997), A 140-kilodalton extracellular protein is essential for the accumulation of *Staphylococcus epidermidis* strains on surfaces, *Infect. Immun.*, 65, 519–524.

54. Peters, G., Locci, R., and Pulverer, G. (1981), Microbial colonization of prosthetic devices II: scanning electron microscopy of naturally infected intravenous catheters, *Zentralbl. Bakteriol. Mikrobiol. Hyg. [B]*, 173, 293–299.

55. Yarwood, J.M., Bartels, D.J., Volper, E.M., and Greenberg, E.P. (2004), Quorum sensing in *Staphylococcus aureus* biofilms, *J. Bacteriol.*, 186, 1838–1850.

56. von Eiff, C., Heilmann, C., and Peters, G. (1999), New aspects in the molecular basis of polymer-associated infections due to Staphylococci, *Eur. J. Clin. Microbiol. Infect. Dis.*, 18, 843–846.

57. Peters, K., Schmidt, H., Unger, R.E., Otto, M., Kamp, G., and Kirkpatrick, C.J. (2002), Software-supported image quantification of angiogenesis in an *in vitro* cell culture system: application to studies of biocompatibility, *Biomaterials*, 23, 3413–3419.
58. Hausman, M.R. and Rinker, B.D. (2004), Intractable wounds and infections: the role of impaired vascularity and advanced surgical methods for treatment, *Am. J. Surg.*, 187 (5A), 44S–55S.
59. Kremery, V. and Barnes, A.J. (2002), Non-*albicans Candida* spp. causing fungaemia: pathogenicity and antifungal resistance, *J. Hosp. Infect.*, 50, 243–260.
60. Reagan, D.R., Pfaller, M.A., Hollis, R.J., and Wenzel, R.P. (1990), Characterization of the sequence of colonization and nosocomial candidemia using DNA fingerprinting and a DNA probe, *J. Clin. Microbiol.*, 28, 2733–2738.
61. Voss, A., Hollis, R.J., Pfaller, M.A., Wenzel, R.P., and Doebbeling, B.N. (1994), Investigation of the sequence of colonization and candidemia in nonneutropenic patients, *J. Clin. Microbiol.*, 32, 975–980.
62. Nucci, M. and Anaissie, E. (2001), Revisiting the source of candidemia: skin or gut? *Clin. Infect. Dis.*, 33, 1959–1967.
63. Rusk, N. (2008), Fighting fire with fire, *Nat. Methods*, 5 (11), 920.

# CHAPTER 4

# BIOFILM RESISTANCE–TOLERANCE TO CONVENTIONAL ANTIMICROBIAL AGENTS

## 4.1. INTRODUCTION

Antibiotic resistance–tolerance is the preeminent obstacle faced by the medical community in regard to biofilm-driven infections. On the other hand, several well-recognized puzzles in microbiology have remained unsolved for decades. These include latent bacterial infections, unculturable microorganisms, persister cells, and biofilm multidrug tolerance (MDT). Approximately 70% of bacteria frequently associated with nosocomial infections are resistant–tolerance to at least one of the drugs commonly used to treat them [1]. Many of these resistance mechanisms are tied to the morphology of the biofilm, which provides a unique shelter for the bacteria to thrive (Fig. 4.1). Exchange of genetic material within biofilms through the conjugation process has been observed to occur at rates up to 1000-fold higher than those found in planktonic populations. Hypermutation, in which bacteria mutate at higher rates to evolve under stressful conditions (e.g., antibiotic challenge), also occurs more frequently within biofilms [2,3]. Therefore, development of resistance mechanisms (e.g., specialized efflux pumps or upregulation of porin proteins on the cell surface) can quickly be selected for and propagated throughout the community.

Whereas, *Pseudomonas aeruginosa* causes an incurable infection in cystic fibrosis (CF) patients [4], *Staphylococcus aureus* and *Staphylococcus epidermidis* infects indwelling medical devices [5]. These are probably the

*Biofilm Eradication and Prevention: A Pharmaceutical Approach to Medical Device Infections*,
By Tamilvanan Shunmugaperumal

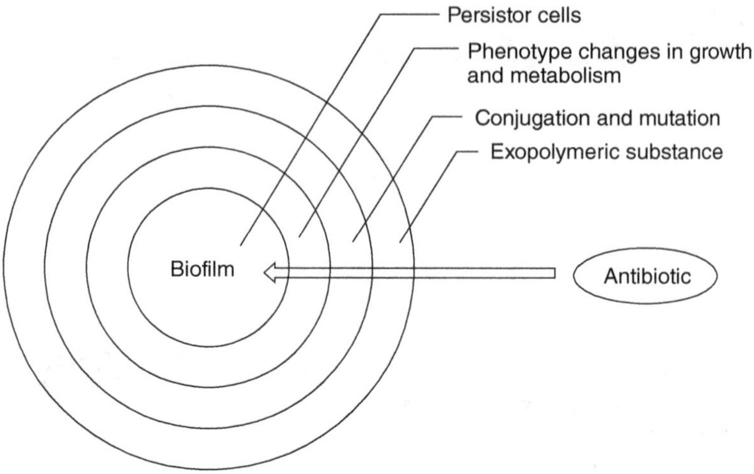


**Figure 4.1.** Conceptual diagram of biofilm resistance–tolerance mechanisms to antimicrobial therapy.

best-known biofilm-producing bacterial organisms. An interesting review by Drenkard [6] was published on the various mechanisms that are connected in *P. aeruginosa* biofilm resistance. A schematic representation of these mechanisms that may be implicated in *P. aeruginosa* biofilm resistance is shown in Fig. 4.2. The following picture of the different mechanisms involved in resistance during biofilm development appears to be most consistent with the current literature. During the early stages of biofilm development, changes in gene expression induced by surface attachment lead to the emergence of a biofilm-specific phenotype that potentially increases biofilm resistance. Later on, the production of the exopolysaccharide matrix contributes to increasing cell survival by delaying antimicrobial penetration. As biofilms mature, the increase in cell density creates gradients of nutrient and oxygen availability leading to a reduction in metabolic activity and growth rate. Furthermore, the increase in cell density also leads to the activation of quorum-sensing (QS) systems. On the other hand, nutrient starvation and oxygen limitation induce the general stress response and up-regulation of efflux pumps. Finally, environmental conditions present in the biofilm also induce or select for phenotypic/persister variants resistant to high concentrations of antimicrobials. Additional studies are still necessary to further elucidate precisely how each of these different mechanisms contributes to the overall resistance displayed by bacterial biofilms.

Both *S. aureus* and *S. epidermidis* have evolved to become highly adaptable human pathogens. Colonization by either species does not usually lead to adverse events. However, once the epithelial and mucosal surfaces have been breached, serious disease can result, ranging from minor skin infections to

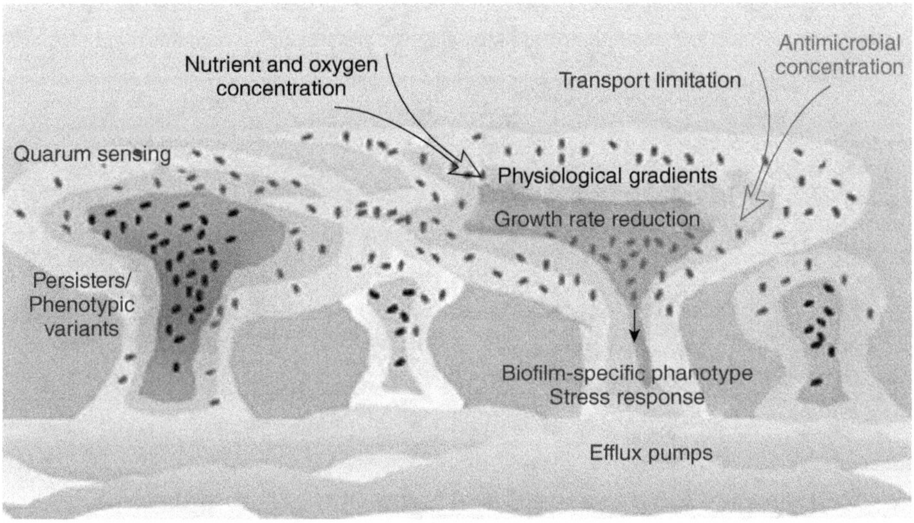


**Figure 4.2.** *Pseudomonas aeruginosa* biofilm resistance. Schematic representation of mechanisms proposed to be involved in *P. aeruginosa* biofilm resistant antimicrobial agents. The increase in bacterial density within biofilm microcolonies (indicated by darkening colors) determines gradients of nutrient and oxygen concentration (indicated by a narrowing arrow). Mechanisms biofilm resistant may include restricted antimicrobial penetration (indicated by a narrowing arrow) mediated by the polysaccharide matrix, and reduction in growth rate and metabolic activity caused by nutrient and oxygen gradients. Biofilm bacteria may also undergo physiological, metabolic, and phenotypic changes leading to a biofilm-specific phenotype. Other resistant mechanisms may include emergence of phenotypic or persister variants (represented as maroon bacteria) within the biofilm population, induction of the general stress response, upregulation of efflux pumps, and activation of quorum-sensing (QS) systems. (See color insert.)

systemic life-threatening infection. While the clinical presentation of staphylococcal infection is not unique, treatment of these infections is increasingly problematic because of the resistance of clinical isolates to an increasing number of antimicrobial agents. Most staphylococcal infections result in acute disease; however, bacterial persistence and recurrent infections are also commonly observed, particularly among patients with indwelling medical devices. The increased use of such devices has resulted in an increase in staphylococcal device-related infection.

Both *S. aureus* and *S. epidermidis* are also common causes of biofilm-mediated prosthetic device-related infection. The polysaccharide adhesion mechanism encoded by the *ica* operon is currently the best understood mediator of biofilm development, and represents an important virulence determinant. More recently, the contributions of other virulence regulators, including the global regulators *agr*, *sarA*, and $\sigma^B$, to the biofilm phenotype also have

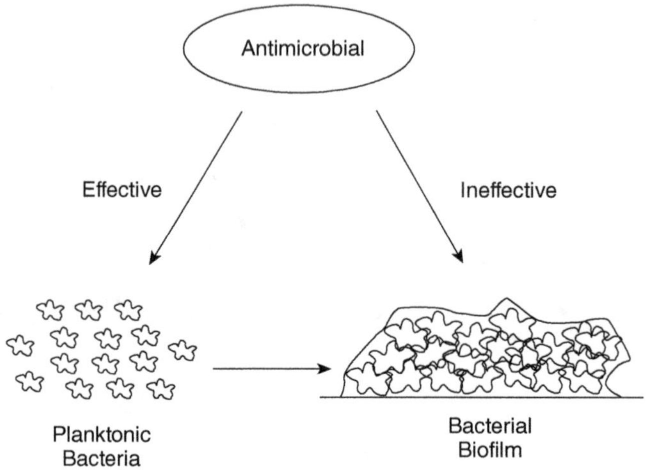


**Figure 4.3.** Effect of antimicrobial agents on two forms of bacteria.

been investigated. Nevertheless, little has changed at the bedside; the clinical and laboratory diagnosis of device-related infection can be difficult, and biofilm resistance frequently results in failure of therapy. A review by Fitzpatrick et al. [7] assesses the way in which advances in the understanding of biofilm genetics may impact on the clinical management of device-related infection.

*In vitro* data indicate that microorganisms in biofilms are substantially more resistant to killing by antimicrobial agents than are planktonic bacteria (Fig. 4.3). In a study of *S. epidermidis* clinical isolates, for example, all isolates were susceptible to vancomycin in the planktonic state; however, when grown as a biofilm, almost three-quarters had minimum bactericidal concentrations (defined as 99.9% reduction in colony-forming units) of >2048 μg mL$^{-1}$ of vancomycin [8]. In a study of *Propionibacterium acnes* prosthetic hip-associated isolates growing in *in vitro* biofilms on poly(methyl methacrylate) (PMMA), for example, all isolates showed considerably greater resistance to cefamandole, ciprofloxacin, and vancomycin in the biofilm than in the planktonic state [9]. Biofilm resistance to antimicrobial agents begins at the attachment phase and increases as the biofilm ages [10]. In a study of *S. epidermidis* biofilms, for example, vancomycin exhibited decreased killing as the biofilm aged from 6 h to 2 days [11]. This observation is paralleled in orthopedic clinical practice where debridement with retention of an infected prosthesis is more successful for early in comparison to late postoperative infection (provided that it is done expeditiously).

The discovery of persisters in biofilms [12,13] has rekindled interest in these unusual cells. The molecular mechanisms that underlie the formation of dormant persister cells are now being unraveled. The number of persisters in a growing population of bacteria rises at mid-log and reaches a maximum of ~1% at stationary state. Similarly, slow-growing biofilms produce substantial numbers of persisters. The ability of a biofilm to limit the access of the immune system components, and the ability of persisters to sustain an antibiotic attack, could then account for the recalcitrance of such infections *in vivo* and for their relapsing nature. Indeed, the presence of persister cells might be important in the aetiology of many recalcitrant infectious diseases.

Yeast *Candida albicans* forms recalcitrant biofilm infections that are tolerant to antibiotics, similarly to bacterial biofilms. *Candida albicans* biofilms produce multidrug tolerant persisters that are not mutants, but rather phenotypic variants of the wild type. Unlike bacterial persisters, however, *C. albicans* persisters were only observed in a biofilm, but not in a planktonic stationary population. Nevertheless, it is interesting to note that *C. albicans* isolates grown as biofilms are more resistant to fluconazole, voriconazole, nystatin, terbenafine, and amphotericin B, than are planktonic forms [14].

In search of the mechanism of persister formation, whether in bacteria or in yeast, the research group of Lewis at Northeastern University, published a review article in *Nature Reviews—Microbiology* journal in 2007 [15]. This subject was again re-examined in a book chapter based on new findings since the previous journal publication by the same author [16]. This chapter will not cover all findings on the mechanism of persister formation because the aforementioned publications already do so. Rather, this chapter highlights the biology of persister cells found in microbial biofilms, biofilm resistance–tolerance to conventional antimicrobial agents, and then offer an insight into their role in infectious disease. But first, let us consider the difference between resistance of regular cells and drug tolerance of persisters. This will provide a useful framework for the subsequent discussion of persisters and their properties.

## 4.2. MULTIDRUG RESISTANCE OF REGULAR CELLS VERSUS MULTIDRUG TOLERANCE OF PERSISTERS

Microorganisms in biofilms are resistant to antimicrobial agents, but not in the classic sense. Conventional–traditional–classic antibiotic resistance (of planktonic bacteria) usually involves inactivation of the antibiotic (e.g., by β-lactamases), modification of targets (e.g., as occurs with vancomycin resistance in enterococci), and exclusion of the antibiotic to their targets (e.g., through efflux) [17]. In addition, this conventional antimicrobial resistance, however, generally is considered distinct from biofilm-associated antimicrobial resistance (Exception, in *P. aeruginosa*, PvrR, a regulatory protein, modulates the phenotypic switch from antimicrobial resistant to antimicrobial-susceptible and apparently also regulates biofilm formation [18]). It is therefore better to define biofilm antibiotic resistance as the ability of biofilm bacteria to survive antibiotic treatment by using its existing complement of genes. This regulation can occur as an innate result of growing in a biofilm or be induced by the antimicrobial agent itself. Indeed, several innate biofilm phenomena and antibiotic-induced factors have been revealed that provide explanations for the ability of bacterial biofilms to survive under antibiotic or other chemical pressures [17]. Hence, biofilm antimicrobial resistance is most likely the result of a complex mixture of these innate and induced factors [19].

The genetic mechanisms of this biofilm antibiotic resistance appear to fall into two general classes: innate resistance factors and induced resistance factors [19]. Innate mechanisms are activated as part of the biofilm developmental pathway, the factors being integral parts of biofilm structure and physiology. Innate pathways include decreased diffusion of antibiotics through the biofilm matrix, decreased oxygen and nutrient availability accompanied by altered metabolic activity, formation of persisters, and other specific molecules not fitting into the above groups. Induced resistance factors include those resulting from induction by the antimicrobial agent itself. Biofilm antibiotic resistance is likely manifested as an intricate mixture of innate and induced mechanisms. Many researchers are currently trying to overcome this extreme biofilm antibiotic resistance by developing novel therapies aimed at disrupting biofilms and killing the constituent bacteria. These studies have led to the identification of several molecules that effectively disturb biofilm physiology, often by interrupting bacterial QS. In this manner, manipulation of innate and induced resistance pathways holds much promise for treatment of biofilm infections [19].

Scientifically, resistance implies a genotypic switch–genetically determined mechanism by which microorganisms resist antimicrobials. In many cases, biofilm bacteria are not resistant, but are tolerant to antimicrobials because of the uniquely protective environment of the biofilm and the accompanying alteration in metabolic rate of such bacterial consortia. This is evidenced when such bacteria are returned to the planktonic state and normal (prebiofilm) sensitivity to antimicrobials is observed [15]. The insensitivity of biofilm bacteria to antibiotics is a function of cell wall composition, surface structure, and phenotypic variation in enzymatic activity [20,21]. It has also been suggested that the negatively charged exopolysaccharide is very effective in protecting bacterial cells from cationic antibiotics by restricting their permeation [22]. A full review of biofilm resistance against antimicrobials is beyond the scope of this chapter [15,17,19].

Numerous mechanisms were proposed for the multidrug resistance (MDR) of regular cells (Table 4.1 [23,24]), and in most cases, there is a fairly good understanding of these processes at the molecular level. It is interesting to note that all of the theoretically logical possibilities of antibiotic resistance seem to have been realized in nature. Importantly, all of these mechanisms accomplish the same aim, which is to prevent the antibiotic from binding–hitting to its target (Fig. 4.4). Each of these resistance mechanisms allows cells to grow at an elevated concentration of antibiotic. Bactericidal antibiotics kill the cell not by inhibiting the target, but by corrupting its function to create a

**TABLE 4.1. Theoretically Possible Mechanisms of MDR in Regular Bacterial Cells[a]**

| |
|---|
| Target modification by mutation |
| Target modification by specialized enzymatic changes |
| Target substitution (e.g., expressing an alternative target) |
| Antibiotic modification or destruction |
| Antibiotic efflux |
| Restricted permeability to antibiotics |

[a]See Refs. [23 and 24].

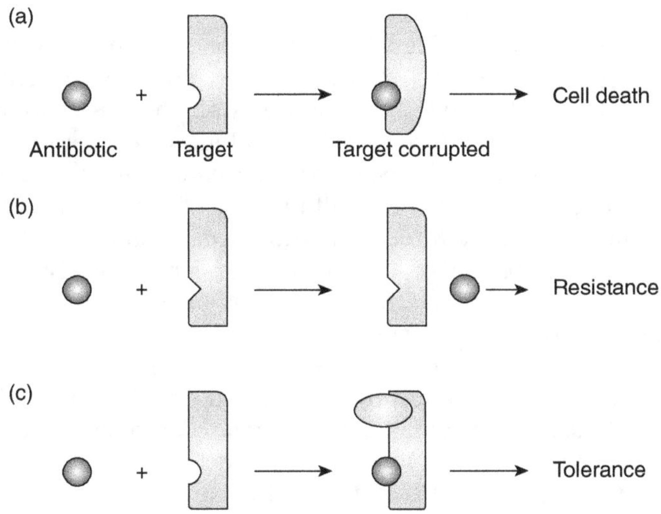


**Figure 4.4.** Resistance versus tolerance to bactericidal antibiotics. (Reproduced with permission from Kim Lewis *Nat. Rev. Microbiol.*, 5, 48–56, 2007 [15]). (**a**) The antibiotic (pink) binds to the target (blue) altering its function, which causes cell death, (**b**) The target of the antibiotic has been altered so that it fails to bind the antibiotic and the cell becomes resistant to treatment with the drug, (**c**) A different molecule (yellow) inhibits the antibiotic target. This prevents the antibiotic from corrupting its functions, resulting in tolerance. (See color insert.)

toxic product. Aminoglycoside antibiotics kill the cell by interrupting translation, which produces misfolded toxic peptides [25]. Beta-lactam antibiotics (e.g., penicillin), inhibit peptidoglycan synthesis, which activates, by an unknown mechanism, autolysin enzymes present in the cell wall [26]. This leads to digestion of the peptidoglycan by autolysins and cell death. Fluoroquinolones inhibit the ligase step of the DNA gyrase and topoisomerase, without affecting the preceding nicking activity. Consequently, the enzymes are converted into endonucleases [27].

In contrast to MDR of regular cells, the MDT of persisters to antibiotics might function by preventing target corruption by a bactericidal agent through the blocking of the antibiotic target(s) (Fig. 4.4). In simple words, tolerance acts by a preventative blocking of the antibiotic targets. If persisters are dormant and have little or no cell-wall synthesis, translation or topoisomerase activity, then the antibiotics will bind to, but will be unable to corrupt, the function of their target molecules. In this way, tolerance could enable resistance to killing by antibiotics, but at the price of nonproliferation.

## 4.3. PERSISTER CELLS

Persisters were described by Bigger in 1944 in one of the first studies on the mechanism of penicillin action [28]. Bigger discovered that penicillin lysed a growing population of *Staphylococcus* spp., but plating this transparent solution on nutrient medium produced surviving colonies. In order to test whether these were mutants, the colonies were grown, treated with penicillin, and the new population again produced a small number of persisters surviving lysis. This experiment was repeated recently with *Escherichia coli* and several different antibiotics. The results produced were similar [29,30]. By the time the mechanism of biofilm resistance to killing was being investigated, Bigger's work was all but forgotten, a curiosity known to few microbiologists. Moyed [31–35] picked up the problem in the 1980s and undertook a targeted search for persister genes. He reasoned that treating a population of *E. coli* with ampicillin would select for mutants with increased production of persisters. After ampicillin application, cells were allowed to recover, and the enrichment process was repeated. This is different from the conventional selection for resistant mutants, where cells that can grow in the presence of antibiotic are favored. After testing for mutants that had the same minimum inhibitory concentration (MIC) to ampicillin (thus not resistant), but survived better, several strains were obtained, and one of them was used to map the mutation to a *hipBA* locus. The mutant appeared to carry a mutation in the *hipA* gene. This *hipA7* allelic strain was found to make 1% persisters in exponential cultures, ~1000 times more than the wild type. Deletion of *hipBA* had no apparent effect on persister formation, suggesting that *hipA7* mutant [36] produced a pleiotropic artifact. Another possibility is that *hipA* is part of a redundant set of genes, and knocking out any single one does not produce a phenotype. Like Bigger's work before him, the studies of Moyed were largely forgotten.

Persister cells have been proposed as an additional innate mechanism for biofilm antibiotic resistance [37]. In the persister theory, a small subpopulation of bacteria, whether in biofilms or planktonic culture, differentiates into dormant, spore-like cells that survive after extreme antibiotic treatment (Figs. 4.5 and 4.6). Differentiation into this dormant state has been hypothesized to be the result of phenotypic variation rather than a stable genetic change [29].

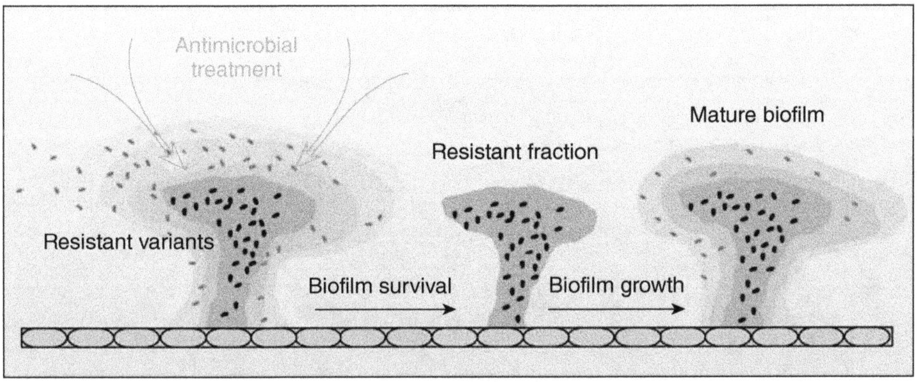


**Figure 4.5.** Resistance mechanism mediated by phenotypic–persister variants. Antimicrobial treatment of bacterial biofilms leads to the eradication of most of the biofilm susceptible population. A small fraction of phenotypic–persister variants (represented as maroon bacteria) survives antimicrobial treatment and is able to start biofilm development once antimicrobial therapy is discontinued. (See color insert.)

Interestingly, the results of recent studies suggest that, while persisters may be phenotypic variants, specific genetic elements are required to form the persister state. Studies by Spoering, Vulíc, and Lewis implicated altered genetic activation of the glycerol-3-phosphate regulated genes *glpD*, *glpABC*, and *plsB* in *E. coli* as a mechanism of persister development [38]. The *glpD* gene was initially found to be important for this developmental pathway because plasmid driven expression of the gene could increase the formation of ampicillin-resistant persisters in the exponential phase by ~10-fold. Mutating the *glpD* gene or other genes involved in glycerol-3-phosphate metabolism, including *glpABC* or *plsB*, decreased tolerance to ampicillin by >100-fold, indicating a role for glycerol-3-phosphate metabolism in persister formation. However, it was not reported whether these mutations altered the growth rate of the cell or the minimum inhibitory concentration for ampicillin. Further, given glycerol-3-phosphate's central metabolic role, these mutations did not provide any direct mechanistic insight into how persisters might be generated.

One mechanism proposed to explain the ability of persisters to resist the action of antibiotics is similar to a mechanism long hypothesized for biofilm resistance, namely, a slowed growth rate. Indeed, persisters exhibit slow or no growth, as observed by microscopy of *E. coli* in a microfluidic device [39]. This decreased growth rate may inhibit antimicrobial action, as discussed in the previous paragraph. However, persisters can survive even after treatment with ofloxacin, which exerts bactericidal activity against nongrowing microorganisms [13,40], suggesting that limited growth rate alone cannot account for the increased antibiotic resistance of persisters. Alternatively, global transcriptional profiling by microarray analysis of persister cells revealed activation of

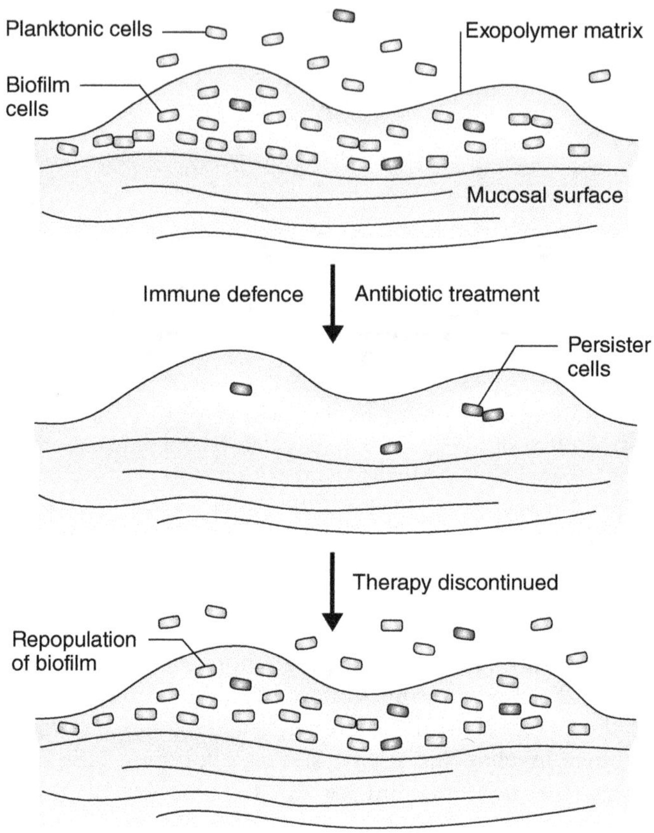


**Figure 4.6.** This figure shows a model of biofilm resistance to killing based on persister survival. Initial treatment with antibiotic kills normal cells (colored green) in both planktonic and biofilm populations. The immune system kills planktonic persisters (colored pink), but the biofilm persister cells (colored pink) are protected from the host defences by the exopolymer matrix. After the antibiotic concentration is reduced, persisters resuscitate and repopulate the biofilm and the infection relapses (Reproduced with permission from Kim Lewis *Nat. Rev. Microbiol.*, 5, 48–56, 2007 [15]). (See color insert.)

numerous stress response pathways [40,41], potentially implicating these cells as hardy, stress-resistant microorganisms.

Another major factor influencing formation of persisters appears to be chromosomal toxin–antitoxin (TA) systems [37], which have previously been associated with programmed cell death in bacteria. Several TA modules were upregulated by microarray analysis of persisters in *E. coli*, including *dinJ/yafQ*, *relBE*, and *mazEF* [41,42]. Overexpression of the *relE* toxin gene, in particular, led to tolerance of high levels of such disparate antibiotics as ofloxacin, cefotaxime, and tobramycin [41]. The *hipBA* TA locus has also been found to be important for formation and maintenance of persisters, and mutation of the *hipA* toxin gene can enrich for persisters within in an *E. coli* population [29,31,41,43]. It has been suggested that these TA modules actually induce stasis of the bacterial cell by inhibiting the activity of a particular cellular machine (e.g., the ribosome [41]). It was proposed that this inert state then prevents the deleterious functions induced by antibiotics. For example, an aminoglycoside cannot induce the formation of misfolded proteins if its target ribosome has been rendered static. In this sense, persister bacteria are considered antibiotic-tolerant rather than antibiotic-resistant [37,41]. Evidence for this induced stasis comes from studies demonstrating that, while overexpression of the *relE* or *chpAK* toxin genes in *E. coli* rapidly reduced colony-forming units (CFUs), subsequent transcription of the *relB* or *chpAI* antitoxins, respectively, led to a restoration of colony formation on agar plates [44]. In other words, the toxin expressing bacteria were nongrowing, yet nondead, and addition of antitoxin resuscitated these cells. Thus, random fluctuations of toxin and antitoxin levels may modulate the formation and awakening of dormant persisters.

Intriguingly, recent work offers a mechanistic explanation for *E. coli* persisters that neither grow nor die in the presence of bactericidal agents and so are MDT [41]. However, no compelling evidence is available to show any of the well-characterized mechanisms uniquely responsible for biofilm resistance. In short, three main mechanisms (hindrance to drug penetration into the biofilm, presence of antibiotic inactivating enzymes, efflux and oxygen deprivation for biofilm resistance) are described to indicate why the biofilm microorganisms show resistance or tolerance against microbial agents. A number of novel drug delivery carriers have been and are being investigated in an attempt to drive antimicrobial agents through biofilms (see in later chapter 11, sections 11.2–11.7 for details).

## 4.4. PERSISTER CONTROVERSIES

Intriguingly, persister research has led to several claims about biofilm antibiotic resistance in opposition to generally accepted biofilm tenets. Specifically disputed is the widely held, and well-supported, hypothesis that biofilms are more resistant to antimicrobial killing than planktonic bacteria. This argument has led some researchers to solely examine planktonic cultures for phenotypic and genotypic analysis of persisters. For example, Spoering and Lewis showed that stationary-phase *P. aeruginosa* was equally or more resistant than biofilm cultures to several antibiotics [13]. This effect was quantified as greater bacterial CFU after 6 h of antibiotic challenge and was hypothesized to be the result of equal or greater persister formation in the planktonic stationary-phase bacteria compared to the biofilm population. Similarly, Harrison et al. [43] discovered that planktonic and biofilm populations of *E. coli* required similar levels of amikacin and ceftrioxone to effect complete eradication of the population in 2h. However, in this latter study, *E. coli* biofilms were more resistant to tobramycin than planktonic phase cells at 2h. Further, increasing the incubation time to 24h revealed a much greater antibiotic resistance of biofilms to all three antibiotics compared to planktonic cells. In other words, planktonic bacteria were more sensitive to lower antibiotic concentrations when treated for longer periods of time. This result leads one to wonder whether increased antibiotic incubation periods could have produced a similar effect in the work by Spoering and Lewis and similar studies [12,13].

An additional concern in these studies is the variance in bacterial numbers between planktonic and biofilm populations at the start of antibiotic treatment. Thus, while Spoering and Lewis [13] found a greater number of surviving stationary planktonic-phase bacteria compared to biofilm bacteria after antibiotic treatment, they also started with a significantly greater number of stationary planktonic phase bacteria then biofilm bacteria. In effect, in the stationary-phase cultures, the units of antibiotic per bacterial cell were markedly decreased relative to biofilm bacteria. This difference might have led to an apparent increase in antibiotic resistance. In a later study of *E. coli* resistance to metal oxyanions, Harrison et al. [43] equalized planktonic and biofilm bacterial numbers before antibiotic challenge and found that this action did not significantly alter the MIC. However, in these conditions for the planktonic bacteria, they reported that "the proportion of surviving cells was smaller than the fraction of survivors recovered from biofilms" [43]. As with the increased incubation time mentioned above, it would be intriguing to determine the effect of starting with similar bacterial numbers using the system as described by Spoering and Lewis [13].

Based on the results of these studies, it may be misleading to consider biofilm antibiotic resistance as a stationary-phase persister phenomenon. Alternatively, perhaps planktonic persisters have differentiated into a more biofilm-like phenotype, although there is no data to support this theory at this time. Recent microarray studies of *E. coli* suggested that the persister transcriptional profile represents a unique physiological state, distinct from exponential- or stationary-phase bacteria [42]. While no comparison was made to biofilms, it is intriguing to speculate that the persister phenotype is similar to the biofilm state. Indeed, the most highly expressed gene in persisters compared to nonpersisters in this microarray analysis was *ygiU*, which is also induced in biofilms and acts as a global regulator influencing biofilm formation [42]. Further, mathematical modeling has predicted a steady accumulation of persisters as a biofilm matures and ages [45]. Thus, despite inconsistencies in persister literature, persister formation remains an intriguing concept as a supporting mechanism of biofilm antibiotic resistance.

In summary, innate formation of persisters might represent a common mechanism used by a wide range of bacteria during biofilm formation. Creation of this tenacious population within the biofilm may drastically inhibit the complete eradication of the biofilm during even prolonged, high-level antibiotic treatment. However, at this stage, it is unclear what relationship, if any, can be drawn between planktonic persisters and biofilm resistance, and furthermore, the mechanisms(s) by which persisters form and/or confer increased antibiotic tolerance. As discussed above, the process of biofilm formation apparently leads to innate mechanisms of antibiotic-resistant bacteria. That is, some mechanisms of resistance appear to be part and parcel of growing in a biofilm. Inhibited diffusion through the matrix, reduced metabolism by nutrient limitation, and formation of dormant persisters all appear to impact the development of a protective environment within the biofilm. Working in combination, these pathways might confer a multilayered network of security for the constituent bacteria. Further exposition of the genetic pathways that lead to these innate phenomena may very well result in improved treatment regimes for disruption and elimination of bacterial biofilms.

## 4.5. PERSISTENT INFECTIONS

Several infections [e.g., syphilis, lyme disease, and tuberculosis (TB)] persist for years in the body in an apparently benign form. In most chronic (persistent) infections, it seems that the pathogen is at least partially shielded from the immune system: *Treponema pallidum* (syphilis) and *Borrelia burgdorferi* (Lyme disease) migrate into the central nervous system (CNS), whereas *Mycobacterium tuberculosis* (TB) is hidden in macrophages or granulomas [46]. Are dormant persister cells responsible for the latent, asymptomatic stages of disease? Significant progress has been made toward understanding the metabolic functions that are required for the persistence of *M. tuberculosis* [47], but no persister genes have been isolated for this, or any other, pathogen that causes latent infection. However, note that *M. tuberculosis* has at least 60 TA loci [48].

### 4.5.1. Single-Compound Antipersister Antibiotic Development

Given the prominent role of tolerance to antibiotics in the aetiology of infectious disease, the need for compounds that could eradicate persisters is obvious. Another important factor to consider is the potential causality between tolerance and antibiotic resistance. A lengthy, lingering infection that is not eradicated owing to tolerance is likely to provide a fertile ground for the emergence of resistant mutants, or for the acquisition of resistance determinants through horizontal gene transfer from other species. A mathematical model predicts that tolerance substantially increases the incidence of resistance [49].

In trying to combat persisters, we might have encountered the ultimate adversary. Indeed, persisters evolved over billions of years to accomplish a single feat: survival. During this time, this cell type has encountered a huge array of harmful compounds, and the inability of any antibiotic in current use to eliminate persisters provides a sobering view of the magnitude of the challenge to those aiming to develop antipersister therapies.

An antipersister drug based on traditional approaches could be produced by combining a conventional antibiotic (e.g., a fluoroquinolone), and an inhibitor of an essential persister protein. As mentioned previously, a mutation in the essential *E. coli* gene encoding PlsB, which increases the $K_m$ (Michaelis constant) of the PlsB protein, results in a 2–3-log drop in the frequency of persister formation [38]. Proteins (e.g., PlsB) that are essential for maintaining the persister state might be attractive targets for antipersister drug development.

However, unlike conventional antibiotic treatment, antipersister therapies face an additional hurdle. The U.S. Food and Drug Administration (FDA) only requires testing against rapidly growing bacteria, and market conditions are excellent for a new conventional broad-spectrum antibiotic. Why then commit resources to the considerably more challenging development of an antipersister therapy?

A disarmingly simple approach to sterilize an infection was first proposed by Bigger in 1944 [28]. The proposal is to kill bacterial cells with a high dose of an antibiotic, then allow the antibiotic concentration to decrease, which will enable persisters to resuscitate and start to grow. If a second dose of antibiotic is administered shortly after persisters start to grow, a complete sterilization might be achieved. This approach is successful *in vitro*, and a *P. aeruginosa* biofilm can essentially be sterilized with two consecutive applications of a fluoroquinolone. Perhaps understandably, this approach has not been received with enthusiasm by specialists in clinical microbiology. The goal of established therapies is to maintain the plasma level of an antibiotic at a maximum concentration, in order to discourage the development of resistance. Most importantly, an optimal pulse-dosing regimen would probably vary from patient to patient. However, it seems that some patients might have inadvertently taken solving the problem of intractable persistent infections into their own hands. Individuals who suffer from persistent infections that require a lengthy therapy are often cured, but why a year-long regimen is better than a month-longer one is unclear. An efficacious fluctuating dose of antibiotics administered serendipitously by the patient might be responsible for persister eradication in these cases. The patients might adjust drug dosing simply through being absent-minded, which sooner or later could produce the perfect drug-administration regimen. Curing persistent infections might therefore result from patient noncompliance. Analyzing how persistent infections are cured might shed light on the likelihood of developing a rational regimen for the pulse-dosing sterilization of infection.

Persister cells can be killed by antiseptics, but these are obviously toxic and largely unsuitable for systemic applications. The recent development of sterile surface materials [50,51] provides an attractive approach for producing a relatively nontoxic antiseptic by covalently attaching the antimicrobial molecule to the surface. This would prevent leaching of the antiseptic and limit contact with tissue cells. There is an obvious problem with this approach: Once attached to the surface, an antimicrobial molecule is immobilized and is unable to reach and kill the pathogen. This problem was solved by linking the antimicrobial compound to a long, flexible polymeric chain that is covalently anchored to the surface of a material. Attaching a long chain of poly-(4-vinyl-*N*-alkylpyridinium bromide) to an amino glass slide (derivatized to contain free-$NH_2$ groups) produced a material that remained largely sterile. Similar immobilized polymers poly [2-(dimethylamino)ethylmethacrylate] [52] and *N*-alkyl-polyethylenimine, rapidly depolarized and killed *S. aureus* or *E. coli* cells that came into contact with the surface, leaving no evidence of surviving persisters [53]. Importantly, in order to be effective, the sterile-surface polymers must be long enough to penetrate across the cell envelope [54,55], as shorter versions were ineffective [54].

The attractive properties of antimicrobial polymers are likely to lead to the development of products with sterile surfaces that will prevent the growth of biofilms on catheters and indwelling devices. These materials, however, do not address the need for systemic sterilizing antibiotics. Is it possible to develop a single-compound antipersister antibiotic? Known target-specific antibiotics do not sterilize an infection. Antiseptics that can kill persisters do not have specific targets, but damage the cell membrane or macromolecules (e.g., DNA and proteins) and are cytotoxic. Considering this, developing a single-molecule sterilizing antibiotic does not seem feasible. However, let us consider the formation of a perfect antibiotic from first principles (Fig. 4.7). The proposed proantibiotic compound is benign, but a bacterial enzyme converts it into a reactive antiseptic compound in the cytoplasm. The active molecule cannot be exported owing to increased polarity, and attaches covalently to many bacterial targets, killing the cell. Irreversible binding of the active antibiotic molecule to cellular targets allows the compound to avoid MDR efflux.

Several existing antimicrobials have properties that closely match those of the ideal proantibiotic. These are isoniazid, pyrazinamide, ethionamide (anti-*M. tuberculosis* drugs), and metronidazole, a broad-spectrum compound that is active against anaerobic bacteria. Once inside the cell, all four compounds convert into active antiseptic-type molecules that covalently bind to their targets. It seems highly pertinent to this discussion that prodrug antibiotics comprise the core of the anti-*M. tuberculosis* drug arsenal.

*Mycobacterium tuberculosis* might form the most intransigent persisters, and excellent bactericidal properties are a crucial feature for any anti-TB antibiotic. Preferred targets have been identified for isoniazid and ethionamide [56], which might indicate a relatively limited reactivity of these compounds. The existence of preferred targets indicates that the prodrug products are not particularly reactive, and that there is considerable scope for the development of better sterilizing antibiotics based on the same principles.

In summary, the entrance of cells into a dormant, persistent state is largely responsible for the MDT of infections. The presence of dormant persisters in biofilms accounts for their tolerance to all known antimicrobials. Persisters are likely to be responsible for latent (chronic) diseases (e.g., TB), which can be suppressed, but not eradicated, with existing antimicrobials. The need to

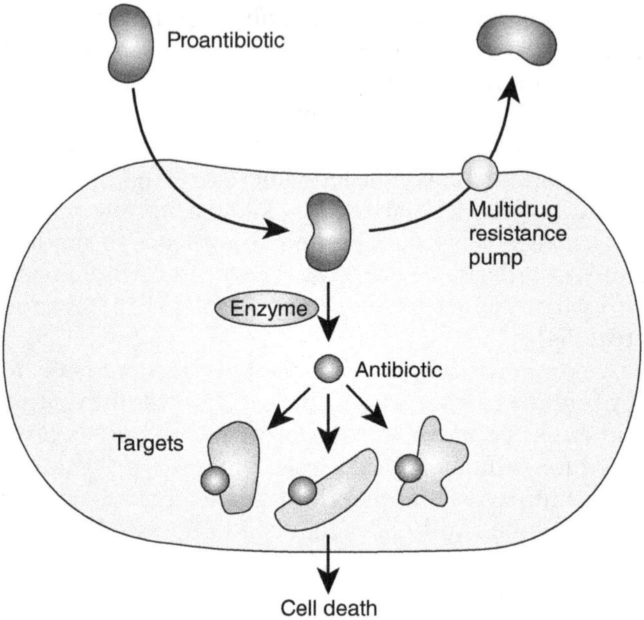


**Figure 4.7. The perfect antibiotic.** The proantibiotic is benign, but a bacterial enzyme converts it into a reactive antibiotic in the cytoplasm. The active molecule does not leave the cytoplasm (owing to increased polarity), and attaches covalently to many targets, thereby killing the cell. Irreversible binding to the targets prevents the antibiotic from multidrug resistance efflux. (Reproduced with permission from Lewis *Nat. Rev. Microbiol.*, 5, 48–56, 2007 [15].) (See color insert.)

develop novel therapeutics capable of killing persister cells and eradicating infections is acute. Finding genes responsible for persister formation and maintenance should lead to drugs that disable persisters and might allow conventional antibiotics to eradicate an infection.

The inability of most bacterial species to grow *in vitro* is well recognized, and it seems likely that uncultivable species might enter into a protective dormant state in unfamiliar environments. Although some important progress has recently been made in the study of persisters and uncultivable bacteria, there is a striking disparity between the importance of the problems outlined here, and the number of professionals working in these fields. This problem is especially prominent in the field of unculturable bacteria, in which the proportion of such species is ≥99%, whereas >99.9% of professionals work on the 1% of cultivable organisms, and only a minority of scientists are studying the uncultivability issue. Similarly, MDT of biofilms is responsible for 65% of all cases of infections in the developed world, but there are probably fewer than 10 scientists working on the mechanism underlying this problem. In addition, out of this tiny band of scientists there are probably fewer than 10 academics, and none at all working in industry, to tackle the problem of persisters. The significance of the problems outlined above is, however, appreciated in the microbiology community. For example, a recent American Society of Microbiology (ASM) session outlined the most significant directions for research and bacterial unculturability and biofilms were ranked among the most important topics [57]. Why then is there a reluctance to study problems that are both exciting and important? Taking on the challenges of such recalcitrant problems is fraught with uncertainty. Solving these problems requires patience and risk-taking, both from researchers and from the agencies that support their work.

## 4.6. PERSISTERS IN YEAST BIOFILMS

Eukaryotic yeasts have a lifestyle that is very similar to that of prokaryotic microorganisms. Not surprisingly, analogous adaptations evolved in these two groups in response to similar environmental challenges through convergent evolution. For example, yeasts form biofilms that, similarly to bacterial biofilms, are responsible for highly recalcitrant infections [58].

The focus of yeast biofilm research has been on *C. albicans*, an important human pathogen that causes oral thrush, relapsing vaginosis, and is a leading cause of morbidity and mortality in immunocompromised individuals. The biofilm forms when single cells attach to a surface and grow into microcolonies, which then merge and produce a complex three-dimensional (3D) structure that is held together by hyphae and an exopolymer matrix. The biofilm contains a mixture of yeast, hyphae, and pseudohyphae. Similarly to bacteria, yeast biofilm exopolymer matrix restricts penetration of immune system components [59], but does not appreciably hinder diffusion of antifungal drugs [60,61].

Genes encoding MDR pumps, MDR1, CDR1, and CDR2, are upregulated upon attachment of *C. albicans* cells to a surface, and this accounts for the resistance of young biofilms to azole antibiotics [62]. However, the high level of drug resistance of mature biofilms (≥48 h) was not affected by deletion of all three of these genes either singly or in combination, including an *mdr1*Δ *cdr1*Δ *cdr2*Δ triple mutant [62–64]. Decreased ergosterol content [62,64] and a diminished level of ergosterol biosynthetic gene expression [65] have been reported in mature *Candida* biofilms and may contribute to drug resistance. Indeed, azoles act by inhibiting ergosterol biosynthesis, and amphotericin B binds to ergosterol. However, ergosterol is unlikely to be involved in the action of echinocandins that inhibit the synthesis of cell wall B-glucan [66], or chlorhexidine, a membrane-active antiseptic that is very effective against bacteria that lack sterols. In essence, the mechanism of *C. albicans* biofilm antifungal resistance remains largely unknown.

The same approach, which was used previously for bacteria to examine biofilm resistance, has been followed for *C. albicans*. A dose-dependent experiment with two highly microbicidal agents, amphotericin B and chlorhexidine (the only compounds that can kill nongrowing yeast cells), showed complete elimination of cells in exponential and stationary planktonic populations. However, a distinctly biphasic killing was observed in a mature biofilm, indicating the presence of persisters [67]. Similarly to bacteria, yeast persisters are not mutants and upon reinoculation, surviving cells reproduced the original wild-type population with a new fraction of persister cells. Staining with fluorescein, which specifically binds to dead yeast cells, showed live persisters within a yeast biofilm treated with amphotericin. These rare live cells were either yeast or pseudohyphae and were morphologically unremarkable. Sorting this stained population showed that dim cells form colonies, while bright ones do not. This method, similarly to the approach described for bacteria, opens the way for obtaining a gene expression profile of yeast persisters.

Quite unexpectedly, *C. albicans* persisters were only apparent in a biofilm culture and not in a nongrowing stationary population. Both biofilm and planktonic stationary cultures produce a mix of cell types, including yeast, pseudohyphae, and hyphae. The specific production of persisters in a biofilm is distinctly different from what is observed in bacteria, where a stationary planktonic culture actually makes more persisters than a biofilm [13]. This probably suggests that the biofilm, and not the planktonic population, is the survival mode of yeast life, and that is where specialized survivor cells are produced.

Dependence of persister production on biofilm formation suggested that these two forms of yeast populations may share part of the same developmental program. A number of genes involved in yeast biofilm formation have been identified, and mutants deleted in these elements were tested for persister production [67]. Surprisingly, all tested mutants appeared to produce normal levels of persisters. Among the tested mutants was the flo8 strain, which does not make hyphae. Its biofilm consists of a simple layer of yeast cells attached to the surface [68], which indicates that attachment is sufficient for persister formation.

A surface contact-dependent Mck1p kinase that affects biofilm formation and invasiveness in *C. albicans* was recently described [64], but a strain deleted in mck1 was able to produce normal levels of persisters as well. Analysis of biofilm mutants suggests that known genes are not involved in persister formation. Persister isolation based on cell sorting opens the possibility of obtaining their transcription profile, which is likely to point to persister genes.

*Candida albicans* in biofilms on poly(vinyl chloride) (PVC) disks has been reported to be 30–2000 times more resistant to fluconazole, amphotericin B, flucytosine, itraconazole, and ketoconazole than planktonic cells [69], and the biofilm structure remained intact at an amphotericin B concentration of 11 times the MIC. Non-*albicans Candida* species were also resistant. *In vitro*, the newer triazoles were also found to be ineffective against *C. albicans* and *C. parapsilosis* biofilms [14]. However, caspofungin has been shown to be effective against *C. albicans* biofilms [70]. Glucan synthesis may thus prove to be an effective target for biofilms. Suggested mechanisms of biofilm resistance include restricted penetration of drugs through the matrix, slow growth of organisms in biofilms accompanied by changes in cell surface composition affecting their susceptibility to drugs, and unique biofilm-associated patterns of gene expression [71,72].

In summary, *Candida* spp. produce biofilms on medical devices to various degrees and the difference in biofilm production and architecture by *Candida* spp. on different device materials is reflected in the different epidemiological trends in *Candida* device-related infections. *In vitro*, *Candida* biofilms are highly resistant to most antifungal agents except caspofungin, thereby posing a therapeutic challenge in managing device-associated *Candida* infections.

## 4.7. INDUCED RESISTANCE FACTORS MECHANISMS

As with any environmental change, antibiotic treatment can alter regulatory patterns within bacteria. Antibiotic treatment can be a harsh stress, even for bacteria within a biofilm. Consequently, one would predict that there must be some antibiotic-regulated genes that influence antibiotic resistance or sensitivity of biofilm bacteria. As previously mentioned, antibiotics can activate regulatory pathways, leading to a profound effect upon the biofilm matrix and achieved biomass [73–76], and it is likely that numerous genetic loci are activated upon treatment with antibiotics. These induced factors may work synergistically with innate factors to enhance survival in the face of strong antimicrobial stresses.

Very little work has been done to identify antibiotic-induced factors in biofilms. However, microarray analyses have yielded some intriguing clues. Imipenem treated *P. aeruginosa* biofilms strongly expressed alginate genes and the chromosomal b-galactosidase gene *ampC* [73]. Expression of *ampC* was restricted to the outer edges of microcolonies, as determined by epifluorescence and confocal scanning laser microscopy of an *ampC-GFP* transcriptional fusion [77]. In a separate study, tobramycin treatment of *P. aeruginosa* biofilms resulted in upregulation of *PA1541* and *PA3920*, two possible antibiotic efflux systems [78]. Although no functional data have been generated, both of these studies identified a number of hypothetical genes that were upregulated or repressed upon antibiotic treatment [73,78], suggesting that many more factors are potentially involved in biofilm resistance than have currently been identified.

Efflux pumps and β-lactamases are some of the key mechanisms used by planktonic bacteria to overcome antibiotic challenge. Previous research has generally disregarded these factors and other planktonically associated systems as not important for biofilm antibiotic resistance. There is much experimental evidence to support this view [17,79]. Nevertheless, as mentioned above, β-lactamases and possibly efflux pumps might exert some influence during a biofilm lifestyle [22,73,77,78]. In reconciling these conflicting observations, it is intriguing to reflect on the numerous hypothetical genes that are differentially regulated during antibiotic exposure. It seems likely that novel orthologs will be discovered that might play a large role in antibiotic resistance specifically within biofilms. For example, *E. coli* alone contains genes for 37 proposed efflux pumps [80]. It is possible that some of these genes are important for survival in a planktonic state, while others may be biofilm-specific. The putative efflux genes mentioned above, which were discovered by microarray analysis of tobramycin treated *P. aeruginosa* biofilms (*PA1541* and *PA3920*), may be examples of such biofilm-specific orthologs [78]. It is also possible that redundant function of similar proteins within the same bacterium may have obscured the activity of previously tested gene products. For example, in another study, a *P. aeruginosa* mutant with deletions in both the *mexAB–oprM* and the *mexCD–oprJ* efflux pumps could not establish biofilms in the presence of azithromycin, while the single mutation constructs of each behaved as wild type [81].

It is clear that much more research is needed to expose additional and/or novel antibiotic-induced factors in biofilms. The multifactorial nature of biofilm antibiotic resistance has hindered identification of these pathways, and much remains to be elucidated about induced factors in biofilm resistance. Discovery of these unknown factors will lead to new and better treatments for biofilm related infections.

## 4.8. FUTURE PERSPECTIVE

Conventional antibiotic resistance mechanisms do not seem to influence biofilm survival, and dispersion of the biofilm bacteria leads to reversion to an antibiotic-sensitive state. These results have led to the identification of several intriguing resistance models, either resulting from an innate property of a biofilm lifestyle or an effect induced by the antimicrobial stress itself. It is tempting to speculate that any one of these models alone (e.g., persister formation) can fully explain biofilm resistance. However, from our current understanding of research findings, none of these phenomena can adequately account for every aspect of the biofilm-resistance phenotype. Further, these models share common features and themes (e.g., decreased metabolic activity seen in anaerobic microcolony environments and with persisters). Also intriguing is the possibility that the biofilm matrix might slow the progress of an antibiotic through the microcolony such that the bacteria can sufficiently activate expression of protective factors in response to the biocide.

The overlap between these resistance paradigms has led some researchers to propose a layered model of biofilm resistance, wherein the outer layers of the microcolonies provide a first-line defense by inhibiting the diffusion of antimicrobial agents, bacteria deeper within the biofilm can be further protected by altered metabolic states, and development of persisters throughout enhances bacterial survival [79]. Throughout, innately expressed as well as antibiotic-induced genes might provide additional protection. In short, biofilm antibiotic resistance results from an overlapping mixture of innate and induced microbial activities, intricately woven together with redundant form and function.

Our understanding of the molecular details of resistance mechanisms of bacterial biofilms is still in its infancy. While many genetic factors have been identified, many more questions remain. Even for many known resistance genes, it is uncertain how they interact with each other to establish a resistance phenotype. Much greater knowledge of genetic responses to antimicrobial agents will facilitate the production of new and better drugs to eradicate biofilms. Manipulation of regulatory and expression networks holds much promise for future treatment of biofilm infections. Indeed, QS inhibitors have demonstrated an exceptional ability to disrupt biofilm structure and act synergistically with a number of antibiotics (see details in Chapter 9).

On the other hand, enhancing and nurturing the impervious nature of beneficial biofilms may lead to improvement, for example, in the production of biologically derived chemicals and bioremediation. Particularly in industrial settings, chemically resistant bacterial biofilms provide a hardy platform for a number of applications involving high concentrations of toxic metals or other chemicals [82]. Studies have revealed the utility of biofilms in the synthesis of ethanol, poly-3-hydroxybutyrate, benzaldehyde, and other chemicals [83–85]. Biofilms have also assisted waste-water treatment, phenol bioremediation, biodegradation of 2,4- and 2,6-dinitrophenol, and bioremediation of toxic metal contamination of environmental sites [86–89]. These applications highlight the usefulness of extremely resistant biofilms in chemical synthesis and breakdown. In either case, whether beneficial or harmful biofilms, elucidation of the genetic mechanisms of innate and induced biofilm resistance, however, holds the key to solving this great mystery.

## 4.9. CASE STUDY

Nontypeable *Haemophilius influenzae* (NTHi) is a Gram-negative, pleomorphic bacterium that colonizes the human nasopharynx. This bacteria can cause a variety of infections including otitis media, sinusitis, conjunctivitis, bronchitis, and pneumonia [90,91]. Recent reports state that β-lactamase-negative ampicillin (AMP)-resistant (BLNAR) strains have increased in some countries [92–94], although their global prevalence remains low [95]. The prevalence of BLNAR strains was reported to be 2.4% in the United States between 2002 and 2003 [96] and 9.3% in Spain between 1998 and 1999 [97]. But the prevalence of BLNAR in Japan has increased rapidly from 5.8% in 2000 to 34.5% in 2004 even in *H. influenzae* type b isolated from patients with meningitis [98]. The result has been an increasing number of cases of otitis media, which are difficult to treat, and an increasing incidence of treatment failures with oral antibiotics in children [99]. However, the pathogenesis of NTHi infection of airway cells involves both an intracellular life cycle, as well as biofilm formation on the surface of the airway epithelia. Studies from Kaczmarek et al. [100] also indicate that otitis media, paranasal sinusitis, and lower respiratory tract infections have become more difficult to treat with oral antibiotic therapy. This finding is due in large part to antibiotic resistance, which points to a need for oral therapy to change along with surgical management of tympanostomy or tympanostomy tubes [101]. But it is not clear whether drug resistance alone causes treatment difficulty and failure in otitis media among young children as NTHi invade human bronchial epithelial cells by macropinocytosis [102] and form biofilms both *in vitro* and *in vivo* [103]. It should be emphasized, however, that resistance in BLNAR strains results from mutations in the ftsl gene-encoding, penicillin-binding protein (*pbp*) 3, which mediates septal peptidoglycan synthesis [98].

To investigate the antimicrobial effect of various antibiotics against NTHi biofilm formation, Kaji et al. [104] conducted a comparative study using both β-lactamase-negative ampicillin (AMP)-susceptible (BLNAS) and AMP-resistant (BLNAR) NTHi strains. In a microtiter biofilm assay, both levofloxacin and gatifloxacin, of the fluoroquinolone antibiotic group, significantly inhibited biofilm formation by BLNAS and BLNAR NTHi in a dose-dependent fashion compared to ampicillin of the penicillin antibiotic group, cefatoxime of the cephalosporin antibiotic group, and both erythromycin and clarithromycin of the macrolides antibiotic group. Furthermore, in flow-cell chamber studies, CLSM counted survival bacteria in mature biofilms that had been treated with gatifloxacin, ampicillin, cefatoxime, and erythromycin [104]. Only gatifloxacin completely killed the BLNAR NTHi isolates within biofilms without regard to the thickness of the biofilm formation. Thus, the result of this study suggests that fluoroquinolones potentially have a role in therapy against diseases caused by both BLNAS and BLNAR NTHi isolates within biofilms [104]. However, since the use of fluoroquinolones in young children is limited in many countries because of adverse neurological side effects, the clinical antimicrobial effectiveness against BLNAS and BLNAR NTHi in biofilms has not been confirmed.

## REFERENCES

1. National Institutes of Health (2006), (Available at http://www.idph.state.ia.us/adper/common/pdf/abx/tab9_niaid_resistance.pdf.) Accessed on 6/25/2009.
2. Blázquez, J. (2003), Hypermutation as a factor contributing to the acquisition of antimicrobial resistance, *Clin. Infect. Dis.*, 37, 1201–1209.
3. Velkov, V.V. (2002), New insights into the molecular mechanisms of evolution: stress increases genetic diversity, *Mol. Biol.*, 36, 209–215.
4. Singh, P.K., Schaefer, A.L., Parsek, M.R., Moninger, T.O., Welsh, M.J., and Greenberg, E.P. (2000), Quorum sensing signals indicate that cystic fibrosis lungs are infected with bacterial biofilms, *Nature*, 407, 762–764.

5. Mack, D., Becker, P., Chatterjee, I., Dobinsky, S., Knobloch, J.K., Peters, G., Rohde, H., and Herrmann, M. (2004), Mechanisms of biofilm formation in *Staphylococcus epidermidis* and *Staphylococcus aureus*: functional molecules, regulatory circuits, and adaptive responses, *Int. J. Med. Microbiol.*, 294, 203–212.

6. Drenkard, E. (2003), Antimicrobial resistance of *Pseudomonas aeruginosa* biofilms, *Microbes Infect.*, 5, 1213–1219.

7. Fitzpatrick, F., Humphreys, H., and O'Gara, J.P. (2005), The genetics of staphylococcal biofilm formation—will a greater understanding of pathogenesis lead to better management of device-related infection? *Clin. Microbiol. Infect.*, 11, 967–973.

8. Khardori, N., Yassien, M., and Wilson, K. (1995), Tolerance of *Staphylococcus epidermidis* grown from indwelling vascular catheters to antimicrobial agents, *J. Ind. Microbiol.*, 15, 148–151.

9. Ramage, G., Tunney, M.M., Patrick, S., Gorman, S.P., and Nixon, J.R. (2003), Formation of *Propionibacterium acnes* biofilms on orthopaedic biomaterials and their susceptibility to antimicrobials, *Biomaterials*, 24, 3221–3227.

10. Knobloch, J.K., Von Osten, H., Horstkotte, M.A., Rohde, H., and Mack, D. (2002), Minimal attachment killing (MAK): a versatile method for susceptibility testing of attached biofilm-positive and-negative *Staphylococcus epidermidis*, *Med. Microbiol. Immunol. (Berl.)*, 191, 107–114.

11. Monzon, M., Oteiza, C., Leiva, J., Lamata, M., and Amorena, B. (2002), Biofilm testing of *Staphylococcus epidermidis* clinical isolates: low performance of vancomycin in relation to other antibiotics, *Diagn. Microbiol. Infect. Dis.*, 44, 319–324.

12. Brooun, A., Liu, S., and Lewis, K. (2000), A dose-response study of antibiotic resistance in *Pseudomonas aeruginosa* biofilms, *Antimicrob. Agents Chemother.*, 44, 640–646.

13. Spoering, A.L. and Lewis, K. (2001), Biofilms and planktonic cells of *Pseudomonas aeruginosa* have similar resistance to killing by antimicrobials, *J. Bacteriol.*, 183, 6746–6751.

14. Kuhn, D.M., George, T., Chandra, J., Mukherjee, P.K., and Ghannoum, M.A. (2002), Antifungal susceptibility of *candida* biofilms: unique efficacy of amphotericin B lipid formulations and echinocandins, *Antimicrob. Agents Chemother.*, 46, 1773–1780.

15. Lewis, K. (2007), Persister cells, dormancy and infectious disease, *Nat. Rev. Microbiol.*, 5, 48–56.

16. Lewis, K. (2008), Multidrug tolerance of biofilms and persister cells, in: Romeo, T. Ed., *Bacterial Biofilms. Current Topics in Microbiology and Immunology 322*, Springer-Verlag, Berlin Heidelberg, pp. 107–131.

17. Patel, R. (2005), Biofilms and antimicrobial resistance, *Clin. Ortho. Relat. Res.*, 437, 41–47.

18. Drenkard, E. and Ausubel, F.M. (2002), Pseudomonas biofilm formation and antibiotic resistance are linked to phenotypic variation, *Nature (London)*, 416, 740–743.

19. Anderson, G.G. and O'Toole, G.A. (2008), Innate and induced resistance mechanisms of bacterial biofilms, in: Romeo, T. Ed., *Bacterial Biofilms. Current Topics*

*in Microbiology and Immunology 322*, Springer-Verlag: Berlin Heidelberg, pp. 85–105.

20. Fux, C.A., Costerton, J.W., Stewart, P.S., and Stoodley, P. (2005), Survival strategies of infectious biofilms, *Trends Microbiol.*, 13, 34–40.
21. Davies, D. (2005), Understanding biofilm resistance to antibacterial agents, *Nat. Rev. Drug Discov.*, 2, 114–122.
22. Anderl, J.N., Franklin, M.J., and Stewart, P.S. (2000), Role of antibiotic penetration limitation of *Klebsiella pneumoniae* biofilm resistance to ampicillin and ciprofloxacin, *Antimicrob. Agents Chemother.*, 44, 1818–1824.
23. Lewis, K., Salyers A., Taber H., and Wax, R. (2002), *Bacterial Resistance to Antimicrobials: Mechanisms, Genetics, Medical Practice and Public Health*, Marcel Dekker, New York.
24. Levy, S.B. and Marshall, B. (2004), Antibacterial resistance worldwide: causes, challenges and responses, *Nature Med.*, 10, S122–S129.
25. Davis, B.D., Chen, L.L., and Tai, P.C. (1986), Misread protein creates membrane channels: an essential step in the bactericidal action of aminoglycosides, *Proc. Natl Acad. Sci. USA*, 83, 6164–6168.
26. Bayles, K.W. (2000), The bactericidal action of penicillin: new clues to an unsolved mystery, *Trends Microbiol.*, 8, 274–278.
27. Hooper, D.C. (2002), in: Lewis, K., Salyers A., Taber H., and Wax, R. Eds., *Bacterial Resistance to Antimicrobials: Mechanisms, Genetics, Medical Practice and Public Health*, Marcel Dekker, New York, pp. 161–192.
28. Bigger, J.W. (1944), Treatment of staphylococcal infections with penicillin, *Lancet*, 497–500.
29. Keren, I., Kaldalu, N., Spoering, A., Wang, Y., and Lewis, K. (2004), Persister cells and tolerance to antimicrobials, *FEMS Microbiol. Lett.*, 230, 13–18.
30. Wiuff, C., Zappala, R.M., Regoes, R.R., Garner, K.N., Baquero, F., and Levin, B.R. (2005), Phenotypic tolerance: antibiotic enrichment of noninherited resistance in bacterial populations, *Antimicrob. Agents Chemother.*, 49, 1483–1494.
31. Moyed, H.S. and Bertrand, K.P. (1983), *hipA*, a newly recognized gene of *Escherichia coli* K-12 that affects frequency of persistence after inhibition of murein synthesis, *J. Bacteriol.*, 155, 768–775.
32. Moyed, H.S. and Broderick, S.H. (1986), Molecular cloning and expression of *hipA*, a gene of *Escherichia coli* K-12 that affects frequency of persistence after inhibition of murein synthesis, *J. Bacteriol.*, 166, 399–403.
33. Scherrer, R. and Moyed, H.S. (1988), Conditional impairment of cell division and altered lethality in *hipA* mutants of *Escherichia coli* K-12, *J. Bacteriol.*, 170, 3321–3326.
34. Black, D.S., Kelly, A.J., Mardis, M.J., and Moyed, H.S. (1991), Structure and organization of *hip*, an operon that affects lethality due to inhibition of peptidoglycan or DNA synthesis, *J. Bacteriol.*, 173, 5732–5739.
35. Black, D.S., Irwin, B., and Moyed, H.S. (1994), Autoregulation of *hip*, an operon that affects lethality due to inhibition of peptidoglycan or DNA synthesis, *J. Bacteriol.*, 176, 4081–4091.
36. Korch, S.B., Henderson, T.A., and Hill, T.M. (2003), Characterization of the hipA7 allele of *Escherichia coli* and evidence that high persistence is governed by (p)ppGpp synthesis, *Mol. Microbiol.*, 50, 1199–1213.

37. Lewis, K. (2005), Persister cells and the riddle of biofilm survival, *Biochemistry (Mosc.)*, 70, 267–274.
38. Spoering, A.L., Vulic, M., and Lewis, K. (2006), *GlpD* and *PlsB* participate in persister cell formation in *Escherichia coli*, *J. Bacteriol.*, 188, 5136–5144.
39. Balaban, N.Q., Merrin, J., Chait, R., Kowalik, L., and Leibler, S. (2004), Bacterial persistence as a phenotypic switch, *Science*, 305, 1622–1625.
40. Kaldalu, N., Mei, R., and Lewis, K. (2004), Killing by ampicillin and ofloxacin induces overlapping changes in *Escherichia coli* transcription profile, *Antimicrob. Agents Chemother.*, 48, 890–896.
41. Keren, I., Shah, D., Spoering, A., Kaldalu, N., and Lewis, K. (2004), Specialized persister cells and the mechanism of multidrug tolerance in *Escherichia coli*, *J. Bacteriol.*, 186, 8172–8180.
42. Shah, D., Zhang, Z., Khodursky, A., Kaldalu, N., Kurg, K., and Lewis, K. (2006), Persisters: a distinct physiological state of *E. coli*, *BMC Microbiol.*, 6, 53–61.
43. Harrison, J.J., Ceri, H., Roper, N.J., Badry, E.A., Sproule, K.M., and Turner, R.J. (2005), Persister cells mediate tolerance to metal oxyanions in *Escherichia coli*, *Microbiology*, 151, 3181–3195.
44. Pedersen, K., Christensen, S.K., and Gerdes, K. (2002), Rapid induction and reversal of a bacteriostatic condition by controlled expression of toxins and antitoxins, *Mol. Microbiol.*, 45, 501–510.
45. Roberts, M.E. and Stewart, P.S. (2005), Modelling protection from antimicrobial agents in biofilms through the formation of persister cells, *Microbiology*, 151, 75–80.
46. Cole, S.T., Eisenach, K.D., McMurray, D.N., and Jacobs, W.R. Jr, eds., *Tuberculosis and the Tubercle Bacillus*, ASM Press, Washington DC, 2005.
47. Gomez, J.E. and McKinney, J.D. (2004), *M. tuberculosis* persistence, latency, and drug tolerance, *Tuberculosis (Edinb)*, 84, 29–44.
48. Pandey, D.P. and Gerdes, K. (2005), Toxin-antitoxin loci are highly abundant in free-living but lost from host associated prokaryotes, *Nucleic Acids Res.*, 33, 966–976.
49. Levin, B.R. and Rozen, D.E. (2006), Non-inherited antibiotic resistance, *Nature Rev. Microbiol.*, 4, 556–562.
50. Tiller, J.C., Liao, C.J., Lewis, K., and Klibanov, A.M. (2001), Designing surfaces that kill bacteria on contact, *Proc. Natl. Acad. Sci. USA*, 98, 5981–5985.
51. Lewis, K. and Klibanov, A.M. (2005), Surpassing nature: rational design of sterile-surface materials, *Trends Biotechnol.*, 23, 343–348.
52. Lee, S.B., Koepsel, R.R., Morley, S.W., Matyjaszewski, K., Sun, Y., and Russell, A.J. (2004), Permanent, nonleaching antibacterial surfaces. 1. Synthesis by atom transfer radical polymerization, *Biomacromolecules*, 5, 877–882.
53. Milovic, N.M., Wang, J., Lewis, K., and Klibanov, A.M. (2005), Immobilized *N*-alkylated polyethylenimine avidly kills bacteria by rupturing cell membranes with no resistance developed, *Biotechnol. Bioeng.*, 90, 715–722.
54. Lin, J., Qiu, S., Lewis, K., and Klibanov, A.M. (2003), Mechanism of bactericidal and fungicidal activities of textiles covalently modified with alkylated polyethylenimine, *Biotechnol. Bioeng.*, 83, 168–172.

55. Morgan, H.C., Meier, J.F., and Merker, R.L. (2000), Method of creating a biostatic agent using interpenetrating network polymers, *US Patent 6,146,688*.
56. Vilcheze, C. Weisbrod, T.R., Chen, B., Kremer, L., Hazbón, M.H., Wang, F., Alland, D., Sacchettini, J.C., and Jacobs, W.R., Jr. (2005), Altered NADH/NAD+ ratio mediates coresistance to isoniazid and ethionamide in mycobacteria, *Antimicrob. Agents Chemother.*, 49, 708–720.
57. Hurst, C.J. (2005), Divining the future of microbiology, *ASM News*, 71, 262–263.
58. Kumamoto, C.A. and Vinces, M.D. (2005), Alternative *Candida albicans* lifestyles: growth on surfaces, *Annu. Rev. Microbiol.*, 59, 113–133.
59. Hoyle, B.D., Jass, J., and Costerton, J.W. (1990), The biofilm glycocalyx as a resistance factor, *J. Antimicrob. Chemother.*, 26, 1–5.
60. Baillie, G.S. and Douglas, L.J. (2000), Matrix polymers of *Candida* biofilms and their possible role in biofilm resistance to antifungal agents, *J. Antimicrob. Chemother.*, 46, 397–403.
61. Samaranayake, Y.H., Ye, J., Yau, J.Y., Cheung, B.P., and Samaranayake, L.P. (2005), *In vitro* method to study antifungal perfusion in *Candida* biofilms, *J. Clin. Microbiol.*, 43, 818–825.
62. Mukherjee, P.K., Chandra, J., Kuhn, D.M., and Ghannoum, M.A. (2003), Mechanism of fluconazole resistance in *Candida albicans* biofilms: phase-specific role of efflux pumps and membrane sterols, *Infect. Immun.*, 71, 4333–4340.
63. Ramage, G., Bachmann, S., Patterson, T.F., Wickes, B.L., and Lopez-Ribot, J.L. (2002), Investigation of multidrug efflux pumps in relation to fluconazole resistance in *Candida albicans* biofilms, *J. Antimicrob. Chemother.*, 49, 973–980.
64. Kumamoto, C.A. (2005), A contact-activated kinase signals *Candida albicans* invasive growth and biofilm development, *Proc. Natl. Acad. Sci. USA*, 102, 5576–5581.
65. Garcia-Sanchez, S., Aubert, S., Iraqui, I., Janbon, G., Ghigo, J.M., and d'Enfert, C. (2004), *Candida albicans* biofilms: a developmental state associated with specific and stable gene expression patterns, *Eukaryot. Cell*, 3, 536–545.
66. Datry, A. and Bart-Delabesse, E. (2006), Caspofungin: mode of action and therapeutic applications, *Rev. Med. Interne.*, 27, 32–39.
67. LaFleur, M.D., Kumamoto, C.A., and Lewis, K. (2006), *Candida albicans* biofilms produce antifungal tolerant persister cells, *Antimicrob. Agents Chemother.*, 50, 3839–3846.
68. Cao, F., Lane, S., Raniga, P.P., Lu, Y., Zhou, Z., Ramon, K., Chen, J., and Liu, H. (2006), The Flo8 transcription factor is essential for hyphal development and virulence *Candida albicans*, *Mol. Biol. Cell*, 17, 295–307.
69. Hawser, S.P. and Douglas, L.J. (1995), Resistance of *Candida albicans* biofilms to antifungal agents *in vitro*, *Antimicrob. Agents Chemother.*, 39, 2128–2131.
70. Ramage, G., Vande Walle, K., Bachmann, S.P., Wickes, B.L., and Lopez-Ribot, J.L. (2002), *In vitro* pharmacodynamic properties of three antifungal agents against preformed *Candida albicans* biofilms determined by time-kill studies, *Antimicrob. Agents Chemother.*, 46, 3634–3636.
71. Kumamoto, C.A. (2002), *Candida* biofilms, *Curr. Opin. Microbiol.*, 5, 608–611.
72. Douglas, L.J. (2003), *Candida* biofilms and their role in infection, *Trends Microbiol.*, 11, 30–36.

73. Bagge, N., Schuster, M., Hentzer, M., Ciofu, O., Givskov, M., Greenberg, E.P., and Hoiby, N. (2004), *Pseudomonas aeruginosa* biofilms exposed to imipenem exhibit changes in global gene expression and beta-lactamase and alginate production, *Antimicrob. Agents Chemother.*, 48, 1175–1187.

74. Hoffman, L.R., D'Argenio, D.A., MacCoss, M.J., Zhang, Z., Jones, R.A., and Miller, S.I. (2005), Aminoglycoside antibiotics induce bacterial biofilm formation, *Nature (London)*, 436, 1171–1175.

75. Rachid, S., Ohlsen, K., Witte, W., Hacker, J., and Ziebuhr, W. (2000), Effect of subinhibitory antibiotic concentrations on polysaccharide intercellular adhesion expression in biofilm-forming *Staphylococcus epidermidis*, *Antimicrob. Agents Chemother.*, 44, 3357–3363.

76. Sailer, F.C., Meberg, B.M., and Young, K.D. (2003), β-Lactam induction of colanic acid gene expression in *Escherichia coli*, *FEMS Microbiol. Lett.*, 226, 245–249.

77. Bagge, N., Hentzer, M., Andersen, J.B., Ciofu, O., Givskov, M., and Hoiby, N. (2004), Dynamics and spatial distribution of beta-lactamase expression in *Pseudomonas aeruginosa* biofilms, *Antimicrob. Agents Chemother.*, 48, 1168–1174.

78. Whiteley, M., Bangera, M.G., Bumgarner, R.E., Parsek, M.R., Teitzel, G.M., Lory, S., and Greenberg, E.P. (2001), Gene expression in *Pseudomonas aeruginosa* biofilms, *Nature (London)*, 413, 860–864.

79. Stewart, P.S. and Costerton, J.W. (2001), Antibiotic resistance of bacteria in biofilms, *Lancet*, 358, 135–138.

80. Pages, J.M., Masi, M., and Barbe, J. (2005), Inhibitors of efflux pumps in Gram-negative bacteria, *Trends Mol. Med.*, 11, 382–389.

81. Gillis, R.J., White, K.G., Choi, K.H., Wagner, V.E., Schweizer, H.P., and Iglewski, B.H. (2005), Molecular basis of azithromycin-resistant *Pseudomonas aeruginosa* biofilms, *Antimicrob. Agents Chemother.*, 49, 3858–3867.

82. Morikawa, M. (2006), Beneficial biofilm formation by industrial bacteria *Bacillus subtilis* and related species, *J. Biosci. Bioeng.*, 101, 1–8.

83. Kunduru, M.R. and Pometto, A.L. 3rd. (1996), Continuous ethanol production by *Zymomonas mobilis* and *Saccharomyces cerevisiae* in biofilm reactors, *J. Ind. Microbiol.*, 16, 249–256.

84. Li, X.Z., Webb, J.S., Kjelleberg, S., and Rosche, B. (2006), Enhanced benzaldehyde tolerance in *Zymomonas mobilis* biofilms and the potential of biofilm applications in fine-chemical production, *Appl. Environ. Microbiol.*, 72, 1639–1644.

85. Zhang, S., Norrlow, O., Wawrzynczyk, J., and Dey, E.S. (2004), Poly(3-hydroxybutyrate) biosynthesis in the biofilm of *Alcaligenes eutrophus*, using glucose enzymatically released from pulp fiber sludge, *Appl. Environ. Microbiol.*, 70, 6776–6782.

86. Lendenmann, U., Spain, J.C., and Smets, B.F. (1998), Simultaneous biodegradation of 2,4-dinitrotoluene and 2,6-dinitrotoluene in an aerobic fluidized-bed biofilm reactor, *Environ. Sci. Technol.*, 32, 82–87.

87. Luke, A.K. and Burton, S.G. (2001), A novel application for *Neurospora crassa*: Progress from batch culture to a membrane bioreactor for the bioremediation of phenols, *Enzyme Microb. Technol.*, 29, 348–356.

88. Nicolella, C., van Loosdrecht, M.C., and Heijnen, J.J. (2000), Waste water treatment with particulate biofilm reactors, *J. Biotechnol.*, 80, 1–33.

89. Singh, P. and Cameotra, S.S. (2004), Enhancement of metal bioremediation by use of microbial surfactants, *Biochem. Biophys. Res. Commun.*, 319, 291–297.
90. Murphy, T.F. and Apicella, M.A. (1987), Nontypeable *Haemophilius influenzae*: a review of clinical aspects, surface antigens, and the human immune response to infection, *Rev. Infect. Dis.*, 9, 1–15.
91. Koyama, J., Ahmed, K., Zhao, J., Saito, M., Onizuka, S., Oma, K., Watanabe, K., Watanabe, H., and Oishi, K. (2007), Strain specific pulmonary defense activity after repeated airway immunizations with non-typeable *Haemophilius influenzae* in a mouse model, *Tohoku J. Exp. Med.*, 211, 63–74.
92. Suzuki, K., Nishimura, T., and Baba, S. (2003), Current status of bacterial resistance in the otolaryngology field: results from the second Nationwide Survey in Japan, *J. Infect. Chemother.*, 9, 46–52.
93. Castanheira, M., Gales, A.C., Pignatari, AC., Jones, R.N., and Sadar, H.S. (2006), Changing antimicrobial susceptibility patterns among *Streptococcus pneumoniae* and *Haemophilus influenzae* from Brazil: Report from the SENTRY Antimicrobial Surviellance Program (1998–2004), *Microb. Drug Resis.*, 2, 91–98.
94. Sunakawa, K. and Farell, D.J. (2007), Mechanisms, molecular and sero-epidemiology of antimicrobial resistance in bacterial respiratory pathogens isolated from Japanese children, *Ann. Clin. Microbiol. Antimicrob.*, 13, 6–7.
95. Hoban, D. and Felmingham, D. (2002), The PROTEKT surveillance study: antimicrobial susceptibility of *Haemophilus influenzae* and *Moraxella catarrhalis* from community-acquired respiratory tract infections, *J. Antimicrob. Chemother.*, 50, 49–59.
96. Heilmann, K.P., Rice, C.L., Miller, A.L., Miller, N.J., Beekmann, S.E., Pfaller, M.A., Richter, S.S., and Doren, G.V. (2005), Decreasing prevalence of beta-lactamase production among respiratory tract isolates of *Haemophilus influenzae* in the United States, *Antimicrob. Agents. Chemother.*, 49, 2561–2564.
97. Marco, F., Garcia-de-Lomas, J., Garcia-Rey, C., Bouza, E., Aguiler, L., Fernandez-Mazarrasa, C., and The Spanish Surveillance Group for Respiratory Pathogens (2001), Antimicrobial susceptibilities of 1730 *Haemophilus influenzae* respiratory tract isolates in Spain in 1998–1999, *Antimicrob. Agents. Chemother.*, 45, 3226–3228.
98. Hasegawa, K., Chiba, N., Kobayashi, R., Murayama, S.Y., Iwata, S., Sunakawa, K., and Ubukata, K. (2004), Rapidly increasing prevalence of beta-lactamase nonproducing, ampicillin-resistant *Haemophilus influenzae* type b in patients with meningitis, *Antimicrob. Agents. Chemother.*, 48, 1509–1514.
99. Starner, T.D., Zhang, N., Kim, G., Apicella, M.A., and McCray, P.B., Jr. (2006), *Haemophilus influenzae* forms biofilms on airway epithelia: implications to cystic fibrosis, *Am. J. Respir. Crit. Care Med.*, 174, 213–220.
100. Kaczmarek, F.S., Gootz, T.D., Dib-Kajj, F., Shang, W., Hallowell, S., and Cronan, M. (2004), Genetic and molecular characterization of beta-lactamase-negative ampicillin-resistant *Haemophilus influenzae* with unusually high resistance to ampicillin, *Antimicrob. Agents Chemother.*, 48, 1630–1639.
101. Casey, J.R. and Pichichero, M.E. (2004), Changes in frequency and pathogens causing acute otitis media in 1995–2003, *Pediatr. Infect. Dis. J.*, 23, 824–828.
102. Ketterer, M.R., Shao, J.Q., Hornick, D.B., Buscher, B., Bandi, V.K., and Apicella, M.A. (1999), Infection of primary human bronchial epithelial cells by *Haemophilus*

*influenzae*: macropinocytosis as a mechanism of airway epithelial cell entry, *Infect. Immun.*, 67, 4161–4170.

103. Hall-Stoodley, L., Hu, F.Z., Gieseke, A., Nistico, L., Nguyen, D., Hayes, J., Forbes, M., Greenberg, D.B., Dice, B., Burrows, A., Wackym P.A., Stoodley, B., Post, J.C., Ehrlich, G.D., and Kerschner, J.E. (2006), Direct detection of bacterial biofilms on the middle-ear mucosa of children with chronic otitis media, *JAMA*, 296, 202–211.
104. Kaji, C., Watanabe, K., Apicella, M.A., and Watanabe, H. (2008), Antimicrobial effect of fluoroquinolones for the eradication of Nontypeable *Haemophilus influenzae* isolates within biofilms, *Tohoku J. Exp. Med.*, 214, 121–128.

## CHAPTER 5

# ANALYTICAL TECHNIQUES USEFUL TO STUDY BIOFILMS

### 5.1. INTRODUCTION

Pipelines, catheters, teeth, plant roots, and the lungs of cystic fibrosis (CF) patients are but a few of the most widely recognized surfaces where the effects of biofilms are readily apparent. The biofilms that form on such surfaces almost invariably house a complex mixture of species, rendering them not particularly amenable to molecular genetic studies. To be able to address questions regarding the molecular basis of biofilm formation, investigators have developed artificial biofilm model systems that are easy to control and reproducible from laboratory to laboratory.

While there are numerous laboratory conditions that favor biofilm formation, investigators have routinely utilized four general systems for the study of biofilms. First among these systems is the flow cell [1,2]. Flow cells are small chambers with transparent surfaces where submerged biofilms can form and be continually fed fresh nutrients. Biofilm flow-cell setups allow the cultivation of biofilms under continuous hydrodynamic conditions. The biofilm flow-cell system consists of five major components: (1) a medium reservoir, (2) a multichannel peristaltic pump, (3) bubble traps, (4) flowcells, and (5) an effluent reservoir. All parts are consecutively connected via Si tubings, splitters, and connectors (Fig. 5.1). The submerged biofilms that form on flow cells are particularly amenable to observation through confocal laser scanning microscopy (CLSM), which allows for the capture of images of biofilm development in

*Biofilm Eradication and Prevention: A Pharmaceutical Approach to Medical Device Infections*, By Tamilvanan Shunmugaperumal

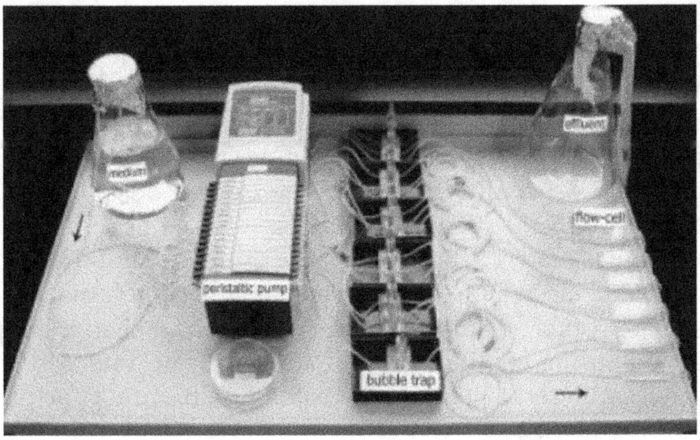


**Figure 5.1.** Major components of biofilm flow cell.

real time. The results obtained using flow cells have provided us with the familiar images of submerged biofilms consisting of mushroom-like structures separated by water-filled channels [1,2].

Investigations of biofilms established under controlled conditions in laboratory setups have provided fascinating insights into the fundamental capacities of bacteria to form multicellular structures. Studies of model systems (e.g., flow-chamber-grown biofilms) have revealed a set of inherent elements of the participating microbes that can facilitate their organization into multicellular communities. For example, among these factors are the production of matrix compounds, cell-surface bound proteins, the secretion of biosurfactants, cellular migration, and regulatory elements (e.g., signal transduction systems, and intra- and extracellular signal messenger molecules. However, flow cells can be cumbersome and are not easily adapted for high-throughput mutant screens.

Submerged biofilms can be studied as well in batch culture under conditions of no flow in microtiter dishes [3,4]. In this system, large numbers of samples can be quickly analyzed. Using the microtiter dish assay system, many investigators have carried out high-throughput screens and identified genes involved in biofilm formation and maintenance in numerous bacterial species [3,5–9]. The floating pellicles that form at the liquid–air interface of standing cultures represent another form of biofilm that is easily studied and adaptable for mutant screens [10–12]. Finally, the colonies that grow on the surface of agar dishes and demonstrate macroscopically complex architecture are now widely recognized as a form of biofilm (reviewed in Branda et al. [2]). This complex colony morphology correlates with production of extracellular matrix and the morphological variation observed in colonies often correlates with the cells' ability to form robust biofilms in other assays. Like pellicles and the biofilms that form on the walls of microtiter dish wells, colonies are amenable to high-throughput screens to identify genes involved in biofilm formation and maintenance.

While all four systems for studying biofilm formation have been successful in broadening our understanding of biofilm development among diverse microorganisms, note that there can be variation among the phenotypes observed as one moves between systems. For example, mutants that exhibit a biofilm defect in one system may have imperceptible or no phenotype in another [3]. The converse also holds true; there are classes of biofilm mutants that do have a reproducible phenotype across all systems (e.g., mutants defective for extracellular matrix production) [11]. Ultimately, of these four general systems, no single one stands out as clearly superior; rather, the methods complement each other. Analyses of the phenotypic changes expressed by different mutants using combinations of several or all four of these systems can greatly aid our understanding of the role that different gene products play in biofilm development.

Individual species of bacteria vary greatly with regard to the environmental conditions under which they will produce maximal amounts of biofilm. These optimal conditions may, in fact, be telling something about the biology and/or ecology of the organism [9,13]. Moreover, many commonly used laboratory strains produce only frail or weak biofilms when compared to wild strains of the same species. In a number of instances, it has been possible to show that this stems from laboratory strains having accumulated numerous mutations over years of passaging through liquid cultures in a process that is referred to as domestication [9,13]. In working with liquid cultures of dispersed populations of cells, it appears to have been unwittingly enriched for strains that have lost some of their potential to form structured multicellular communities while growing rapidly in liquid culture.

Our understanding of biofilms has developed as the methods for biofilm examination and characterization have evolved. Much of the early investigative work on biofilms relied heavily on the scanning electron microscope (SEM). This technique utilizes graded solvents (alcohol, acetone, and xylene) to gradually dehydrate the specimen prior to examination, since water of hydration is not compatible with the vacuum used with the electron beam. This dehydration process results in significant sample distortion and artifacts; the extracellular polymeric substances, which are ~95% water, will appear more as fibers than as a thick gelatinous matrix surrounding the cells.

The use of transmission electron microscope (TEM) and specific polysaccharide stains (e.g., ruthenium red) allowed researchers both to identify the nature of these extracellular fibers in biofilms and to better elucidate their association with the cells. Electron microscopy has been used for the examination and characterization of biofilms on medical devices [14,15] and in human infections [16,17]. Because of its excellent resolution properties, the electron microscope will, in spite of its limitations, continue to be an important tool for the biofilm scientist.

Enumerating the number microbes attached to surfaces by colony counts is a traditional way that was first developed to study biofilm morphology.

The second technique involves the analyses of the surface topography of the biofilm both superficially and internally. Mass spectrometric and atomic force microscopic methods in conjunction with CLSM are routinely used nowadays to envision the biofilm structure. Especially, the use of the CLSM overcomes the difficulty in obtaining clear photomicrographs of the often densely packed microorganisms under hydrated conditions [18]. The CLSM can document biofilm morphology and physiology in four dimensions under *in situ* conditions.

The use of both CLSM and epifluorescence microscopy requires that the organisms in the biofilms be stained with fluorescent stains. These stains are designed to emit light at specific wavelengths and can be used to probe specific cellular functions. For example, nucleic acid stains [e.g., DAPI (4′6′-diamidino-2-phenylindole), acridine orange, and Syto 9] will stain the DNA and RNA of all cells regardless of their viability. Other stains have been developed for probing cell viability. Propidium iodide is taken up only by cells with damaged cytoplasmic membranes, and 5-cyano-2,3-ditolyl tetrazolium chloride is taken up and reduced to 5-cyano-2,3-ditolyl tetrazolium chloride-formazan only by cells that have a functioning cytochrome system. The use of a suite of such stains allows the biofilm researcher to quantify all the cells and determine which ones are viable.

Fluorescent antisera and fluorescent *in situ* hybridization (FISH) probes may enable us to identify specific organisms within a mixed biofilm community. Green fluorescent protein (GFP), a constitutively produced, plasmid-mediated molecule, can allow biofilms to be examined noninvasively, without fixation or staining [19].

This chapter intends to give an overview of these already developed analytical techniques, which are useful in the study of biofilm morphology.

## 5.2. CONVENTIONAL PLATE COUNTING

Common laboratory methods used to study the development of biofilms on surfaces rely on the removal of attached cells from the surfaces and subsequent quantification by conventional plate counts. These are indirect methods of studying bacterial biofilm formation on surfaces. These techniques are also abrasive to attached cells and may result in injury, which could, in turn, result in viable but nonculturable bacteria. Thus, most of these techniques incorporate a resuscitation step of several minutes to allow for cell recovery [20]. Methods that have previously been used to remove attached bacteria from surfaces for subsequent enumeration by plate counts include surface scraping of stainless steel coupons [21], vortexing of Si tubing [22], polystyrene [23] and colon tissue [24]; sonication of poly(vinyl chloride) (PVC) [25,26] and polycarbonate coupons [27], leaves [28], enamel disks [29], Si catheter sections [30] and goethite particles [31], and shaking stainless steel coupons or glass wool with beads [20,32–35]. Although these methods allow for enumeration of bacteria attached to surfaces by colony counts, they do not reveal the *in situ* structure of a biofilm.

Table 5.1 lists several of the methods that have been used by clinical microbiologists for the recovery and measurement of clinically relevant biofilms on indwelling medical devices. For most of these techniques, a determination of the recovery efficiency of the method (i.e., the percentage of cells that are actually recovered from the biofilm) is needed. Methods that allow a determination of biofilm cell count in the implanted device without necessitating device removal (e.g., the endoluminal brush technique) could provide a distinct advantage for the clinical practitioner, potentially alleviating the need for device removal when the device is found not to contain intraluminal biofilms. These methods all rely on the quantification of biofilm cells as a measurement of total biofilm accumulation. Other methods have been used by biofilm researchers for measuring biofilms, including total protein [43], absorbance at either 550 [44] or 950 nm [45], tryptophan fluorescence [46], endotoxin [47], and total adenosine triphosphate (ATP) [48]. Any of these methods could be investigated for the measurement of clinically relevant biofilms.

At this point, it should be obvious that any method that sets out to estimate the efficacy of a treatment against biofilms should use biofilms and not planktonic cells to do so. Standard CLSI (Clinical and Laboratory Standards Institute)/NCCLS (National Committee for Clinical Laboratory Standards) broth microdilution methods for susceptibility testing cannot accurately estimate antimicrobial efficacy against biofilms, because these techniques are based on the exposure of planktonic organisms to the antimicrobial agent. However, a number of apparatuses have been developed for this purpose (Table 5.2). All of the model systems presented have been shown to provide useful information on biofilm processes, and several of these systems have been used to determine the efficacy of various antimicrobial agents against biofilm-associated organisms. Key parameters that may affect the rate and extent of biofilm formation in a model system, and that therefore should be considered in model system design, are given in Table 5.3.

The modified Robbin's device (MRD) has provided important information regarding biofilm physiology and antibiotic susceptibility [55]. Morck et al. [56] demonstrated an important correlation between the antibiotic susceptibilities of biofilms *in vitro* using the MRD and the efficacy of antibiotic treatment *in vivo*.

The *in vitro* study of bacterial biofilms can be done in flow chambers like the MRD (Fig. 5.2) [57]. It consists of an artificial multiport sampling catheter containing 25 evenly spaced sampling ports devised so that Si disks attached to sampling plugs are located on the inner surface, without disturbing flow characteristics. The sampling plugs can be removed and replaced aseptically. Inoculation is made by passing an exponential phase culture of *P. aeruginosa* through the flow catheter for several hours followed by sterile medium. Biofilms develop over time and may be sampled by removal of the disks and analyzed appropriately [57]. By means of the CLSM, it is possible to examine

**TABLE 5.1. Methods That Have Been Used for Measurement of Biofilms on Catheters**

| Method | Basic Protocol | Advantage | Limitation(s) | Reference |
|---|---|---|---|---|
| Roll-plate | Roll the catheter tip over the surface of a blood agar plate | Easy to use | Examines only catheter outer surface, inaccurate | 36 |
| Vortex, then viable count | Catheter section in PBS[a] is vortexed then cultured on different media | Measures intraluminal and extraluminal biofilm | Recovery efficiency unknown | 37 |
| Sonicate, vortex, then viable count | Catheter section in TSB,[b] sonicate then vortex, then culture on blood agar | Measures intraluminal and extraluminal biofilm | Recovery efficiency unknown | 38 |
| Sonicate, vortex, homogenize, then viable count | Catheter section in PBS,[a] sonicate/vortex repeatedly, then homogenize and culture on blood agar | Recovery efficiency Determined | Measures intraluminal biofilm only | 39 |
| Acridine orange direct staining | Following roll-plate method, catheter section is stained with acridine orange | Allows direct examination of catheter | Method does not allow quantification | 40 |
| Endoluminal brush | Brush is introduced into the implanted catheter, removed, placed into PBS, sonicated, and plated | Allows examination of indwelling catheter | Effect of procedure on patient and recovery efficiency unknown | 41 |
| Alginate swab | Swab introduced into the implanted catheter, removed, then streaked over a blood agar plate | Allows examination of indwelling catheter | Effect of procedure on patient and recovery efficiency unknown | 42 |

[a]Phosphate-buffered saline, PBS.
[b]Trypticase soy broth, TSB.

**TABLE 5.2. Apparatuses That Have Been Used for Growing and Testing Biofilms**

| Apparatus | Organism(s) Tested | Flow Dynamics | Substratum | Method for Removing and Quantifying Biofilm | Reference |
|---|---|---|---|---|---|
| Modified Robbins device | *Pseudomonas pseudomallei* | Batch/mixing | Silastic disks | Method of removal not given; viable count | 49 |
| Calgary biofilm device | *P. aeruginosa, Staphylococcus aureus, Escherichia coli* | Batch/mixing | Plastic pegs | Sonicate peg, then viable count | 50 |
| Disk reactor | Gram-negative bacteria | Batch/mixing | Teflon coupons | Sonicate, vortex, homogenize, then viable or direct count | 51 |
| CDC[a] biofilm reactor | Gram-negative bacteria | Continuous/ open system | Needleless connectors (plastic) | Sonicate, vortex, homogenize, then viable or direct count | 52 |
| Perfused biofilm fermentor | *Candida albicans* | Continuous/ open system | Cellulose-acetate filters | Shake in sterile water, then viable count | 53 |
| Model bladder | Gram-negative bacteria | Continuous/ open system | Urinary catheters | Direct examination by SEM[b] or TEM[c] or by chemical analysis | 54 |

[a]Centers for Disease Control = CDC.
[b]Scanning electron microscopy = SEM.
[c]Transmission electron microscopy = TEM.

**TABLE 5.3. Factors to Consider in the Development of a Model Biofilm System**

| Medium | Inoculum | Hydrodynamics | Substratum |
|---|---|---|---|
| Composition, temperature, presence of antimicrobial agents | Identity of organism, No. of cells | Flow rate, presence of shear, batch versus open system, retention time | Roughness, chemistry, conditioning films |

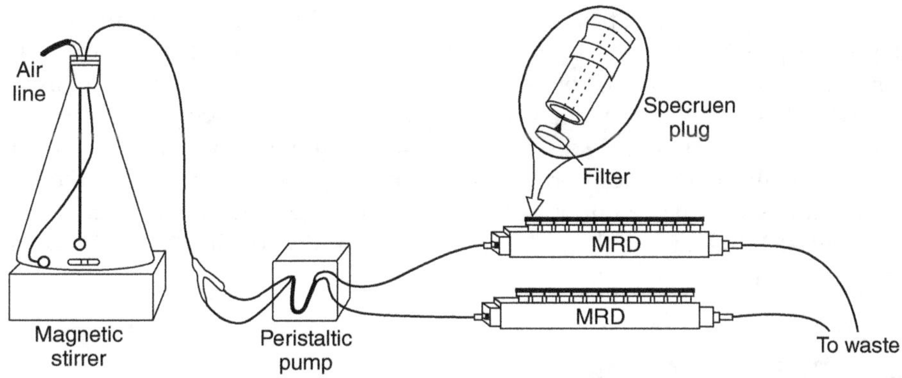


**Figure 5.2.** Diagrammatic representation of two parallel MRD and the apparatus used to provide a continuous flow of bacteria and the culture medium. (Reprinted from Høiby et al. [137] with permission.)

living, fully hydrated biofilms, and the activity of genes in individual bacterial cells can then be studied by utilizing the molecular biology technique of GFP labeled cells [58].

Animal models of biofilm bacteria (e.g., *P. aeruginosa* in the lungs) can be established in rats and mice by intratracheal inoculation of bacteria in agar or alginate beads. The histopathologic changes, the immune response of the body, and the efficacy of antibiotic therapy can then be followed [59,60].

While the MRD has proven to be an effective model of biofilm formation, it is not suited for rapid antibiotic susceptibility testing in a clinical laboratory setting. A model that is becoming increasingly used for high-throughput screening of antimicrobial compounds–biofilm susceptibility is the Calgary Biofilm Device (CBD)/MBEC (minimum biofilm eradication concentration) plate, where polycarbonate pegs are used in a 96-well format to form statistically identical biofilms, can be snapped off, sonicated, and counts performed [50,61]. The CBD produced 96 equivalent biofilms, making it the first assay system truly amenable to antibiotic susceptibility testing for adherent bacterial populations. The CBD requires no pumps or tubing, making the process much simpler to set up than the MRD, and eliminates a major source of possible contamination. The availability of multiple testing sites greatly reduced the time required to determine the antibiotic susceptibilities of biofilms from weeks with the MRD to 3 days with the CBD. The CBD is also amenable to automation because it is built on the typical platform for 96-well plates. Furthermore, in order to validate the CBD as a tool for studying biofilm organization, three proof-in-principle experiments were carried out by Harrison et al. [62]. First, the biofilm structures of several bacterial strains were evaluated in rich and minimal media, and the images presented show that bacteria and fungi adopted a diverse range of structural conformations that were dependent on strain genetics as well as growth conditions. Second, Live/Dead® staining was used to illustrate that dead cells were a frequent, albeit variable, component of biofilms. Last of all, the staining of extracellular polymers using fluorophore-conjugated lectins was used to show that this component of the biofilm matrix was unevenly distributed throughout the surface-adherent communities. Collectively, these proof-in-principle experiments were designed to simultaneously illustrate important caveats in the interpretation of microscopy data, but also to validate the CBD as a potential tool for the study of biofilm structure–function relationships [62].

## 5.3. MODERN METHODS

To study biofilm formation and its architecture, microscopic imaging techniques like SEM or CLSM are also widely utilized [18,63–69].

### 5.3.1. Confocal Laser Scanning Microscope

The CLSM is a "high-tech" epifluorescence microscope that creates a thin (~0.3 μm) plan of focus in which out-of-focus light has been blocked, traditionally by optical barriers (confocal apertures: "pinholes"), but also now by the physics of light absorption (multiphoton microscopy). Laser light is scanned across the specimen to provide intense, deeply penetrating excitation energy. Intrinsic fluorophores (biomolecules, e.g., GFP or chlorophyll) or exogenous probes (e.g., fluorescently labeled antibodies) are excited, the resulting fluorescence is detected by photomultiplier tubes, and a digital image is produced. A "stack" (series of digital XY optical sections) is automatically collected through computer-controlled alteration of the microscope's stage in the Z dimension. Sections in the XZ plane (saggital sections) can also be acquired. The stack can be computer processed to create, for example, three-dimensional (3D) reconstructions.

Because biofilms are complex 3D structures, the analysis of them is not trivial. While microbial single cells easily can be monitored using a conventional microscope, biofilms require additional resolution in the direction vertical to the substratum (the $z$-axis).

Three basic designs of CLSM are commercially available. "Real-time" instruments are optimized for video-rate data collection at the expense of axial and lateral resolution. These devices are used primarily for study of physiological phenomena on time scales of seconds or less and are generally not useful for precise optical sectioning (morphological–structural studies). Therefore, they have not yet found significant application to biofilm work. The other types of CLSM are point-scanning devices that offer high lateral and axial resolution, but slower image collection times. The "single-photon" CLSM has been available from several manufacturers for ~25 years and is by far the most used instrument in biological confocal microscopy. "Multiphoton" CLSMs, which use tunable infrared (IR) lasers to provide excitation energy "packets" (multiples of photons of longer wavelengths absorbed by lower-wavelength-excitable fluors), are in their first generation of commercialization and theoretically offer several advantages over their single-photon counterparts: lower sample damage through photobleaching or heating; deeper penetration of excitation light; and absence of light-obstructing pinholes (the optical section is created solely by point of focus of the laser beam). Both performance and reliability issues (e.g., the lasers can be difficult to tune and to stabilize) and laser price currently limit the market for such instruments. Although these issues will be resolved with time, as they were for single-photon devices, the longer excitation wavelengths of the multiphoton systems result in a twofold poorer theoretical resolution than in single-photon instruments.

In most biological disciplines, the CLSM has been used primarily to examine fixed specimens that have been stained with fluorescently labeled antibodies or organelle-specific probes. In contrast, bacterial biofilm studies began with the study of living biofilms as they grew on the surfaces of glass-walled perfusion chambers (flow cells [70,71]), thereby permitting semicontinuous nondestructive documentation of temporal changes in biofilm structure–physiology. Conventional microscopy of multilayered biofilms is possible, however, through cryoembedding (frozen samples are embedded in a cryomounting media, e.g., TissueTek) or ambient temperature embedding in a plastic resin, followed by microtome cutting; micrographs are made of the individual (optically clearer) sections. This lower-cost (a research-grade cryostat, ~\$20,000, together with a high-quality epifluorescence microscope, ~\$25,000, vs the ~\$200,000 price of a CLSM) approach to obtaining biofilm images can produce sections as thin as 2 μm: only ~10-fold thicker than what can be achieved by confocal microscopy (optical sectioning), but still of lower resolution. Fixation and dehydration required by sectioning techniques can affect 3D relationships. Finally, physical sectioning, even when performed with a flawless technique, can disrupt the fine structure of the biofilms, thereby making spatial analysis difficult.

Although the confocal microscope has permitted examination of intact biofilms, an embedding step is still advantageous in certain casts. Embedded biofilms are much more robust and easy to handle. This is especially important for techniques in which multiple washing steps are required (e.g., nucleic acid hybridization). Agarose embedding is both simple and gives good results when used with confocal microscopy [72]. Disadvantages arc that the agarose solution must be warm and that the final preparations are rather fragile due to the low (0.1%) agarose concentration. The DNA sequencing grade acrylamide (20%) is an excellent embedding medium [1,58,73]. These preparations are easily handled and, for example, are suitable for 16S ribosomal probing or analysis of GFP expression. Furthermore, acrylamide is virtually nonfluorescent. Biofilms embedded by either of these methods may be fixed, dehydrated, or still living.

The earliest applications of the CLSM to biofilm research were descriptive. However, application of confocal microscopy to 3D localization of non-enzyme reporter systems, notably GFP, has brought significant advances to the understanding of gene activity in biofilms. Ribosome content is an indicator of growth potential (and is closely correlated to growth rate in exponentially growing cultures), measurements of ribosome-hybridization-signal intensity have therefore been used for *in situ* determination of cellular activity [74]. The physiological state of cells in biofilms can be assessed using specially designed promoter–reporter systems that fluoresce only in actively dividing cells. A recent study demonstrated the growth activity and potential of *Pseudomonas* cells within a biofilm. Briefly, an expression cassette that contained GFP under control of a ribosomal ribonucleic acid (rRNA) promoter (growth-phase regulated; active only when cells are growing) was constructed, and the growth phase of individual cells in the developing biofilm was monitored. Ribosome synthesis rate and content were determined *in situ* giving a snapshot of the cells' physiology. Furthermore, microautoradiography was used in combination with ribosomal probing to visualize activity–identity relations in microbial communities [75].

Confocal microscopy has been adeptly applied to the examination of free-radical-based disinfection [76]. A nonfluorescent compound was employed that became fluorescent after reaction with hydroxyl radicals or superoxide and excitation with 365-nm light. A time series of saggital sections showed that free radicals were generated within the biofilm during exposure to potassium monopersulfate, presumably from the action of catalysts incorporated into the substratum. Radical generation was suggested to be the cause of a depth-dependent killing of biofilm bacteria exposed to the persulfate.

A new technology, stimulated emission depletion (STED), has been developed into commercial products, increasing the optical resolution even further by using two photon excitation in combination with quenching of nearby fluorescence that could otherwise deteriorate the image [77,78]. However, STED is currently of limited use since only a few fluorophores are suitable for this particular laser excitation.

### 5.3.2. Fluorescent Labeling of Biofilm Cells

Confocal microscopy and derived methods require the specimen to be fluorescent. The biofilm must therefore either be autofluorescent by means of indigenous fluorescent molecules, or the biofilm cells must express a fluorescent protein (e.g., GFP [79]), individual biofilm cells, or other components of the multicellular structure must be stained. Therefore, it is important to recognize that confocal microscopy (or any epifluorescence technique) cannot generally be used to visualize unstained (nonfluorescent) material, (e.g., the exopolymer matrix in which biofilm cells are embedded). Hence, conclusions regarding presence–absence–changes in the extracellular matrix when that matrix was not specifically stained are based primarily on assumptions. Exopolymer and its relationship to biofilm structure–function can, however, be explored using appropriate techniques (e.g., fluorescent lectin staining [80,81]).

Early biofilm studies by the Caldwell group employed a simple, yet efficient way of detecting the biomass in flow cells. The void volume (i.e., the liquid phase) was supplemented with a solution of fluorescein isothiocyanate (FITC), leaving the biomass unstained. The resulting images were "negatives" and the biofilm could be rendered as the dark portions of the images [69,82]. This gave sufficiently high resolution to determine cell sizes and spatial relations. More recently developed stains (e.g., the Syto stains) (Invitrogen, Carlsbad, CA), can efficiently stain cells in virtually any color of the rainbow. In combination with propidium iodide (PI), it is further possible to specifically stain live and dead cells. The dye Syto 9 will stain all cells green regardless if they are dead or alive, while it is generally assumed that only cells with a damaged membrane will be stained by the red PI dye, indicating dead cells. Recent results suggest that PI might be of limited use as a cell viability indicator for some environmental bacterial species [83]. Therefore, for each individual bacterial species, a fine tuning of the dye combinations is recommended prior to performing the actual experiments using mixtures of live and dead cells of known ratios [84]. Recently, the assumption that PI only targets dead cells was confirmed also for biofilm cells that had been exposed to a membrane damaging agent, using cell sorting of harvested biofilm cells and regrowth test of the separated green (GFP-tagged) and red (PI-stained) cells on nutrient agar. Only the cells that were not labeled red with PI were able to grow [85]. Stains targeting the extracellular matrix (e.g., lectins) [86,87] or calcofluor white [88,89] can also be employed to visualize the surrounding of the biofilm cells. In addition, the extracellular DNA component of the matrix can be visualized by the use of different DNA binding fluorophores [90].

If genetic manipulation of the biofilm cells is possible, chromosomal tagging of cells with a gene cassette encoding the GFP can be a useful option [91]. Alternatively, plasmids encoding for the GFP might be introduced into the cells prior to biofilm examinations. Depending on the construction, this fluorescent tagging can be used as a simple labeling to verify the location of the cells in a biofilm, or, by selecting suitable variants of *gfp* genes and promoters, it can be used for monitoring gene expression in biofilms. Such tagging of biofilm cells has been done to monitor metabolic–physiological activity in biofilms by introducing constructs encoding for Gfp derivatives with a short half-life, placed under transcriptional control of a ribosomal promoter [92]. For example, the *gfp*[AGA] gene, encoding for a GFP with a short half-life, was placed under transcriptional control of the ribosomal promoter rrnBP1 and introduced into either *E. coli* or *Pseudomonas* spp. wild-type strains. Cells that have a high metabolic–physiological activity can be expected to exhibit high gfp[AGA] expression and emit a high-fluorescent signal, whereas cells that have low metabolic–physiological activity can be expected to exhibit a low or have no expression of the fluorescent protein [92,93]. Further, using GFP variants with different emission spectra (e.g., the CFP-cyan fluorescent protein), YFP (yellow fluorescent protein), and RFP (red fluorescent protein), the spatial distribution of either cells in a multispecies biofilm can be determined, or of a number of (mutant) strains of a single species tagged with different colors [94–97].

Another way of fluorescently labeling biofilm cells is through the use of FISH, where specific probes hybridize to the 16S rRNA in the cells. For FISH a DNA probe is designed to match a distinct region of the cell's rRNA. The probes can be conjugated to a fluorescent dye (e.g., FITC or Rhodamine) or to an enzyme (e.g., horseradish peroxidase), which deposits fluorescent molecules. It might be challenging to introduce larger conjugates (e.g., the horseradish peroxidase) into cells of thicker biofilms without destroying biofilm cells due to harsh permeabilization procedures. Therefore FISH involving probes with larger conjugates might preferentially be applied on thin sections of thick biofilms. The number of ribosomes present in a given cell is proportional to the growth potential of the cell, and FISH labeling can consequently also be used to determine the growth status of a biofilm cell [98]. However, under certain conditions (e.g., stress), cells might have increased numbers of ribosomes, although their actual growth rate is low. The probe design can be adjusted so that the probe only labels a single species by targeting a so-called variable rRNA region, or a probe can label all cells belonging to the same domain or phylum by choosing a more conserved region. An example of such a probe is the widely used EUB338, which can hybridize to virtually all bacteria [99]. However, the growing knowledge on rRNA encoding sequences has revealed that probes formerly believed to be universal fail to be able to hybridize to species or entire phyla of microorganisms in the realm they originally were thought to cover completely [100].

### 5.3.3. Biofilm Studies Using an Atomic Force Microscope

Comparatively few reports are, however, available for biofilm examination using atomic force microscopy (AFM) alone [101,102]. Moreover, one of the most promising techniques for the characterization of microbial cell adhesion and cell–surface interactions is atomic force spectroscopy (AFS). This method uses AFM operating in force mode, which offers both imaging capabilities and quantitative measurements of forces between the AFM tip and the sample. A review published by Dufrene [103] described the fundamentals and applications of AFM–AFS relevant to biofilm systems. A recently reported investigation combined AFS with confocal force microscopy to demonstrate, *in situ*, interaction between a living bacterium, *E. coli*, and the silicon nitride surface of the AFM tip [104]. The study revealed that an outer-membrane protein of *E. coli* extending outside the cell wall was responsible for cell adhesion to the tip surface. Few interesting reports on the application of AFM in biofilm characterization pertinent to medical devices are deployed here.

The use of the endotracheal (ET) tube (Fig. 5.3) in the management of a patient in the intensive care unit (ICU) allows oxygenation and positive pressure ventilation. Even though intubation of the upper-respiratory tract overcomes host defences (e.g., the cough reflex and mucociliary escalator) [105],

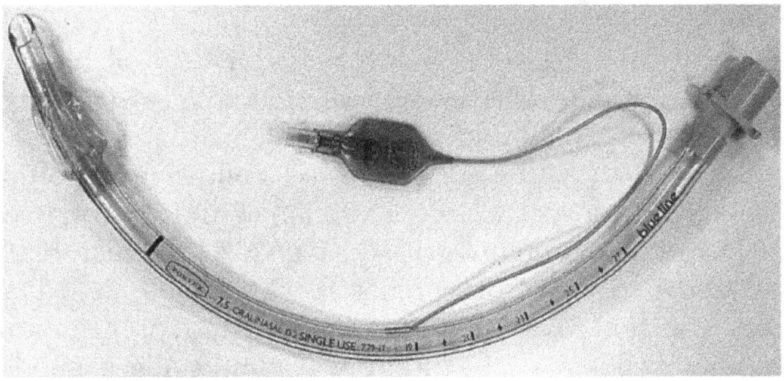

**Figure 5.3.** The endotracheal tube.

the consequent coloniation of ET tubes, which are commonly fabricated from PVC, with bacterial pathogens, is a common occurrence in conjunction with subsequent microbial biofilm formation [106–108]. Dissemination of this biofilm into the lower-respiratory tract carries pathogenic populations of microorganisms to compromised regions of the lung, where the onset of infection commences [109,110]. Such is the importance of the ET tube and microbial biofilm in the pathogenesis of nosocomial pneumonia that the phrase "ventilator-associated pneumonia" has entered the medical vocabulary. Gorman et al. [111] examined the sequential steps involved in the formation of biofilm on PVC by AFM and the concomitant development of resistance to an antibiotic (ceftazidime) and to a nonantibiotic antimicrobial agent (hexetidine). Biofilm surface microrugosity (surface roughness or smoothness analysis by the AFM) was then measured as $R_q$, the root mean square of the vertical dimension of the surface [111]. Alternatively, microrugosity may also be defined as the number of atomic levels at the surface. *Staphylococcus aureus* and *P. aeruginosa* isolated from ET tube biofilm were employed. The surface microrugosity of bacteria growing in sessile mode on PVC decreased significantly ($p < 0.05$) over the period of 4, 24, 48 h, and 5 days. The progressive accretion of bacterial glycocalyx was readily visualized in micrographs leading to a smoother surface topography with time.

By using AFM, the influence of host environmental conditions in the oropharynx and trachea on the cell surface characteristics of *Candida* spp. (*C. albicans*, *C. tropicalis*, and *C. krusei*) and their adhesion to the medical device biomaterials, PVC and Si, was studied by Jones et al. [112]. In addition, the effects of the salivary conditioning film on the surface properties of these biomaterials were reported. Treatment of Si and PVC with saliva significantly altered the surface properties, notably reducing both the advancing and receding contact angles and, additionally, the microrugosity. These effects may contribute to the decreased adherence of saliva-treated microorganisms to these biomaterials. Furthermore, it is important that physiological conditions should be employed whenever biocompatibility of oropharyngeal biomaterials is under investigation [112].

Another study from the same research group [113] describes the physicochemical properties of hexetidine-impregnated PVC ET biomaterials and their resistance to microbial adherence (*S. aureus* and *P. aeruginosa*). Increasing the concentration of hexetidine decreased the tensile properties and hydrophobicity, yet increased PVC microrugosity. Following hexetidine release (21 days), the surface properties were similar to those of native PVC. The resistance of hexetidine-containing PVC (1 or 5%) to microbial adherence (following defined periods of drug release) was greater than that of native PVC and was constant over the examined period of hexetidine release. Interestingly, microrugosity of 12 different commercially available catheters were determined (values are given in Table 5.4) in a separate study and no specific relationship to the chemical nature of the biomaterial was seen by Jones et al. [114].

Carbon, in a range of forms [including diamond-like carbon (DLC)], has been used in many biomedical applications over the past two decades [115]. Diamond-like carbon, in particular, exhibits many useful properties, including low friction, chemical inertness, biological compatibility, smoothness, and resistance to wear [116,117]. The low friction, resistance to wear, and smoothness of DLC films are advantages in biomedical applications where wear occurs on implements used for repeated high-friction tasks [116,118]. Many researchers have studied the biocompatibility of DLC. For example, Thomson et al. [119] reported that no adverse effects were observed following contact of mouse

**TABLE 5.4. Surface Roughness (Microrugosity) of the Urethral Catheters Determined by AFM**

| Catheter Name | Mean Microrugosity (Rq, nm) (mean ± s.e.m) |
|---|---|
| Silicone-containing bismuth subcarbonate (low concentration) | 59.92 ± 14.72 |
| Silicone-containing barium sulfate | 39.48 ± 6.92 |
| Silicone-containing bismuth subcarbonate (high concentration) | 56.02 ± 14.80 |
| Silicone-containing bismuth trioxide | 82.56 ± 7.85 |
| Per-Q-cath | 43.11 ± 9.06 |
| V-Caths | 56.78 ± 13.46 |
| Polyurethane-containing bismuth subcarbonate | 109.79 ± 27.80 |
| Landmark | 141.06 ± 21.27 |
| Polyurethane-containing barium sulfate | 85.66 ± 14.65 |
| L-Cath | 66.44 ± 16.67 |
| Picc Shield | 5.45 ± 1.25 |
| Centermark | 93.26 ± 23.38 |

peritoneal macrophages and mouse fibroblasts in culture with DLC. Similarly, Matthew et al. [120] and Dowling et al. [121] reported that DLC coated steel and cobalt–chrome implants, respectively, were biocompatible both *in vivo* and *in vitro*, while Jones et al. [122] demonstrated the hemocompatibility of DLC coatings. Despite the emerging use of DLC as a coating for medical devices, few studies have examined the resistance of DLC coatings onto medical polymers to both microbial adherence and encrustation. In a recent study by Jones et al. [123], amorphous DLC of a range of refractive indices (1.7–1.9) and thicknesses (100–600 nm) was deposited onto polyurethane, a model polymer, and the resistance to microbial adherence (*E. coli*; clinical isolate) and encrustation examined using *in vitro* models. This study also utilizes the AFM to see the surface roughness of selected samples of uncoated and DLC coated polyurethane [123]. Films of lower thicknesses (100 and 200 nm; of defined refractive index, 1.8), exhibited the greatest resistance to encrustation and to microbial adherence. In conclusion, this study has uniquely illustrated both the microbial antiadherence properties and resistance to urinary encrustation of DLC coated polyurethane.

Since the extracellular polymeric substances (EPS) themselves form the dynamic biofilm matrix [124], an in-depth analysis on the physicochemical properties of the EPS should be performed. The majority of analytical studies on EPS have focused on an analysis of the structure and function of polysaccharides, often ignoring the presence and possible significance of other types of macromolecules. In most cases, carbohydrates, including neutral sugars, acidic polysaccharides, and aminosugars, are indeed the most abundant chemical species and, therefore, are proposed to act as structural elements responsible for the mechanical stability of biofilms. However, one must be cautious when claiming the importance of a certain type of polysaccharides in biofilm development. An investigation has demonstrated that the exopolysaccharide alginate, produced by mucoidal strains of *P. aeruginosa*, is, in fact, not required to form the biofilms of two predominant environmental phenotypes of nonmucoidal strains of *P. aeruginosa* [125]. These two strains, PA14 and PAO1, have traditionally been used to study biofilms, and in several reports alginate has been assumed to be a key EPS component mediating cell adhesion of the above strains. Evidence that a polysaccharide different than alginate facilitates attachment of *P. aeruginosa* PA14 and PAO1 strains to surfaces would probably have considerable implications in developing strategies for preventing biofilm development by these bacteria.

In contrast to the abundant literature on polysaccharides, reports on EPS proteins are scarce. Yet, extracellular enzyme activities are readily observed in biofilms. It has been proposed that exoenzymes ought to be considered an integral part of the EPS matrix [124]. The most powerful tool for the characterization of proteins is mass spectrometry (MS). Analytical MS has undergone rapid development with respect to ionization methods, instrumentation, and the ability to detect large molecules and deal with complex mixtures. Undisputedly, MS has the potential to monitor microbiological systems and to elucidate the molecular mechanisms and interrelationships in such systems. Reaching a deeper understanding of the interfacial chemistry governing biofilm process is likely to require the extensive use of modern MS. The remarkable progress in MS is due primarily to two ionization methods developed in the late 1970s, namely, electrospray ionization (ESI) and matrix assisted laser desorption ionization (MALDI) [126]. It also found that this ionization method has unique advantages for the MS characterization of complex biological mixtures (e.g., EPS).

In the most common use of this technique, the ions are desorbed inside a vacuum chamber and mass analyzed in a time-of-flight (TOF) mass spectrometer. The capabilities of modern instruments [e.g., TOF, ion trap, and Fourier transform mass spectrometers (FTMS)] were unknown only a few years ago. Tandem MS (MS/MS), which consists of two mass analyzers in series, is a powerful instrument for the analysis of complex materials and, for example, quadrupole–TOF (Q–TOF) have been successfully applied to the analysis of biological materials. Several applications of MALDI and ESI of interest to medical and environmental microbiology's have been demonstrated and/or are under active development. These include bacterial identification, monitoring bacterial protein expression and post-translational modifications in response to environmental or metabolic stimuli, and measuring bacterial cell growth. However, the application of modern MS to biofilms research, including biofouling and biocorrosion, remains virtually unexplored.

To prove that MS can help to elucidate the composition of complex biological materials with minimum sample preparation and with high sensitivity, some studies have focused on the characterization of organic and inorganic components using EPS, as well as mixtures of model compounds, using ESI and laser spray ionization (LSI) MS. The EPS used in these studies were recovered from cultures of a recently described, new strain of a sulfate-reducing bacterium *Desulfovibrio alaskensis* [127].

Apart from AFM–AFS and MS, other techniques of surface science and numerous forms of light and electron microscopy are valuable tools to understand the nature of microbially influenced interfacial processes. These tools are extensively used for that purpose [128–131]. A comprehensive list of various analytical techniques useful for the characterization of biofilm on inanimate surfaces including medical devices is shown in Table 5.5.

Colloid probe AFM, which is another version of AFS, has recently been used to demonstrate differences in the adhesive properties of three *E. coli* strains to the surface of glass [132]. The study employed a gradient force analysis method. The authors proposed that the values of adhesion (sticking) coefficients between bacterial cells and glass were correlated solely with the length of the exopolymers on surfaces of an *E. coli* wall. Biological force microscopy (BFM), offers a great advantage for quantitative measurements of the adhesion characteristics of microorganisms on surfaces with varied physicochemical properties. A recent example is the use of BFM to study living marine diatoms of *Navicula* genus as a bioprobe. The adhesive properties of

**TABLE 5.5. Techniques Used in the Study of Biofilm**

| Technique | Abbreviation |
|---|---|
| Atomic force spectroscopy | AFS |
| Atomic force microscopy | AFM |
| Biological force microscopy | BFM |
| Biological force spectroscopy | BFS |
| Confocal force microscopy | CFM |
| Mass spectrometry | MS |
| Tandem MS | MS/MS |
| Fourier transform mass spectrometry | FTMS |
| Electrospray ionization | ESI |
| Matrix-assisted laser desorption ionization | MALDI |
| Laser spray ionization | LSI |
| Time-of-flight (TOF) mass spectrometry | TOFMS |
| Quadruple/TOF | Q-TOF |
| X-Ray diffraction | XRD |
| X-Ray photoelectron spectrometry | XPS |
| Near-edge X-ray absorption fine structure spectroscopy | NEXAFS |
| Energy dispersive X-ray analysis | EDX |
| Scanning transmission X-ray microscopy | STXM |
| Secondary ion mass spectrometry | SIMS |
| Surface-enhanced laser desorption ionization | SELDI |

the EPS associated with individual diatom cells to two physicochemically different materials, mica and a fouling-release Si elastomer, Intersleek, were quantitatively characterized [133]. By using different diatoms, as well as the same diatom in different growth stages, it was found that the work of detachment of a single diatom from either type of surface strongly depended on the individual *Navicula* cell, but not on its growth stage. Generally, the adhesion forces to the hydrophobic intersleek surface and the hydrophilic mica surface were of similar strength. The study concluded that the adhesion of *Navicula* to surfaces with different physicochemical properties is governed by the macromolecular specificity of diatom EPS. However, the EPS macromolecules that mediate diatom cell adhesion have not been identified, and curve profiles strongly suggested that both proteins and polysaccharides are the likely bioadhesives.

### 5.3.4. Emerging Techniques

The "holy grail" of biofilm infections is an "early-warning" diagnostic method that would allow for noninvasive detection of the early stages of tissue or biomedical implant infection [134] and an expedient response. Such diagnostics are only now just emerging.

Currently, only upon the onset of a cyclical fever in an implant recipient, will a patient receive a battery of blood tests meant to detect any infecting microorganisms; such as, colony-forming plate count assays that typically take anywhere from 48–72h and are only capable of detecting planktonic not sessile cells. Emergence of polymerase chain reaction (PCR) techniques have shortened the time period, but they also sample body fluids (blood, saliva, urine), which will not provide an accurate estimate of the actual biofilm flora colonizing an implant.

The environmental scanning electron microscope (ESEM) is a low-vacuum (near atmospheric pressure) SEM that can be used to image hydrated samples. While the images arc essentially topographical, the technique has been used to demonstrate the degree of exopolymer hydration in manganite-reducing biofilms through comparison with dehydrated samples [135].

Scanning probe microscopy (SPM) comprises a suite of instruments that move a cantilevered probe across a surface in contact or near-contact modes. Force microscopy measures the attraction–repulsion of the probe tip at resolutions up to the atomic level (atomic force microscopy). Variations in tip shape–composition, or in the way tip–surface interaction is modulated, give different resolutions and modes of interaction (force modulation microscopy, electrostatic force microscopy, magnetic force microscopy, phase detection microscopy). Although the images from these techniques are topographical renderings of the tip–surface interaction (meaning that the topographical information actually represents numerical physicochemical data), common applications are simply topographical (obtaining the shape of the surface is the final goal of the imaging process). The SPM has produced images of dried bacteria that resemble those acquired with SEM [136], but with greater detail at similar magnification (especially in 3D) and without extensive sample preparation (e.g., critical point drying and coating). Gross antibiotic-induced changes in the cell wall of a bacterium have been demonstrated [137]. Since water is integral to biological structures, SPM is now being applied to hydrated living biological specimens [138], although tip contamination must be considered [139]. In an elegant application of force microscopy to bacterial structure, the influence of lipopolysaccharide composition on relative strength-of-interaction between cells and substrata has been quantified in aqueous solution [140].

Chemical microscopy (Raman–IR microscopy) provides spatially resolved (line scanning, global imaging) chemical information on the specimen (i.e., spectroscopy through the microscope). As in any microscope technique, spatial resolution is finally limited by the wavelengths of light used. Infrared microscopy (a broadband absorption technique) has historically used wavelengths (~2800–10,000nm) too long for resolution of a single bacterial cell; spatial resolution is at best tens of micrometers. Raman microscopy (a light-scattering technique that measures small shifts in photon energy around a single wavelength) using visible or near-IR lasers, however, can resolve single bacterial cells [141]. Furthermore, several commercial confocal Raman microscopes exist that theoretically allow the chemical nature of biofilms to be investigated in the axial as well as the lateral dimensions.

Correlative microscopy is the use of two or more techniques to investigate the same (or parallel) specimens. Relatively easily performed with fixed specimens, CLSM images have been combined with electron micrographs to examine biofilm architectural features [72]. The approach can also be employed with living biofilms; for example, Palmer and co-workers [142,143] combined photon-counting microscopy (to measure luciferase activity, i.e., gene expression) with laser confocal microscopy to analyze, first, viability of *Vibrio* species [142] and, second, membrane potential in genetically engineered *P. putida* cells [143]. The potential for combination of the scanning probe or Raman methods with epifluorescence techniques (including confocal) promises major advances in our understanding of biofilm structure–function.

Xiong et al. [144] reports a rapid, continuous method for real-time monitoring of biofilm development, both *in vitro* and in a mouse infection model, through noninvasive imaging of bioluminescent bacteria (Fig. 5.4). Two important biofilm-forming bacterial pathogens, *S. aureus* and *P. aeruginosa*, were made bioluminescent by insertion of a complete *lux* operon. These bacteria produce significant bioluminescent signals for both *in vitro* studies and in an *in vivo* model, allowing effective real-time assessment of the physiological state of the biofilms. *In vitro* viable counts and light output correlated well for 10 days or longer, provided that the growth medium was replenished every 12 h. Recovery of the bacteria from the catheters of infected animals showed that the bioluminescent signal corresponded to the CFU and that the *lux* constructs were highly stable even after many days *in vivo*. The IVIS™ imaging system, commercially available from the Xenogen Corporation, has been used to follow *in vivo* either infections of soft tissue or medical biomaterial implants or to assess the subsequent efficacy of a single or multiple antibiotic treatment. As a research tool, the IVIS camera system and microscope can detect the emitted photons from both live and dead bacteria.

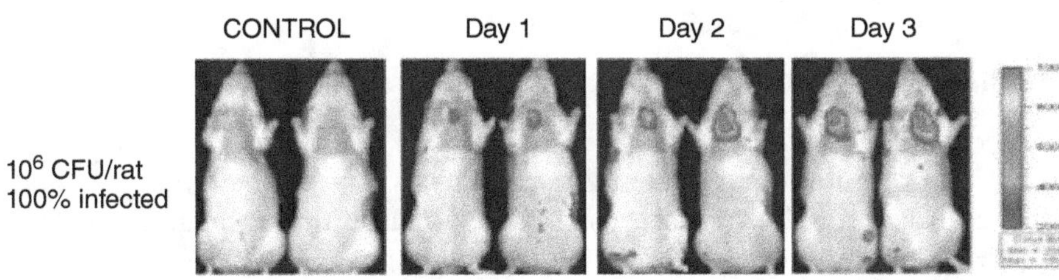


**Figure 5.4.** Real-time monitoring of *S. aureus* Xen29 in an experimental-rat endocarditis model. Two representative animals infected intravenously with either normal saline (control) or $10^6$ CFU of *S. aureus* strain are shown. The animals were imaged ventrally, with their chest area shaved, to avoid background signal from animal hair. The process of infection was monitored daily by detecting photon emission around the region of interest (heart area) over a 6-day course. (See color insert.)

An alternative to bioluminescence is the noninvasive detection of fluorescent bacteria using multiphoton laser scanning microscopy. Currently, there are no prokaryotic dyes for *in vivo* use available commercially. As bacterial expression of fluorescent proteins (e.g., GFP and its variants) does not require any substrate for endogenous production, it is the method of choice for prokaryotic labeling. Unfortunately, use of GFP is restricted to only those bacterial species that can produce the fluorescent protein, which eliminates the tracking of some key pathogens (e.g., *S. epidermidis* or obligates anaerobes). Månsson et al. [145,146] reports on the application of multiphoton microscopy for noninvasive *in vivo* imaging of ongoing infection. *Escherichia coli* GFP uropathogenic strains were injected by micropuncture directly into the proximal tubule of a nephron in an exteriorized kidney. Then, the rat is immediately placed on the stage of a multiphoton microscope, with the injected kidney placed in a saline-filled glass-bottom Petri dish (far from noninvasive!). Using multiphoton microscopy, the progression of infection was visualized in the anesthetized rat for an extended period of time. In a similar review by Roux et al. [147], *Cryptococcus neoformans* (a yeast labeled with fluorescein isothiocyanate, FITC) was injected into the tail vein of a mouse. Fluorescence images of the pathogens arriving in the brain of the infected animal were recorded 10 min later, using a multiphoton microscope.

While both of the imaging methods above are capable of detecting either luminescent or fluorescent bacteria *in vivo*, they are both dependent on the use of bacteria that have been manipulated *in vitro* then purposely injected to a subject to mimic an infection. Unfortunately, real infections are not caused by recombinant bacteria that glow. However, multiphoton microscopy does have the potential, especially in the near-IR wavelength range, to detect the onset of biofilm formation provided one could label the biofilm cells *in situ*. Hypothetically, one could imagine periodically injecting an implant patient with immunoconjugated quantum dots [148,149] that are specific to potential implant-colonizing bacteria. The circulating immuno-quantum dots would locate targeted biofilm bacteria thus allowing early stage multiphoton detection of an implant infection.

### 5.3.5. Biofilm Image Analysis

The recorded microscopic images can be used immediately or processed further for presentation or quantitative analysis. The images that originate from a confocal microscope are usually grayscale bitmap images, one from each focal plane and one for each detection channel (color). An image of a 30-μm thick biofilm recorded with a step size of 0.5 μm in three channels will result in $30/0.5 \times 3 = 180$ individual images. Most microscope software pack the images in containers (e.g., the LIF file format) for Leica and the LSM format for Zeiss microscopes. Furthermore the images will be relatively large. In the example above, the standard resolution of 256 graytones and $512 \times 512$ pixels will result in a file with the size of 45 MB. Special software is required to handle these files and to render the beautiful biofilm representations. While several packages are available, a few seem to dominate the market: Imaris (Bitplane, Bern, Switzerland), Amira (Visage Imaging, Carlsbad, CA), Volocity (Improvision, Coventry, the UK), Voxblast (Vaytek, Fairfield, IA), and Metamorph (Molecular Devices, Sunnyvale, CA) [150, for a review]. Designed for the visualization of larger eukaryotic cells, most 3D presentation packages are not optimized for the small cell sizes involved in microbial biofilms. The average size for a Gram-negative bacterium is $1 \times 2$–$3\,\mu m$. The optical resolution of a confocal image recorded with one-photon excitation is at best $0.48\,\mu m$ [151]. This means that a sampling of images with a smaller step size than this will not provide more information. Consequently, a single bacterium will not appear in more than one or two independent focal plane images, making a 3D reconstruction of the single cell difficult. However, the biofilm as a whole is much larger and can be rendered in 3D, although it may not be possible to locate the individual cells. Typically, a step size of $1\,\mu m$ is used for bacterial biofilms [94]. The visualization software packages also include tools for cleaning up the recordings, such as filters for noise and cross-talk (the situation where one fluorophore is recorded in the detection channel of another fluorophore). The main features of these softwares are, however, their capability to visualize the spatial organization of the recorded data. The rendering can be in perspective 3D, or as 2D images in all three axes, $x$–$y$, $x$–$z$, and $y$–$z$. Sequential recordings over time can also be rendered, and animated, providing a four-dimensional (4D) dataset, $x$–$y$–$z$–$t$. Such data sets can quickly be very large (sometimes several gigabytes) and it sets new requirements for the software and hardware. Special detector systems mounted on the confocal microscope might also facilitate the simultaneous separation of a number of fluorescent spectra (lambda mode configurations) originating from different fluorescent molecules or proteins with overlapping emission spectra, increasing the possibilities for multifluorescent labeling of biofilms [95].

Quantitative analysis of 3D images can be challenging and several groups have developed special software packages for this purpose. The algorithms start by determining the extent of the biomass, by thresholding each focal plane image. This step is crucial and much effort has been put into optimizing it. It can be done either manually, semi- or fully automated. Some of the first attempts to provide robust quantification software were done by Yang et al. [152] and Heydorn et al. [153] with the programs ISA and COMSTAT, respectively. Both extract a number of parameters, which can be used to characterize the biofilm: biomass (pixels occupied by biomass), biofilm height, height distribution, roughness coefficient, and diffusion distances, to mention a few. Both programs are developed in MATLAB (MathWorks, Natik, MA). The COMSTAT program utilizes a command-line interface within the MATLAB shell, which is required for operation, whereas ISA and its successor ISA3D are compiled programs that do not require the MATLAB package. Other quantification software is available (e.g., the web-based PHILIP) [154], a program that has a higher level of automation than the ISA and COMSTAT packages. Further developments of PHILIP have taken the automation and robustness toward the threshold function to a new level [155,156]. The COMSTAT software is now available in a complete reprogrammed version 2, which is running on the platform independent software foundation Java. This new version of COMSTAT software uses the same thresholding algorithms as the PHILIP software and also incorporates a number of new features (e.g., a user programmable plug-in interface for end user defined image processing functions and a wider range of image formats).

### 5.3.6. Identification of Biofilm in Human Chronic Diseases: The Chronic Rhinosinusitis Case

The mucosa of the nasal airway and that of the paranasal sinuses is increasingly accepted as a single pathophysiological unit, and thus the term "rhinosinusitis" [157] is commonly used. A continuum also exists for pathology between the upper and lower respiratory tracts [158]. However, distinct clinical entities still exist with localized rhinosinusitis and more diffuse panrespiratory disease. The categorization of chronic rhinosinusitis (CRS) by a pathophysiologic mechanism is still largely undefined.

Standard imaging techniques can describe bacterial biofilm morphology well, but offers little to describe the species present, gene expression, or phenotype. Attempts to culture the bacteria, away from the environment from which they were sampled will inevitably led to a change in phenotype. The use of species specific FISH enables the *in situ* identification of bacteria and their EPS.

Harvey and Lund [159] have undertaken a systematic review of the published literature for biofilms and their role in CRS. Both Medline (1966–2006) and Embase (1988–2006) were searched until November 2006, which yielded 652 articles, 13 of which provided original research of biofilms in CRS. There were 7 studies demonstrating biofilm morphology in mucosal samples from human CRS patients. One study showed similar evidence for biofilms in an animal model of CRS. The FISH techniques with CLSM were employed in one study to demonstrate biofilm formation *in situ* by *S. pneumoniae*, *S. aureus*, *H. influenza*, and *P. aeruginosa*. *In vitro* biofilm forming capacity of microbiological samples, after culture, was assessed in two studies. Correlation with a clinical outcome was also made in these papers. One study demonstrated biofilm growth in removed frontal sinus stents.

Recent FISH studies on biopsies of sinus mucosa in CRS [160] and repeated in middle ear mucosa [161], represent species specific *in situ* identification of biofilms in disease states. However, these techniques still require mucosa *ex vivo*, are expensive, limited in assessing polymicrobial communities, and require skill found at only a few centers.

Molecular-based bacterial detection techniques that utilize PCR (e.g., PCR cloning) will allow nonculturable bacteria to be defined [162]. Denaturing gradient gel electrophoresis (DGGE) and temperature-gradient gel electro-

**TABLE 5.6. Identifying Biofilm Through COMSTAT[a] and Image Structure Analyzer (ISA)[b]**

| Detection Techniques for Biofilms |
|---|
| **Imaging** |
| Scanning Electron Microscopy (SEM) |
| Transmission Electron Microscopy (TEM) |
| Confocal laser scanning microscopy (CLSM) |
| **Imaging Adjuncts (with CLSM)** |
| LIVE/DEAD Stain–BacLight Bacterial Viability Kit |
| Fluorescent *in situ* hybridization (FISH) with species specific probes |
| **Image processing (defines structure, mean thickness, roughness, substratum coverage and surface to volume ratio)** |
| COMSTAT image processing script for MATLAB[a] |
| Image Structure Analyzer (ISA)[b] |
| **Identification of non-planktonic bacteria** |
| PCR cloning |
| Denaturing gradient gel electrophoresis (DGGE) |
| Temperature gradient gel electrophoresis (TGGE) |

[a]See Ref. [162].
[b]See Ref. [163].

phoresis (TGGE) may also help to better define the microbial community within sinuses. Finally, computer aided analysis of CLSM enables conceptualization of the complex 3D biofilm structure [163]. A summary of identification techniques through COMSTAT [162] and Image structure analyzer (ISA) [163] is present in Table 5.6.

## 5.4. CONCLUDING REMARKS

In conclusion, the gold standard in biofilm research is an approach, which involves flow-cell technology in combination with CLSM. Unlike other techniques, this particular methodology allows getting insight into details of developmental processes, spatial organization, and function of laboratory-grown biofilms in real time under continuous and noninvasive conditions down to the single-cell level [164].

Confocal laser scanning microscopy monitoring of fluorescently color-coded bacteria, grown in mixed-strain biofilms under continuous hydrodynamic conditions, has provided intriguing insights into the spatiotemporal developmental processes, in some cases down to the single-cell level [94,95]. Detailed microscopic examinations have revealed that, even in monospecies biofilms, a number of physiologically distinct cell subpopulations exist and differentiate during biofilm development [93]. Intriguingly, recent investigations have revealed that these distinct cell subpopulations exhibit differential sensitivity toward antimicrobial agents [85,93]. One cell subpopulation, which exhibited high metabolic–physiological activity, was sensitive toward conventional antimicrobial agents, whereas the second subpopulation was refractory [93]. However, the second cell subpopulation exhibited increased sensitivity toward a number of membrane-targeting compounds, and hence a combined antimicrobial treatment using a conventional antibiotic and a membrane-targeting compound was able to kill both cell subpopulations of the biofilm [93]. These, and other studies, highlight the importance of studying the characteristics of biofilm cells at the subpopulation, and single-cell level.

Future studies, facilitated by sophisticated approaches and new technologies, will increase our understanding of microbial life in multicellular communities. Investigations of spatiotemporal gene expression in biofilms by the use of fluorescent reporter genes and CLSM will continue, and expand through the use of combinations of gene expression markers. Cell subpopulations will be isolated from complex communities (e.g., by the use of microdissection, microfluidic devices, or cell sorting), and subjected to further analysis [e.g., RT (reverse transcriptase)-PCR measurements of specific gene expression or DNA array analysis of global gene expression] [165]. Moreover, new technologies are emerging that might enable analysis of global transcription profiles of single biofilm cells [166,167]. Studies of the metabolic functions of subpopulations of biofilm cells or single cells in a biofilm community, might be done by the use of NanoSIMS (nanometer-scale secondary-ion mass spectrometry) or related techniques, which have recently been successfully employed on microbial cells from the environment and microbial cells associated with animals [168,169]. Insight into the features of distinct cell subpopulations and single cells in microbial communities will provide a more complete understanding of the microbial multicellular lifestyle, and were open up new strategies to manipulate harmful biofilms to restore and maintain human well-being.

## REFERENCES

1. Christensen, B., Sternberg, C., Andersen, J., Palmer, R. Jr., Nielsen, A., Givskov, M., and Molin, S. (1999), Molecular tools for the study of biofilm physiology, in: Doyle, R.J. Ed., *Biofilms*, Academic Press, London, UK, pp. 20–42 [Abelson, J., Simon, M.I. (Series Eds.), *Methods in Enzymology Series* Vol. 310].
2. Branda, S.S., Vik, S., Friedman, L., and Kolter, R. (2005), Biofilms: the matrix revisited, *Trends Microbiol.*, 13, 20–26.
3. O'Toole, G.A. and Kolter, R. (1998), Initiation of biofilm formation in *Pseudomonas fluorescens* WCS365 proceeds via multiple, convergent signalling pathways: a genetic analysis, *Mol. Microbiol.*, 28, 449–461.
4. O'Toole, G.A., Pratt, L.A., Watnick, P.I., Newman, D.K., Weaver, V.B., and Kolter, R. (1999), Genetic approaches to study of biofilms, *Methods Enzymol.*, 310, 91–109.

5. O'Toole, G.A. and Kolter, R. (1998), Flagellar and twitching motility are necessary for *Pseudomonas aeruginosa* biofilm development, *Mol. Microbiol.*, 30, 295–304.
6. Pratt, L.A. and Kolter, R. (1998), Genetic analysis of *Escherichia coli* biofilm formation: roles of flagella, motility, chemotaxis and type I pili, *Mol. Microbiol.*, 30, 285–293.
7. Watnick, P.I. and Kolter, R. (1999), Steps in the development of a *Vibrio cholerae* El Tor biofilm, *Mol. Microbiol.*, 34, 586–595.
8. Watnick, P.I., Lauriano, C.M., Klose, K.E., Croal, L., and Kolter, R. (2001), The absence of a flagellum leads to altered colony morphology, biofilm development and virulence in *Vibrio cholerae* O139, *Mol. Microbiol.*, 39, 223–235.
9. Valle, J., Toledo-Arana, A., Berasain, C., Ghigo, J.M., Amorena, B., Penades, J.R., and Lasa, I. (2003), SarA and not sigmaB is essential for biofilm development by *Staphylococcus aureus*, *Mol. Microbiol.*, 48, 1075–1087.
10. Guvener, Z.T. and McCarter, L.L. (2003), Multiple regulators control capsular polysaccharide production in *Vibrio parahaemolyticus*, *J. Bacteriol.*, 185, 5431–5441.
11. Friedman, L. and Kolter, R. (2004), Genes involved in matrix formation in *Pseudomonas aeruginosa* PA14 biofilms, *Mol. Microbiol.*, 51, 675–690.
12. Enos-Berlage, J.L., Guvener, Z.T., Keenan, C.E., and McCarter, L.L. (2005), Genetic determinants of biofilm development of opaque and translucent *Vibrio parahaemolyticus*, *Mol. Microbiol.*, 55, 1160–1182.
13. Branda, S.S., Gonzalez-Pastor, J.E., Ben-Yehuda, S., Losick, R., and Kolter, R. (2001), Fruiting body formation by *Bacillus subtilis*, *Proc. Natl. Acad. Sci. USA.*, 98, 11621–11626.
14. Raad, I., Costerton, W., Sabharwal, U., Sacilowski, M., Anaissie, W., and Bodey, G.P. (1993), Ultrastructural analysis of indwelling vascular catheters: a quantitative relationship between luminal colonization and duration of placement, *J. Infect. Dis.*, 168, 400–407.
15. Stickler, D., Morris, N., Moreno, M-C., and Sabbuba, N. (1998), Studies on the formation of crystalline bacterial biofilms on urethral catheters, *Eur. J. Clin. Microbiol. Infect. Dis.*, 17, 649–652.
16. Ferguson, D.J.P., McColm, A.A., Ryan, D.M., and Acred, P. (1986), A morphological study of experimental staphylococcal endocarditis and aortitis. II. Interrelationship of bacteria, vegetation and cardiovasculature in established infections, *Br. J. Exp. Pathol.*, 67, 679–686.
17. Nickel, J.C. and Costerton, J.W. (1992), Coagulase-negative staphylococcus in chronic prostatitis, *J. Urol.*, 147, 398–401.
18. Palmer Jr, R.J. and Sternberg, C. (1999), Modern microscopy in biofilm research: confocal microscopy and other approaches, *Curr. Opin. Biotech.*, 10, 263–266.
19. Bloemberg, G.V., O'Toole, G.A., Lugtenberg, B.J.J., and Kolter, R. (1997), Green fluorescent protein as a marker for *Pseudomonas* spp., *Appl. Environ. Microbiol.*, 63, 4543–51.
20. Lindsay, D. and von Holy, A. (1997), Evaluation of dislodging methods for laboratory-grown bacterial biofilms, *Food Microbiol.*, 14, 383–390.
21. Jeong, D.K. and Frank, J.F. (1994), Growth of *Listeria monocytogenes* at 10°C in biofilms with microorganisms isolated from meat and dairy processing environments, *J. Food Prot.*, 57, 576–586.

22. Anwar, H., Strap, J.L., and Costerton, J.W. (1992), Eradication of biofilms cells of *Staphylococcus aureus* with tobramycin and cephalexin, *Can. J. Microbiol.*, 38, 618–625.
23. Buck, J.W. and Andrews, J.H. (1999), Localized, positive charge mediates adhesion of *Rhodosporidium toruloides* to barley leaves and polystyrene, *Appl. Environ. Microbiol.*, 65, 2179–2183.
24. Pryde, S.E., Richardson, A.J., Stewart, C.S., and Flint, H.J. (1999), Molecular analysis of the microbial diversity present in the colonic wall, colonic lumen and cecal lumen of a pig, *Appl. Environ. Microbiol.*, 65, 5372–5377.
25. Green, P.N. and Pirrie, R.S. (1993), A laboratory apparatus for the generation and biocide efficacy testing of Legionella biofilms, *J. Appl. Bacteriol.*, 74, 388–393.
26. Doolittle, M.M., Cooney, J.J., and Caldwell, D.E. (1995), Lytic infection of *Escherichia coli* biofilms by bacteriophage T4, *Can. J. Microbiol.*, 41, 12–18.
27. Leriche, V. and Carpentier, B. (1995), Viable but non-culturable *Salmonella typhimurium* in single and binary species biofilms in response to chlorine treatment, *J. Food Prot.*, 58, 1186–1191.
28. Morris, C.E., Monier, J-M., and Jacques, M-A. (1998), A technique to quantify the population size and composition of the biofilm component in communities of bacteria in the phyllosphere, *Appl. Environ. Microbiol.*, 64, 4789–4795.
29. Pratten, J., Barnett, P., and Wilson, M. (1998), Composition and susceptibility to chlorhexidine of multispecies biofilms of oral bacteria, *Appl. Environ. Microbiol.*, 64, 3515–3519.
30. Stickler, D.J., Morris, N.S., McLean, R.J.C., and Fuqua, C. (1998), Biofilms on indwelling urethral catheters produce quorum-sensing signals molecules *in situ* and *in vitro*, *Appl. Environ. Microbiol.*, 64, 3486–3490.
31. Gauthier, V., Redercher, S., and Block, J-C. (1999), Chlorine inactivation of Sphingomonas cells attached to geothite particles in drinking water, *Appl. Environ. Microbiol.*, 65, 355–357.
32. Bloomfield, S.F., Arthur, M., Begun, K., and Patel, H. (1993), Comparative testing of disinfectants using proposed European surface test methods, *Lett. Appl. Microbiol.*, 17, 119–125.
33. Oosthuizen, M.C., Steyn, B., Lindsay, D., Brözel, V.S., and von Holy, A. (2001), Novel method for the proteomic investigation of a dairy-associated Bacillus cereus biofilm, *FEMS Microbiol. Lett.*, 194, 47–51.
34. Lindsay, D., Brözel, V.S., Mostert, J.F., and von Holy, A. (2002), Differential efficacy of a chlorine dioxide-containing sanitizer against single species and binary biofilms of a dairy-associated Bacillus cereus and a *Pseudomonas fluorescens* isolate, *J. Appl. Microbiol.*, 92, 352–361.
35. Lindsay, D., Brözel, V.S., and von Holy, A. (2006), Biofilm/spore response in *Bacillus cereus* and *Bacillus subtilis* during nutrient limitation, *J. Food Prot.*, 69, 1168–1172.
36. Maki, D.G., Weise, C.E., and Sarafin, H.W. (1977), A semiquantitative culture method for identifying intravenous-catheter-related infection, *New Engl. J. Med.*, 296, 1305–1309.
37. Tenney, J.H., Moody, M.R., Newman, K.A., Schimpff, S.C., Wade, J.C., Costerton, J.W., and Reed, W.P. (1986), Adherent microorganisms on lumenal surfaces of long-term intravenous catheters, *Arch. Intern. Med.*, 146, 1949–1954.

38. Sherertz, R.J., Raad, I.I., Belani, A., Koo, L.C., Rand, K.H., Pickett, D.L., Straub, S.A., and Fauerbach, L.L. (1990), Three-year experience with sonicated vascular catheter cultures in a clinical microbiology laboratory, *J. Clin. Microbiol.*, 28, 76–82.

39. Donlan, R.M., Murga, R., Bell, M., Toscano, C.M., Carr, J.H., Novicki, T.J., Zuckerman, C., Corey, L.C., and Miller, J.M. (2001), Protocol for the detection of biofilms on needleless connectors attached to central venous catheters, *J. Clin. Microbiol.*, 39, 750–753.

40. Zufferey, J., Rime, R., Francioli, P., and Bille, J. (1988), Simple method for rapid diagnosis of catheter-associated infection by direct acridine orange staining of catheter tips, *J. Clin. Microbiol.*, 26, 175–177.

41. Kite, P., Dobbins, B.M., Wilcox, M.H., Fawley, W.N., Kindon, A.J.L., Thomas, D., Tighe, M.J., and McMahon, M.J. (1997), Evaluation of a novel endoluminal brush method for *in situ* diagnosis of catheter related sepsis, *J. Clin. Pathol.*, 50, 270–282.

42. Cercenado, E., Ena, J., Rodriguez-Creixems, M., Romero, I., and Bouza, E. (1990), A conservative procedure for the diagnosis of catheter-related infections, *Arch. Intern. Med.*, 150, 1417–1420.

43. Mittelman, M.W., Kohring, L.L., and White, D.C. (1992), Multipurpose laminar-flow adhesion cells for the study of bacterial colonization and biofilm formation, *Biofouling*, 6, 39–51.

44. Jacobs, L., DeBruyn, E.E., and Cloete, T.E. (1996), Spectrophotometric monitoring of biofouling, *Water Sci. Technol.*, 34, 533–540.

45. Taylor, R.J. (1996), Efficacy of industrial biocides against bacterial biofilms, Ph.D. thesis, University of Birmingham, Birmingham, UK.

46. Angell, P., Arrage, A.A., Mittelman, M.W., and White, D.C. (1993), On line, non-destructive biomass determination of bacterial biofilms by fluorometry, *J. Microbiol. Methods*, 18, 317–327.

47. Rioufol, C., Devys, C., Meunier, G., Perraud, M., and Goullet, D. (1999), Quantitative determination of endotoxins released by bacterial biofilms, *J. Hosp. Infect.*, 43, 203–209.

48. Walter, R.W. and Cooke, L.M., (1997), Presented at the National Association of Corrosion Engineers Annual Conference, Paper No. 410.

49. Vorachit, M., Lam, K., Jayanetra, P., and Costerton, J.W. (1993), Resistance of *Pseudomonas pseudomallei* growing as a biofilm on silastic disks to ceftazidime and cotrimoxazole, *Antimicrob. Agents Chemother.*, 37, 2000–2002.

50. Ceri, H., Olson, M.E., Stremick, C., Read, R.R., Morck, D., and Buret, A. (1999), The Calgary Biofilm Device: new technology for rapid determination of antibiotic susceptibilities of bacterial biofilms, *J. Clin. Microbiol.*, 37, 1771–1776.

51. Donlan, R. M., Murga, R., and Carson, L. (1999), Growing biofilms in intravenous fluids, in: Wimpenny, J., Gilbert, P., Walker, J., Brading, M., and Bayston, R. eds., *Biofilms: the good, the bad, and the ugly*, Bioline, Cardiff, Wales, pp. 23–29.

52. Murga, R., Miller, J.M., and Donlan, R.M. (2001), Biofilm formation by gram-negative bacteria on central venous catheter connectors: effect of conditioning films in a laboratory model, *J. Clin. Microbiol.*, 39, 2294–2297.

53. Baillie, G.S. and Douglas, J. (1998), Effect of growth rate on resistance of *Candida albicans* biofilms to antifungal agents, *Antimicrob. Agents Chemother.*, 42, 1900–1905.

54. Stickler, D.J., Morris, N.S., and Winters, C. (1999), Simple physical model to study formation and physiology of biofilms on urethral catheters, *Methods Enzymol.*, 310, 494–501.

55. Raad, I., Darouiche, R., Hachem, R., Sacilowski, M., and Bodey, G.P. (1995), Antibiotics and prevention of microbial colonization of catheters, *Antimicrob. Agents Chemother.*, 39, 2397–2400.

56. Morck, D.W., Lam, K., McKay, S.G., Olson, M.E., Prosser, B., Ellis, B.D., Cleeland, R., and Costerton, J.W. (1994), Comparative evaluation of fleroxacin, ampicillin, trimethoprim-sulfamethoxazole, and gentamicin as treatments of catheter-associated urinary tract infections in a rabbit model, *Int. J. Antimicrob. Agents*, 4, S21–S27.

57. Kharazmi, A., Giwercman, B., and Høiby, N. (1999), Robbins Device in Biofilm Research, in: Doyle, R.J. Ed., *Biofilms*, Academic Press, London, UK, pp. 207–215 [Abelson, J., Simon, M.I. (Series Eds.), *Methods in Enzymology Series Vol. 310*].

58. Møller, S., Sternberg, C., Andersen, J.B., Christensen, B.B., Ramos, J.L., Givskov, M., and Molin S. (1998), *In situ* gene expression in mixed-culture biofilms: evidence of metabolic interactions between community members, *Appl. Environ. Microbiol.*, 64, 721–732.

59. Johansen, H. and Høiby, N., (1999), Animal models of chronic pneumonia (rodents), in: Zak, O. and Sande, M. Eds., *Handbook of Animal Models of Infection*, Academic Press, London, pp. 517–526.

60. Moser, C., Johansen, H.K., Song, Z.J., Hougen, H.P., Rygaard, J., and Hoiby, N. (1997), Chronic *Pseudomonas aeruginosa* lung infection is more severe in Th-2 responding BALB/c mice compared to Th-1 responding C3H/HeN mice, *APMIS*, 105, 838–842.

61. Olson, M.E., Ceri, H., Morck, D.W., Buret, A.G., and Read, R.R. (2002), Biofilm bacteria: formation and comparative susceptibility to antibiotics, *Can. J. Vet. Res.*, 66, 86–92.

62. Harrison, J.J., Ceri, H., Yerly, J., Stremick, C.A., Hu, Y., Martinuzzi, R., and Turner, R.J. (2006), The use of microscopy and three-dimensional visualization to evaluate the structure of microbial biofilms cultivated in the Calgary Biofilm Device, *Biol. Proced. Online*, 8, 194–215.

63. Nancharaiah, Y.V., Venugopalan, V.P., Wuertz, S., Wilderer, P.A., and Hausner, M. (2005), Compatibility of the green fluorescent protein and a general nucleic acid stain for quantitative description of a Pseudomonas putida biofilm, *J. Microbiol. Methods*, 60, 179–187.

64. Neu, T.R., Woelfl, S., and Lawrence, J.R. (2004), Three-dimensional differentiation of photo-autotrophic biofilm constituents by multi-channel laser scanning microscopy (single-photon and two-photon excitation), *J. Microbiol. Methods*, 56, 161–172.

65. Ahmed, K., Gribbon, P.N., and Jones, M.N. (2002), The application of confocal microscopy to the study of liposome adsorption onto bacterial biofilms, *J. Liposome Res.*, 12, 285–300.

66. Neu, T.R., Kuhlicke, U., and Lawrence, J.R. (2002), Assessment of fluorochromes for two-photon laser scanning microscopy of biofilms, *Appl. Environ. Microbiol.*, 68, 901–909.

67. Adair, C.G., Gorman, S.P., Byers, L.M., Gardner, T., and Jones, D.S. (2000), Confocal laser scanning microscope examination of microbial biofilms, in: An, Y.H. and Friedman R.J. Eds., *Principles, Methods and Applications*, Humana Press, Totowa, NJ, pp. 249–259.

68. Bradshaw, D.J., Marsh, P.D., Schilling, K.M., and Cummins, D. (1996), A modified chemostat system to study the ecology of oral biofilms, *J. Appl. Bact.*, 80, 124–130.

69. Lawrence, J.R., Korber, D.R., Hoyle, B.D., Costerton, J.W., and Caldwell, D.E. (1991), Optical sectioning of microbial biofilm, *J. Bact.*, 173, 6558–6567.

70. Palmer, R. Jr. (1999), Microscopy flowcells: perfusion chambers for real-time study of biofilms, in: Doyle, R.J. Ed., *Biofilms*, Academic Press, London, UK, pp. 160–165. [Abelson, J. and Simon, M.I. (Series Eds.), *Methods in Enzymology Series, Vol. 310*].

71. Kolenbrander, P., Andersen, R., Kazmerzak, K., Wu, R., and Palmer, R, Jr. (1999), Spatial organization of oral bacteria in biofilms, in: Doyle, R.J. Ed., *Biofilms*, Academic Press, London, UK, pp. 322–334 [Abelson, J. and Simon, M.I. (Series Eds.), *Methods in Enzymology Series, Vol. 310*].

72. Masso-Deyà, A., Whallon, J., Hickey, R., and Tiedje, J. (1995), Channel structures in aerobic biofilms of fixed-film reactors treating contaminated groundwater, *Appl. Environ. Microbiol.*, 61, 769–777.

73. Sternberg, C., Christensen, B., Møller, S., Andersen, J., and Molin, S. (1996), Localisation of cells active in gene expression and gene transfer in intact bacterial biofilms, in: Hastings, J., Kricka, L., and Stanley, P. Eds., *Bioluminescence and Chemiluminescence: Molecular Reporting with Photons*, John Wiley & Sons, Inc. Chichester, UK, pp. 371–374.

74. Paulsen, L., Ballard, G., and Stahl, D. (1993), Use of rRNA fluorescence *in situ* hybridization for measuring the activity of single cells in young and established biofilms, *Appl. Environ. Microbiol.*, 59, 1359–1360.

75. Lee, N., Nielsen, P., Andreasen, K., Juretschko, S., Nielsen, J., Schleifer, K-H., and Wagner, M. (1999), Combination of fluorescent in situ hybridization and micro-autoradiography, a new tool for structure-function analyses in microbial ecology, *Appl. Environ. Microbiol.*, 65, 1289–1297.

76. Wood, P., Caldwell, D., Evan, E., Jones, M., Korber, D., Wolfaardt, G., Wilson, M., and Gilbert, P. (1998), Surface-catalysed disinfection of thick *Pseudomonas aeruginosa* biofilms, *J. Appl. Microbiol.*, 84, 1092–1098.

77. Dyba, M., Keller, J., and Hell, S.W. (2005), Phase filter enhanced STED-4Pi fluorescence microscopy: Theory and experiment, *New J. Phys.*, 134, 1–21.

78. Harke, B., Keller, J., Ullal, C.K., Westphal, V., Schonle, A., and Hell, S.W. (2008), Resolution scaling in STED microscopy, *Opt. Exp.*, 16, 4154–4162.

79. Chalfie, M., Tu, Y., Euskirchen, G., and Ward, W.W. (1994), Green fluorescent protein as a marker for gene expression, *Science*, 263(5148), 802–805.

80. Møller, S., Korber, D., Wolfaardt, G., Molin, S., and Caldwell, D. (1997), Impact of nutrient composition on a degradative biofilm community, *Appl. Environ. Microbiol.*, 63, 2433–2438.

81. Wolfaardt, G., Lawrence, J., Robarts, R., and Caldwell, D. (1998), *In situ* characterization of biofilm exopolymers involved in the accumulation of chlorinated organics, *Microb. Ecol.*, 35, 213–223.
82. Caldwell, D.E., Korber, D.R., and Lawrence, J.R. (1992), Imaging of bacterial cells by fluorescence exclusion using scanning confocal laser microscopy, *J. Microbiol. Methods*, 15, 249–261.
83. Shi, L., Günther, S., Hübschmann, T., Wick, L.Y., Harms, H., and Müller, S. (2007), Limits of propidium iodide as a cell viability indicator for environmental Bacteria, *Cytometry A*, 71A, 592–598.
84. Molecular Probes. Sheet mp07007. Available at http://probes.invitrogen.com/media/pis/.
85. Haagensen, J.A., Klausen, M., Ernst, R.K., Miller, S.I., Folkesson, A., Tolker-Nielsen, T., and Molin, S. (2007), Differentiation and distribution of colistin-and sodium dodecyl sulfate-tolerant cells in *Pseudomonas aeruginosa* biofilms, *J. Bacteriol.*, 189, 28–37.
86. Lawrence, J.R., Neu, T.R., and Swerhone, G.D.W. (1998), Application of multiple parameter imaging for the quantification of algal, bacterial and exopolymer components of microbial biofilms, *J. Microbiol. Methods*, 32, 253–261.
87. Strathmann, M., Wingender, J., and Flemming, H.C. (2002), Application of fluorescently labelled lectins for the visualization and biochemical characterization of polysaccharides in biofilms of *Pseudomonas aeruginosa*, *J. Microbiol. Methods*, 50, 237–248.
88. Yang, L., Haagensen, J.A.J., Jelsbak, L., Johansen, H.K., Sternberg, C., Høiby, N., and Molin, S. (2008), *In situ* growth rates and biofilm development of *Pseudomonas aeruginosa* populations in chronic lung infections, *J. Bacteriol.*, 190, 2767–2776.
89. Stewart, P.S., Murga. R., Srinivasan, R., and de Beer, D. (1995), Biofilm structural heterogenecity visualized by three microscopic methods, *Water Res.*, 8, 2006–2009.
90. Allesen-Holm, M., Barken, K.B., Yang, L., Klausen, M., Webb, J.S., Kjelleberg, S., Molin, S., Givskov, M., and Tolker-Nielsen, T. (2006), A characterization of DNA release in *Pseudomonas aeruginosa* cultures and biofilms, *Mol. Microbiol.*, 59, 1114–1128.
91. Lambertsen, L., Sternberg, C., and Molin, S. (2004), Mini-Tn7 transposons for site- specific tagging of bacteria with fluorescent proteins, *Environ. Microbiol.*, 6, 726–732.
92. Sternberg, C., Christensen, B.B., Johansen, T., Na, T., Andersen, J.B., Givskov, M., and Molin, S. (1999), Distribution of bacterial growth activity in flow-chamber biofilms, *Appl. Environ. Microbiol.*, 65, 4108–4117.
93. Pamp, S.J., Gjermansen, M., Johansen, H.K., and Tolker-Nielsen, T. (2008), Tolerance to the antimicrobial peptide colistin in *Pseudomonas aeruginosa* biofilms is linked to metabolically active cells, and depends on the pmr and mexAB-oprM genes, *Mol. Microbiol.*, 68, 223–240.
94. Klausen, M., Aaes-Jorgensen, A., Molin, S., and Tolker-Nielsen, T. (2003), Involvement of bacterial migration in the development of complex multicellular structures in *Pseudomonas aeruginosa* biofilms, *Mol. Microbiol.*, 50, 61–68.

95. Pamp, S.J., and Tolker-Nielsen, T. (2007), Multiple roles of biosurfactants in structural biofilm development by *Pseudomonas aeruginosa*, *J. Bacteriol.*, 189, 2531–2539.

96. Tolker-Nielsen, T., Brinch, U.C., Ragas, P.C., Andersen, J.B., Jacobsen, C.S., and Molin, S. (2000), Development and dynamics of *Pseudomonas* sp. Biofilms, *J. Bacteriol.*, 182, 6482–6489.

97. Kjaergaard, K., Schembri, M.A., Ramos, C., Molin, S., and Klemm, P. (2000), Antigen 43 facilitates formation of multispecies biofilms, *Environ. Microbiol.*, 2, 695–702.

98. Møller, S., Pedersen, A.R., Poulsen, L.K., Arvin, E., and Molin, S. (1996), Activity and three-dimensional distribution of toluene-degrading *Pseudomonas putida* in a multispecies biofilm assessed by quantitative *in situ* hybridization and scanning confocal laser microscopy, *Appl. Environ. Microbiol.*, 62, 4632–4640.

99. Stahl, D.A. and Amann, R. (1991), Development and application of nucleic acid probes, in: Stackebrandt, E. and Goodfellow, M., Eds., *Nucleic Acid Techniques in Bacterial Systematics*, John Wiley & Sons, Inc., New York, pp. 205–248.

100. Amann, R. and Fuchs, B.M. (2008), Single-cell identification in microbial communities by improved fluorescence *in situ* hybridization techniques, *Nat. Rev. Microbiol.*, 6, 339–348.

101. Gunning, P.A., Kirby, A.R., Parker, M.L., Gunning, A.P., and Morris, V.J. (1996), Comparative imaging of *Pseudomonas putida* bacterial biofilms by scanning electron microscopy and both DC and AC non-contact atomic force microscopy, *J. Appl. Bact.*, 81, 276–282.

102. Bremer, P.J., Geesey, C.G., and Drake, B. (1992), Atomic force microscope examination of the topography of a hydrated bacterial biofilm on a copper surface, *Curr. Microbiol.*, 24, 223–230.

103. Dufrene, Y.F. (2004), Using nanotechniques to explore microbial surfaces, *Nature Rev. Microbiol.*, 2, 451–460.

104. Lower, B.H., Yongsuthon, R., Vellano III, F.P., and Lower, S.K. (2005), Simultaneous force and fluorescence measurements of a protein that forms a bond between a living bacterium and a solid surface, *J. Bacteriol.*, 187, 2127–2137.

105. Levine, S.A. and Niederman, M.S. (1991), The impact of tracheal intubation on host defenses and risks for nosocomial pneumonia, *Clin. Chest. Med.*, 12, 523–543.

106. Sottile, F.D., Marrie, T.J., Prough, D.S., Hobgood, C.D., Gower, D.J., Webb, L.X., Costerton, J.W., and Gristina, A.G. (1996), Nosocomial pulmonary infection: possible etiologic significance of bacterial adhesion to endotracheal tubes, *Crit. Care. Med.*, 14, 265–270.

107. Gorman, S.P., Adair, C., O'Neill, F., Goldsmith, C., and Webb, H. (1993), Influence of selective decontamination of the digestive tract on microbial biofilm formation on endotracheal tubes from artificially ventilated patients, *Eur. J. Clin. Microbiol. Infect. Dis.*, 12, 9–17.

108. Rubenstein, J.S., Kabat, T.K., Shulman, S.T., and Yogev, R. (1992), Bacterial and fungal colonisation of endotracheal tubes in children: a prospective study, *Crit. Care Med.*, 20, 1544–1549.

109. Adair, C.G., Gorman, S.P., Feron, B.M., Byers, L., Jones, D.S., Goldsmith, C.E., Moore, J.E., Kerr, J.R., Curran, M.D., Hogg, G., Webb, C.H., McCarthy, G.J., and

Milligan, K.R. (1999), Implications of endotracheal tube biofilm for ventilator-associated pneumonia, *Int. Care Med.*, 25, 1072–1076.

110. Inglis, T.J.J. (1995), New insights into the pathogenesis of ventilator-associated pneumonia, *J. Hosp. Infect.*, 108, 409–413.
111. Gorman, S.P., McGovern, J.G., Woolfson, A.D., Adair, C.G., and Jones, D.S. (2001), The concomitant development of poly(vinyl chloride)-related biofilm and antimicrobial resistance in relation to ventilator-associated pneumonia, *Biomaterials*, 22, 2741–2747.
112. Jones, D.S., McGovern, J.G., Adair, C.G., Woolfson, A.D., and Gorman, S.P. (2001), Conditioning film and environmental effects on the adherence of *Candida* spp. to silicone and poly(vinylchloride) biomaterials, *J. Mater. Sci. Mater. Med.*, 12, 399–405.
113. Jones, D.S., McGovern, J.G., Woolfson, A.D., Adair, C.G., and Gorman, S.P. (2002), Physicochemical characterization of hexetidine-impregnated endotracheal tube poly(vinyl chloride) and resistance to adherence of respiratory bacterial pathogens, *Pharm. Res.*, 19, 818–824.
114. Jones, D.S., Garvin, C.P., and Gorman, S.P. (2004), Relationship between biomedical catheter surface properties and lubricity as determined using textural analysis and multiple regression analysis, *Biomaterials*, 25, 1421–1428.
115. Lettington, A.H. (1993), Applications of diamond-like carbon thin films, *Philos. Trans R. Soc. London*, 342, 287–296.
116. Bull, S.J. (1995), Tribology of carbon coatings: DLC, diamond and beyond, *Diamond Relat. Mater.*, 4, 827–836.
117. Monaghan, D.P., Laing, K.C., and Logan, P.A. (1993), How to deposit DLC successfully, *Mater. World*, 1, 347–349.
118. Bull, S.J. and Chalker, P.R. (1995), High performance diamond and diamond-like coatings, *J. Miner. Met. Mater. Soc.*, 47, 16–19.
119. Thomson, L.A., Law, F.C., Rushton, N., and Franks, J. (1991), Biocompatibility of diamond-like carbon coating, *Biomaterials*, 12, 37–40.
120. Matthew, A., Myer, B., and Rushton, N. (2001), *In vitro* and *in vivo* investigations into the biocompatibility of diamond-like carbon (DLC) coatings for orthopedic applications, *J. Biomed. Mater. Res.*, 58, 319–328.
121. Dowling, D.P., Donnelly, K., and O'Brien, T.P. (1996), Application of diamond like carbon films as hermetic coatings on optical fibres, *Diamond Relat. Mater.*, 5, 492–495.
122. Jones, M.I., McColl, R., Grant, D.M., Parker, K.G., and Parker, T.L. (2000), Protein adsorption and platelet attachment and activation, on TiN, TiC, and DLC coatings on titanium for cardiovascular applications, *J. Biomed. Mater. Res.*, 52, 413–421.
123. Jones, D.S., Garvin, C.P., Dowling, D., Donnelly, K., and Gorman, S.P. (2006), Examination of surface properties and *in vitro* biological performance of amorphous diamond-like carbon-coated polyurethane, *J. Biomed. Mater. Res. Part B: Appl. Biomater.*, 78B, 230–236.
124. Wingender, J., Neu, T.R., and Flemming, H-C. (1999), What are bacterial extracellular polymeric substances? in: Wingender, J., Neu, T.R., and Flemming H-C. Eds., *Microbial Extracellular Polymeric Substances: Characterization, Structure and Function*, Springer-Verlag, New York, pp. 1–15.

125. Wozniak, D.J., Wyckoff, T.J.O., Starkey, M., Keyser, R., Azadi, P., O'Toole, G.A., and Parsek, M.R. (2003), Alginate is not a significant component of the extracellular polysaccharide matrix of PA14 and PAO1 *Pseudomonas aeruginosa* biofilms, *Proc. Natl. Acad. Sci. USA*, 100, 7907–7912.

126. Hiraoka, K. (2004), Laser spray: electric field-assisted matrix-assisted laser desorption/ionization, *J. Mass Spectrom.*, 39, 341–350.

127. Feio, M.J., Zinkevich, V., Beech, I.B., Llobet-Brossa, E., Eaton, P., Schmitt, J., and Guezennec, (2004), *Desulfovibrio alaskensis* sp. nov., a sulphate-reducing bacterium from a soured oil reservoir, *Int. J. Syst. Evol. Microbiol.*, 54, 1747–1752.

128. Benzerara, K., Yoon, T.H., Menguy, N., Tyliszczak, T., and Jr Brown, G.E. (2005), Nanoscale environments associated with bioweathering of a Mg-Fe-pyroxene, *Proc. Natl. Acad. Sci. USA*, 102, 979–982.

129. Toner, B., Fakra, S., Vilalobos, M., Warwick, T., and Sposito, G. (2005), Spatially resolved characterization of biogenic manganese oxide production within a bacterial biofilm, *Appl. Environ. Microbiol.*, 71, 1300–1310.

130. Beech, I.B. (2004), Corrosion of technical materials in the presence of biofilms-current understanding and state-of-the-art methods of study, *Int. Biodeter. Biodegr.*, 53, 177–183.

131. Omoike, A., Chorover, J., Kwon, K.D., and Kubicki, J.D. (2004), Adhesion of bacterial exopolymers to a-FeOOH: Inner-sphere complexation of phosphodiester groups, *Langmuir*, 20, 11108–11114.

132. Li, H. and Logan, B.E. (2004), Analysis of bacterial adhesion using a gradient force analysis method and colloidal probe atomic force microscopy, *Langmuir*, 20, 8817–8822.

133. Arce, F.F., Avci, R., Beech, I.B., Cooksey, K.E., and Wiggelsworth-Cooksey, B. (2004), A live bioprobe for studying diatom-surface interactions, *Biophys. J.*, 87, 4284–4297.

134. Bauer, T.W., Parvizi, J., Kobayashi, N., and Krebs, V. (2006), Diagnosis of periprosthetic infection, *J. Bone Joint Surg. Am.*, 88, 869–882.

135. Larson, I., Little, B., Nealson, K., Ray, R., Stone, A., and Tian, J. (1998), Magnetite reduction by *Shewanella putrifaciens* MR-4, *Am. Mineralogist.*, 83, 1564–1572.

136. Umeda, A., Saito, M., and Amako, K. (1998), Surface characteristics of Gram-negative and Gram-positive bacteria in an atomic force microscope image, *Microbiol. Immunol.*, 42, 159–164.

137. Braga, P. and Ricci, D. (1998), Atomic force microscopy: application to investigation of *Escherichia coli* morphology before and after exposure to cefodizime, *Antimicrob. Agents Chemother.*, 42, 18–22.

138. Kuznetsov, Y., Malkin, A., and McPherson, A. (1997), Atomic force microscopy studies of living cells: visualization of motility, division, aggregation, transformation, and apoptosis, *J. Struct. Biol.*, 120, 180–191.

139. Schaus, S. and Henderson, E. (1997), Cell viability and probe-cell membrane interactions of XR1 glial ceils imaged by atomic force microscopy, *Biophys. J.*, 73, 1205–1214.

140. Razatos, A., Ong, Y-L., Sharma, M., and Georgiou, G. (1998), Molecular determinants of bacterial adhesion monitored by atomic force microscopy, *Proc. Nat. Acad. Sci. USA*, 95, 11059–11064.

141. Colarusso, P., Whitley, A., Levm, I., and Lewis, E. (1999), Raman microscopy and imaging of inorganic and biological materials with liquid crystal tunable filters, in: Morris, M.D. Ed., *Biomedical Applications of Raman Spectroscopy, Proceedings of SPIE*, SPIE Press, Bellingham, Washington, Vol. 3608.
142. Phiefer, C., Palmer, R, Jr., and White, D. (1999), Comparison of relative photon flux from single cells of the bioluminescent marine bacteria *Vibrio fiscberi* and *Kbrio harveyi* using photon-counting microscopy, *Luminescence*, 14(3):147–151.
143. Palmer, R Jr., Applegate, B., Burlage, R., Sayler, G., and White, D. (1999), Heterogeneity of gene expression and activity in bacterial biofilms. In Bioluminescence and Chemical Luminescence: Perspectives for the 21sf Century, in: Roda, A., Pazzagli, M., Kricka, L.J., and Stanley, P.E. Eds., *Proceedings of the 10th International Symposium or Biolummescence and Chemiluminescence*, September 4–8: John Wiley & Sons, Inc., Bologna, Chichester, UK, pp. 609–612.
144. Xiong, Y.Q., Willard, J., Kadurugamuwa, J.L., Yu, J., Francis, K.P., and Bayer, A.S. (2005), Real-time *in vivo* bioluminescent imaging for evaluating the efficacy of antibiotics in a rat *Staphylococcus aureus* endocarditis model, *Antimicrob. Agents Chemother.*, 49, 380–387.
145. Månsson, L.E., Melican, K., Molitoris, B.A., and Richter-Dahlfors, A. (2007), Progression of bacterial infections studied in real time; novel perspectives provided by multiphoton microscopy, *Cell Microbiol.*, 9, 2334–2343.
146. Månsson, L.E., Melican, K., Boekel, J., Sandoval, R.M., Hautefort, I., Tanner, G.A., Molitoris, B.A., and Richter-Dahlfors, A. (2007), Real-time studies of the progression of bacterial infections and immediate tissue responses in live animals, *Cell Microbiol.*, 9, 413–424.
147. Roux, P., Münter, S., Frischknecht, F., Herbomel, P., and Shorte, S.L. (2004), Focusing light on infection in four dimensions, *Cell Microbiol.*, 6, 333–343.
148. Su, X.L. and Li, Y. (2004), Quantum dot biolabeling coupled with immunomagnetic separation for detection of *Escherichia coli* O157: H. *Anal. Chem.*, 76, 4806–4810.
149. Yang, L. and Li, Y. (2006), Simultaneous detection of *Escherichia coli* O157:H7 and *Salmonella typhimurium* using quantum dots as fluorescence labels, *Analyst*, 131, 394–401.
150. Neu, T.R. and Lawrence, J.R. (2005), One-photon versus two-photon laser scanning microscopy and digital image analysis of microbial biofilms, *Methods Microbiol.*, 34, 89–136.
151. Hell, S.W., Reiner, G., Cremer, C., and Stelzer, E.H.K. (1993), Aberrations in confocal fluorescence microscopy induced by mistakes in refractive index, *J. Microsc.*, 169, 391–405.
152. Yang, X., Beyenal, H., Harkin, G., and Lewandowski, Z. (2000), Quantifying biofilm structure using image analysis, *J. Microbiol. Methods*, 39, 109–119.
153. Heydorn, A., Ersbøll, B.K., Hentzer, M., Parsek, M.R., Givskov, M., and Molin, S (2000), Experimental reproducibility in flow-chamber biofilms, *Microbiology*, 146, 2409–2415.
154. Xavier, J.B., White, D.C., and Almeida, J.S. (2003), Automated biofilm morphology quantification from confocal laser scanning microscopy imaging, *Water Sci. Technol.*, 47, 31–37.

155. Merod, R.T., Warren, J.E., McCaslin, H., and Wuertz, S. (2007), Toward automated analysis of biofilm architecture: Bias caused by extraneous confocal laser scanning microscopy images, *Appl. Environ. Microbiol.*, 73, 4922–4930.

156. Mueller, L.N., de Brouwer, J.F., Almeida, J.S., Stal, L.J., and Xavier, J.B. (2006), Analysis of a marine phototrophic biofilm by confocal laser scanning microscopy using the new image quantification software PHLIP, *BMC Ecol.*, 6, 1.

157. Lanza, D.C. and Kennedy, D.W. (1997), Adult rhinosinusitis defined, *Otolaryngol. Head Neck Surg.*, 117, S1–S7.

158. Bousquet, J., Van Cauwenberge, P., Khaltaev, N., Aria Workshop, G., and World Health, O. (2001), Allergic rhinitis and its impact on asthma, *J. Allergy Clin. Immunol.*, 108, S147–S334.

159. Harvey, R.J. and Lund, V.J. (2007), Biofilms and chronic rhinosinusitis: systematic review of evidence, current concepts and directions for research, *Rhinology*, 45, 3–13.

160. Sanderson, A.R., Leid, J.G., and Hunsaker, D. (2006), Bacterial biofilms on the sinus mucosa of human subjects with chronic rhinosinusitis, *Laryngoscope*, 116, 1121–1126.

161. Hall-Stoodley, L., Hu, F.Z., Gieseke, A., Nistico, L., Nguyen, D., Hayes, J., Forbes, M., Greenberg, D.B., Dice, B., Burrows, A., Wackym P.A., Stoodley, B., Post, J.C., Ehrlich, G.D., and Kerschner, J.E. (2006), Direct detection of bacterial biofilms on the middle-ear mucosa of children with chronic otitis media, *JAMA*, 296, 202–211.

162. Heydorn, A., Nielsen, A.T., Hentzer, M., Sternberg, C., Givskov, M., Ersboll, B.K., and Mølin, S. (2000), Quantification of biofilm structures by the novel computer program COMSTAT, *Microbiol.*, 146, 2395–2407.

163. Beyenal, H., Donovan, C., Lewandowski, Z., and Harkin, G. (2004), Three-dimensional biofilm structure quantification, *J. Microbiol. Methods*, 59, 395–413.

164. Pamp, S.J., Sternberg, C., and Tolker-Nielsen, T. (2009), Insight into the microbial multicellular lifestyle via flow-cell technology and confocal microscopy, *Cytometry Part A*, 75A, 90–103.

165. Lenz, A.P., Williamson, K.S., Pitts, B., Stewart, P.S., and Franklin, M.J. (2008), Localized gene expression in *Pseudomonas aeruginosa* biofilms, *Appl. Environ. Microbiol.*, 74, 4463–4471.

166. Zhong, J.F., Chen, Y., Marcus, J.S., Scherer, A., Quake, S.R., Taylor, C.R., and Weiner, L.P. (2008), A microfluidic processor for gene expression profiling of single human embryonic stem cells, *Lab. Chip.*, 8, 68–74.

167. Marcus, J.S., Anderson, W.F., and Quake, S.R. (2006), Microfluidic single-cell mRNA isolation and analysis, *Anal. Chem.*, 78, 3084–3089.

168. Behrens, S., Losekann, T., Pett-Ridge, J., Weber, P.K., Ng, W.O., Stevenson, B.S., Hutcheon, I.D., Relman, D.A., and Spormann, A.M. (2008), Linking microbial phylogeny to metabolic activity at the single-cell level by using enhanced element labeling-catalyzed reporter deposition fluorescence in situ hybridization (EL-FISH) and NanoSIMS, *Appl. Environ. Microbiol.*, 74, 3143–3150.

169. Lechene, C.P., Luyten, Y., McMahon, G., and Distel, D.L. (2007), Quantitative imaging of nitrogen fixation by individual bacteria within animal cells, *Science*, 317(5844), 1563–1566.

# PART II

# BIOFILM-RELATED INFECTIONS IN VARIOUS HUMAN ORGANS (NONDEVICE-RELATED CHRONIC INFECTIONS)

## CHAPTER 6

# BIOFILM-RELATED INFECTIONS IN OPHTHALMOLOGY

### 6.1. INTRODUCTION

The majority of bacteria live in biofilms and not as individual microorganisms. About 65% of human infections are related to biofilm biophysiology. Indeed, most biofilms enter the body either through the skin, digestive tract, ocular surfaces, respiratory system, or mouth [1]. Therefore, the major objective of Part II is to describe biofilm-related infections occurring in ocular tissues, oral cavity, topical skin region (chronic wound), and lung with cystic fibrosis (CF) condition.

Unlike physicians and scientists whose interests lie buried deep in the viscera, ophthalmologists and optometrists have the benefit of easy access to the eye. Consequently, many of the earliest prosthetic devices were developed for the eye, ranging from contact lenses to lens implants. While these innovations provide wide-ranging benefits for patients, clinicians became aware that they could be associated with infections, and they instituted clinical practices (e.g., soaking scleral buckles in antibiotics before implantation and prescribing disposable contact lenses).

*Biofilm Eradication and Prevention: A Pharmaceutical Approach to Medical Device Infections*,
By Tamilvanan Shunmugaperumal

## 6.2. STRUCTURE OF THE EYE

The ocular globe can be divided into anterior and posterior segments (Fig. 6.1). The anterior segment consists of external cornea, conjunctiva, aqueous humor, iris-ciliary body, and lens. All these parts in the ocular cavity sometimes are referred to as extra-ocular tissues. The cornea and the lens consist of avascular and transparent structures. These structures obtain most of their necessary nutrients from aqueous humor. The cornea also partially depends on the tear fluid for essential nutrients (e.g., amino acids and vitamins). The iris–ciliary body and conjunctiva are highly vascular tissues. The aqueous humor is a dynamic watery fluid that is continuously secreted by the ciliary body and drained out by a trabecular meshwork returning to the blood stream through schlemm's canal. It maintains a pressure that imparts the convex shape to the front of the globe. The posterior chamber primarily consists of outer-sclera, choroid, retina, and the vitreous humor. These parts of the ocular cavity are also called intra-ocular tissues. The sclera is an avascular tissue and acts as an outer-protective layer. Underneath the sclera is a highly vascular choroid that supplies nutrients to both outer-sclera and inner-retina. The innermost layer is retina that is primarily responsible for image formation and thus the vision. The blood–retinal barrier is comprised of retinal pigment epithelium and the endothelium of retinal blood vessels. Unlike the aqueous humor, vitreous humor is a clear watery viscous fluid that is replaced at a very slow rate. The primary purpose of the vitreous humor is to provide a cushioned support for the rest of ocular structures, as well as a clear unobstructed path of light to the retina.

The epithelial surface of the eye is continuously exposed to potential pathogens, but rarely becomes infected. This is due in large part to the tear

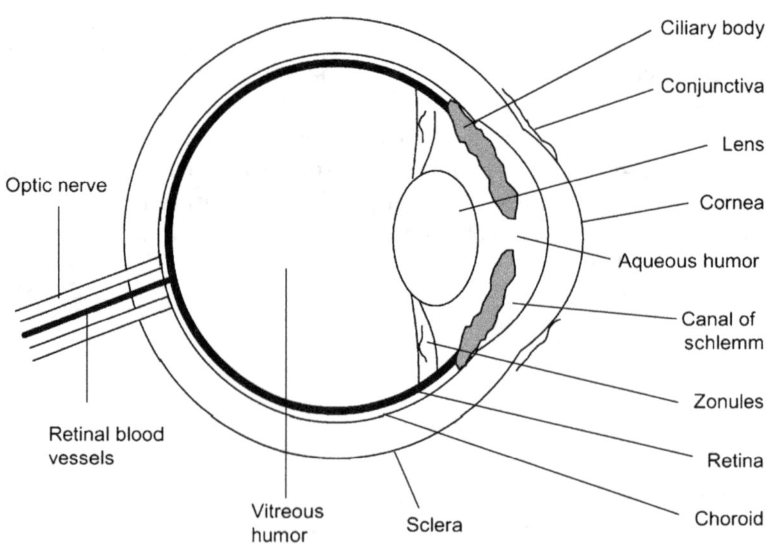


**Figure 6.1.** Structure of the human eye.

film that contains a number of antimicrobial components that are renewed during tearing. To help prevent corneal or conjunctival infection, the anterior eye harbors a variety of antimicrobial defenses that do not manifest damaging inflammatory immunopathological mechanisms [2]. Tear proteins [e.g., lactoferrin (Lf), lysozyme, and complement] comprise the nonadaptive antimicrobial factors along with anatomical barriers and mucous secretions. Lactoferrin is an 82-kDa protein [3] produced in the acinar cells of the lacrimal gland [4]. The Lf found in most secretions is almost entirely as apo-Lf [5]. Thus it has the ability to tightly bind any free Fe and effectively compete with bacteria for this essential cofactor. Lactoferrin has been shown to bind free Fe in tears, reducing the availability of Fe required for bacterial growth [6]. The ability to acquire Fe *in vivo* is thought to be an essential requirement for colonization and invasion by pathogenic bacteria [7]. Indeed, Lf has been shown to inhibit the growth of a number of bacterial species implicated in adverse events in tear film. Lactoferrin possesses a high isoelectric point with positively charged amino acids clustered at the N-terminus. This overall positive charge at physiological pH allows Lf to interact with the negatively charged surface components of bacteria. The Lf can directly bind both Gram-positive and -negative bacteria perturbing bacterial membranes. Due to the cationic nature of the N-terminus, Lf binding to bacteria can be a nonspecific interaction due to the negative charge on bacterial membranes [8]. Direct interactions between the lipopolysaccharide (LPS) from Gram-negative cell membranes and the Lf have been investigated. Human Lf binds to the lipid A region of LPS with high affinity [9] resulting in a concomitant increase in membrane permeability. This action is due to lactoferricin (Lfc), a peptide obtained from Lf by enzymatic cleavage, which is active not only against bacteria, but even against fungi, protozoa, and viruses [10].

Interestingly, *Staphylococcus epidermidis*, the most common bacterial isolate from tears, appears to be susceptible to Lf only in the presence of lysozyme [11], a situation that would occur naturally in tears. The mechanism of the synergy between Lf and lysozyme has been studied. Leitch and Willcox [11] found initial Lf binding to cell-bound lipoteichoic acid (LTA) was required prior to bacteria becoming susceptible to lysozyme. It was proposed that, on binding to the anionic LTA of *S. epidermidis*, the cationic protein Lf decreases the overall negative charge on the bacterial surface, allowing greater accessibility of lysozyme to the underlying peptidoglycan [12].

The involute interaction of Lf with bacteria can effect host defense through competition for free iron, binding of bacterial LPS, and inhibition of bacterial biofilm formation or adversely can result in inadvertent promotion of pathogenesis through bacterial competition for Lf iron reserves. Thus, working in concert with lysozyme and/or topical applications of antibiotics, Lf may increase susceptibility to these bactericidal agents by increasing penetration into the biofilm. In addition, through its unique combination of antimicrobial action and anti-inflammatory activities, Lf in the tear film plays an important role in the maintenance of ocular health [10].

## 6.3. OCULAR INFECTIONS

Viruses, bacteria, and fungi are the three most common pathogens causing various ocular disorders. The infections can be either local, where only the ocular tissues are infected or systemic, where the infection spreads to various other organs.

Immunocompromised patients are more prone to these infections. While some infections are self-limiting and benign, others are associated with serious consequences, (e.g., vision loss). In spite of various anti-infective agents being readily available, delivery of these compounds to the target tissue is still a challenging task. While the anterior chamber infections are treated mostly by topical ophthalmic drops, intravitreal and subconjuntival injections are widely indicated for the treatment of posterior chamber infections.

## 6.4. BACTERIAL INFECTIONS

Ocular tissues in both anterior and posterior segments are prone to a wide variety of bacterial infections. Both the Gram-positive and -negative species can infect the ocular structures. Most of these infections occur either after a surgery or during trauma. *Staphylococcus* species are the most common bacterial pathogens responsible for various ocular bacterial infections. These infections produce inflammation, a process generated mostly by the immune system. The symptoms include redness from an increased blood supply, swelling from fluid accumulating in the tissues, pain from nerve irritation, and ultimately vision loss. Even though the infections are caused by different bacterial species, most of them are treated with wide spectrum antibiotics. Antibiotics may be either bactericidal (kills the bacteria) or bacteriostatic (inhibit the growth of bacteria). These agents are divided into various classes based on their chemical structure (Table 6.1). Most of them exert their action by either inhibiting the cell wall synthesis or by blocking the protein synthesis. Most of the antibiotics have broad-spectrum activity, where they kill or inhibit a wide range of Gram-positive or -negative bacteria.

The delivery of most of these antibiotics to the anterior chamber is also a difficult task, as most of them are the substrates for efflux proteins like P-glycoprotein (P-gp) present on the cornea [13]. These efflux pumps on the cornea restrict the absorption of these antibiotics into cornea, resulting in reduced ocular bioavailability [14]. Various strategies have been implemented to improve the corneal permeability of these antibiotics. Both the addition of efflux pump inhibitors and prodrug design have been shown to be viable strategies in enhancing the corneal permeability and ocular bioavailability of these antibiotics [15]. In a transporter-mediated prodrug strategy, the efflux substrate is modified chemically such that it will have diminished affinity toward the efflux transporter, but high affinity for an influx transporter. In this way, it can circumvent the P-gp mediated efflux. Similar strategies have

**TABLE 6.1. Classification of Antibiotic Used to Treat Various Extra- and Intraocular Aetiological Conditions**

| Class | Mechanism of Action | Examples | Notes |
|---|---|---|---|
| Aminoglycosides | Inhibit bacterial protein synthesis (translation) | Streptomycin, gentamycin, amikacin | Bactericidal and particularly active against aerobic and Gram-negative bacteria. |
| β-Lactums | Inhibit bacterial cell wall synthesis by binding penicillin-binding proteins (PBPs) located on the bacterial cell wall. | Penicillins, cephalosporins, carbapenems, monobactums | (1) Contains Betalactum ring (2) Bactericidal and active against Gram-positive bacteria |
| Macrolides | Inhibit bacterial protein synthesis (translation) | Erythromycin, azithromycin, clarithromycin | Effective against aerobic and anaerobic Gram-positive organisms |
| Fluroquinolines | Inhibit bacterial DNA synthesis by inhibiting bacterial topomerase enzymes | Norfloxacin, ciprofloxacin | (1) Broad-spectrum antibiotics (2) Relatively less active against staphylococci and streptococci. |
| Tetracyclines | Inhibit bacterial protein synthesis (translation) | Tetracyclines, doxycycline | (1) Broad-spectrum antibiotics (2) Contraindicated with preparations containing vitamins and minerals. |
| Sulfonamides | Block metabolic process (e.g., synthesis of folic acid), which is necessary for synthesis of nucleic acids | Cotrimoxizole, trimethaprim | (1) Synthetic antibacterial with sulfonamide group (2) Broad-spectrum antibiotics |
| Lincosamides | Inhibit bacterial protein synthesis (translation) | Clindamycin, lincomycin | (1) Effective against anaerobic Gram-positive bacteria (2) Bacteriostatic |

**TABLE 6.1.** ***Continued***

| Class | Mechanism of Action | Examples | Notes |
|---|---|---|---|
| Polypeptides | Inhibit bacterial cell wall synthesis | Bacitracin, polymixin | (1) Bacitracin is active against Gram-positive organisms (2) Polymixin is a cationic polypeptide that is active against Gram-negative species |
| Chloramphenicol | Inhibit bacterial protein synthesis (translation) | Chloramphenicol | (1) Bacteriostatic and a broad-spectrum antibiotics (2) No longer used due to bone marrow toxicity and resistance |
| Rifamycins | Inhibit bacterial DNA dependent RNA polymerase and thus blocks the RNA synthesis | Rifamycin, rifabutin | (1) Active against Gram-positive and some Gram-negative species (2) Rifamycin is rarely used alone due to bacterial resistance |
| Glycopeptides | Inhibit bacterial cell wall synthesis, prevent the transport of cell wall precursors from cytoplasm to the cell wall | Vancomycin | (1) Effective against anaerobic Gram-positive bacteria |

been employed for delivery of these antibiotics either intravitreally or subconjuntivally for the treatment of posterior chamber infections. In addition, bacterial resistance is also a major obstacle that could lead to treatment failure [16]. Antibiotic resistance in bacteria may be an inherent trait of the organism (e.g., a particular type of cell wall structure) that renders it naturally resistant, or it may be acquired by means of mutation in its own DNA or by acquisition of resistance-conferring DNA from another source [17]. This problem can be overcome by appropriate use of antibiotic regimens and constant development of novel antibacterial agents that are not substances for efflux proteins.

## 6.5. BLEPHARITIS

Blepharitis is one of the most common bacterial infections of the eyelids and lachrymal apparatus and is caused by *S. epidermis* and *S. aureus*. If left untreated it can lead to keratitis and chronic conjunctivitis that are described below.

## 6.6. BACTERIAL KERATITIS

The infection develops in a vascular cornea. It is commonly caused by *S. aureus*, *S. pneumoniae*, and Gram-negative coliform bacteria. The infection can lead to corneal ulceration. The cornea, as well as the eyelids, becomes very edematous. The most noticeable feature is rapid progression of the disease that can result in corneal disintegration. Contact lens also serves as an exogenous risk factor of ulcerative bacterial keratitis, as it can affect the precorneal tear film and corneal epithelium [18]. Usage of extended wear contact lenses can increase the risk [19]. If left untreated it can lead to perforation of cornea that can result in permanent blindness. The common symptoms include severe pain, redness, decreased vision, and photophobia.

At the initial stages of infection, broad-spectrum antibiotics may be recommended. These antibiotics should be effective against most of the Gram-positive and -negative organisms. The treatment should be initiated by the "shot gun" approach, which is a combination of cephalosporin and amino glycosides. To achieve the therapeutic concentrations of antibiotics at the site of action, "fortified" antibiotics are also recommended [20,21]. Such form of treatment can assist in achieving higher concentrations in the cornea in a relatively short time period. A loading regimen consisting of 2% tobramaycin and 5% cefamandole may be indicated. This regimen is continued until the specific bacterial strain is identified and then treatment may be gradually shifted to a single drug.

Fluoroquinolones (e.g., ciprofloxacin, ofloxacin, lomefloxacin, and norfloxacin) have shown excellent activity against both Gram-negative and -positive bacteria and are used frequently for the treatment of bacterial keratitis [22]. Ciprofloxacin hydrochloride (0.3%) and ofloxacin (0.3%) have been shown to be effective in treatment of bacterial keratitis [23]. Subconjunctival delivery of antibiotics might achieve higher concentrations. Collagen shields are produced when the contact lenses, made out of animal collagen, are soaked in antibiotics, this approach can be used to prolong corneal drug levels [24].

## 6.7. CONJUNCTIVITIS

Conjunctivitis is characterized by irritation, itching, and tearing. Bacterial conjunctivitis usually unilateral usually does not spread to the contralateral eye. Staphylococcus species are the most common bacterial pathogens responsible for this disorder. Other species include *Streptococcus pneumoniae*, *Chlamydio trachomatis*, and *Haemophilus influenza*. In children, it is primarily caused by *H. influenza*, *S. pneumoniae*, and *Moraxella catarrhalis* [25]. Conjunctivitis can be classified as infectious and noninfectious types. Infectious conjunctivitis, also called pinkeye, is caused either by bacteria or viruses. It is highly contagious. Noninfectious conjunctivitis can be caused by allergies, pollen, and underlying diseases. The common symptoms include red eye, pain, swelling of eyes, blurred vision, itching, and a gritty feeling. Acute conjunctivitis is self-limiting, but it usually takes up to 3 weeks to treat it completely. Infection is often present in one eye and gradually spreads to the other. No single antibiotic can cover all possible pathogens responsible for conjunctivitis. Since most of the causative organisms are Gram-positive bacteria it would be appropriate to choose an antibiotic that can cover a maximum number of Gram-positive species [26]. The most commonly used broad-spectrum antibiotics include erythromycin and bacitracin-polymyxin B ointments. Aminoglycosides like gentamicin, neomycin, and tobramycin have more of Gram-negative coverage and are also relatively more toxic to cornea [26]. The bacteriostatic agent sulfacetamide (10%) is also commonly prescribed for topical use. The antibiotics tetracycline and chloramphenicol are topically used as well. However, tetracycline is available only as an ointment. In addition, fluroroquinoline derivatives are also commonly prescribed to treat bacterial conjunctivitis.

## 6.8. HYPER-ACUTE BACTERIAL CONJUNCTIVITIS

This severe, vision-threatening ocular infection requires immediate medical attention. The symptoms are characterized by a yellow–green purulent discharge that can reaccumulate even after being wiped off [27]. The infection is frequently caused by *Neisseria gonorrehoeae and N. meningitides*, the former being the most common organisms. If left untreated it may lead to severe perforation and ulceration of cornea leading to partial or total vision loss. It is usually treated with systemic antibiotics. Cephalosporin and spectinimycin are recommended [26].

## 6.9. CHRONIC BACTERIAL CONJUNCTIVITIS

This ailment commonly is caused by staphylococcus species [28]. It often develops in conjunction with blepharitis. The symptoms include itching, burning, and morning eyelash crusting [26]. Some patients may even have recurrent styes and chalazia (lipogranulomas) of the eyelid margin, which is due to chronic inflammation of meibomian glands (meibomianitis). Treatment is similar to normal conjunctivitis, which includes antibiotics (e.g., fluoroquinolines).

## 6.10. OCULAR CHLAMYDIAL INFECTIONS

Ocular *Chlamydia trachomatis* occur in two forms, that is, trachomatis (chronic keratoconjunctivtis) and inclusive conjunctivitis. Trachoma is the most common cause of ocular morbidity and blindness [26]. Inclusive keratoconjunctivitis is a sexually transmitted disease. It is a common cause of conjunctivitis in neonates. Infants are generally exposed to *C. trachomatis* from the infected cervix of mother. It is usually treated with antibiotics and if necessary they are administered systemically [29]. Oral azithromycin is also recommended as a cost-effective means of controlling endemic trachoma [30,31]. Novel vaccine delivery systems also have been evaluated for trachoma treatment [32].

## 6.11. ENDOPHTHALMITIS

This infection involves the inflammation of intraocular tissues (aqueous and vitreous humor). It can be caused by bacteria, fungi, and viruses. Most cases of endophthalmitis are caused by bacteria, usually following ocular surgery. The most common causative organisms are *S. aureus*, *S. epidermidis*, and several other streptococcus species. The symptoms depend on the type of endophthalmitis. Postoperative endophthalmitis occurs mostly after cataract surgery. It can occur by the infection being caused either during the surgery or because of the surgical equipment that is not sterile [33]. Depending on the severity, it can also lead to loss of vision. The symptoms include red eye, swollen eyelids, and blurred vision. Post-traumatic endophthalmitis is caused by penetrating eye injuries. It is also associated with red eye and swollen eyelids. Hematogenous endophthalmitis occurs when the systemic infection spreads to the eye and settles in the ocular tissues. It causes a mild decrease in vision and floaters in the eye.

Since infection is caused by bacteria, corticosteroids and flouroquinoline derivatives are usually recommended. The route of administration depends on the type and severity of the disease. The regimen is administered by the intravenous route to treat hematogenous endophthalmitis. Topical antibiotics are given if there is a wound infection on the ocular segment. Corticosteroids are given to decrease inflammation [34]. If the vision loss is severe, vitrectomy needs to be performed, where a portion of the infected vitreous humor is replaced with sterile saline or other compatible fluid [35].

## 6.12. OCULAR IMPLANTS AND POSTOPERATIVE ENDOPHTHALMITIS

Serious bacterial infections of the eye are often associated with abiotic prosthetic materials, (e.g., intraocular lenses, IOL), contact lenses, and scleral buckles [36]. All of these abiotic prosthetic materials are otherwise called ocular implants. An ocular implant is a medical device designed to replace or augment a damaged or missing visual organ. An IOL is an artificial aid surgically implanted in the living eye to restore sharpness of vision, frequently after the removal of a cataract. It has to be transparent, tolerable material of an appropriate size, shape, and power for the individual patient. Somehow it also has to remain in place.

In 1765, Casanova's Memoirs record a meeting with an Italian itinerant oculist, Tadini, at a dinner in Warsaw. Tadini showed him some polished crystal spheres in a box and claimed he could implant them underneath the cornea to replace the crystalline lens: "A substance which I can place in the cornea to supply the loss of the crystalline matter". The great lover's reply was "There's a great difference between a tooth and the crystalline humour; and though you may have succeeded in putting an artificial tooth into a gum, this treatment will not do with the eye". (*Memoirs of Jacques Casanova de Seingalt*, Vol. 6b). A fellow guest, an unnamed German ophthalmologist, ridiculed the claim in print and Tadini never mentioned the spheres again, but Casanova may have been responsible for conveying the idea to the Court at Dresden where, according to a publication of 1795 by the Swiss surgeon Rudolph Schiferli (1773–1837), the Court Oculist Casaamata had tried to insert a glass lens underneath a corneal wound, but this had fallen into the bottom of the eye. In other words, this was the first failed attempt to correct aphakia via artificial lens implantation.

In 1948, the British ophthalmic surgeon, (Sir) Harold Ridley (1906–2001) of St. Thomas' Hospital and the Moorfields Eye Hospital, consulted with John Pike (1902–1983), the senior optical specialist at the Rayner Optical Company about designing and manufacturing an implantable lens. He had been stimulated in this by a chance remark of a medical student, F.S. (Steve) Perry, who had commented during a cataract extraction the year before that it was a pity that the diseased lens could not be replaced. Ridley had the humility and the foresight to take on board this idea from a student. The first operation, on a 45-year old female, took place at St. Thomas' on 29 November 1949 though, as a two stage process, the IOL was only implanted permanently 3 months later, on 8 February 1950. The procedure was highly controversial and apparently St. Thomas' was chosen because it provided better security.

Ridley discounted glass as being too heavy of a material. The first IOL was made of a perspex called Transpex 1. This material was a specially recreated plastic, copying that used in the windows of RAF fighter aircraft during the Second World War. Doctors had observed that when these windows shattered, the eye injuries suffered by airmen, although otherwise horrific, were that bit less severe due to the inert properties of the perspex within the body, which therefore experienced no secondary inflammation. The Pharma company (ICI) resumed production of this form of perspex at Ridley's request. They made all of his lenses and charged only ~£1 each. At 100 mg in weight, they were ~20 times heavier than modern intraocular lenses. In the 1950s, by now called Perspex CQ (Clinical Quality), the material was made using a compression moulding process. The lens blanks were then profiled and polished by hand.

The first few operations were conducted out of the glare of publicity and when the wider ophthalmological profession became aware of the revolutionary procedure, opinion was sharply divided as to its wisdom. As the lenses had to be sterilized in a cetrimide solution at the point of insertion, several patients suffered severe postoperative reactions. By 1957, all lenses were supplied by the manufacturer presterilized in sodium hydroxide. Early patients were often left highly myopic. When Ridley performed his last operation in 1964, it was judged that 20% of his patients had experienced long-term failure and the main obstacle remained that of secure positioning. By the late 1960s, fixation lenses were devised using loops of nylon 66, but even then only a few hundred IOLs were supplied each year. Nylon also degraded when it absorbed water. During the 1970s, demand from the United States spurred the growth of the IOL manufacturing industry in Britain and elsewhere.

**Some Early Types of IOL:**

Ridley lens (1949–1950).

Strampelli rigid anterior chamber lens (1954).

Epstein collar stud lens—South African design utilizing the iris.

Choyce anterior chamber lens (MK1 1956)—an English modification of Strampelli's design. The MKVIII of 1963 proved especially popular. Peter Choyce (1919–2001) modified his anterior lens design 9 times in 23 years. In July 1966, he founded the "Implant Club" with a membership of 16 international practitioners.

Binkhorst Iris Clip lens (1957)—with four fixation loops, so-called because the lens resembled a paper clip.

1959–The J-shaped fixation made by placing a cut in the flexible nylon haptic loop was first designed in Barcelona, but the idea only really caught on in an American posterior chamber lens of 1975.

Fyodorov lens (1966)—a Russian modification of the Binkhorst. Svyatoslav Fyodorov (Fedorov) used fibers for his external haptic loops drawn from the patient's own achilles tendon.

Boberg-Ans semiflexible anterior chamber lens—Danish design.

Pearce posterior chamber lens (1975).

The IOLs were designed with flanges (later fenestrated) by Copeland, but these became associated with chronic uveitis. "Claw" designs were tried though sometimes an additional suture was deemed necessary to hold the lens in place. Fantastic shapes were produced, notably the "Pregnant 7".

In the 1980s, design developments were influenced by viscoelastic technology. Posterior chamber lenses became much more popular and extracapsular cataract extraction surgery was refined. Ridley even underwent a successful operation himself, in his old hospital of St. Thomas'. Since 1993, their standard of manufacture has been governed by the European Medical Devices Directive. The perspex material now included an anti-UV (ultraviolet) absorbing agent preventing light-induced damage to the back of the eye. Several multinational optical corporations acquired IOL manufacturing divisions.

The implantation of modern IOLs requires minimally invasive surgery. A Si folding lens can be inserted through a tiny incision or injected into the eye. The technology has close parallels with the contact lens industry and soft hydrophilic materials (e.g., acrylic) have been adapted for IOL use. Today there are in excess of 1500 designs and it is estimated that >200 million people have benefited from an intraocular lens.

Three main types of IOL are commonly used following cataract extraction: an anterior chamber IOL positioned in front of the iris, but behind the cornea; an iris clip lens straddling the pupil and a posterior chamber IOL, now the lens of choice, situated behind the iris within or on the capsular bag [37]. Poly(methyl methacrylate) (PMMA) is a suitable material for IOL implants due to its high light transmittance, low specific gravity, relatively high refractive index (1.49), and desirable mechanical properties. It is chemically inert, biostable, and can be lathe cut, milled, polished, and sterilized. The standard IOL consists of a central optic supported by haptics, projections from the main body of the lens. The central optic part of IOL is made up of PMMA, Si, hydrogel–hydrophilic acrylic, and hydrophobic acrylic, whereas the haptic part of IOL is build with polypropylene, PMMA rigid, polyacrylic, polyimide (elastimide), and polyvinylidene fluorides. While standard IOLs are monofocal and as such focused in one position, the development of multifocal IOL allows the eye to be in focus for far and near objects simultaneously. The brain then chooses which image to ignore [38]. However, the quality of the vision is reduced compared with a standard monofocal lens. Consequently, accommodating IOL has been developed to overcome these issues. A further type of available is the toric lens, the shape of which is designed to overcome astigmatism and thus obviate the requirement for surgical manipulation of the cornea. Further developments include the wavefront and light-adjustable lens. The former is designed to provide better contrast sensitivity by correcting for an average range of higher-order aberrations that alter vision. The light-adjustable lens, when implanted, can be altered with laser light to change its power. Blue-filtering lenses are also available, as it is thought that blue light may be an etiological factor in age-related macular degeneration. Thus development in cataract surgery have concentrated on designing implants that return the eye to its precataract state, meaning that patients no longer require spectacles after surgery to see near objects. Research into these new technologies will continue as there is a move toward producing lenses that meet the exact individual requirements of the patient.

## 6.13. PHAKIC, APHAKIC, AND PSEUDOPHAKIC IOLs

Phakia is the presence of the natural crystalline lens and aphakia is the absence of the natural crystalline lens, either from natural causes or because it has been removed. Pseudophakia is the substitution of the natural crystalline lens with a synthetic lens. Pseudophakic IOLs (PIOLs) are used in cataract surgery. The root of these words comes from the Greek word phakos "lens". The IOLs have been used since 1999 for correcting larger errors in myopic (near-sighted), hyperopic (far-sighted), and astigmatic eyes. This type of IOL is also called PIOL (phakic intraocular lens), and the crystalline lens is not removed. The PIOLs can be either spheric or toric. The latter is used for astigmatic eyes. The difference is that toric PIOLs have to be inserted in a specific angle, or the astigmatism will not be fully corrected, or it can even get worse. According to placement site in the eyes phakic, IOLs can be divided to the following:

1. Angle supported PIOLs: Those IOLs are placed in the anterior chamber. They are notorious for their negative impact on the corneal endothelial lining, which is vital for maintaining a healthy dry cornea.
2. Iris supported PIOLs: This type is gaining more and more popularity. The IOL is attached by claws to the mid-peripheral iris by a technique called enclavation. It is believed to have a lesser effect on corneal endothelium.
3. Sulcus supported PIOLs: These IOLS are placed in the posterior chamber in front of the natural crystalline lens. They have special vaulting so as not to be in contact with the normal lens. The main complications with this type are their tendency to cause cataracts and/or pigment dispersion.

Most PIOLs have not yet been approved by the U.S. Food and Drug Administration (FDA), but many are under investigation, and some of the risks the FDA found during a 3-year study of the Artisan lens, produced by Ophtec USA Inc., are (1) a yearly loss of 1.8% of the endothelial cells, (2) a 0.6% risk of retinal detachment, (3) a 0.6% risk of cataract (other studies have shown a risk of 0.5–1.0%), and a 0.4% risk of corneal swelling.

## 6.14. ACCOMMODATIVE IOLs

One of the major disadvantages of conventional IOLs is that it is primarily focused for distance vision. Though patients who undergo a standard IOL implantation no longer experience clouding from cataracts, they are unable to accommodate, or change focus from near to far, far to near, and in distances in between. Accommodating IOLs interact with cilliary muscles and zonules, using hinges at both ends to "latch on" and move forward and backward inside the eye using the same mechanism for normal accommodation. These IOLs have a 4.5-mm square-edged optic and a long-hinged plate design with polyimide loops at the end of the haptics. The hinges are made of an advanced Si called BioSil that was thoroughly tested to make sure it was capable of unlimited flexing in the eye [39]. There are many advantages to accommodating IOLs. For example, light comes from and is focused on a single focal point, reducing halos, glares, and other visual aberrations. Accommodating IOLs provide excellent vision at all distances (far, intermediate, and near), projects no unwanted retinal images, and produces no loss of contrast sensitivity or central system adaptation. Accommodating IOLs have the potential to eliminate or reduce the dependence on glasses postcataract surgery. For some, accommodating IOLs may be a better alternative to refractive lens exchange (RLE) and monovision [40].

The FDA approved Eyeonics Inc.'s accommodating IOL, Crystalens AT-45, in November 2003. Bausch & Lomb acquired Crystalens in 2008, and introduced a newer model called Crystalens HD in 2008. Crystalens is the only FDA approved accommodating IOL currently on the market and it is approved in the United States and Europe [41].

NuLens Ltd. is currently in patient trials with a new Accommodative IOL technology with the potential to provide over 10 diopters of accommodative power. With an IOL that sits on top of the collapsed capsular bag, the NuLens Accommodative IOL may be the first intraocular lens to provide real, comfortable, and lasting accommodation for near, intermediate, and far distances.

## 6.15. MICROBIAL ADHESION AND IOL: *IN VITRO* AND *IN VIVO* STUDIES

Postoperative endophthalmitis remains a serious sight-threatening complication of intraocular surgery. Despite improvements in prophylaxis, surgical techniques, and treatments, endophthalmitis represents a therapeutic emergency that often causes severe and persistent visual impairment [42,43].

*Staphylococcus epidermidis* is currently recognized as a major etiological agent of endophthalmitis after cataract surgery [44,45]. This species, which is part of the ocular and periocular surface flora, is characterized by its ability to adhere to polymer surfaces (e.g., IOL). The binding of bacteria, primarily during the implantation process or immediately after, is the first step in IOL colonization [44,46–48]. It is followed by bacterial accumulation in multilayered cell clusters embedded in an exopolysaccharide matrix, leading to the formation of a confluent structured biofilm. Biofilm formation on polymer surfaces is a complex process that depends on bacterial cell surface characteristics, the nature of the polymer material, and environmental factors (e.g., medium composition and flow conditions) [49–51]. Few different studies have tried to determine which biomaterial was the stickiest. Comparison between these studies is difficult, and even impossible because conditions varied. The strains, the incubation times, and the quantitative or qualitative methods to determine bacterial adhesion all varied.

In 1994, Cusumano et al. [52] published an scanning electron microscope (SEM) study to quantify bacterial growth on the surface of IOL. They used a coagulase-negative staphylococcus strain, but did not specify which one. All cultures were maintained under standard conditions at 38 °C for 5 days. At the end of the incubation time, all IOL were removed and fixed for SEM. Silicone, hydrogel and PMMA IOLs were tested. But no specifications were given about the manufacturers and model numbers of the IOLs, since many of their Si properties differ [53]. Depending on the material used, mean bacterial counts increased in the following order: PMMA, hydrogel, and Si. All the differences were statistically significant. Gabriel et al. [53] found similar results in 1998. They compared the relative adherence of a clinical strain of *Pseudomonas aeruginosa* to four IOLs made of PMMA, hydrophobic acrylic, and Si. The IOLs were incubated for 2 and 18 h. Two methods were used: radiolabel and SEM. Adhesion was greater on Si than on PMMA, and it was greater on PMMA than on hydrophobic acrylic. The differences between the three groups were statistically significant. Pinna et al. [54] showed the contrary with another strain, as they proved that hydrophobic acrylic IOLs were more prone to *S. epidermidis* adherence than PMMA IOLs [54]. This trend was found by using bacterial counting after 3 min incubation and by using SEM. However, at 90 min, hydrophobic acrylic IOLs had a lower viable bacterial count than did the PMMA IOLs. Discrepancies depending on the technique used and the length of incubation might be explained by the fact that the culture method can detect only viable bacteria, whereas in SEM the total number of bacteria includes both live and dead microorganisms. Moreover, hydrophobic acrylic material may have a toxic time-dependent effect on bacterial cells, which may eventually reduce bacterial adhesion to the lens. As for Ng et al. [44], who used a culture method, adherence of *S. epidermidis* to hydrogel IOL was significantly weaker than its adherence to PMMA IOLs, after an incubation time of 60 min. Garcia-Saenz et al. [55] compared the *in vitro* adherence of slime-producing and non-slime-producing *S. epidermidis* to different IOLs. The lenses were incubated for 18 h with bacteria, then sonicated and vortexed. Quantitative cultures were performed. Slime-negative strains of *S. epidermidis* adhered to all IOLs, but at a lower level than slime-positive strains. The most adherent lenses were hydrophobic acrylic and heparinized PMMA. The least adherent IOLs were hydrogel, PMMA, and fluorine PMMA. Adhesion was average for Si. This study is not in keeping with others, demonstrating that bacterial adhesion to heparinized PMMA is weaker than to PMMA [56–59]. Apparently, the long, hydrophilic heparin chains that were fixed on the lens surface could capture neighboring water molecules, forming a hydrophilic boundary layer all around the lens. This, in turn, decreased cellular and bacterial binding [57,58]. However, heparin bears negative charges that might inhibit adherence [56,57]. Perhaps the heparin chains coating the PMMA is not dense enough in places allowing some hydrophobic areas to be exposed. Finally, an *in vitro* study to determine the adherence of *S. epidermidis* to IOLs made of five different biomaterials was performed [47]. Adhesion was weakest on hydrogel and strongest on the Si polymer. The pairwise comparison of PMMA, heparinized PMMA, and hydrophilic acrylic was not significant.

An *in vitro* model that more closely resembled intraocular conditions, especially by mimicking aqueous humor composition and intraocular hydrodynamic conditions, was designed recently by Baillif et al. [60]. The model allowed the study of *S. epidermidis* biofilm formation (from the primary attachment phase to the biofilm maturation phase) on IOL. Moreover, it was the first time the entire process of biofilm development on IOL surfaces was examined. Very recently, Baillif et al. [61] again used the same dynamic model to compare *S.epidermidis* adherence and growth on four commercially available IOL biomaterials: PMMA, Si, hydrophilic acrylic, and hydrophobic acrylic (Table 1.2), and concluded that bacterial adhesion to and biofilm development on the IOL surface depended on the characteristics of the biomaterials (Fig. 6.2). Furthermore, the results suggest that hydrophobic IOL (e.g., Si or hydrophobic acrylic IOL) are more permissive to *S. epidermidis* adhesion and growth than hydrophilic IOLs. Moreover, adhesion is also affected by the nature of the

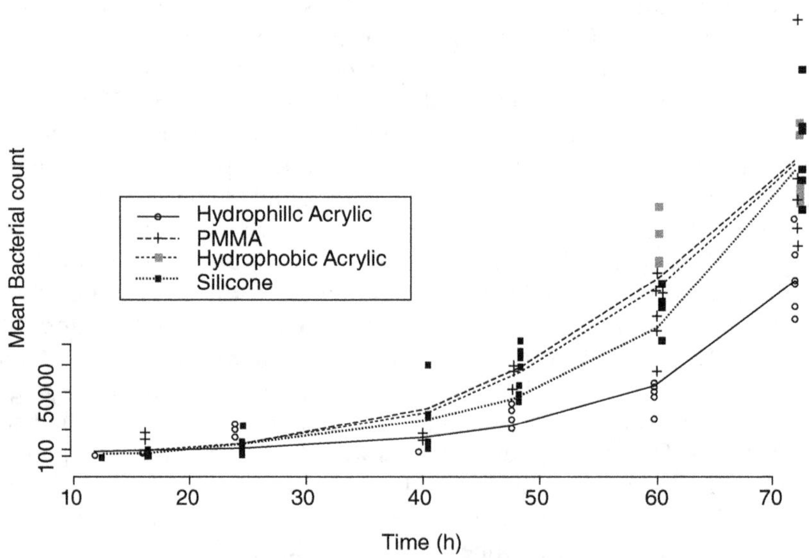


**Figure 6.2.** Bacterial adherence of *S. epidermidis* to IOLs of four biomaterials (polynomial poisson regression on the square root of bacterial counts). (Redrawn with permission from Baillif et al. *J. Cataract. Refract. Surg.*, 34, 153–158, 2008 [61].)

surrounding medium. Its influence seems essential, making it difficult to extrapolate the *in vitro* results to the clinical situation. Since this medium is very difficult to model because of its complexity, an *in vivo* study seemed essential.

Kodjikian et al. [62] published the first study concerning *in vivo* evolution of bacteria adhering to five different intraocular materials. The study was performed with 120 IOLs (PMMA, heparinized PMMA, Si, hydrophobic acrylic, and hydrogel) and used a *S. epidermidis* reference strain (ATCC 14990) that produces a great amount of slime. The domestic pig was chosen as a suitable animal model of endophthalmitis, after a bibliographical analysis and a personal study of its aqueous humor composition. This study involved 135 domestic pigs. The authors knew that identical amounts of bacteria could not be bound to each material because there had been significant differences in their prior *in vitro* study [47]. So, bacterial concentration was adjusted spectrophotometrically to $10^8$ CFU $mL^{-1}$ in order to reach a minimum bacterial count of ~$10^5$ CFU per IOL. Lenses were incubated in bacterial suspension for 1 h at 37 °C. After removing crystalline lens (by manual extracapsular extraction) under aseptic conditions and general anesthesia, previously infected IOLs were implanted into the anterior chamber. To explant the lenses, the authors first enucleated the involved eye. After washing with povidone iodine 5% solution and rinsing with sterile balanced salt solution (BSS), the authors performed a large corneal incision to remove the IOLs with sterile forceps under a laminated flux hood. Animals were killed after 24 h (24H group), 72 h (72H group), and 1 week (1W group). Bacterial counting was conducted on every IOL removed. For each of the five lens materials, 18 IOLs were implanted before being removed to measure the amount of remaining bacteria (hereafter termed CFU IOL) and 6 were used to control the precise amount of adhering bacteria before implantation (termed CFU control). Results were expressed as ratios (Fig. 6.3). A positive ratio indicated that bacterial growth took place on the IOL surface, whereas a negative one meant that the count of bound bacteria had decreased between IOL implantation and removal. Only two materials, Si and heparinized PMMA [heprin surface modified (HSM) PMMA], presented a globally positive ratio, (i.e., bacterial growth on their surface). The others had a globally negative ratio, showing a decline of the bacterial population colonizing their surface. Moreover, only hydrogel showed an immediate negative ratio at 24 h. The ratio of attached bacteria per area unit found on hydrogel, fluorine PMMA, acrylic, heparinized PMMA, and finally on the Si polymer, went in increasing order. Comparing pairs of materials gave statistically significant differences, except between hydrogel and fluorine PMMA. These results are in agreement with some *in vitro* studies, concluding that intermediate hydrophobicity is an important promoting factor of bacterial binding [63–65]. Bacteria adhere less to IOLs composed of hydrophilic (e.g., hydrogel) or very hydrophobic (e.g., fluorine PMMA) materials than to intermediate hydrophobic (e.g., Si) materials [52,55]. Hydrogel and fluorine PMMA showed *in vitro* low bacterial adhesion [55,58]

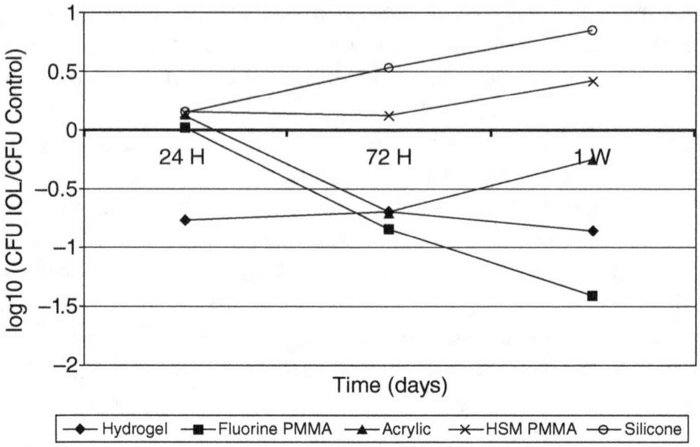


**Figure 6.3.** *In vivo* evolution of bacteria adhering to IOLs in pig eyes at three different time periods. Results are expressed as $\log_{10}$ CFU IOL/CFU control 50 mm$^{-2}$.

and so logically *in vivo* bacterial decline. Modifying the surface of PMMA IOLs with fluorine thus seems to be a better way of reducing bacterial adhesion than coating it with heparin. According to some previous studies [57,58,66], HSM IOLs provide a highly hydrated surface that modifies some structural fatty acids of *S. epidermidis*, reducing bacterial adhesion. However, the present work showed that HSM PMMA allowed bacterial growth, proving that bacteria adhered rather firmly. This result may relate to the fact that heparin behaves as an adhesion receptor in some *Staphylococcus* species [67]. A biomaterial unsuitable for *in vivo* growth of bacteria would be very useful in clinical practice. The IOLs made of such a material would most likely prevent the development of endophthalmitis, unlike other biomaterials that allow bacterial growth on their surfaces. Colonization of the IOL surface may lower intrinsic and extrinsic defenses of the eye, making them unable to fully eradicate the adherent bacteria, thus causing an infection. These results from Kodjikian et al. [62] suggest that the risk of endophthalmitis after cataract extraction followed by IOL implantation under antibiotic prophylaxis may be lower with IOLs made of less sticky material (e.g., hydrogel and fluorine PMMA). Recent epidemiological studies of endophthalmitis rates with various IOL biomaterials have reached similar conclusions that the implantation of Si IOL might be associated with increased rates of endophthalmitis [68,69].

The perfect biomaterial that could prevent postoperative endophthalmitis does not yet exist. Globally, hydrophilic materials and hydrophobic acrylic seem to be less sticky than Si or PMMA, but the clinical relevance remains to be proven! There are elementary precautions to be taken to reduce the risk of postoperative endophthalmitis. At the insertion of an IOL, care must be taken to minimize contact with the external eye. Moreover, the external eye must be prepared carefully just before operation, sterile adhesive barriers must be used to keep the eyelashes away from the field of surgery, and the IOL packaging must be opened only at the last moment. Further biomaterial development should respect the following goals: development of a material that reduces or does not allow bacterial adhesion (discovering new biomaterials or working on antibiotic surface treatments), development of a material that is injectable (and not foldable) to prevent contact with the ocular surface, development of a material that is biocompatible, safe and not toxic, and development of a material that prevents lens epithelial proliferation. Studies will also continue to involve a new type of IOLs with new implantation techniques (e.g., the twin-capsulorhexis IOL), implanted in the bag-in-the-lens-technique and accommodating Si IOL.

## 6.16. CONTACT LENS

Contact lenses have been classified according to material of construction, design, wear schedule, and frequency of disposal. Soft contact lenses are made of either hydrogel or Si and are designed to allow oxygen to diffuse through the lens material to provide oxygen to the cornea. Hard contact lenses are constructed of PMMA and move with each blink, allowing oxygen-containing tears to flow underneath the lens [70]. Bacteria adhere readily to both types of lenses [70–73].

Miller et al. [74] examined initial attachment of *P. aeruginosa* to hydrophilic contact lenses (hydrogels) and found that the rate of adherence varied depending on water content and polymer composition. Though these were initial adhesion studies, only 2 h in duration, they observed extracellular matrix polymers by transmission electron microscopy (TEM) and Rh red staining. The degree of attachment was found to depend on a number of factors, including the nature of the substrate, pH, electrolyte concentration, ionic charge of the polymer, and bacterial strain tested. Their results showed that there was greater adherence to hydrophobic surfaces and to lenses composed of nonionic polymers.

Stapleton et al. [73] also observed greater adhesion of *P. aeruginosa* to low-water-content nonionic lenses than to ionic lenses. They found that maximal adhesion occurred after 45 min and did not increase for contact periods as long as 24 h. Miller et al. [74] also showed that *P. aeruginosa* adhesion was enhanced by mucin, lactoferrin, lysozyme, immunoglobulin A, bovine serum albumin (BSA), and mixtures of these molecules, though exposure to human tears resulted in both an increase and decrease in adherence depending on the lens formulation tested. These investigators noted that the data would not allow an accurate prediction of how these molecules would perform under *in situ* conditions.

Organisms that have been shown to adhere to contact lenses include *P. aeruginosa*, *S. aureus*, *S. epidermidis*, *Serratia* spp., *Escherichia coli*, *Proteus* spp., and *Candida* spp. [70]. An established biofilm (extensive exopolymer matrix) was demonstrated by SEM of a lens removed from a patient with keratitis caused by *P. aeruginosa* [42]. McLaughlin-Borlace et al. [75] also provided evidence of biofilms on the surfaces of 20 contact lens samples collected from patients with a clinical diagnosis of microbial keratitis. In several cases, the biofilms contained multiple species of bacteria or bacteria and fungi.

Biofilms have also been shown to develop on contact lens storage cases [70,75,76]. In fact, the lens case has been implicated as the primary source of organisms for contaminated lens disinfectant solutions and lenses [76]. One study found that 80% of asymptomatic lens users had contaminated storage cases [76]. These investigators found that bacterial biofilms were present on 17 of 20 storage cases examined, a significantly greater percentage than the percentage of lenses containing biofilms. They also isolated the identical organism from the lens case and the corneas of infected patients for 9 of 12 samples examined. Additionally, studies have found that the protozoan *Acanthamoeba* may be a component of these biofilms [70,76]. These organisms feed on the biofilm bacteria and may also be a cause of microbial keratitis.

Several studies have compared the efficacy of contact lens storage and cleaning solutions against bacterial biofilms on lens storage cases. Wilson et al. [76] compared quaternary ammonium compounds, chlorhexidine-gluconate, and hydrogen peroxide ($H_2O_2$). He found that 3% $H_2O_2$ was most effective in inactivating bacterial biofilm organisms that were 24-h old (*Serratia marcescens*, *P. aeruginosa*, *S. epidermidis*, or *Streptococcus pyogenes*). Biofilms of *C. albicans*, on the contrary, were highly resistant to all treatments, including $H_2O_2$. Another study found that sodium salicylate was effective in decreasing initial bacterial adherence to lenses and cases [77]. However, one study found that biofilms could be detected on contact lenses removed from patients with microbial keratitis whose lens storage cases were treated according to the manufacturer's instructions with disinfectants (e.g., $H_2O_2$ and chlorine-release systems) [75]. Gandhi et al. [78] showed that *S. marcescens* could grow in chlorhexidine disinfectant solutions. Further research is needed to determine the efficacy of disinfectant solutions against model system biofilms and natural biofilms on contact lenses that have been removed from patients with an active infection.

## 6.17. MISCELLANEOUS BIOFILM-MEDIATED OCULAR INFECTIONS

Bacteria are capable of producing a wide array of molecules that can mediate human diseases independent of the presence of live organisms. This can occur through an immune reaction to a bacterial antigen, the direct effects of a toxin on host tissues, or the response of host tissues to a signaling molecule. Such molecules are produced by bacteria in biofilms, as well as planktonic cells. However, because bacteria in biofilms can persist in niches that do not support planktonic cells, biofilms may facilitate the contamination of the medical environment with these molecules. This mechanism can be illustrated by examples involving endotoxin release from biofilms. Endotoxin or lipopolysaccharide is a heat stable component derived from the outer-cell membrane of Gram-negative bacteria. Endotoxin's effects on host tissues are numerous and mediated in part by eliciting the production of numerous cytokines and by stimulating the coagulation and complement cascades [79,80].

Diffuse lamellar keratitis (DLK) is a complication of laser *in situ* keratomileusis (LASIK) in 0.2–3% of patients [81]. Several authors have described outbreaks of DLK that were associated with contamination of sterilizer reservoirs with Gram-negative organisms [82,83]. The outbreaks ended when proper decontamination and biofilm control procedures were introduced. Because, by definition, DLK does not involve an active bacterial infection of the cornea, it has been inferred that these outbreaks of DLK resulted from the inoculation of bacterial antigens or toxins, but not live organisms, under the LASIK flap, causing a sterile inflammatory response. Accordingly, Holzer et al. [84] demonstrated in a rabbit model of LASIK that interface inoculation of lipopolysaccharide from *P. aeruginosa* resulted in DLK in 94% of eyes that were not treated with anti-inflammatory agents. Lower rates of DLK were observed in animals treated with anti-inflammatory medication, indicating that DLK is, in part, immune-mediated. Both the outbreak data and animal work suggest that DLK is likely the result of an inflammatory response to molecules derived from bacteria. It is not known whether or not growth in a biofilm influences the ability of Gram-negative bacteria to produce or release endotoxin. However, by facilitating the survival of a bacterial community under conditions that would not have likely supported planktonic growth, biofilm formation can make it possible for endotoxin to contaminate a medical environment.

Reyes et al. [85] described an analogous outbreak of febrile reactions after cardiac catheterization. Bacterial contamination of a medical water supply occurred, which resulted in high levels of endotoxin in the water. Catheters were flushed with this water prior to gas sterilization. Although live organisms were not transmitted to patients, endotoxin persisted on the catheters despite sterilization and found its way into catheterized patients, causing an acute febrile illness. Patients had negative blood cultures, so it was deduced that the epidemic was a consequence of the sterile, but endotoxin contaminated, catheters.

Endotoxin has also been detected in contact lens cases contaminated with Gram-negative bacteria [86]. Bacterial biofilm formation on contact lenses and contact lens cases has been documented [75]. It is beyond the scope of this chapter to review in detail all the ways in which endotoxin can participate in contact lens associated keratitis, but Khatri et al. [87] suggested that it may cause a sterile keratitis associated with contact lens wear. *In vitro* studies of rabbit corneal epithelium demonstrated increased production of proinflammatory eicosanoid compounds in the presence of endotoxin [88]. Immortalized human corneal epithelial cells show "increased monolayer permeability" and disruption of epithelial tight junctions when exposed to endotoxin [89]. Thus, endotoxin derived from biofilms on contact lenses or contact lens cases can both stimulate inflammation directly in the cornea and render the cornea more susceptible to bacterial infection. Again, the ability of Gram-negative bacteria in biofilms to survive in hostile conditions allows endotoxin to enter the medical environment and, thus, come in contact with human tissues.

Punctal plugs can also provide a niche for biofilm formation. Yokoi et al. [90] reported a case of conjunctivitis associated with punctal plugs. The punctal plugs were observed to be covered with white material, which, when cultured, revealed a mixed infection of *Staphylococcus heamolyticus* and *Candida tropicalis*. Scanning electron microscopy of the plug demonstrated bacteria embedded in an extracellular matrix, consistent with a biofilm. Soukasian [91] found culture-positive punctal plugs in 19 of 20 explanted plugs, with *P. aeruginosa* being the most frequently isolated organism. Although the ocular surface is commonly exposed to *P. aeruginosa* owing to the ubiquity of this microbe in the environment, the conjunctiva rarely becomes colonized with this organism. The abiotic substrate of the punctal plug, like that of the contact lens, may provide a site where bacteria can persist in a biofilm relatively protected from host mucosal defenses while shedding organisms onto the ocular surface. Ultimately, an acute or chronic infection of the ocular surface can develop.

In the eye, the examination of pathologic specimens of corneas from patients with chronic microbial keratitis has revealed bacteria growing in a biofilm in patients with infectious crystalline keratopathy (ICK) [92]. Fixation of the corneal tissue in the presence of ruthenium red demonstrates bacteria encased in an exopolysaccharide matrix in patients with ICK, but not in the corneas of other patients with chronic keratitis. Infectious crystalline keratopathy often has a prolonged clinical course, is difficult to culture, and is poorly responsive to antibiotic treatment, all features consistent with a biofilm infection. Further investigation is required to determine whether or not biofilm formation on biotic surfaces plays a role in other corneal infections.

An additional example of biofilm formation on the ocular surface in the absence of an abiotic substrate has been recently reported. Mihara et al. [93] described a patient with a history of pterygium surgery 7 years earlier who presented with eye pain and a white–yellow mass. Clinically, the mass was considered to be a calcification, but when it was removed and examined microscopically, a biofilm composed of Gram-positive bacilli and associated with neutrophils was identified. Although no definite conclusions can be made as to the conditions that allowed biofilm to form in this patient, her previous history of pterygium surgery requiring a scleral graft and the use of mitomycin C suggests it might have been the result of abnormal tissue on the ocular surface. This abnormal tissue might have provided a more permissive environment for biofilm formation, either because of alterations in its structure or as a result of impaired mucosal immunity. In either case, it suggests that under appropriate conditions bacterial biofilm formation can occur on the eye in the absence of abiotic materials.

## 6.18. SCLERAL BUCKLE INFECTIONS

Several studies have documented a high rate of biofilm formation on scleral buckles. One report noted that 65% of buckles removed for infection or extrusion had evidence of bacterial biofilms [94]. Another report found biofilms on 5 of 28 buckles removed for all causes, but only 3 of these had clinical evidence of infection [95]. Biofilm formation on scleral buckles again illustrates the ability of abiotic materials to provide a niche for biofilm formation. These bacterial communities can participate in disease in a number of ways. First, they provide a reservoir of available bacteria, which may cause a clinical infection in a susceptible host. Additionally, some cases of buckle extrusion may represent immune-mediated tissue damage due to a chronic bacterial colonization of the buckle. In this model, the bacterial biofilm does not produce a suppurative infection, but does induce a chronic and destructive inflammatory response [96]. The biofilm on the scleral buckle is resistant to elimination by the immune system; nonetheless, it provides chronic immune stimulus through the ongoing release of planktonic cells and antigens. Over time, this may induce immune mediated tissue damage, separate from that caused by direct interactions between bacteria and host tissue.

From an understanding of the biofilm concept, ophthalmologists will incorporate it when biofilm-related diseases arise in their clinical practice on various ocular infections. Biofilms can play a role in disease as a source of biologically active molecules (endotoxin production), as a source of infection, and by forming directly on ocular tissues.

## REFERENCES

1. Jakob, H.G., Morneff-Lipp, M., Bach, A., von Pückler, S., Windeler, J., Sonntag, H., and Hagl, S. (2000), The endogenous pathway is a major route for deep sternal wound infection, *Eur. J. Cardiothorac. Surg.*, 17, 154–160.
2. Chandler, J.W. and Gillette, T.E. (1983), Immunologic defense mechanisms of the ocular surface, *Ophthalmology*, 90, 585–591.
3. Jensen, O.L., Gluud, B.S., and Birgens, H.S. (1985), The concentration of lactoferrin in tears during post-operative ocular inflammation, *Acta Ophthalmol. (Copenh.)*, 63, 341–345.
4. Gillette, T.E. and Allansmith, M.R. (1980), Lactoferrin in human ocular tissues, *Am. J. Ophthalmol.*, 90, 30–37.
5. Makino, Y. and Nishimura, S. (1992), High-performance liquid chromatographic separation of human apolactoferrin and monoferric and diferric lactoferrins, *J. Chromatogr.*, 579, 346–349.
6. Kijlstra, A. (1990), The role of lactoferrin in the nonspecific immune response on the ocular surface, *Reg. Immunol.*, 3, 193–197.
7. Yu, R.H. and Schryvers, A.B. (2002), Bacterial lactoferrin receptors: insights from characterizing the Moraxella bovis receptors, *Biochem. Cell Biol.*, 80, 81–90.

8. Ling, J.M. and Schryvers, A.B. (2006), *Perspectives* on interactions between lactoferrin and bacteria, *Biochem. Cell Biol.*, 84, 275–281.
9. Appelmelk, B.J., An, Y.Q., Geerts, M., Thijs, B.G., de Boer, H.A., MacLaren, D.M., de Graaff, J., and Nuijens, J.H. (1994), Lactoferrin is a lipid A binding protein, *Infect. Immun.*, 62, 2628–2632.
10. Flanagan, J.L. and Willcox, M.D.P. (2009), Role of lactoferrin in the tear film, *Biochimie*, 91, 35–43.
11. Leitch, E.C. and Willcox, M.D. (1998), Synergic antistaphylococcal properties of lactoferrin and lysozyme, *J. Med. Microbiol.*, 47, 837–842.
12. Leitch, E.C. and Willcox, M.D. (1999), Elucidation of the antistaphylococcal action of lactoferrin and lysozyme, *J. Med. Microbiol.*, 48, 867–871.
13. Dey, S., Patel, J., Anand, B.S., Jain-Vakkalagadda, B., Kaliki, P., Pal, D., Ganapathy, V., and Mitra, A.K. (2003), Molecular evidence and functional expression of P-glycoprotein (MDR1) in human and rabbit cornea and corneal epithelial cell lines, *Invest. Ophthalmol. Visual Sci.*, 44, 2909–2918.
14. Dey, S., Gunda, S., and Mitra, A.K. (2004), Pharmacokinetics of erythromycin in Rabbit corneas after single-dose infusion: role of P-glycoprotein as a barrier to *in vivo* ocular drug absorption, *J. Pharmacol. Exp. Therapeut.*, 311, 246–255.
15. Katragadda, S., Talluri, R.S., and Mitra, A.K. (2006), Modulation of P-glycoprotein-mediated efflux by prodrug derivatization: an approach involving peptide transporter-mediated influx across rabbit cornea, *J. Ocul. Pharmacol. Ther.*, 22, 110–120.
16. Livermore, D.M. (2003), Bacterial resistance: origins, epidemiology, and impact, *Clin. Infect. Dis.*, 36(Suppl 1), S11–S23.
17. Barker, K.F. (1999), Antibiotic resistance: a current perspective, *Br. J. Clin. Pharmacol.*, 48, 109–124.
18. Vallas, V., Stapleton, F., and Willcox, M.D. (1999), Bacterial invasion of corneal epithelial cells, *Aust. New Zeal. J. Ophthalmol.*, 27, 228–230.
19. Tabbara, K.F., El-Sheikh, H.F., and Aabed, B. (2000), Extended wear contact lens related bacterial keratitis, *Br. J. Ophthalmol.*, 84, 327–328.
20. Lauffenburger, M.D. and Cohen, K.L. (1993), Topical ciprofloxacin versus topical fortified antibiotics in rabbit models of *Staphylococcus* and *Pseudomonas keratitis*, *Cornea*, 12, 517–521.
21. Lin, C.P. and Boehnke, M. (2000), Effect of fortified antibiotic solutions on corneal epithelial wound healing, *Cornea*, 19, 204–206.
22. Mah, F.S. (2004), Fourth-generation fluoroquinolones: new topical agents in the war on ocular bacterial infections, *Curr. Opin. Ophthalmol.*, 15, 316–320.
23. Prajna, N.V., George, C., Selvaraj, S., Lu, K.L., McDonnell, P.J., and Srinivasan, M. (2001), Bacteriologic and clinical efficacy of ofloxacin 0.3% versus ciprofloxacin 0.3% ophthalmic solutions in the treatment of patients with culture-positive bacterial keratitis, *Cornea*, 20, 175–178.
24. Callegan, M.C., Engel, L.S., Clinch, T.E., Hill, J.M., Kaufman, H.E., and O'Callaghan, R.J. (1994), Efficacy of tobramycin drops applied to collagen shields for experimental *Staphylococcal keratitis*, *Curr. Eye Res.*, 13, 875–878.
25. Weiss, A., Brinser, J.H., and Nazar-Stewart, V. (1993), Acute conjunctivitis in childhood, *J. Pediat.*, 122, 10–14.

26. Morrow, G.L. and Abbott, R.L. (1998), Conjunctivitis, *Am. Fam. Physician*, 57, 735–746.
27. Mannis, M.J. and Sugar, J. (1995), Syphilis, serologic testing, and the setting of standards for eye banks, *Am. J. Ophthalmol.*, 119, 93–95.
28. Boustcha, E. and Nicolle, L.E. (1995), Conjunctivitis in a long-term care facility, *Infect. Control Hosp. Epidemiol.*, 16, 210–216.
29. Sommer, A. (2005), Systemic antibiotics for community-wide trachoma control, *Arch. Ophthalmol.*, 123, 687–688.
30. Tabbara, K.F., Abu-el-Asrar, A., al-Omar, O., Choudhury, A.H., and al-Faisal, Z. (1996), Single-dose azithromycin in the treatment of trachoma. A randomized, controlled study, *Ophthalmology*, 103, 842–846.
31. Schachter, J., West, S.K., Mabey, D., Dawson, C.R., Bobo, L., Bailey, R., Vitale, S., Quinn, T.C., Sheta, A., Sallam, S., Mkocha, H., Mabey, D., Faal, H. (1999), Azithromycin in control of trachoma, *Lancet*, 354, 630–635.
32. Igietseme, J., Eko, F., He, Q., Bandea, C., Lubitz, W., Garcia-Sastre, A., and Black, C. (2005), Delivery of Chlamydia vaccines, *Expert Opin. Drug Delivery*, 2, 549–562.
33. Field, D. and Merrick, E. (2006), Postoperative endophthalmitis: caution is the watchword, *J. Perioperat. Pract.*, 16, 16–20.
34. Elston, R.A. and Chattopadhyay, B. (1991), Postoperative endophthalmitis, *J. Hospit. Infect.*, 17, 243–253.
35. Das, T. and Sharma, S. (2003), Current management strategies of acute post-operative endophthalmitis, *Semin. Ophthalmol.*, 18, 109–115.
36. Zegans, M.E., Shanks, R.M., and O'Toole, G.A. (2005), Bacterial biofilms and ocular infections, *Ocul. Surf.*, 3, 73–80.
37. Lloyd, A.W., Faragher, R.G.A., and Denyer, S.P. (2001), Ocular biomaterials and implants, *Biomaterials*, 22, 769–785.
38. Woodcock, M., Shah, S., and Smith, R.J. (2004), Recent advances in customizing cataract surgery, *Br. J. Med.*, 328, 92–96.
39. Slade, Stephen. "Accommodating IOLs: Design, Technique, Results". Rev. Ophthalmol.. 2005. 20 March 2009. (Available at http://www.revophth.com/index.asp?page=1_751.htm.)
40. "Crystalens Accommodating IOL." USA Eyes. 2008. Council of Refractive Surgery Quality Assurance. 20 March 2009. (Available at http://www.usaeyes.org/lasik/faq/crystalens-2.htm.)
41. Segre, Liz. "Intraocular Lenses (IOLs): New Advances Including AcrySof ReStor, Tecnis, ReZoom, and Crystalens." All About Vision. 2009. Access Media Group LLC. 20 March 2009. (Available at http://www.allaboutvision.com/conditions/iol.htm.)
42. Wejde, G., Montan, P., Lundström, M., Stenevi, U., and Thorburn W. (2005), Endophthalmitis following cataract surgery in Sweden: national prospective survey 1999–2001, *Acta Ophthalmol. Scand.*, 83, 7–10.
43. Kresloff, M.S., Castellarin, A.A., and Zarbin, M.A. (1998), Endophthalmitis, *Surv. Ophthalmol.*, 42, 193–224.
44. Ng, E.W.M., Barrett, G.D., and Bowman, R. (1996), *In vitro* bacterial adherence to hydrogel and poly(methyl methacrylate) intraocular lenses, *J. Cataract Refract. Surg.*, 22, 1331–1335.

45. Jansen, B. and Peters, G. (1991), Modern strategies in the prevention of polymer-associated infections, *J. Hosp. Infect.*, 19, 83–88.
46. Kodjikian, L., Burillon, C., Lina, G., Roques, C., Pellon, G., Freney, J., and Renaud, F.N. (2003), Biofilm formation on intraocular lenses by a clinical strain encoding the ica locus: a scanning electron microscopy study. *Invest. Ophthalmol. Vis. Sci.*, 44, 4382–4387.
47. Burillon, C., Kodjikian, L., Pellon, G., Martra, A., Freney, J., and Renaud, F.N. (2002), *In vitro* study of bacterial adherence to different types of intraocular lenses, *Drug Dev. Ind. Pharm.*, 28, 95–99.
48. Griffiths, P.G., Elliot, T.S.J., and McTaggart, L. (1989), Adherence of *Staphylococcus epidermidis* to intraocular lenses, *Br. J. Ophthalmol.*, 73, 402–406.
49. Cramton, S.E. and Götz, F. (2004), Biofilm development in staphylococcus. in: Ghannoum, M. and O'Toole, G.A., Eds., *Microbial Biofilms*, ASM Press, Washington, DC, pp. 64–84.
50. An, Y.H. and Friedman, R.J. (1998), Concise review of mechanisms of bacterial adhesion to biomaterial surfaces, *J. Biomed. Mater. Res.*, 43, 338–348.
51. Katsikogianni, M. and Missirlis, Y.F., Concise review of mechanisms of bacterial adhesion to biomaterials and of techniques used in estimating bacteria-material interactions, (Available at, http://www.ecmjournal.org/journal/papers/vol008/pdf/v008a05.pdf.) Accessed on March 16, 2009.
52. Cusumano, A., Busin, M., and Spitznas, M. (1994), Bacterial growth is significantly enhanced on foldable intraocular lenses, *Arch. Ophthalmol.*, 112, 1015–1016.
53. Gabriel, M.M., Ahearn, D.G., Chan, K.Y., and Patel, A.S. (1998), *In vitro* adherence of *Pseudomonas aeruginosa* to four intraocular lenses, *J. Cataract Refract. Surg.*, 24, 124–129.
54. Pinna, A., Zanetti, S., Sechi, L.A., Usai, D., Falchi, M.P., and Carta, F. (2000), *In vitro* adherence of *Staphylococcus epidermidis* to polymethyl methacrylate and ACRYSOF intraocular lenses, *Ophthalmology*, 107, 1042–1046.
55. Garcia-Saenz, M.C., Arias-Puente, A., Fresnadillo-Martinez, M.J., and Matilla-Rodriguez, A. (2000), *In vitro* adhesion of *Staphylococcus epidermidis* to intraocular lenses, *J. Cataract Refract. Surg.*, 26, 1673–1679.
56. Abu el-Asrar, A.M., Shibl, A.M., Tabbara, K.F., and al-Kharashi, S.A. (1997), Heparin and heparin-surface-modification reduce *Staphylococcus epidermidis* adhesion to intraocular lenses, *Int. Ophthalmol.*, 21, 71–74.
57. Portoles, M., Refojo, M.F., and Leong, F.L. (1993), Reduced bacterial adhesion to heparin-surface-modified intraocular lenses, *J. Cataract Refract. Surg.*, 19, 755–759.
58. Arciola, C.R., Caramazza, R., and Pizzoferrato, A. (1994), *In vitro* adhesion of *Staphylococcus epidermidis* on heparin-surfacemodified intraocular lenses, *J. Cataract Refract. Surg.*, 20, 158–161.
59. Prosdocimo, G., Grandesso, S., and Amici, G. (1997), Influence of optic and haptic materials on the adherence of *Staphylococcus epidermidis* to intraocular lenses: a pilot study, *Eur. J. Ophthalmol.*, 7, 241–244.
60. Baillif, S., Casoli, E., Marion, K., Roques, C., Pellon, G., Hartmann, D.J., Freney, J., Burillon, C., and Kodjikian, L. (2006), A novel *in vitro* model to study staphylococcal biofilm formation on intraocular lenses under hydrodynamic conditions, *Invest. Ophthalmol. Vis. Sci.*, 47, 3410–3416.

61. Baillif, S., Ecochard, R., Casoli, E., Freney, J., Burillon, C., and Kodjikian, L. (2008), Adherence and kinetics of biofilm formation of *Staphylococcus epidermidis* to different types of intraocular lenses under dynamic flow conditions, *J. Cataract Refract. Surg.*, 34, 153–158.
62. Kodjikian, L., Burillon, C., Chanloy, C., Bostvironnois, V., Pellon, G., Mari, E., Freney, J., and Roger, T. (2002), *In vivo* study of bacterial adhesion to five types of intraocular lenses, *Invest. Ophthalmol. Vis. Sci.*, 43, 3717–3721.
63. Ruther, P. and Vincent, R. (1980), The adhesion of microorganisms to surfaces, physico-chemical aspects, in: Berkeley, R.C.W., Melling, J., Rutter, P.R., and Vincent, B., Eds., *Microbial. Adhesion to Surfaces*, Ellis Horwwod, London, pp. 79–91.
64. Magnusson, K.E. (1982), Hydrophobic interaction—a mechanism of bacterial binding, *Scand. J. Infect. Dis. Suppl.*, 33, 32–36.
65. Pascual, A., Fleer, A., Westerdaal, N.A., and Verhoef, J. (1986), Modulation of adherence of coagulase-negative staphylococci to Teflon catheters *in vitro*, *Eur. J. Clin. Microbiol.*, 5, 518–522.
66. Paulsson, M., Gouda, I., Larm, O., and Ljungh, A. (1994), Adherence of coagulase-negative staphylococci to heparin and other glycosaminoglycans immobilized on polymer surfaces, *J. Biomed. Mater. Res.*, 28, 311–317.
67. Rostand, K.S. and Esko, J.D. (1997), Microbial adherence to and invasion through proteoglycans, *Infect. Immunol.*, 65, 1–8.
68. Wong, T.Y. and Chee, S-P. (2004), Risk factors of acute endophthalmitis after cataract extraction: a case-control study in Asian eyes, *Br. J. Ophthalmol.*, 88, 29–31.
69. ESCRS Endophthalmitis Study Group. (2007), Prophylaxis of postoperative endophthalmitis following cataract surgery: results of the ESCRS multicenter study and identification of risk factors, *J. Cataract Refract. Surg.*, 33, 978–988.
70. Dart, J.K.G. (1996), Contact lens and prosthesis infections, in: Tasman, W. and Jaeger, E.A., Eds., *Duane's Foundations of Clinical Ophthalmology*, Lippincott–Raven, Philadelphia, pp. 1–30.
71. Miller, M.J. and Ahearn, D.G. (1987), Adherence of *Pseudomonas aeruginosa* to hydrophilic contact lenses and other substrata, *J. Clin. Microbiol.*, 25, 1392–1397.
72. Stapleton, F. and Dart, J. (1995), *Pseudomonas* keratitis associated with biofilm formation on a disposable soft contact lens, *Br. J. Ophthalmol.*, 79, 864–865.
73. Stapleton, F., Dart, J.K., Matheson, M., and Woodward, E.G. (1993), Bacterial adherence and glycocalyx formation on unworn hydrogel lenses, *J. Br. Contact Lens Assoc.*, 16, 113–117.
74. Miller, M.J., Wilson, L.A., and Ahearn, D.J. (1988), Effects of protein, mucin, and human tears on adherence of *Pseudomonas aeruginosa* to hydrophilic contact lenses, *J. Clin. Microbiol.*, 26, 513–517.
75. McLaughlin-Borlace, L., Stapleton, F., Matheson, M., and Dart, J.K.G. (1998), Bacterial biofilm on contact lenses and lens storage cases in wearers with microbial keratitis, *J. Appl. Microbiol.*, 84, 827–838.
76. Wilson, L.A., Sawant, A.D., and Ahearn, D.G. (1991), Comparative efficacies of soft contact lens disinfectant solutions against microbial films in lens cases, *Arch. Ophthalmol.*, 109, 1155–1157.
77. Farber, B.F., Hsi-Chia, H., Donnenfield, E.D., Perry, H.D., Epstein, A., and Wolff, A. (1995), A novel antibiofilm technology for contact lens solutions, *Ophthalmology*, 102, 831–836.

78. Gandhi, P.A., Sawant, A.D., Wilson, L.A., and Ahearn, D.G. (1993), Adaptation and growth of *Serratia marcescens* in contact lens disinfectant solutions containing chlorhexidine gluconate, *Appl. Environ. Microbiol.*, 59, 183–188.

79. Manocha, S., Feinstein, D., Kumar, A., and Kumar, A. (2002), Novel therapies for sepsis: antiendotoxin therapies, *Expert Opin. Investig. Drugs*, 11, 1795–1812.

80. Holzheimer, R.G. (2001), Antibiotic induced endotoxin release and clinical sepsis: a review, *J. Chemother.*, 13, 159–172.

81. Melki, S.A. and Azar, D.T. (2001), LASIK complications: etiology, management, and prevention, *Surv. Ophthalmol.*, 46, 95–116.

82. Holland, S.P., Mathias, R.G., Morck, D.W., Chiu, J., and Slade, S.G. (2000), Diffuse lamellar keratitis related to endotoxins released from sterilizer reservoir biofilms, *Ophthalmology*, 107, 1227–1233, discussion 1233–1224.

83. Mamalis, N. (2003), Diffuse lamellar keratitis, *J. Cataract. Refract. Surg.*, 29, 1849–1850.

84. Holzer, M.P., Solomon, K.D., Vargas, L.G., Sandoval, H.P., Kasper, T.J., Vroman, D.T., and Apple, D.J. (2002), Diffuse lamellar keratitis. Postoperative prophylactic treatment with corticosteroids in an experimental animal study, *Ophthalmologe*, 99, 849–853.

85. Reyes, M.P., Ganguly, S., Fowler, M., Brown, W.J., Gatmaitan, B.G., Friedman, C., and Lerner, A.M. (1980), Pyrogenic reactions after inadvertent infusion of endotoxin during cardiac catheterizations, *Ann. Intern. Med.*, 93, 32–35.

86. Fulk, G.W., Davis, R.D., and Abbey, M.M. (1997), Endotoxin concentration in contact lens storage cases, *J. Am. Optom. Assoc.*, 68, 296–300.

87. Khatri, S., Lass, J.H., Heinzel, F.P., Petroll, W.M., Gomez, J., Diaconu, E., Kalsow, C.M., and Pearlman, E. (2002), Regulation of endotoxin induced keratitis by PECAM-1, MIP-2, and toll-like receptor 4, *Invest. Ophthalmol. Vis. Sci.*, 43, 2278–2284.

88. Vafeas, C.P., Mieyal, A., Urbano, F., Falck, J.R., Chauhan, K., Berman, M., and Schwartzman, M.L. (1998), Hypoxia stimulates the synthesis of cytochrome P450-derived inflammatory eicosanoids in rabbit corneal epithelium, *J. Pharmacol. Exp. Ther.*, 87, 903–910.

89. Yi, X., Wang, Y., and Yu, F.S. (2000), Corneal epithelial tight junctions and their response to lipopolysaccharide challenge, *Invest. Ophthalmol. Vis. Sci.*, 41, 4093–4100.

90. Yokoi, N., Okada, K., Sugita, J., and Kinoshita, S. (2000), Acute conjunctivitis associated biofilm formation on a punctual plug, *Jpn. J. Ophthalmol.*, 44, 559–560.

91. Soukiasian, S. The microbial flora from explanted punctal plugs. OMIG Abstract. (Available at http://eyemicrobiology.upmc.com/2003Abstracts/2003Abs18.htm.)

92. Fulcher, T.P., Dart, J.K., McLaughlin-Borlace, L., Howes, R., Matheson, M., and Cree, I. (2001), Demonstration of biofilm in infectious crystalline keratopathy using ruthenium red and electron microscopy, *Ophthalmology*, 108, 1088–1092.

93. Mihara, E., Shimizu, M., Togue, C., and Inoue, Y. (2004), Case of a large, movable bacterial concretion with biofilm formation on the ocular surface, *Cornea*, 23, 513–515.

94. Holland, S.P., Pulido, J.S., Miller, D., Ellis, B., Alfonso, E., Scott, M., and Costerton, J.W. (1991), Biofilm and scleral buckle associated infections. A mechanism for persistence, *Ophthalmology*, 98, 933–938.
95. Asaria, R.H., Downie, J.A., McLaughlin-Borlace, L., Morlet, N., Munro, P., and Charteris, D.G. (1999), Biofilm on scleral explants with and without clinical Infection, *Retina*, 19, 447–450.
96. Costerton, J.W., Stewart, P.S., and Greenberg, E.P. (1999), Bacterial biofilms: a common cause of persistent infections, *Science*, 284, 1318–1322.

# CHAPTER 7

# BIOFILM-RELATED INFECTIONS IN THE ORAL CAVITY

## 7.1. INTRODUCTION

Most oral bacteria live symbiotically in biofilms. This symbiotic association gives the bacteria different communal properties than individual planktonic bacteria. Importantly, this difference is significant for periodontal infections, decay, and possible systemic correlations that reportedly are related to oral biofilm agents [1]. Dentists can play a significant role in controlling pathogenic microorganisms with both oral and systemic ramifications by understanding the interrelationships of the biofilm infectious pathology and how these biofilm organisms and their group products relate to systemic reactions. In addition, because dentists understand more about biofilm biology as it relates to human infections, they are also better able to realize how these associations affect healthy and compromised immune systems.

Management of oral biofilms allows dentists to help control the pathogens responsible for periodontal disease and decay. Increasing evidence indicates that the oral system is a portal for pathogenic microorganisms. This situation is cumulative with systemic effects that can overcome an individual's resistance threshold, culminating in systemic sequela [2,3]. New evidence indicates that controlling these oral pathogens has systemic benefits, as oral pathology is related to cardiovascular and respiratory diseases, diabetes, and systemic inflammatory responses, as well as low birth weight and preterm deliveries.

*Biofilm Eradication and Prevention: A Pharmaceutical Approach to Medical Device Infections*,
By Tamilvanan Shunmugaperumal

Note that some insurance companies now cover periodontal scaling for gingivitis and periodontal disease for pregnant women and patients at risk for pregnancy [4].

## 7.2. BIOFILM FORMATION IN ORAL CAVITY

When a healthy tooth is cleaned thoroughly, it is coated rapidly by a complex mixture that includes glycoproteins, acidic praline-rich proteins, mucins, bacterial debris, exoproducts, and sialic acids. There are three steps by which oral biofilm forms on this surface and eventually attaches to tooth surfaces. First, a conditioning film or pellicle is deposited on the tooth enamel. Next, primary colonizers experience cell-to-surface attachment. Finally, cell–cell interactions occur between the mid-to-late colonizers and the early colonizers. Maturation of the biofilm requires a cell–cell interaction called coaggregation, which Whitaker et al. [5] defined as "the recognition and adhesion between genetically distinct bacteria". Glucan-like molecules on the surfaces of bacteria recognize a series of lipoproteins (lipoprotein receptor antigen I family), proteins, and extracellular components with which they may coaggregate [6].

## 7.3. BIOFILM-RELATED DISEASES IN ORAL CAVITY

It has been estimated that 36.8% (43 million) of adult Americans have periodontal disease, making it one of the world's most prevalent chronic diseases [7]. Periodontal disease, which encompasses gingivitis and periodontitis, results from plaque-associated bacterial infections, which cause inflammation in the surrounding tissue. Dental caries (dental decay) is a destructive condition of the dental hard tissues that, if unchecked, can progress to inflammation and death of vital pulp tissue, with eventual spread of infection to the periapical area of the tooth and beyond. Plaque that becomes mineralized with calcium and phosphate ions is termed calculus or tartar. The disease process involves acidogenic plaque bacteria, including *Streptococcus mutans*, *Streptococcus sobrinus*, and *Lactobacillus* spp. [8], whereas periodontal diseases can involve both the soft and hard tissues and are the most common inflammatory destructive conditions that affect humans. They are initiated by components of the plaque that develops on the hard root surface adjacent to the soft tissues of the supporting periodontium and may be confined to the gingiva (gingivitis) or extend to the deeper supporting structures with destruction of the periodontal ligament and the alveolar bone that supports the teeth (periodontitis). Such loss of attachment, with associated periodontal pocket formation, may ultimately lead to loosening and loss of the affected teeth. *Porphyromonas gingivalis*, *Prevotella intermedia*, and *Aggregatibacter actinomycetemcomitans* are regarded as the major pathogens in advancing periodontitis [9].

Within the subgingival crevice, the primary source of nutrients for the developing biofilm is gingival crevice fluid, a serum exudate that bathes the gingival crevice. This fluid provides proteins, glycoproteins, and other nutrients. Bacterial nutrients may also originate from saliva and the host diet (especially fermentable carbohydrates) [10]. Though there is a constant flow of air through the oral cavity, the tooth surface rapidly becomes anaerobic on colonization with microorganisms. Marsh [10] noted that redox potential ($E_h$) fell from >+200 to −30 mV within 2 days of colonization and to −150 mV after 7 days. The $E_h$ of the gingival crevice is usually lower than that of other sites around a healthy tooth. Bradshaw et al. [11] used a model system oral biofilm and demonstrated that anaerobes increased in proportion to aerobes with increasing biofilm age. They showed that mixed cultures can protect obligate anaerobes in the biofilms from the toxic effects of oxygen.

## 7.4. BIOFILM IN ORAL CAVITY AND SYSTEMIC INFLAMMATION

Oral bacteria have different effects on the host organism. Some live in a symbiotic commensal relationship, while others live in a sycophant association. Gram-negative anaerobic bacteria that live in deeper periodontal pockets produce endotoxins, biofilm products, peptides, polysaccharides, acids, and toxins, all of which are capable of evoking an inflammatory host response. These bacterial products can also affect neutrophils, host immune system responses, and individual cellular responses [12]. Furthermore, the bacterial invasion and host modification suggest a host immune activation with the formation of antigens, endotoxins and inflammatory mediators, all of which have an effect on cardiovascular disease, β-cell functions of the pancreas, atherosclerosis, stroke, and other systemic involvements.

The initial host response to this bacterial effect involves an infiltration of polymorphonuclear leukocytes, macrophages, and lymphocytes. These cells and other periodontal connective tissue cells synthesize and locally release Interleukin 1, 6, 8 (IL-1, IL-6, IL-8), tumor necrosis factor alpha (TNF-α), prostaglandin E (Pg E), and multiple matrix metalloproteinases (MMPs), excessive amounts of which are responsible for the disruption and destruction of tissue [12]. Both chewing and dental procedures have increased bacteremia. According to Geerts et al. [13], some patients with periodontal disease (compared with healthy controls) reported a fivefold increase in systemic endotoxemia as a result of chewing, which led the authors to conclude that localized periodontitis inflammation can spread into systemic circulation.

Because the surface area of the human periodontal ligament is 8.0–20 $cm^2$, it is likely that bacteria and byproducts enter through the ulcerated pocket epithelium into the systemic circulatory system. The increase in proinflammatory markers may be related to the direct release into the bloodstream of the polymorphonuclear (PMN) leukocyte components or monocyte and/or macrophage components or the formation of acute-phase proteins (C-reactive protein, serum Amyloid A, and fibrinogen) in the liver as a response to the bacterial challenge [14]. Directly or indirectly, the proinflammatory agents activate host enzymes IL-1, TNF-α, IL-6, and PgE, all of which influence the chemotaxis of PMNs. These agents also increase gingival blood vessel permeability, bone resorption, and tissue repair [15].

Inflammation is the initial response of the immune system to infection, irritation, or injury. Inflammation is characterized by redness, heat, swelling, pain, and organ dysfunction [16]. Research indicates that patients suffering with systemic inflammation have additional risks of systemic effects. A 2005 study revealed that patients with elevated systemic inflammatory markers were associated with a significantly higher risk of cardiovascular death, even when traditional cardiovascular risk factors were controlled [17]. It was proposed that systemic inflammation was the cause of the resultant atherosclerosis, although the authors stated that additional research was needed for a better understanding of the cause–effect relationship.

In 2004, Gan et al. [18] reported associations between chronic obstructive pulmonary disease (COPD) and systemic inflammation. The C-reactive protein levels increased in COPD patients, as did plasma fibrinogen levels and serum TNF-α. In addition, circulating leukocytes were higher in COPD patients than among control subjects. All statistical variances were at confidence levels of 95% or more [18]. Other research has determined that inflammatory cytokines increase as COPD exacerbations occur, suggesting a notable association between pulmonary and systemic inflammation [19].

The World Health Organization (WHO) has clearly delineated the route of infection from oral pathogens in the upper respiratory system [20]. In a 2003 study, Scannapieco et al. [20] examined how oral pathogens could participate in the pathogenesis of respiratory infection by reviewing five studies that investigated upper respiratory infections and oral treatments. All five studies treated the oral pathogens and all reported that the upper-respiratory infections decreased [20]. Upon entering the alveolus, either the biofilm bacteria or their reactions are able to modify the region to enable bacterial growth [21].

Systemic infections that are triggered by a septic or infectious event may play a role in acute lung injuries and acute respiratory distress syndrome. A 2002 study by Takala et al. [22] found that patients with acute lung injuries had significantly higher concentrations of serum IL-8, IL-6, and soluble IL-2 than control patients. Patients with an acute infection (e.g., pancreatitis) had higher serum inflammatory markers (IL-8, IL-6, IL-2) than control patients. This study demonstrated that acute infections and the host responses play a role in acute lung injuries [22].

Diabetic patients are known to have a higher incidence of periodontal disease. Several studies have concluded that the extent and magnitude of periodontal disease is increased in uncontrolled diabetics with elevated inflammatory markers [23]. In 2006, it was reported that poorly controlled diabetics demonstrated a higher collagenase level than either controlled diabetics or control subjects [24]. The high levels of collagenase are associated with degradation of the periodontal tissues.

Mahamed et al. [25] demonstrated that diabetic mice inoculated with *Actinobacillus actinomycetemcomitans* exhibited significantly more alveolar bone loss when compared to both preinoculated and control mice. Kesavalu et al. [26] reported that mice who received subcutaneous inoculations with *P. gingivalis* and *A. actinomycetemcomitans* induced host changes in the formation of IL-1β, IL-6, and TNF-α. It appears that IL-1β caused β-cell changes that inhibited insulin secretion and stimulated nitric oxide synthase [26].

Recently, Steer et al. [27] found that IL-1 caused insulin secretion to decrease further by stimulating β-cell necrosis through the production of nitric oxide, which induces the β-cell death. This finding was confirmed by Arnush et al. [28], who found that cellular damage stimulated IL-1 release by islet cell macrophages, decreased insulin production, and stimulated nitric oxide synthase expression.

Treating periodontal disease has an effect on the diabetic markers. Schara et al. [29] reported on diabetic patients who received a full-month of disinfection; at 3 and 9 months, the patients had a significantly lower plaque index, less bleeding on probing, a reduction in probing depth, and gain of clinical attachment. There was also a significant reduction in the serum level of HbA1c at 3 months, but this reduction disappeared at 6 and 12 months, suggesting that diabetic patients must receive full-month disinfection every 3 months.

In 2001, Iwamoto et al. [30] reported that patients who were treated with local antimicrobial therapy once a week for 1 month experienced fewer bacteria in the pocket ($p > 0.01$), significantly reduced TNF-α levels ($p > 0.015$), and significantly reduced HbA1c values ($p > 0.007$). Kol and Palatella [31] found that topical application of doxycycline enhanced wound healing and decreased reparative response time in diabetic mice.

Other studies have investigated the role of inflammation in myocardial infarction (MI), atherosclerosis, and stroke. Systemic pathogens are suspected as a potential trigger for atherosclerotic plaque inflammation and have been associated with cerebrovascular symptomatology in patients with carotid disease [32]. Several inflammatory mediators are involved in the atherosclerotic lesion. The cytokine TNF-α inflammatory mediator induces the expression of MMP, which is responsible for protein breakdown. The MMPs attack the cap of the plaque and may cause it to rupture or form a thrombosis, which may lead to MI and strokes [32].

Researchers have identified genetic transformations that enable the oral pathogen *P. gingivalis* to invade and infect human arterial cells [33]. Kuramitsu et al. [34] discovered *Porphyromonas gingivalis* was able to increase endothelial cell expression of monocyte chemoattractant protein-1. *Porphyromonas gingivalis* was able to increase monocyte recruitment and intercellular adhesion molecule 1 (ICAM-1) through an increased expression of monocyte chemoattractant protein-1. The increased ICAM-1 facilitates attachment of monocytes to endothelial cells. It also increases the endothelial cellular production of elastase or gelatinase (the forms of an MMP), which fosters plaque rupture [34]. Additional factors [e.g., heat shock protein 60 (HSP60) from bacterial and host cells] mediate endothelial cell inflammatory expression of intercellular and vascular cell adhesion molecules, both of which activate the secretion of monocyte–macrophage, IL-6, and TNF-α [35]. A 2002 study by Yamazaki et al. [35] helped to explain the periodontal–cardiovascular link by noting that atherosclerosis patients showed the highest antibody levels of host and *P. gingivalis* HSP60.

Certain bacteria (e.g., *A. actinomycetemcomitants* and *P. gingivalis*) have been implicated in host-related infections. According to a 2000 study by Haraszthy et al. [36], 80% of endoarterectomy specimens tested positive for periodontal pathogens, while more than 59% had two or more periodontal pathogens. *Porphyromonas gingivalis* is capable of inducing low-density lipoprotein (LDL) into a foam-cell formation. This step was accomplished by promoting LDL binding to macrophages and promoting macrophage modification of LDL to foam-cell formation, resulting in the pathogenesis of atherosclerosis. This pathogenesis was accomplished by the increasing IL-6 serum level in response to an increase in IL-1 and TNF factors to protect against tissue damage [34].

The IL-6 increases the synthesis of C-reactive protein, which in turn increases the rate of phagocytosis of bacteria. The IL-6 also increases the synthesis of fibrinogen while decreasing albumin and transferring levels. This acute phase reaction increases fever, erythrocyte sedimentation rate, and secretion of glucocorticoids in addition to activating the complement and clotting cascade [37]. This reaction may help to explain platelet coagulation at arterial sites that contain *P. gingivalis* and *A. actinomycetemcomitans*.

A host of cardiovascular interrelationships that involve periodontal disease and systemic involvement have been reported in the literature. Coronary disease and strokes were more common among patients with seropositive antibody levels for *P. gingivalis* [38,39]. Desvarieux et al. [40] reported a direct relationship between the presence of five periodontal pathogens and the thickness of the tunica intima and tunica media in the carotid artery. Jain et al. [41] found a positive correlation between the severity of periodontal disease and the extent of aortic lipid deposits, reporting that aortic lipid deposits were induced in rabbits through a high-fat-content diet and induced periodontitis. Recently, Gibson et al. [42] reported on mice infected with *P. gingivalis* and noted that the mice exhibited an increased atherosclerotic plaque formation that was prevented by immunization against *P. gingivalis*. The authors found *P. gingivalis* in the blood of the animals and the aortic arch tissues, as well as increased atherosclerotic plaque. Those animals immunized for *P. gingivalis* did not incur *P. gingivalis*-accelerated atherosclerosis.

According to the literature, periodontal pathogens have been found in the human circulatory system. Kozarov et al. [43] found viable *A. actinomycetemcomitants* in the human carotid arteries. A 2003 study involving a DNA probe analysis of aneurysm repair tissue reported that 50 of 56 patients (90%) tested positive for bacterial DNA in the aneurysm tissue [44]. Recently, Fiehn et al.

[45] found periodontal pathogenic microbes in atherosclerotic plaques in patient's carotid and femoral arteries, although only patients with a history of MI showed elevated bacterial levels. The presence of these periodontal pathogenic microbes in the atherosclerotic arteries of the MI patients led the authors to suggest a relationship between the number of bacteria and systemic health (MI) issues [45]. This presence may relate to an increase in collagenase (MMP 1 and 13) and elastase–gelatinase (MMP 2, 9, and 12) in aortic aneurysm tissues. Here, the bacteria invade the artery wall and induce a localized inflammation with associated cytokines, which causes host cells to stimulate MMPs of elastin and connective tissues, leading to the aortic aneurysms.

Treating periodontal disease has a profound effect on patient's systemic inflammatory markers. D'Aiuto et al. [46] found that patient's C-reactive protein and IL-6 levels were reduced significantly after tetracycline products were placed in periodontal pockets. The authors concluded that the periodontal pathogens caused the systemic marker increase and that reducing the periodontal disease reduced the systemic markers. However, only 79% of the patients in this study experienced a decrease in inflammatory markers once the periodontal therapy was delivered. Taylor et al. [47] treated periodontal disease with full-mouth extractions and found the patient's C-reactive protein levels, plasminogen activator inhibitor-1, fibrinogen, and white and plasma cell counts all were reduced significantly. Rahman et al. [48] found that extracting teeth and replacing them with implants also reduced C-reactive protein levels, suggesting that treating periodontal disease modified the patient's C-reactive inflammatory markers significantly. Figure 7.1 shows how the bacterial products have a direct effect, producing cytokines and chemokines that have direct and indirect localized and systemic effects [49].

## 7.5. BIOFILM IN ORAL CAVITY AND IMMUNE SYSTEM RESPONSE

The host immune system is able to protect against most planktonic bacteria invasions through one of three mechanisms:

1. Phagocytosis of invading microorganisms by blood cells.
2. Proteolytic reactions leading to localized responses (clotting, opsonization).
3. Synthesis of antimicrobial peptides [50].

Bacteria that live in biofilms are remarkably resistant to host defenses and therapy with conventional antibiotics [51,52] because mixed biofilm populations differ from their planktonic counterparts in both genotypic diversity and phenotypic gene expression [53,54]. Polymorphonuclear (PMN) leukocytes are the first line of defense against infection [55]. Biofilms may have an adverse effect on PMN function. Jesatitis et al. [52] found that neutrophils that settle on biofilms lacked pseudopods, had impaired motility, and became enveloped

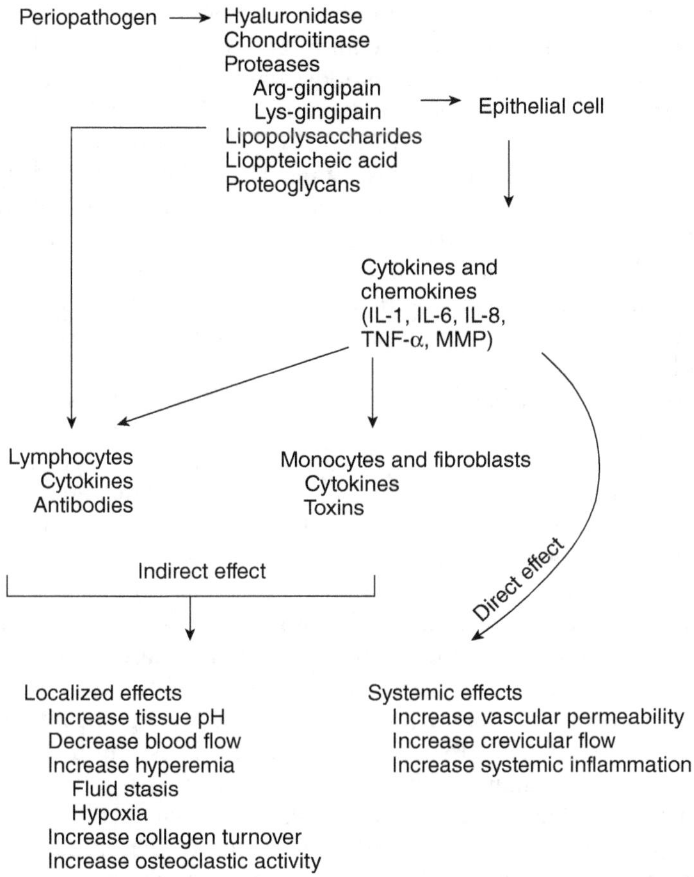


**Figure 7.1.** The bacterial products have a direct effect, producing cytokines and chemokines that have direct and indirect localized and systemic effects.

in the biofilm as planktonic bacteria were released by the biofilm. The PMN also were found to become degranulated with little increase in hydrogen peroxide ($H_2O_2$) production and a diminished oxidative potential [52]. Evidence by Ward et al. [56] illustrated that the immune system of a vaccinated rabbit had no effect on the growth of bacteria in biofilms implanted in the animal, which demonstrated that microorganisms are able to detach from biofilms, and could overcome the host immune system and cause an infection. Certain extracellular bacterial components have been found to interfere with host macrophage phagocytic activitiy [57]. Also, biofilm bacteria have been found unable to be eliminated by phagocytosis by opsonized antibodies in cystic fibrosis (CF) patients [58]. One study found that biofilm cells were less sensitive to death by human PMN cells, leading Yasuda et al. [59] to conclude that biofilm organisms are resistant to oxygen species produced by PMN, and detached biofilm cells may be able to evade the host phagocytic activity in the blood stream and initiate a bloodstream infection. Geerts et al. [13] reported that not only do bacteria enter the bloodstream during mastication, but also endotoxin levels increased in the bloodstream fourfold after mastication. They further found that endotoxin levels of severe periodontitis patients were greater than mild or moderate disease states [13]. Kinane et al. [60] reported on an increased incidence of bacteremia induced from conventional periodontal procedures. Forner et al. [61] stated that the crucial nature of periodontal treatment is the prevention of bacteremia associated with oral procedures, whereas Misra et al. [62] stipulated that there is an increased possibility of bacteremia being more frequent and affecting children with congenital heart defects and subsequent endocarditis.

The host periodontal sulcus responds in predictive ways to associated plaque in localized periodontitis. Verderame et al. [63] discovered epithelial cavitations and ulcerations in response to the microorganisms entrapped in the region. The development of epithelial breakdown and ulcerations might help explain findings by Matheny et al. [64], who, while evaluating the microcirculatory dynamics associated with human gingivitis, discovered a significant increase in the number of blood vessels visible in microscopic fields. Kerdvongbundit et al. [65] evaluated inflammatory changes in the microcirculatory and micromorphologic dynamics of human gingiva before and after conventional treatment (scaling and root planing). Blood flow measured with laser doppler flowmetry demonstrated a statistically significant blood flow increase when the gingival tissues were inflamed. These returned to normal after treatment and remained stable for 3 months post-treatment. Cimasoni of the University of Geneva, School of Dentistry, Department of Periodontics, demonstrated the close proximity and relationship of the bloodstream to the periodontal pocket (unpublished note). The increased exposure of the bloodstream during infection helps explain how periopathogenic cells might become involved systemically or have systemic effects on the host immune system as the host responds to these pathogens.

Research is showing that certain of these pathogens are able to invade human cells [66], thus making pathogen recognition and control more difficult [67]. This cellular invasion and difficulty in recognition places patients with a medical or immune compromise at greater risk from biofilm pathogens [68]. Lü and Jacobson [69] demonstrated that patients with immunodeficiencies, therefore, are more susceptible to infection and experience a greater degree of infection than patients with competent immune systems. Primary immunodeficiencies include humor immunities (affecting B-cell differentiation or antibody production), T-cell defects, combined B- and T-cell defects, phagocytic disorders, and complement deficiencies. These multiple disorders of the host immune systems often involve multiple infections with unusual or opportunistic organisms. These infections occur despite aggressive treatments, with the host experiencing a failure to thrive or grow, which often is associated with a positive family history [70].

## 7.6. NORMAL HOST RESPONSES VERSUS BIOFILM INFECTION

Patients with a normal immune system are able to counter most planktonic bacterial infections on their mucosal surfaces [71]. Figure 7.2 shows the mononuclear phagocyte system in conjunction with various subsets of tissue macrophages that are present in the human body.

Bone marrow is the ultimate source of blood cells, including those destined to become immune cells. The lymphocytes comprise a majority of the immune cells that originate from stem cells and are comprised of T cells, which mature in the thymus and B cells [72]. Lymphocytes travel via the bloodstream and also through the lymphatic vessels as fluid and cells are exchanged between these two systems [73]. Within the lymphatic system are small lymph nodes that contain specialized compartments where immune cells encounter foreign particles [74]. To work effectively, the cells of the immune system must communicate either by physical contact or by chemical messengers (e.g., cytokines) [75].

The main type of lymphocytes are the B and T cells. The B cells work chiefly by fabricating and secreting antibodies in response to specific antigens. These antibodies attach to the antigens and mark the antigen for destruction [76]. Antibodies belong to a large group of molecules known as immunoglobulins, which play different roles in immune system function. Immunoglobulin G (IgG) works to coat microbes, accelerating their recognition and uptake by other cells in the immune system [77]. Immunoglobulin M (IgM) is effective in killing bacteria, whereas immunoglobulin A (IgA) functions via secretions in the digestive tract, tears, and saliva to help guard against entry infections [78]. These and a host of other immunoglobulins are generally effective in

Promonocyte (bone marrow)

↓

Monocyte (blood)

↓

Macrophages (tissues) highly phagocytic

| | |
|---|---|
| Connective tissue | (histiocyte) |
| Liver | (Kupffer cell) |
| Lung | (alveolar macrophage) |
| Spleen | (free and fixed macrophages, sinusoidal lining cell) |
| Lymph node | (free and fixed macrophage) |
| Bone marrow | (macrophages, sinusoidal lining cell) |
| Serous cavity | (peritoneal macrophage) |
| Bone tissue | (osteoclast) |
| Nervous system | (microglia) |

**Figure 7.2.** Mononuclear phagocyte system.

helping manage the host challenge to bacterial infections. According to the National Cancer Institute, in patients with normal immune system antibodies, the antibodies bind to and inactivate bacterial toxins. The antibodies bind to the antigen an make it recognizable to phagocytic cells (opsonization), activating the complement cascade, blocking the antigen from cell invasion, and binding to the cell, which makes it possible for killer immune cells to destroy the pathogens [79]. Results of activation of the complement cascade include stimulating mast cells and basophiles to release granulocytic chemicals, neutrophil attractants and opsonizing compounds, and to generate membrane attractant complexes [C1q, C3, C4, C5, C5–C9(MAC)], factor B and Fb, factor H, and properdine, some of which serve to break down pathogen membranes [80].

Unlike B cells, T cells do not recognize antigens, but their surfaces contain antibody-like receptors that are involved in immune response. Memory T cells are required to maintain immunity, while regulatory T cells help keep the immune system in check to prevent inflammation autoimmunity [81]. Killer T cells directly attack foreign cells, using molecules on their surface to recognize small fragments (antigens) and launch an attack to kill the foreign cell [82]. These surface receptors [major histocompatibility complex (MHC)] molecules are proteins used for nonself-recognition. These MHC proteins present significant problems with transplants as almost all cells are covered with MHC proteins and the donor–recipient pattern must be a close match [83]. Natural killer (NK) cells are armed with granules filled with potent chemicals that are attracted to cells lacking self-MHC molecules and attach to other types of foreign cells. These NK cells have the potential to bind to many types of foreign cells, and then deliver their chemical barrage to kill the pathogens [84].

The T-cell cytokines help to regulate monocyte–macrophage function [85]. Monocytes are phagocytic cells that circulate in the blood and appear to respond to specific cytokines (IL-10), which cause the cells to migrate into tissues, where they develop into macrophages [86]. Macrophages, in response to a host of signals, scavenge and rid the body of worn out cells (PMN) and other debris [87]. The National Cancer Institute has stated that granulocytes that include basophils produce chemicals (e.g., histamines) and are able to destroy planktonic bacteria. However, granulocytes also contribute to inflammation and some allergic reactions. These same authors stated that eosinophils release granulocytic chemicals into surrounding tissues to destroy pathogens [79], while neutrophils are phagocytic cells that are often the first line of defense, as these cells respond to cytokines as well as produce cytokines and ingest planktonic bacteria. They also use a series of enzymes and $H_2O_2$ (superoxide) to kill the ingested pathogens [88]. One of the products produced in a PMN is lactoferrin, which is a multifunctional, antimicrobial (bactericidal) protein. Lactoferrin also is produced on acinar cells of the pancreas, stomach, salivary gland, and other organs [89].

For an immune system to function in a normal manner, its components must be able to create and exchange proteinaceous cytokines, which enable the system to communicate. Some cytokines activate certain immune cells, whereas others turn off specific cells. When stimulated by infection, T helper cells fabricate IL-2, which serves to increase the number of infection-fighting cells and causes them to mature [90]. Specific cytokines attract specific cells, whereas injured cells also produce chemokines that attract or stimulate immune cells and/or are factors in inflammation and the regulation of immune responses [91]. Complement proteins are free-circulating inactive agents that serve to complement antibody-coating antigen complexes. When activated (typically by an antibody), a chain reaction of complements occurs with the end result usually puncturing a hole in the pathogen cell wall, thus increasing the pathogen susceptibility to phagocytosis and/or activation of attractants for phagocytosis [92].

Biofilms complicate the immune system recognition and control. Leid et al. [93] demonstrated that the exopolysaccharide matrix of *Pseudomonas aeruginosa* caused a biofilm to be refractory to the host immune system, but cells unable to form the biofilm matrix were susceptible to the host immune system. Wagner et al. [94] found that when PMN were exposed to biofilms, they experienced a significant alteration of function as they underwent transdifferentiation. These authors found that PMN lost the ability to respond to CD62L and the upregulation of CD14, as well as the expression of CD83, which resulted from the effects of the infection on the host cells. This finding helped to explain how biofilm-affected PMN lost their chemotactic activity, while the production of superoxide and opsonized chemicals (cytotoxic, proteolytic, and collagenolytic) potentially continued and was enhanced. This combination provides a possible explanation of how the affected PMN might contribute to tissue destruction and eventually to tissue lyses.

In conclusion, hypothesis involving periodontal pathology and systemic involvement have much research support, but remain unproven. However, dentists must be aware of these possibilities and should take corrective actions for long-term treatment of periodontal disease that will decrease pathogenic bacteria and provide a homecare system for the patient to maintain oral and oral-systemic health. Treating periodontal disease can lead to improvements in systemic inflammatory markers, as well as a decrease in upper respiratory infections and other systemic effects, suggesting that dentists may be able to assist patients in improving their systemic conditions. Moreover, biofilm disease affects the host immune system in a variety of ways. Various biofilm components cause different immune system responses. Treatment of the biofilm is an integral aspect to decrease the immune system responses, along with decreasing the host systemic inflammation.

## 7.7. PREVENTION STRATEGIES OF PERIODONTAL DISEASES

The prevention of dental caries and the periodontal diseases is traditionally targeted at the mechanical or nonspecific control of dental plaque, as this is the precipitating factor. However, the individual response of the host and other confounding factors can influence disease initiation and progression. Like the other chronic infections (e.g., osteomyelitis secondary to prosthesis-related orthopedic infections), which are optimally treated by debridement and this type of removal of affected tissue and associated biofilms is the standard of care for periodontitis as well. Therefore, common treatments for periodontal disease aim to cure inflamed tissue, reduce bacteria, and eliminate the periodontal pocket. Scaling (removal of calculus and plaque), root planing (removal of necrotic tooth tissue on root surface), and surgery (to remove tissue and reduce pocket depth) have been used in the mechanical treatment of periodontal diseases. However, these procedures are time consuming and demanding on the patient. Recent therapies for treating periodontitis have incorporated various antibiotic and antimicrobial agents. The control of periodontitis is rooted in the removal of established biofilms (plaque) from the subgingival areas, in combination with supplemental antimicrobial agents. Quirynen et al. [95] found that chlorhexidine rinses after mechanical cleaning significantly improved gum health, as measured by a reduction in probing depth of the gingival crevice. Kinniment et al. [96] found that pathogens (e.g., *P. gingivalis* and *Fusabacterium nucleatum*) were inhibited within laboratory oral biofilms by treatment with chlorhexidine, in support of the findings by Quirynen.

Reynolds et al. [97] found that subgingival irrigation with chlorhexidine during ultrasonic scaling provided a significant improvement in probing depth compared to that of the untreated control group. Jeong et al. [98] found that root planing plus a mixture of tetracycline and citric acid containing gel was most effective in decreasing pocket depth. In this case, the root planing consisted of mechanically removing plaque and calculus from the exposed root surfaces. Citric acid acted as a chelating agent to remove mineral deposits on the root surfaces.

Due to the inadequacies of both peroral administration of antimicrobial agents and the use of antibacterial mouthwashes [99,100], recent treatments have focused on the use of controlled release intrapocket antimicrobial drug delivery systems. Examples of these include films [101–103], gels [7,104,105], and semisolids [106–109].

In the development of implantable drug delivery systems for the treatment of periodontal diseases, several key physicochemical properties may be defined. These include, ease of administration into and prolonged retention within the periodontal pocket, controlled release of antimicrobial agent into the crevicular fluid and biodegradation–bioerosion, the latter property facilitating removal of the delivery system and reattachment of the gingiva [106,109]. However, the design of currently available systems is suboptimal. In a publication, the physicochemical properties of gels composed of a novel polymeric complex between poly(methylvinylether-*co*-maleic anhydride) and polyvinylpyrrolidone were described [110]. In particular, several of these systems exhibited rheological properties that were deemed suitable as platforms for topical drug delivery systems. Therefore, in a subsequent paper, Jones et al. [111] examined the physicochemical properties of gels composed of poly(methylvinylether-*co*-maleic anhydride) and polyvinylpyrrolidone and containing tetracycline, designed for the treatment of periodontal disease. Tetracyclines are commonly used for the treatment of periodontal disease and in this study tetracycline was chosen as a representative member of this class of antimicrobial agent [7,99]. In particular, the textural (mechanical) and flow and oscillatory rheological properties and release of tetracycline from these systems are described, due to the applicability of these properties to the clinical and nonclinical performance of periodontal drug delivery systems [99,108,109]. Therefore, antimicrobial approaches, including the use of antimicrobial agents, represent a valuable complement to mechanical plaque control. Such strategies should ideally prevent plaque biofilm formation without affecting the biological equilibrium within the oral cavity, which is inhabited by up to 1000 different species of bacteria at $10^8$–$10^9$ bacteria $mL^{-1}$ saliva or $mg^{-1}$ dental plaque [112].

## 7.8. NOVEL ANTIMICROBIAL THERAPIES FOR DENTAL PLAQUE-RELATED DISEASES

Table 7.1 lists novel strategies developed so far to control oral infection–biofilm. Control of dental plaque-related diseases has traditionally relied on nonspecific removal of plaque by mechanical means (summarized from the review of Allaker and Douglas [113]). Maintenance of oral hygiene often includes use of chemical agents. However, increasing problems of resistance to synthetic antimicrobials have encouraged the search for alternative natural products. Plants are the source of >25% of prescription and over-the-counter preparations, and the potential of natural agents for oral prophylaxis will therefore be considered. Targeted approaches may be directed at the black-pigmented anaerobes associated with periodontitis. Such pigments provide an opportunity for targeted phototherapy with high-intensity monochromatic light. Studies to date have demonstrated selective killing of *P. gingivalis* and *P. intermedia* in biofilms. Functional inhibition approaches, including the use of protease inhibitors, are also being explored to control periodontitis. Replacement therapy, by which a resident pathogen is replaced with a nonpathogenic bacteriocin-producing variant, is currently under development with respect to *S. mutans* and dental caries. Selective light-activated killing, functional inhibition of specific virulence factors, and microbial replacement therapy offer a more targeted approach, whereas a plant-based product may well have a more general disruptive effect on the oral microbiota. However, there is a growing acceptance of natural therapies as complementary to mainstream healthcare.

This increasing consumer demand for effective and safe oral care products will help to further drive the need to investigate plants, including a systematic study of known medicinal plants, as potential sources of novel compounds to control oral infections.

**TABLE 7.1. Novel Strategies Developed So Far to Control Oral Infection–Biofilm**

| Novel Strategies | Examples |
|---|---|
| Plant-based therapy | Phytochemicals (e.g., simple phenols and phenolic acids, quinones, flavones, flavonoids and flavonols, tannins and coumarins, terpenoids–essential oils, alkaloids, and lectins–antimicrobial peptides). |
| Plant-derived substances | Extracts of miswak, tea tree oil, peppermint oil, green tea and manuka honey, eucalyptus, lavandula, sage, and rosmarinus oils,<br>Listerine™ (Essential Oils Rinse) contains the active ingredients thymol, eucalyptol, methyl salicylate and menthol and has been in widespread use for many years. Thymol and eucalyptol are antimicrobial, while methyl salicylate and menthol act as a cleaning agent and local anaesthetic, respectively.<br>Nigerian chewing stick (*Fagara zanthoxyloides*), |
| Oral microbiota modifiation therapy | 1. *Probiotics*<br>Probiotics as defined by the WHO are live microorganisms which, when administered in adequate amounts, confer a health benefit on the host. Introduction of microorganisms as a therapeutic tool for the prevention and treatment of dental caries and periodontal disease could possibly act in the following manner within the oral environment. (a) Direct interactions within dental plaque possibly include the disruption of plaque biofilm formation through competition for binding sites on host tissues and other bacteria, and competition for nutrients and (b) indirect probiotic actions within the oral cavity, including the modulation of aspects of both innate and specific immune function.<br>2. *Replacement therapy.*<br>3. *Phage therapy.*<br>4. *Prebiotics like* human milk contains oligosaccharides that have prebiotic characteristics. |
| Light-activated killing | *Selective killing using* metal-chelating groups like porphyrins<br>1. *Microbial proteases*<br>There are four main classes of proteases; (a) serine proteases (e.g., trypsin-like, elastase), (b) cysteine proteases (e.g., gingipains), (c) aspartic proteases (e.g., *Candida albicans* Saps), and (d) metalloproteases (e.g., microbial keratinases). |
| Functional inhibition | 2. *Protease inhibitors*<br>The main classes of inorganic or synthetic inhibitors are chelators (EDTA),[a] oxidizing agents, thiol-blocking agents, heavy metal ions, methane thiosulfonates, and organomercurials.<br>3. *Porphyromonas gingivalis gingipains*<br>There are two major gingipains, arginine-specific gingipain (RgpA and B) and a lysine-specific gingipain (Lgp). |

[a]Ethylene diamine tetraacetic acid = EDTA.

## 7.9. PERIODONTAL DEVICES

Therapeutic approaches for the treatment of periodontitis include mechanical or surgical methods and administration of systemic antibiotics. However, for systemic administration the drugs must be given in high doses to maintain an effective concentration in gingival crevicular fluid (GCF). High doses of antibiotics cause side effects (e.g., gastrointestinal disorders, development of resistant bacteria, and suprainfection). Systemic therapy has a low-benefit/high-risk ratio [114,115]. With advances in understanding of the etiology and pathogenesis of periodontal disease, attention has been focused on local drug delivery systems. Topical administration of antibacterial agents in the form of mouthwashes is ineffective in controlling disease progression since only a limited amount of drug actually accesses the periodontal pocket. Moreover, the drug is constantly flushed due to a very high fluid clearance rate (an estimated 40 replacements of the fluid an hour within a 5-mm pocket) [116]. Local drug delivery to the pocket in the form of subgingivally placed systems has numerous advantages. Periodontal diseases are localized in the immediate environment of the pocket, which is easily accessible for the insertion of a delivery device using a syringe or tweezers, depending on the physical form of the delivery system. The critical period of exposure of the pocket to the antibacterial drug is between 7 and 10 days [117]. Maintenance of a sustained high drug concentration can be achieved by correct planning, taking into account the high fluid clearance rate. Sustained release devices in the form of fibers, powders, strips, pastes, gels, and ointments have been reported. Some systems undergo a phase change from a liquid to an *in situ* forming solid. These systems have the advantage of syringeable delivery and an implant with good retention. Intrapocket delivery systems can be divided into degradable and nondegradable systems. These intrapocket delivery systems following their insertion into the periodontal pocket, release antimicrobial agents above the minimum inhibitory concentration for a sustained period of time. Thus intrapocket devices have a high-benefit/low-risk ratio [118]. However, nonbiodegradable systems must be removed or discharged from the pocket subsequent to the accomplishment of their drug release function.

## 7.10. NONDEGRADABLE DEVICES

Nondegradable cellulose acetate fibers loaded with tetracycline were first reported in 1983 by Goodson et al. [119]. Various studies on this system showed that it was unable to deliver sustainable levels of tetracycline [119,120], chlorhexidine [121], or metronidazole [122] to become clinically useful, although it eventually led to the development of the commercially available Actisite™ (Alza Corp. Palo Alto, CA) delivery system that is composed of a monolithic ethylene vinyl acetate fiber loaded with 25% tetracycline. The fiber can be placed around the circumference of the tooth to the depth of the pocket and folded upon itself to completely fill the pocket. The drug concentration in the pocket was found constant until the removal of the fiber 10 days after insertion [116]. This system's main disadvantages include a reported 23% risk of extrusion of the fiber from the pocket [123] and need for removal.

Nondegradable film or slab-based devices made from poly(methyl methacrylate) (PMMA) and ethylcellulose have been reported. The PMMA slabs have been formed by mixing various antibiotic agents (tetracycline, metronidazole, or chlorhexidine) during self-polymerization of PMMA similar to the procedure to make bone cements (see Section 11.1), only cured as sheets under high pressure and cut into the desired shape. *In vitro* studies showed that therapeutic levels of all three drugs may be achieved for a period of 2 weeks and that these depend on the nature of the drug and its initial concentration [124]. Clinical studies showed various degrees of efficacy, although they did not evolve to clinical use [125]. The second system is created by dissolving ethylcellulose and drug (chlorhexidine [126,127], metronidazole [128], or minocycline [129]) in either ethanol or chloroform followed by solvent evaporation and cutting of the films to shape. The most extensively studied systems, which contain chlorhexidine, have shown promising clinical results in the maintenance of periodontal pockets over a 2-year period [130].

## 7.11. DEGRADABLE DEVICES

Biodegradable systems are usually polymeric or protein in nature and undergo natural degradation following exposure to gingival fluid components. Various film-based devices have been described. The first degradable systems to be developed were based on hydoxypropylcellulose loaded with various agents: tetracycline, chlorhexidime, and ofloxacin. Similar to other degradable applications described in the literature, fast release of the drug occurs from the film within 2 h, followed by maintenance of tetracycline within the pocket for 24 h after insertion. Several modifications have been made to address the rapid degradation and short duration of drug release. For example, incorporation of methacrylic acid copolymer particles into the film has been reported to prolong the release of ofloxacin *in vitro* and *in vivo* for 7 days [130,131].

Polyhydroxybuteric films loaded with 25% tetracycline or metronidazole have been used clinically. These films demonstrated an improvement in clinical and microbiological parameters, although they suffer from rapid degradation in their mechanical properties. Therefore, they required several consecutive placements every 4 days during the trial [132].

A degradable device based on hydrolyzed gelatin cross-linked by formaldehyde as reported by Steinberg et al. [133] has evolved into the commercial Perio-chip™ (Perio Products Ltd., Jerusalem, Israel). A different commercially available system, Elysol (Dumex, Copenhagen, Denmark), is based on a water-free mixture of melted glycerol monooleate and metronidazole to which sesame oil was added to improve its flow properties in the syringe. The gel flows deep into the periodontal pocket and readily adapts to root morphology. When it comes in contact with water it sets in a liquid crystalline state. The matrix is degraded as a result of neutrophil and bacterial activity within the pocket [134]. Effective doses of metronidazole within the pocket are maintained for 24–36 h. Another antibiotic gel, Atridox™ (Block Drug Corporation Inc., Jersey City, NJ) has a solution formulation that is composed of two separate syringes that are coupled together. One syringe contains 8.5% w/w doxycycline hyclate and the other 37% w/w poly(D,L-lactide) (PDLLA). They are dissolved in a biocompatible carrier of 63% w/w *N*-methyl-2-pyrrolidone, which quickly hardens into a waxlike substance upon contact with the cervicular fluid. The system slowly releases doxycycline into the surrounding tissue for 7 days. This system has been approved by the FDA.

In one study of tetracycline incorporation into halloysite for the treatment of periodontitis, an initial burst of tetracycline release was followed by a dramatic reduction in release. Coating with the cationic polymer chitosan reduced the burst release from 45 to 30% and the total release over 9 days from 88 to 78% [135]. A key concern was delivery of the halloysite to the gingival pocket and its subsequent retention. Combination therapy based on amoxycillin and metronidazole in conventional dosage forms has been widely investigated in clinical dental practice due to its activity against a wide range of anaerobes, facultative, and aerobic bacteria. The combination of amoxycillin and metronidazole has synergistic action and covers a wide range of microflora, with metronidazole inhibiting the anaerobes and amoxycillin inhibiting the facultative aerobic bacteria. Both the drugs are bactericidal in nature and are administered systemically. This is important for complete elimination of subgingivally occurring periodontal pathogens [136]. A biodegradable intrapocket device containing amoxycillin and metronidazole was prepared using 69.29 mg each of amoxycillin and metronidazole, 2.0% diethyl phthalate plasticizer, and 750-mg poly(lactic-*co*-glycolic acid) (PLGA) [137]. The device was optimized on the basis of evaluation parameters (e.g., weight variation, content uniformity, surface pH, and *in vitro* and *in vivo* release studies). The films showed sustained *in vitro* release for a period of 16 days. *In vivo* release studies showed that drug concentrations were maintained above the MIC value for the entire period of the release studies. The samples from this study were capable of inhibiting the growth of most test strains. The combination of amoxycillin and metronidazole in the carrier polymer PLGA not only showed an extended spectrum of antimicrobial activity, but also showed a synergistic effect against *Eubacterium limosum*, a metronidazole-resistant strain.

Zilberman and co-workers [138–140] developed and studied metronidazole-loaded 50/50 poly(D,L-lactic-*co*-glycolic acid) (PDLGA), 75/25 PDLGA, and poly(L-lactic acid) (PLLA) films. These structured films were prepared using the solution-casting technique [140]. Concentrated solutions and high solvent evaporation rates were used in order to obtain most of the drug within the bulk. These films are designed to be inserted into periodontal pockets and treat infections during the metronidazole controlled-release phase, for at least

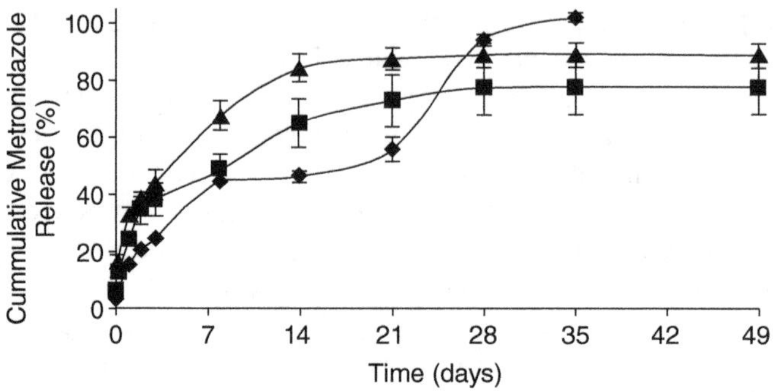


**Figure 7.3.** The effect of copolymer type on metronidazole release profile from films loaded with 10%wt drug. Host polymers: ◆-50/50 PDLGA, ■-75/25 PDLGA, ▲-PDLLA. The experiments were performed in triplicate and the results are presented as means ± standard deviations (With permission from Zilberman and Elsner *J. Control. Rel.*, 130, 202–215, 2008 [138].)

1 month. The effects of copolymer composition and drug content on the release profile, on cell growth, and on bacterial inhibition were investigated. The metronidazole release profiles from films containing 10% drug are presented in Fig. 7.3. Although the 50:50 PDLGA film degrades faster than the 75:25 PDLGA and PDLLA films, the rate of drug release from the latter two films loaded with 10% metronidazole was faster than from the former, due to differences in drug location–dispersion within the film. The drug crystals appear to be located mainly on the surface of the PDLLA and 75:25 PDLGA films, whereas in the 50:50 PDLGA films the drug was located in the bulk and also on the surface. These results indicate that the copolymer composition affects the release profile, while the drug content did not show any significant effect on the shape of the release curves. Human gingival cells and rat mesenchymal bone marrow cells have demonstrated normal *in vitro* growth on the drug-eluting films. The released drug also exhibited effectiveness against *Bacteroides fragilis*. The microbiological inhibition kinetics showed that metronidazole cumulative release during 3 days succeeded in totally inhibiting bacterial growth after 2 days [139].

## 7.12. BIOFILM PROBLEMS IN DENTAL UNIT WATERLINES AND ITS PRACTICAL CONTROL

Dental chair units (DCUs) contain integrated systems that provide the instruments and services for a wide range of dental procedures. These DCUs use water to cool and irrigate DCU supplied instruments (e.g., conventional dental handpieces, high-speed turbine dental handpieces, threeway air–water syringes, and ultrasonic scalers) and tooth surfaces during dental treatment, as the heat generated during instrument operation can be injurious to teeth [141]. The

DCU supplied water is also used for oral rinsing by patients (water supplied via the cup filler outlet) and to wash out the DCU spittoon, or cuspidor, after oral rinsing (water supplied via the bowl-rinse outlet). An intricate network of interconnected narrow-bore tubes called dental unit waterlines (DUWLs) supplies water to all of the DCU supplied instruments, cup-filler, and bowl-rinse outlets [142,143].

In a typical modern DCU, the waterline network consists of many meters of plastic DUWL tubing having an internal diameter of 1–2 mm. Water flow within these narrow bore tubes is laminar and thus the flow at the lumen surfaces is almost negligible compared with that at the center of the lumen. A conditioning pellicle of chemicals mainly from the supply water builds on this inner-face over time providing an easier attachment substrate for microorganisms [144,145]. Thus, microorganisms in DCU supply water attach to the internal surfaces of the DUWL tubing and form microcolonies that eventually give rise to biofilms. These DCUs connected to municipal water supplies usually contain low numbers of several bacterial species that eventually give rise to multispecies biofilm in DUWLs [146–148]. Water stagnation within DUWLs when the equipment is not being used encourages the proliferation of biofilm. The DUWL biofilm matrix also contains both inorganic and organic material derived from supply water and dead microorganisms. From here, planktonic forms of microorganisms and pieces of biofilm are shed to seed biofilm formation elsewhere in the waterline network or are transferred directly into the mouths of patients during dental procedures. Dental handpieces and ultrasonic scalers also aerosolize such biofilm components. These aerosols and fine droplets can enter the lungs of patients and dental healthcare staff [149,150]. Thus, DUWL biofilm acts as a reservoir for ongoing contamination of DUWL output water, and can act as a potential source of cross-infection. Sterilization of the handpieces, syringes, and associated instruments attached to DUWLs has no impact at all on biofilm within DUWLs [142]. Microbial contamination of DUWL output water is a universal problem and all untreated DUWLs in standard DCUs are subject to contamination and will harbor resident biofilms.

Modern DCUs are categorized as medical devices under the European Union Medical Devices Directive [151]. Microbial contamination of a diverse range of medical devices has been shown to be an important cause of cross-contamination and cross-infection, especially in healthcare environments [142]. A single DCU can be in used in the treatment of many patients each day and microbial contamination of specific component parts can be a significant potential source of cross-infection [142,152]. This becomes quite significant where immunocompromised, oral surgery, or endodontic patients are treated [142]. The DCU components that come into direct contact with the patient's oral cavity are of particular concern, including dental unit handpieces, ultrasonic scalers, three-in-one air–water syringes, and suction hoses. Aerosols, splashes, and contact contamination contribute to a microbially contaminated environment in the vicinity of a DCU. Output water provided by a DCU may also be of concern as a potential source of infection as it comes directly from the DCUs extensive network of DUWLs, which can harbor biofilms, and enters the oral cavity of the patient during treatment. Furthermore, aerosols and droplets produced by dental instruments connected to DUWLs may be inhaled by patients and dental healthcare personnel [141,152–154].

Many studies over the last 40 years demonstrated that DUWL output water is often contaminated with high densities of microorganisms, predominantly Gram-negative aerobic heterotropic environmental bacteria, including *Legionella* and *Pseudomonas* species. Untreated DUWLs host biofilms that permit microorganisms to multiply and disperse through the water network and that are aerosolized by DCU instrument use, thus exposing patients and staff to these microorganisms, to fragments of biofilm, and bacterial endotoxins. Yeasts, fungi, and amoebae may also be present in DUWL output water [155]. *Legionella* bacteria live within a variety of amoebae and protozoa commonly found in soil and water, and are often found in association with biofilms, including DUWL biofilms. There is no definitive published evidence, so far, that any patient has ever contracted legionellosis following exposure to contaminated DUWL output water. However, many studies certainly have reported the presence of *legionellae* in DUWLs [150,156–158]. In 1995, Atlas et al. [159] reported the death of a dentist in California resulting from Legionnaire's disease, which was possibly due to exposure to DUWL output water. Occupational exposure to aerosols of waterborne bacteria generated by dental instruments attached to DUWLs may lead to colonization of dental staff and also cause a higher prevalence of antibodies to *Legionella* [159,160].

Dental unit waterline output water is also a potent source of bacterial endotoxin, composed of lipopolysaccharide released from the cell walls of dead Gram-negative bacteria, and levels ranging from 500 to 2560 endotoxin units (EU) $mL^{-1}$ have been reported [157,161]. In contrast, the maximum level of endotoxin permissible in sterile water for irrigation in the United States is $0.25\,EU\,mL^{-1}$. Endotoxin can cause localized inflammation, fever, and shock in susceptible individuals. Interestingly, in medical devices that are prone to biofilm contamination and endotoxin build-up (e.g., humidifiers), a hypersensitivity pneumonitis triggered by contaminating endotoxin is well documented [157]. Inhaled endotoxin can precipitate reactive airway symptoms [162] and asthma severity is directly correlated with the concentration of endotoxin [163]. Furthermore, results from a single, large, practice-based cross-sectional study reported a temporal association between occupational exposure to contaminated DUWL output water with aerobic counts of $>200\,CFU\,mL^{-1}$ at $37\,°C$ and development of asthma in the subgroup of dentists in whom asthma arose following the commencement of dental training [149]. Finally, Putnins et al. [161] suggested that endotoxin present in DUWL output water might stimulate the release of proinflammatory cytokines in gingival tissue during surgery and adversely affect healing.

Dental instruments that are connected to DUWLs and that are used in the patient's mouth (e.g., turbine and conventional handpieces, air–water syringes, and ultrasonic scalers) should contain integrated antiretraction valves or devices that prevent backflow or back siphonage of fluids from the oral cavity into the DUWLs [143]. The need for antiretraction devices has been highlighted by many studies demonstrating that oral fluids can be retracted into DUWLs during DCU instrument use. The detection of oral bacterial species and other human-derived microorganisms in DUWL output water has provided convincing evidence for likely failure of antiretraction devices [146,164–166]. Moreover, an Italian study of 54 DCUs, comprising 18 different models by 6 different DCU manufacturers reported an antiretraction device failure rate of 74% (40–54 DCUs tested) [164]. Thus, retraction of oral fluids (e.g., saliva and blood) during use of dental instruments attached to DUWLs can add to the range of microorganisms present in DUWL biofilms, and therefore in DUWL output water, as well as increasing the potential for transmission of pathogenic microbes. To minimize the potential impact of antiretraction device failure, the current CDC guidelines for infection control in dental healthcare settings recommend that DCU handpieces should be operated to discharge water and air for a minimum of 20–30s after each patient session [167]. All dental handpieces connected to DUWLs should be cleaned, lubricated, and sterilized by autoclaving after each patient use.

Reservoir bottles in DCUs can easily become contaminated with skin organisms (e.g., *Staphylococcus epidermidis* and *Staphylococcus aureus*), the latter a significant human pathogen, which introduces additional human microorganisms into DUWLs [168]. To avoid this, reservoir bottles should be handled carefully and should be cleaned and disinfected regularly. Preferably, reservoir bottles that can be sterilized by autoclaving after cleaning should be used.

Over the last two decades, numerous approaches, both chemical and nonchemical based, for reducing the microbial density in DUWL output water have been proposed, but none that is both efficient at eliminating biofilm, compatible in the long term with the material components of DUWL networks and dental instruments attached to DUWLs, as well as being safe for patients, has been universally adopted [143]. One widely used procedure for reducing the microbial burden in DUWL output water involves flushing DUWLs with fresh water [169,170]. However, whereas this procedure does somewhat reduce the density of microbes in output water, it does not remove biofilm and is ineffective as a means of controlling the quality of DUWL output water [148,171,172]. Another procedure used to improve the quality of DUWL output water involves point-of-use microbial filters at the ends of DUWLs near the instrument attachment sites. These can be very effective, but have to be changed regularly as they become clogged readily and thus add to ongoing maintenance expense [173–175]. Microbial filters attached to the DCU supply water line suffer from similar drawbacks. Filters have no effect whatsoever on existing biofilms in DUWLs. The use of sterile, deionized, or distilled water in independent bottle reservoirs also has no effect on resident biofilms in DUWLs.

The most efficient means of achieving good quality DUWL output water is regular treatment disinfection of DUWLs with a chemical, biocide, or cleaning agent that removes biofilm from DUWLs effectively, resulting in good quality output water [143,146,176–179]; for a list of DUWL treatment agents, see the recent reviews by Walker and Marsh [178] and Coleman et al. [180]. Dental unit waterlines treatment agents are generally divided into two categories, including agents for intermittent DUWL treatment (e.g., once weekly), and agents for continuous or residual DUWL treatment. Note, many DUWL treatment agents have not been developed or endorsed by DCU manufacturers, but rather have been developed by other manufacturers in response to an evident market need. Thus, there is significant potential for incompatibility of DUWL treatment agents with components of the DUWL network, as well as with instruments connected to this network [143].

In the case of residual DUWL treatment agents, there is a lack of independent studies in the literature on potential interactions of such agents and their byproducts on oral tissues and teeth. A number of studies reported that some DUWL treatment agents (e.g., 3 ppm sodium hypochlorite; a 1:10 dilution of Listerine mouthrinse; bio 2000, a 0.12% chlorhexidine gluconate- and 12% ethanol-containing product; and 0.224% BioClear, a citric acid containing product) may adversely affect bonding of composite material to both enamel and dentine [181,182]. With the extended and more widespread use of DUWL treatment agents, it is likely that such adverse effects may become clinically relevant in the case of residual DUWL treatments. In 2002, a clinical laboratory study reported clogging of DUWLs by the accumulation of disinfectant deposits in three of six DCUs treated with the alkaline hydrogen peroxide DUWL treatment agent Sterilex Ultra [146]. Clogging became evident after the fourth consecutive week of once-weekly treatment in the three DCUs, and in one of these, after 14 weeks it became impossible to aspirate water or treatment agent through the air–water syringe waterline, which had to be replaced. Additionally, the pH of DUWL output water in these DCUs remained persistently alkaline (e.g., pH 8.4) for several days post-DUWL treatment. This was in contrast to the DCU supply water (pH 7.0) and DUWL output water from other DCUs (pH 7.0) in the same clinic, which were treated with a different $H_2O_2$ treatment agent not associated with DUWL clogging [143,146]. These findings suggested that residual DUWL treatment agent was present in DUWL output water in the DCUs that exhibited clogging for a considerable time after treatment. Another recent study from the same research laboratory on the long-term efficacy of Planosil (a $H_2O_2$ and Ag ion-containing DUWL treatment agent) in Planmeca Prostyle Compact DCUs, identified several episodes of failure to disinfect DUWLs due to adverse effects on a variety of DCU components [179]. After 6 months of continuous once weekly (15-h overnight) DUWL treatment with Planosil, some episodes of DUWL disinfection failure were directly linked with blockage of and/or leakage from disinfectant intake valves, and corrosion of Al components of disinfectant delivery containers. Valve leakage was linked with damage to an internal glue seal, whereas valve blockage was caused by a combination of dislodged glue and oxidized Al deposits. Protracted exposure to $H_2O_2$ with its strong oxidizing properties was identified as the most likely cause of the adverse affects on the DCUs concerned. The problems were completely resolved in collaboration with the DCU manufacturer, Planmeca, who developed replacement valves and disinfectant delivery containers that were resistant to damage corrosion following prolonged exposure to Planosil [179]. This study highlights the importance of investigating the long-term effects that DUWL treatment agents can have on DCU components and also highlights the important role that DCU manufacturers have in ongoing research and development to identify problems and to continually improve their DCUs.

Modern DCUs are equipped with a suction system that has a variety of purposes. Primarily, the suction system is used to remove oral fluids and debris from the oral cavity during dental procedures and also to minimize aerosol release into the dental clinic environment during the use of dental instruments attached to DUWLs, especially high-speed turbine handpieces [152]. Oral fluids and spent DUWL output water removed by DCU suction hoses and from the DCU cuspidor is eventually released as waste water following particle removal, dental amalgam removal, and disinfection. Special amalgam separators are used to remove amalgam particles generated during the placement and removal of amalgam dental restorations in patients as amalgam contains Hg [143]. The performance of amalgam separators can vary considerably as can the total Hg concentration and total dissolved (ionic) Hg concentration in DCU waste water [183]. A study from the United States reported that iodine-releasing resin cartridges, an effective residual treatment used to control biofilm growth in DUWLs by continuous release of low levels of iodine into DUWL output water, may have a harmful affect on the environment by mobilizing Hg from dental amalgam in DCUs with the release of highly toxic dissolved Hg into the environment from DCU waste water [184]. However, the findings of this study were challenged by other authors who claimed that chloramine used to disinfect municipal water was more likely to have caused the increase in Hg levels rather than iodine [185]. Other studies reported that a range of disinfectants and cleaning agents used to treat DCU waste water lines also cause the release of Hg from dental amalgam when tested in the laboratory. Strong chlorine-containing agents were reported to cause the release of more Hg than other products [186,187]. These findings suggest that it is conceivable that DUWL treatment agents that contain Cl could also mobilize Hg from dental amalgam collected in amalgam filters, traps, and separators, as well as in DCU waste water lines and pipes and release it into the environment.

Various electrochemically activated (ECA) solutions, also called superoxidized water, anolyte, and various other terms, have been used in recent years as a residual treatment to control biofilm in DUWLs. These studies showed that such solutions can be very effective [188–191]. However, some ECA solutions can have the potential for adverse effects on DUWLs and DCU instruments connected to them, following extended use if the parameters for electrochemical activation are suboptimal or if the product used is too concentrated [143]. The literature concerning ECA solutions in all their varieties is confusing due to the use of differing terminologies, technologies, and end products. An ECA solution is generated by passing water containing dilute salt solutions (usually dilute sodium chloride, NaCl) or other mineral solutions through an electrochemical cell designed to generate two streams of activated solution. One is a negatively charged antioxidant solution (catholyte) and the other is a positively charged oxidant solution (anolyte) [143]. Much early work on disinfection using anolyte solutions, used generators that produced an extremely acidic form that could also cause corrosion and harm to some materials. Data from this research group has shown that even short-term exposure of DUWLs to insufficiently dilute anolyte from a number of sources could cause some DUWLs to deteriorate rapidly and cause corrosion damage to other DCU components [143]. When using an ECA solution to control biofilm in DUWLs, it is essential that the ECA generator is capable of consistent quality output at neutral pH. The ECA product concentration in DUWL output water does not need to exceed 1–2 ppm free available chlorine, as it is so effective. This requires accurate dosing into DUWL supply water, as anolyte produced by ECA generators is usually much more concentrated (e.g., 200 ppm).

Very little applied research has been undertaken to investigate whether the materials used to manufacture DUWLs can influence biofilm formation. One Japanese study reported that DUWLs composed of polyvinylidene fluoride were effective in inhibiting biofilm formation and reducing bacterial density in DUWL output water [192]. Another Italian study reported that the aerobic heterotrophic bacterial plate count at 22 °C from polytetrafluorethylene was lower than output water from DUWLs made from polyethylene [193]. These findings indicate that the development of novel DUWL materials with antimicrobial and/or antibiofilm properties is a potentially very productive area for research on DUWL biofilm control. Delivering DCU supply water using Cu pipes may also be beneficial in improving the microbial quality of DCU supply water, as Cu pipework has been shown to possess significant antimicrobial advantages over drinking water pipework of anoter composition [194,195].

A new generation of DCUs with integrated DUWL cleaning systems that facilitate and simplify control of biofilms in DUWLs by cleaning and disinfection with consequent consistent good-quality DUWL output water has been developed [143,146,179,196]. The effectiveness of the Planmeca Waterline Cleaning System (WCS™), a semiautomated DUWL cleaning system developed by the Finnish DCU manufacturer Planmeca, to control DUWL biofilm in two separate Planmeca Prostyle Compact DCUs over a 20-week period using the $H_2O_2$ and Ag ion-containing disinfectant Planosil was reported [146]. The WCS™ was found to be very effective at eliminating DUWL biofilm in these DCUs when used with Sanosil and consistently provided output water with bacterial densities below the American Dental Association (ADA) recommended level of ≤200 CFU $mL^{-1}$ of aerobic heterotrophic bacteria for up to 7-days postdisinfection. The microprocessor-controlled WCS™ was originally developed to be retrofitted to existing Planmeca DCUs having a municipal mains water supply. In recent years, the WCS™ is provided as an integrated DUWL cleaning system in new Planmeca Prostyle Compact DCUs [179]. In a more recent study from the same laboratory, the ability of the WCS™ to maintain the microbiological quality of DUWL output water below the ADA recommended level of ≤200 CFU $mL^{-1}$ of aerobic heterotrophic bacteria was investigated over a longer period (18 months) with a much larger number of DCUs (10 DCUs) using the $H_2O_2$ and Ag ion-containing DUWL disinfectants Planosil and Planosil Forte [179]. A recent study from Poland on DUWL disinfection with a different $H_2O_2$ and Ag ion-containing disinfectant (Oxygenal 6) reported that isolates of *Sphingomonas paucimobilis* were significantly more prevalent (80%) in DUWL output water postdisinfection compared with output water predisinfection (10%) [197]. Planosil was reformulated by the DCU manufacturer Planmeca as a more concentrated form of $H_2O_2$ and Ag ions (Planosil Forte) and when used once weekly was found to maintain bacterial density in output water below the ADA standard for all 10 DCUs during the 17 consecutive weeks studied [179]. In this regard, it is interesting to note that Planmeca recently developed a more advanced microprocessor-controlled DUWL cleaning system called the Water Management System (WMS™), a fully integrated and automated DUWL cleaning system that requires minimal effort on the part of the user [196]. The WMS™ is more advanced and automated than the WCS™ and also contains many additional features, including an air gap. Studies with a Planmeca Compact i DCU demonstrated that the WMS™ consistently provided DUWL output water that passed the ADA quality standard of ≤200 CFU $ml^{-1}$ for up to 7 days after once weekly disinfection with Planosil Forte during a test period of 40 consecutive weeks [196]. However, in WCS™ and WMS™, the consistent provision of good quality DUWL output water was dependent on meticulous implementation of the disinfection protocol by staff undertaking DUWL disinfection [179].

In the Dublin Dental Hospital (Ireland), all 103 DCUs with which the hospital is equipped are supplied with water from a central 8000-L storage tank supplied with potable quality mains water. In June 2006, during a period of warm weather, ongoing routine monitoring detected a bacterial bloom of *Pseudomonas fluorescens* in the 8000-L DCU storage tank, which developed over the course of 1-week between weekly samplings, where the bacterial density rose to >100,000 CFU $mL^{-1}$. Concomitantly, routine weekly testing of DUW output water from several sentinel DCUs showed bacterial densities >100,000 CFU $mL^{-1}$ despite the once-weekly DUWL disinfection regime with Planosil. This incident necessitated disconnecting all of the hospital's DCUs from the tank supply and providing each with fresh potable quality water in clean independent reservoir bottles until the contamination problems with the tank could be resolved. These findings highlighted the necessity for effective control of water quality throughout the DCU supply water network in dental hospitals and multi-DCU clinics, not only within DUWLs.

Electrochemically activated technology was pioneered as a specialized discipline of electrochemistry by Prof. Vitold Bakhir in the 1970s in the former Soviet Union [198–200]. Electrochemically activated solutions were generated by passing a dilute salt solution through an electric field in a flow-through electrolytic module (FEM), segregating the ions formed and producing two oppositely charged solutions possessing altered physical and chemical properties [201]. The activation process changes the state of the salt solution from a stable to a metastable state. The positively charged solution (anolyte) typically has a redox value of +600 mV, and is composed of a mixture of unstable mixed oxidants (predominantly hypochlorous acid) in a physically excited state that is highly microbicidal and able to penetrate biofilms. The negatively charged antioxidant solution (catholyte) has detergent-like properties, typically a pH of 11, a redox value of −600 mV, and contains predominantly sodium hydroxide (NaOH) in an excited state. These active ion species and free radicals are short-lived with a half-life of typically <48 h [201]. Activation also generates electrically and chemically active microbubbles of electrolytic gas, 0.2–0.5 mm in diameter, which enhance the redox potential of anolyte and catholyte [201]. Different kinds of ECA solutions can be generated by different types of ECA generators, each with specific properties and applications [201]. Over the years, several generations of FEM were developed, the FEM-3 being one of the more recent [199,202]. The ECA solution production with consistent quality and characteristics was difficult to achieve prior to the development of FEM-3 technology. The FEM-3-based ECA technology outside Russia is now owned by and has been further refined by the Trustwater Group, Clonmel, Ireland. Anolyte (Ecasol™) produced by Trustwater Group generators has a neutral pH, very much in contrast to anolyte produced by earlier generators from other manufacturers, which was often acidic and corrosive [199,201,202].

A centralized and automated biofilm management system for mains water distribution networks supplying water to DCUs that consistently maintains DUWL supply and output water at better than potable quality simultaneously in 103 DCUs in a Dental Hospital setting was developed (Fig. 7.4) [203,204]. The first part of the system consists of automated filtration of mains water to yield DCU supply water of consistent chemical composition, despite fluctuations in the chemical composition of the inward mains supply. The second part of the system automatically disinfects filtered mains water using dilute (i.e., 2.5 ppm) Ecasol™, a neutral ECA anolyte solution, and over the 100-week study period maintained the average microbiological quality of DUWL supply and output water at <1 and 18.1 CFU $mL^{-1}$, respectively, of aerobic heterotrophic bacteria, significantly better than potable water. The system requires minimal human intervention, is cost-effective, environmentally friendly, and provides a robust solution to the universal problem of biofilm formation in DUWLs. The current system is ideally suited to Dental Hospitals and large dental clinics equipped with many DCUs, but has immense potential for managing water quality in other healthcare facilities, including acute hospitals.

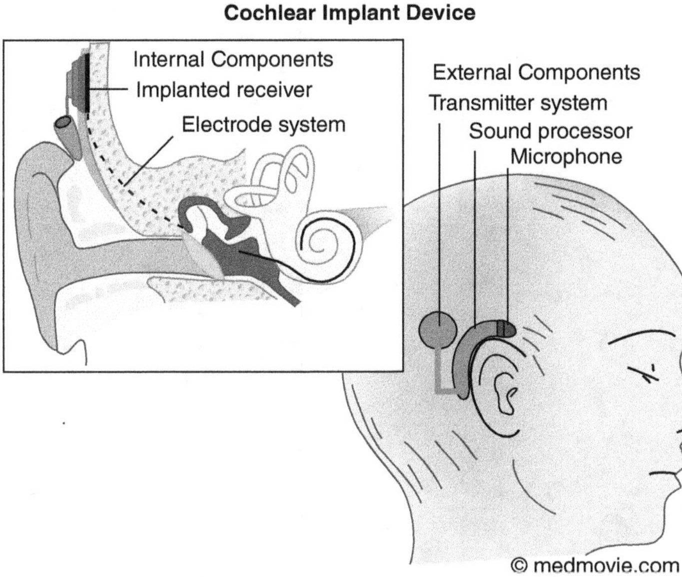


**Figure 2.1.** External and internal components of cochlear implants.

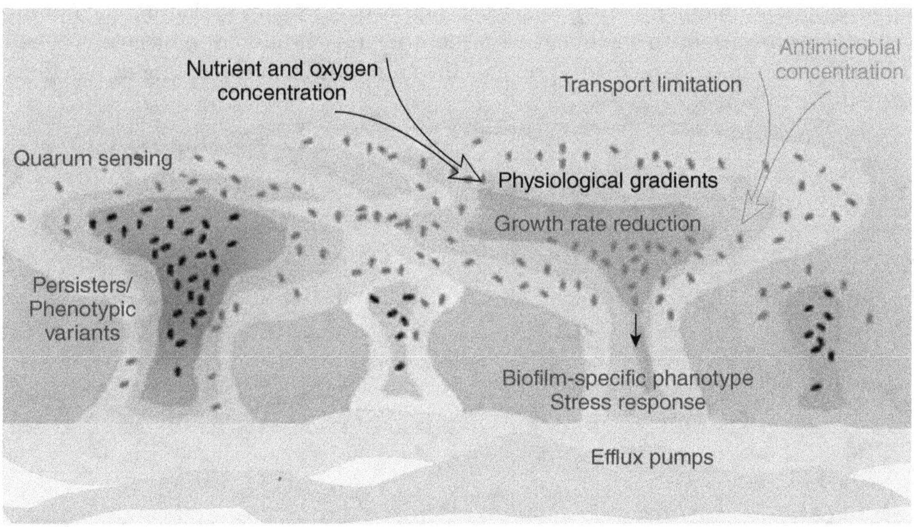


**Figure 4.2.** *Pseudomonas aeruginosa* biofilm resistance. Schematic representation of mechanisms proposed to be involved in *P. aeruginosa* biofilm resistant antimicrobial agents. The increase in bacterial density within biofilm microcolonies (indicated by darkening colors) determines gradients of nutrient and oxygen concentration (indicated by a narrowing arrow). Mechanisms biofilm resistant may include restricted antimicrobial penetration (indicated by a narrowing arrow) mediated by the polysaccharide matrix, and reduction in growth rate and metabolic activity caused by nutrient and oxygen gradients. Biofilm bacteria may also undergo physiological, metabolic, and phenotypic changes leading to a biofilm-specific phenotype. Other resistant mechanisms may include emergence of phenotypic or persister variants (represented as maroon bacteria) within the biofilm population, induction of the general stress response, upregulation of efflux pumps, and activation of quorum-sensing (QS) systems.

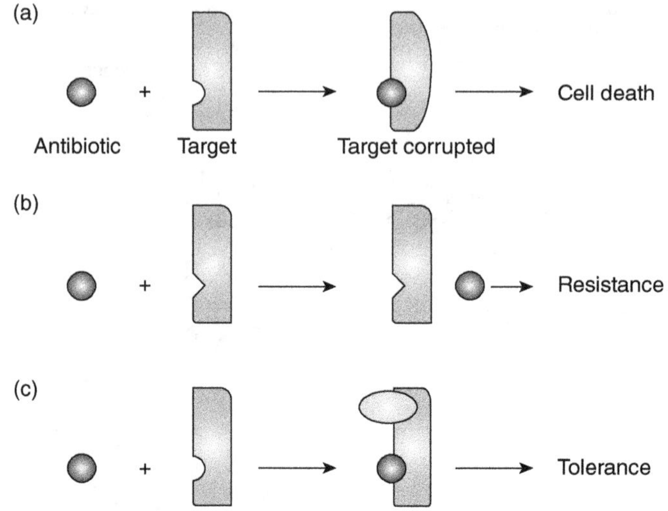


**Figure 4.4.** Resistance versus tolerance to bactericidal antibiotics. (Reproduced with permission from Kim Lewis *Nat. Rev. Microbiol.*, 5, 48–56, 2007 [15].) (**a**) The antibiotic (pink) binds to the target (blue) altering its function, which causes cell death, (**b**) The target of the antibiotic has been altered so that it fails to bind the antibiotic and the cell becomes resistant to treatment with the drug, (**c**) A different molecule (yellow) inhibits the antibiotic target. This prevents the antibiotic from corrupting its functions, resulting in tolerance.

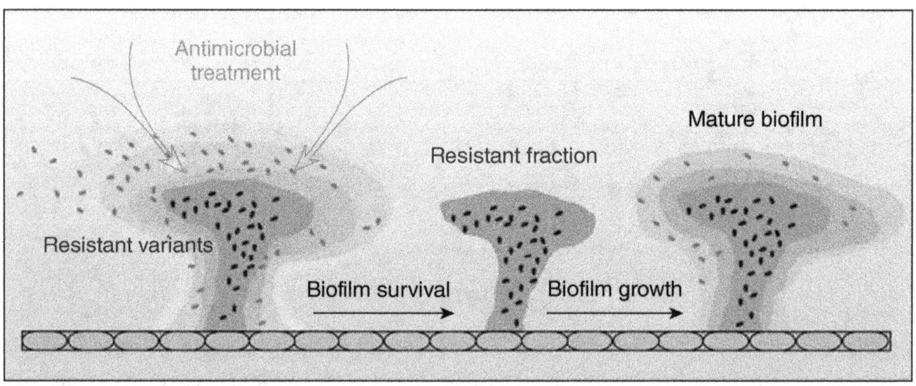


**Figure 4.5.** Resistance mechanism mediated by phenotypic–persister variants. Antimicrobial treatment of bacterial biofilms leads to the eradication of most of the biofilm susceptible population. A small fraction of phenotypic–persister variants (represented as maroon bacteria) survives antimicrobial treatment and is able to start biofilm development once antimicrobial therapy is discontinued.

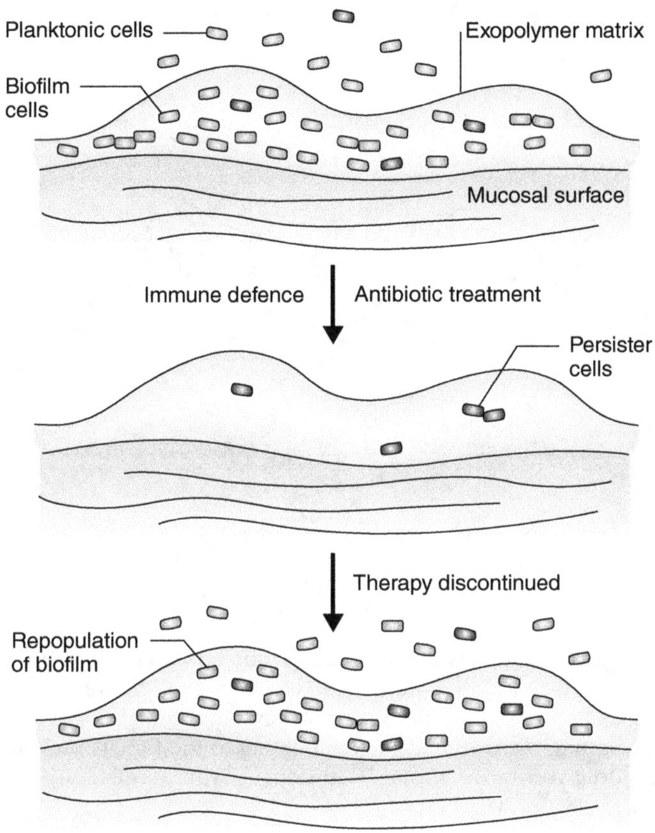


**Figure 4.6.** This figure shows a model of biofilm resistance to killing based on persister survival. Initial treatment with antibiotic kills normal cells (colored green) in both planktonic and biofilm populations. The immune system kills planktonic persisters (colored pink), but the biofilm persister cells (colored pink) are protected from the host defences by the exopolymer matrix. After the antibiotic concentration is reduced, persisters resuscitate and repopulate the biofilm and the infection relapses (Reproduced with permission from Kim Lewis *Nat. Rev. Microbiol.*, 5, 48–56, 2007 [15].)

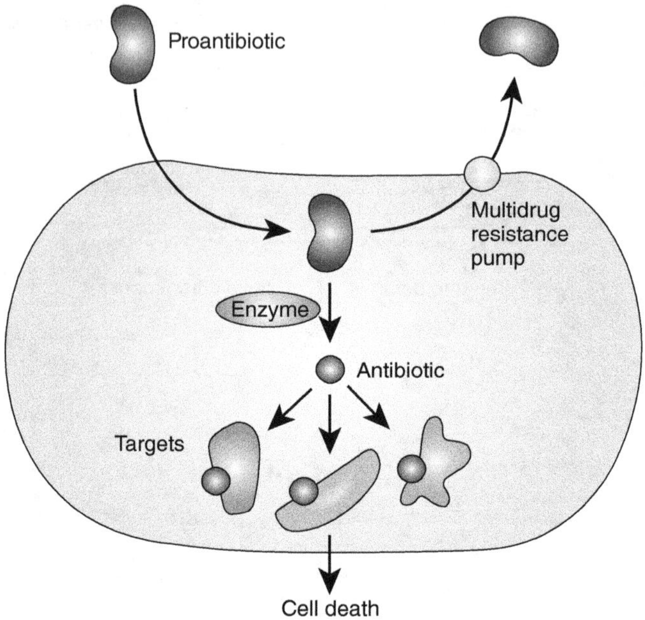


**Figure 4.7. The perfect antibiotic.** The proantibiotic is benign, but a bacterial enzyme converts it into a reactive antibiotic in the cytoplasm. The active molecule does not leave the cytoplasm (owing to increased polarity), and attaches covalently to many targets, thereby killing the cell. Irreversible binding to the targets prevents the antibiotic from multidrug resistance efflux. (Reproduced with permission from Lewis *Nat. Rev. Microbiol.*, 5, 48–56, 2007 [15].)

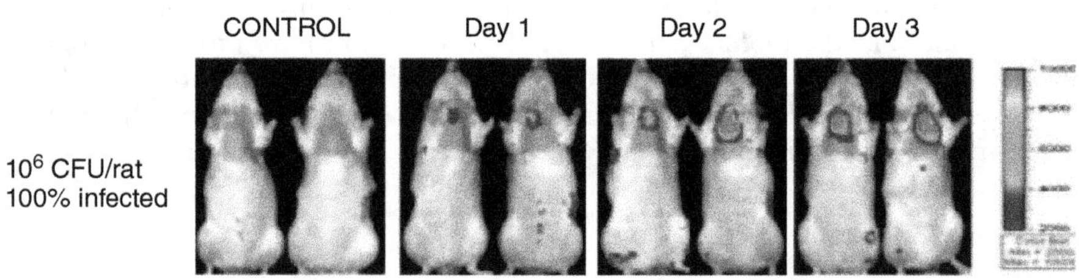


**Figure 5.4.** Real-time monitoring of *S. aureus* Xen29 in an experimental-rat endocarditis model. Two representative animals infected intravenously with either normal saline (control) or $10^6$ CFU of *S. aureus* strain are shown. The animals were imaged ventrally, with their chest area shaved, to avoid background signal from animal hair. The process of infection was monitored daily by detecting photon emission around the region of interest (heart area) over a 6-day course.

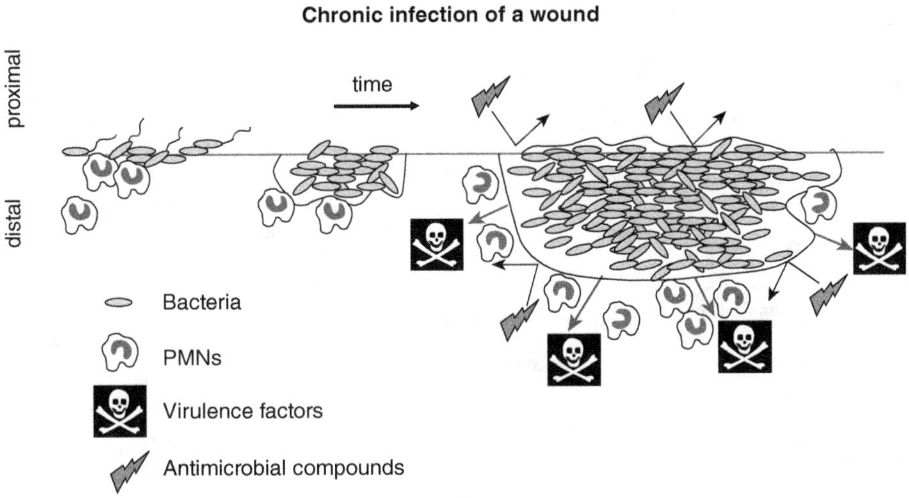


**Figure 8.1.** Development of a biofilm in a chronic infected wound.

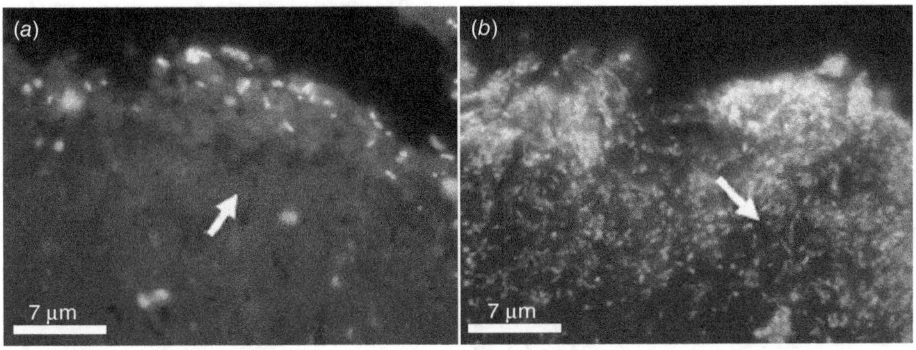


**Figure 8.2.** (*a*) Colonizing planktonic *P. aeruginosa* in a chronic wound; arrows show single cells. (*b*) Shows a larger collection of bacteria in a chronic wound infected with *P. aeruginosa*, with the arrow indicating a microcolony.

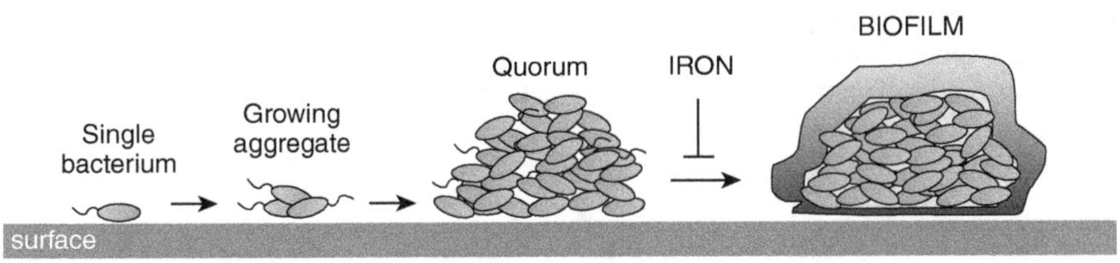


**Figure 8.4.** Schematic of the bacterial-biofilm formation process and its inhibition by high concentrations of Fe. The biofilm is depicted as a cut-away image.

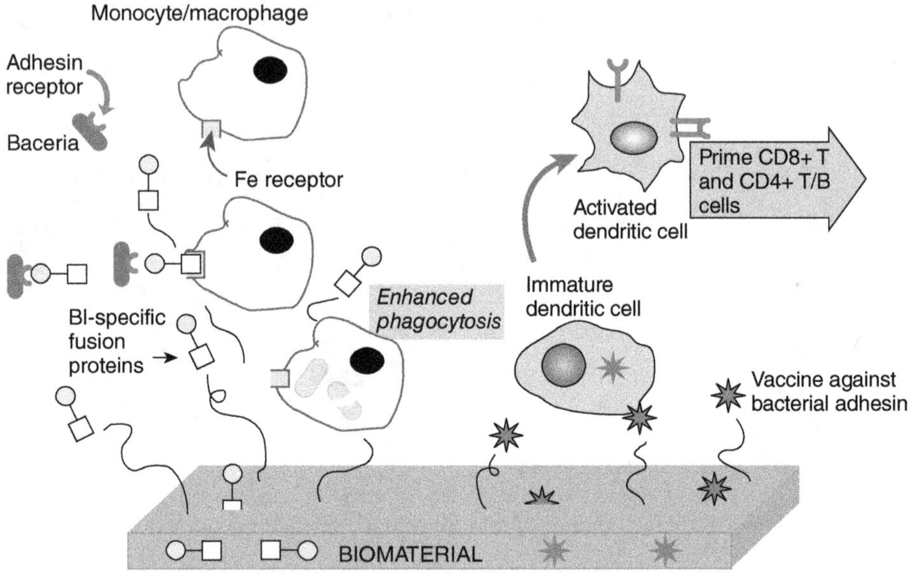


**Figure 9.4.** Hypothetical biomaterial engineered to enhance short- and long-term infection immune response. (Adapted from Bryers [314].)

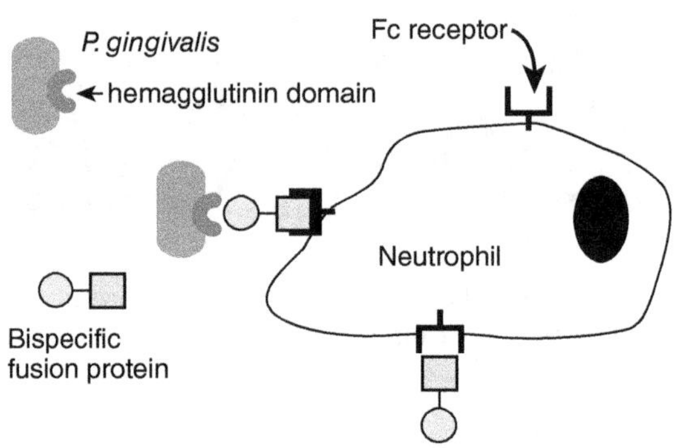


**Figure 9.6.** Use of bispecific fusion proteins to opsonize pathogenic bacteria and enhance phagocytosis. Fc receptor is an antibody possessing its binding specificity known as the Fc (fragment, crystallizable) region.

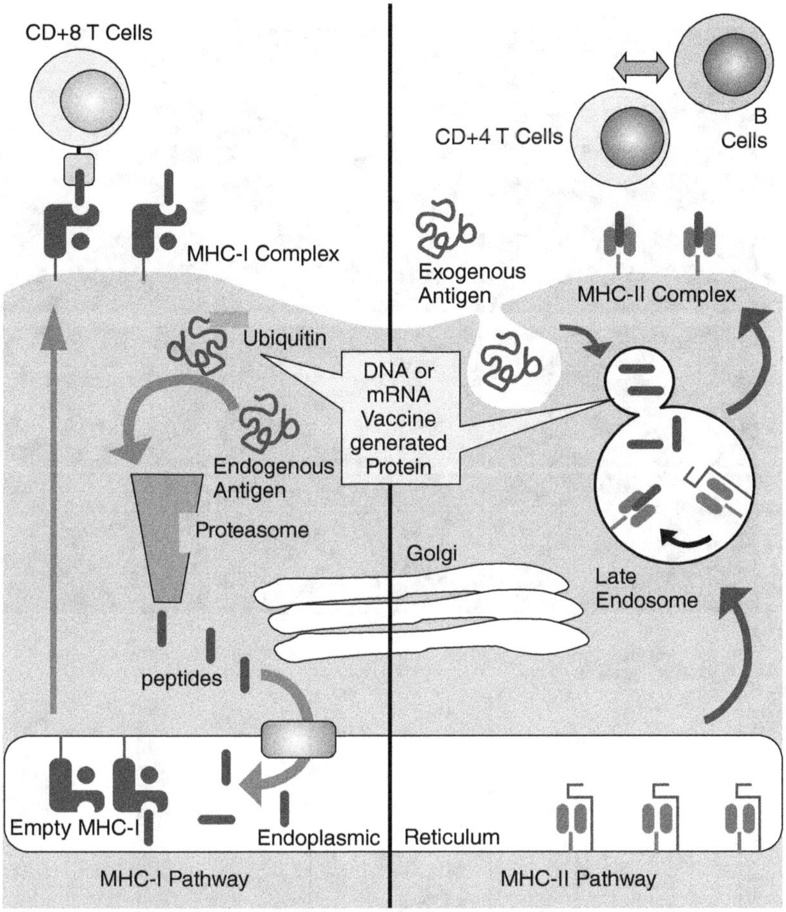


**Figure 9.7.** Antigen presentation and pathways of vaccine response. Plasmid DNA or messenger ribonucleic acid (mRNA) is taken up by dendritic cells for intracellular expression of antigen. Antigen can be secreted (not shown) and subsequently taken up by another DC as an exogenous antigen. Antigen expressed intracellularly by a dendritic cell or taken up through cross-priming is presented by MHC-I to CD8+ T-cells (cytotoxic leukocytes; CTLs). Antigen taken in exogenously or directed by DNA or mRNA trafficking signals are processed by the MHC-II pathway and presented to CD4+ TH cells, which can subsequently secrete: soluble cytokine signals (e.g., IL-12) back to the dendritic cell, proliferative signals (e.g., IL-2 and IFN-$\gamma$) to Tc cells, or signals directed toward B-cells (e.g., IL-4) to induce B-cell proliferation and antibody secretion.

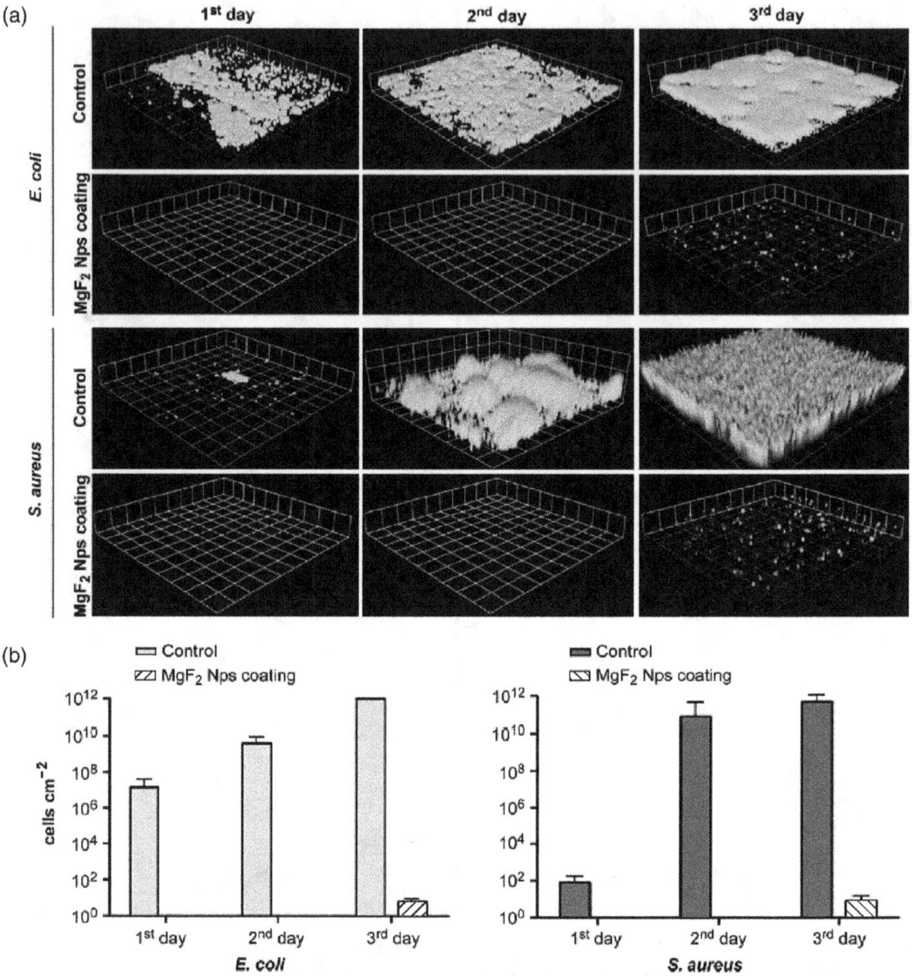


**Figure 11.3.** Extended antibiofilm activity of $MgF_2$.Nps coatings on glass surfaces. (a) Confocal laser scanning microscope (CLSM) images of *E. coli* and *S. aureus* following biofilms formation over the course of three consecutive days on uncoated and $MgF_2$. Nps coated surfaces. Green and red staining represents, respectively, live and dead bacterial cells. In all images 1 unit equals 13.8 mm. (b) Viable count of the biofilm cells. (control refers to the biofilm development on uncoated surface). (Reproduced with permission Lellouche et al. *Biomaterials*, 30, 5969—5978, 2009 [206].)

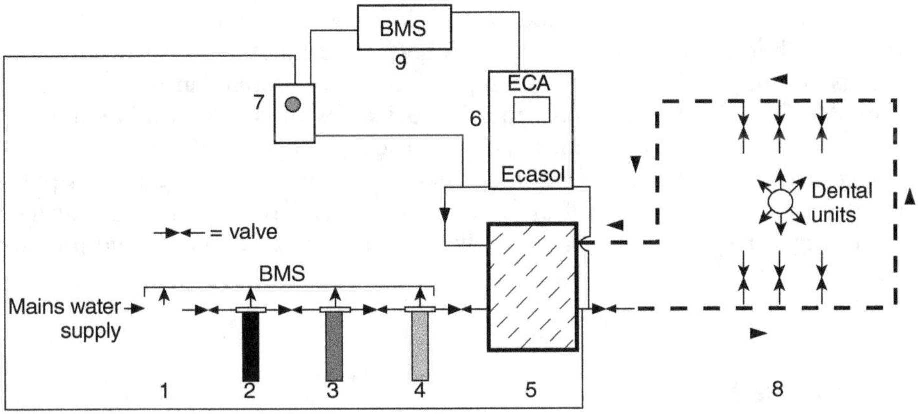


**Figure 7.4.** Schematic diagram of the integrated and automated prefiltration and Ecasol™ disinfection system for maintaining the quality of DUWL supply and output water at better than potable quality at the Dublin Dental Hospital. Key: (1) self-cleaning particle filter; (2) water-softening unit; (3) granular activated carbon filter with 15% (v/v) KDF-85 media; (4) KDF-55 filter; (5) 8000-L processed water storage tank; (6) Trustwater Model 120 EcasolTM generator; (7) Hach chlorine analyzer, (8) recirculating ring main supplying water to DCUs with the direction of flow indicated by arrowheads; (9) connections to buildings management system (BMS) computer and remote site monitoring station, (Reproduced with permission O'Donnell et al. *J. Dent.*, 37, 748–762, 2009 [204].)

The problem of microbial biofilm formation in DUWLs and poor quality DUWL output water has been recognized for more than four decades. In an age where inadequate infection control and prevention in healthcare facilities is seldom absent from the media, it is vital that dental healthcare professionals endeavor to maintain their DCU output water quality at a level that would at least satisfy the ADA standard of $\leq$200 CFU mL$^{-1}$ of aerobic heterotrophs, or preferably, potable water standards. Achieving this objective has been difficult to meet consistently, mainly because of the absence of standards or legislation, but also because DCU manufacturers have been slow to address this issue by DCU engineering and design changes and by the provision of specific guidance on DUWL disinfection. Fortunately, this situation has begun to change with some DCU manufacturers developing and continuing to improve effective integrated and automated DUWL disinfection systems for use with specified DUWL treatment agents that are effective in the long-term and compatible with their DCUs.

Similarly, there are many needs for additional research in periodontology, including the development of biomarkers of current and future disease activity. Effective community- and population-based means of prevention need to be investigated, and although current treatments are generally quite effective in arresting disease progression and restoring some degree of lost periodontal support, further study is needed to develop and test innovative treatment strategies that are less invasive, more cost effective, and take advantage of our increasing understanding of tissue regeneration and repair on the molecular level. As observational studies report associations of common inflammatory periodontal disease with various systemic diseases, large, multicenter randomized controlled trials should be undertaken to investigate the effect of periodontal treatment on risk of systemic diseases and disorders (e.g., adverse pregnancy outcomes, cardiovascular disease, and stroke, diabetes, and pulmonary disease).

## REFERENCES

1. Marsh, P.D. and Martin, M.V. (1999), *Oral Microbiology*, 4th ed., Butterworth–Heinemann, Burlington, MA.
2. Shanies, S. and Hein, C. (2006), The significance of periodontal infection in cardiology, *Grand Rounds*, 1, 24–33.
3. Chun, Y.H., Chun, K.R., Olguin, D., and Wang, H.L. (2005), Biological foundation for periodontitis as a potential risk factor for atherosclerosis, *J. Perio. Res.*, 40, 87–95.
4. Keller, D. (2006), Management of periodontitis for HIV–AIDS patients: Case study, *Dent. Today*, June, 110–113.
5. Whittaker, C.J., Klier, C.M., and Kolenbrander, P.E. (1996), Mechanisms of adhesion by oral bacteria, *Annu. Rev. Microbiol.*, 50, 513–552.
6. Davey, M.E. and O'Toole, G.A. (2000), Microbial biofilms: From ecology to molecular genetics, *Microbiol. Mol. Biol. Rev.*, 64, 847–867.
7. Southard, G.L. and Godowski, K.C. (1998), Subgingival controlled release of antimicrobial agents in the treatment of periodontal disease, *Int. J. Antimicrob. Agents*, 9, 239–253.
8. Allaker, R.P. and Hardie, J.M. (1998), Oral infections. *Topley and Wilson's Microbiology and Microbial Infections*, Vol. 3, 9th ed., Arnold, London, pp. 373–390.
9. Slots, J., Bragd, L., Wikstrom, M., and Dahlen, G. (1986), The occurrence of *Actinobacillus actinomycetemcomitans*, *Bacteroides gingivalis* and *Bacteroides intermedius* in destructive periodontal disease in adults, *J. Clin. Periodontol.*, 13, 570–577.
10. Marsh, P.D. (1995), Dental plaque, in: Lappin-Scott, H.M. and Costerton, J.W., Eds., *Microbial Biofilms*, Cambridge University Press, Cambridge, UK, pp. 282–300.
11. Bradshaw, D.J., Marsh, P.D., Allison, C., and Schilling, K.M. (1996), Effect of oxygen, inoculum composition and flow rate on development of mixed culture oral biofilms, *Microbiology*, 142, 623–629.
12. Kornman, K.S., Page, R.C., and Tonetti, M.S. (1997), The host response to microbial challenge in periodontitis: Assembling the players, *Periodontol.* 2000, 14, 33–53.

13. Geerts, S.O., Nys, M., De, M.P., Charpentier, J., Albert, A., Legrand, V., and Rompen, E.H. (2002), Systemic release of endotoxins induced by gentle mastication: Association with periodontitis severity, *J. Periodontol.*, 73, 73–78.

14. Loos, B.G., Graandijk, J., Hoek, F., Wertheim-van Dillen, P.M., and van der Velden, U. (2000), Elevation of systemic markers related to cardiovascular disease in the peripheral blood of periodontitis patients, *J. Periodontol.*, 71, 1528–1534.

15. Steel, D.M. and Whithead, A.S. (1994), The major acute phase reactants: C- reactive protein, serum Amyloid P and serum Amyloid A component, *Immunol. Today*, 15, 81–88.

16. Butterfield, T.A., Best, T.M., and Merrick, M.A. (2006), The dual roles of neutrophils and macrophages in inflammation: A critical balance between tissue damage and repair, *J. Athl. Train.*, 41, 457–465.

17. Maradit-Kremers, H., Nicola, P.J., Crowson, C.S., Ballman, K.V., and Gabriel, S.E. (2005), Cardiovascular death in rheumatoid arthritis: A population-based study, *Arthritis Rheum.*, 52, 722–732.

18. Gan, W.Q., Man, S.F., Senthilselvan, A., and Sin, D.D. (2004), Association between chronic obstructive pulmonary disease and systemic inflammation: A systemic review and a meta-analysis, *Thorax*, 59, 574–580.

19. Wouters, E.F. (2005), Local and systemic inflammation in chronic obstructive pulmonary disease, *Proc. Am. Thorac. Soc.*, 2, 26–33.

20. Scannapieco, F.A., Bush, R.B., and Paju, S. (2003), Associations between periodontal disease and risk for nosocomial bacterial pneumonia and chronic obstructive pulmonary disease. A systemic review, *Ann. Periodontol.*, 8, 54–69.

21. What are key characteristics of biofilm? (Available at http://www.erc.montana.edu/biofilmbook/MODULE01/Mod01_Blue/Mod01_S04_Blue.htm.) Accessed on 6th February 2009.

22. Takala, A., Jousela, I., Takkunen, O., Kautiainen, H., Jansson, S.E., Orpana, A., Karonen, S.L., and Repo, H. (2002), A prospective study of inflammation markers in patients at risk of indirect acute lung injury, *Shock*, 17, 252–257.

23. Iacopino, A.M. (2001), Periodontitis and diabetes interrelationships: Role of inflammation, *Ann. Periodontol.*, 6, 125–137.

24. Safkan-Seppala, B., Sorsa, T., Tervahartiala, T., Beklen, A., and Konttinen, Y.T. (2006), Collagenases in gingival cervical fluid in Type 1 diabetes mellitus, *J. Periodontol.*, 77, 189–194.

25. Mahamed, D.A., Marleau, A., Alnaeeli, M., Singh, B., Zhang, X., Penninger, J.M., and Teng, Y.T. (2004), G(-) anaerobes-reactive CD4+ T-cells trigger RANKL- mediated enhanced alveolar bone loss in diabetic NOD mice, *Diabetes*, 54, 1477–1486.

26. Kesavalu, L., Chandrasekar, B., and Ebersole, J.L. (2002), *In vivo* induction of proinflammatory cytokines in mouse tissue by *Porphyromonas gingivalis* and *Actinobacillus actinomycetemcomitans*, *Oral Microbiol. Immunol.*, 12, 177–180.

27. Steer, S.A., Scarim, A.L., Chambers, K.T., and Corbett, J.A. (2006), Interleukin-1 stimulates beta-cell necrosis and release of the immunological adjuvant HMGB1, *PLo. S. Med.*, 3, e17.

28. Arnush, M., Heitmeier, M.R., Scarim, A.L., Marino, M.H., Manning, P.T., and Corbett, J.A. (1998), IL-1 produced and released endogenously within human islets inhibits beta cell function, *J. Clin. Invest.*, 102, 516–526.

29. Schara, R., Medvescek, M., and Skaleric, U. (2006), Periodontal disease and diabetes metabolic control: A full-month disinfection approach, *J. Int. Acad. Periodontol.*, 8, 61–66.

30. Iwamoto, Y., Nishimura, F., Nakagawa, M., Sugimoto, H., Shikata, K., Makino, H., Fukuda, T., Tsuji, T., Iwamoto, M., and Murayama, Y. (2001), The effect of antimicrobial periodontal treatment on circulating tumor necrosis factor-alpha and glycated haemoglobin level in patients with Type 2 diabetes, *J. Periodontol.*, 72, 774–778.

31. Kol, R. and Palattella, A. (2006), The use of doxycycline in periodontology. Historic *in vivo* study on mice affected by diabetes mellitus, *Minerva Stomatol.*, 55, 77–86.

32. Emsley, H.C. and Tyrrell, P.J. (2002), Inflammation and infection in clinical stroke, *J. Cereb. Blood Flow Metab.*, 22, 1399–1419.

33. Rodrigues, P.H. and Progulske-Fox, A., (2005), *Gene expression* profile analysis of *Porphyromonas gingivalis* during invasion of human coronary artery endothelial cells, *Infect. Immun.*, 73, 6169–6173.

34. Kuramitsu, H.K., Qi, M., Kang, I.C., and Chen, W. (2001), Role of periodontal bacteria in cardiovascular diseases, *Ann. Periodontol.*, 64, 41–47.

35. Yamazaki, K., Ohsawa, Y., Tabeta, K., Ito, H., Ueki, K., Oda, T., Yoshie, H., and Seymour, G.J. (2002), Accumulation of human heat shock protein 60-reactive T cells in the gingival tissues of periodontitis patients, *Infect Immun.*, 70, 2492–2501.

36. Haraszthy, V.I., Zambon, J.J., Trevisan, M., Zeid, M., and Genco, R. (2000), Identification of periodontal pathogens in atheromatous plaques, *J. Periodontol.*, 71, 1554–1560.

37. Sole, M. My favourite immune protein: Interleukin-6. Accessed on January 16, 2009. (Available at www.bio.davidson.edu/Courses/Immunology/Students/Spring2003/Sole/myfavprotein.htm.)

38. Pussinen, P.J., Jousilahti, P., Alfthan, G., Palosuo, T., Asikainen, S., and Salomaa, V. (2003), Antibodies to periodontal pathogens are associated with coronary heart disease, *Atheroscler. Thromb. Vasc. Biol.*, 23, 1250–1254.

39. Pussinen, P.J., Alfthan, G., Rissanen, H., Reunanen, A., Asikainen, S., and Knekt, P. (2004), Antibodies to periodontal pathogens and stroke risk, *Stroke*, 35, 2020–2023.

40. Desvarieux, M., Demmer, R.T., Rundek, T., Boden-Albala, B., Jacobs, D.R., Jr., Sacco, R.L., and Papapanou, P.N. (2005), Periodontal microbiota and carotid intima-media thickness: The Oral Infections and Vascular Disease Epidemiology Study (INVEST), *Circulation*, 111, 576–582.

41. Jain, A., Batista, E.L., Jr., Serhan, C., Stahl, G.L., and Van Dyke, T.E. (2003), Role for periodontitis in the progression of lipid deposition in an animal model, *Infect. Immun.*, 71, 6012–6018.

42. Gibson, F.C. 3rd., Hong, C., Chou, H.H., Yumoto, H., Chen, J., Lien, E., Wong, J., and Genco, C.A. (2004), Innate immune recognition of invasive bacteria accelerates atherosclerosis in apolipoprotein E-dependent mice, *Circulation*, 109, 2801–2806.

43. Kozarov, E.V., Dorn, B.R., Shelburne, C.E., Dunn, W.A., Jr., and Progulske-Fox, A. (2005), Human atherosclerotic plaque contains viable invasive Actinobacillus actinomycetemcomitants and *Porphyromonas gingivalis*. *Arterrioscler. Thromb. Vasc. Biol.*, 25, e17–18.

44. Marques da Silva, R., Lingaas, P.S., Geiran, O., Tronstad, L., and Olsen, I. (2003), Multiple bacteria in aortic aneurysms, *J. Vasc. Surg.*, 38, 1384–1389.

45. Fiehn, N.E., Larsen, T., Christiansen, N., Holmstrup, P., and Schroeder, T.V. (2005), Identification of periodontal pathogens in atherosclerotic vessels, *J. Periodontol.*, 76, 740–748.

46. D'Aiuto, F., Nibali, L., Parkar, M., Suvan, J., and Tonetti, M.S. (2005), Short-term effects of intensive periodontal therapy on serum inflammatory markers and cholesterol, *J. Dent. Res.*, 84, 269–273.

47. Taylor, B.A., Tofler, G.H., Carey, H.M., Morelkopp, M.C., Philcox, S., Carter, T.R., Elliott, M.J., Kull, A.D., Ward, C., and Schenck, K. (2006), Full-month tooth extraction lowers systemic inflammatory and thrombotic markers of cardiovascular risk, *J. Dent. Res.*, 85, 74–78.

48. Rahman, A.U., Rashid, S., Noon, R., Samuel, Z.S., Lu, B., Borgnakkle, W.S., and Williams, R.C. (2005), Prospective evaluation of the systemic inflammatory marker C-reactive protein in patients with end-stage periodontitis getting teeth replaced with dental implants: A pilot investigation, *Clin. Oral. Implants Res.*, 16, 128–131.

49. Costerton, J. and Keller, D. (2007), Oral peripathogens and systemic effects, *Gen. Dent.*, 210–215. (Available at www.agd.org.)

50. Hoffmann, J.A., Kafatos, F.C., Janeway, C.A., Jr., and Ezekowitz, R.A. (1999), Phylogenetic perspectives in innate immunity, *Science*, 284, 1313–1318.

51. Donlan, R.M. (2000), Role of biofilms in antimicrobial resistance [published erratum appears in: ASAIO *J. Pathol.* 2001; 47(1):99], *ASAIO J.*, 46, S47–S52.

52. Jesatitis, A.J., Franklin, M.J., Berglund, D., Sasaki, M., Lord, C.I., Bleazard, J.B., Duffy, J.E., Beyenal, H., and Lewandowski Z. (2003), Compromised host defense of *Pseudomonas aeruginosa* biofilms: characterization of neutrophils and biofilm interactions, *J. Immunol.*, 171, 4329–4339.

53. Sauer, K., Camper, A.K., Ehrlich, G.D., Costerton, J.W., and Davies, D.G. (2002), *Pseudomonas aeruginosa* displays multiple phenotypes during development as a biofilm, *J. Bacteriol.*, 184, 1140–1154.

54. Boles, B.R., Thoendel, M., and Singh, P.K. (2004), Self-generated diversity produces "insurance effects" in biofilm communities, *Proc. Natl. Acad. Sci. USA*, 101, 16630–16635.

55. Components of the immune system. November 2005. The Merck Manuals Online Medical Library. (Available at http://www.merck.com/mmpe/sec13/ch163b.html.) Accessed on September 29, 2009.

56. Ward, K.H., Olson, M.E., Lam, K., and Costerton, J.W. (1992), Mechanism of persistent infection associated with peritoneal implants, *J. Med. Microbiol.*, 36, 406–413.

57. Shiau, A.L. and Wu, C.L. (1998), The inhibitory effect of *Staphylococcus epidermidis* slime on the phagocytosis of murine peritoneal macrophages is interferon-independent, *Microbiol. Immunol.*, 42, 33–40.

58. Meluleni, G.J., Grout, M., Evans, D.J., and Pier, G.B. (1995), Mucoid *Pseudomonas aeruginosa* growing in a biofilm in vitro are killed by opsonic antibiodies to the mucoid exopolysaccharide capsule but not by antibodies produced during chronic lung infection in cystic fibrosis patients, *J. Immunol.*, 155, 2029–2038.

59. Yasuda, H., Ajiki, Y., Aoyama, J., and Yokota T. (1994), *Interaction* between human polmorphonuclear leucocytes and bacteria released from *in vitro* bacterial biofilm models, *J. Med. Microbiol.*, 41, 359–367.

60. Kinane, D.F., Riggio, M.P., Walker, K.F., MacKenzie, D., and Shearer, B. (2005), Bacteraemia following periodontal procedures, *J. Clin. Periodontol.*, 32, 708–713.

61. Forner, L., Larsen, T., Kilian, M., and Holmstrup, P. (2006), Incidence of bacteremia after chewing, tooth brushing and scaling in individuals with periodontal inflammation, *J. Clin. Periodontol*, 33, 401–407.

62. Misra, S., Percival, R.S., Devine, D.A., and Duggal, M.S. (2007), A Pilot study to assess bacteraemia associated with tooth brushing using conventional, electric or ultrasonic tooth brushes, *Eur. Arch. Paediatr. Dent.*, 8 (Suppl. 1), 42–45.

63. Verderame, R.A., Cobb, C.M., Killoy, W.J., and Drisko, C.L. (1989), Scanning electron microscopic examination of pocket wall epithelium and associated plaque in localized juvenile periodontitis, *J. Clin. Periodontol.*, 16, 234–241.

64. Matheny, J.L., Abrams, H., Johnson, D.T., and Roth, G.I. (1993), Microcirculatory dynamics in experimental human gingivitis, *J. Clin. Periodontol.*, 20, 578–583.

65. Kerdvongbundit, V., Vongsavan, N., Soo-Ampon, S., and Hasegawa, A. (2003), Microcirculation and micromorphology of healthy and inflamed gingivae, *Odontology*, 91, 19–25.

66. Rudney, J.D., Chen, R., and Sedgewick, G.J. (2001), Intracellular *Actinobacilus actinomycetemcomitans* and *Porphyromonas gingivalis* in buccal epithelial cells collected from human subjects, *Infect. Immun.*, 69, 2700–2707.

67. Zhang, Y., Wang, T., Chen, W., Yilmaz, Ö., Park, Y., Jung, Y-II., Hackett, M., and Lamont, R.J. (2005), Differential protein expression by *Porphyromonas gingivalis* in response to secreted epithelial cell components, *Proteomics*, 5, 198–211.

68. Sullivan, K.J., Goodwin, S.R., Sandler, E., and Joyce, M. (2005), Critical care of the pediatric hematopoietic stem cell transplant recipient in 2005, *Pediat. Transplant.*, 9 (Suppl. 7), 12–24.

69. Lü, F.X. and Jacobson, R.S. (2007), Oral mucosal immunity and HIV/SIV infection, *J. Dent. Res.*, 86, 216–226.

70. Cooper, M.A., Pommering, T.L., and Korányi, K. (2003), Primary immunodeficiencies, *Am. Fam. Physician*, 68, 2001–2008.

71. Ganz, T. (2001), Antimicrobial proteins and peptides in host defense, *Semin. Respir. Infect.*, 16, 4–10.

72. Geissmann, F., Jung, S., and Littman, D.R. (2003), Blood monocytes consist of two principal subsets with distinct migratory properties, *Immunity*, 19, 71–76.

73. Michel, C.C. (1997), Starling: the formulation of his hypothesis of microvascular fluid exchange and its significance after 100 years, *Exp. Physiol.*, 82, 1–30.

74. Mc Dowell, J. and Windelspecht, M. (2004), *The Lymphatic System*, Greenwood Press, Westport, CT.

75. Cuk, M., Radoseviç-Strasiç, B., Milin, C., Kirigin, M., and Rukavina, D. (1987), Lymphoid system as a regulator of morphostasis and hormonal modulation of these functions, *Ann. N. Y. Acad. Sci.*, 496, 104–107.

76. Shachar, I. and Flavell, R.A. (1996), Requirement for invariant chain in B cell maturation and function, *Science*, 274, 106–108.

77. IgG, March 2008, Dalhousie University Faculty of Medicine Immunology Bookcase. (Available at http://pim.medicine.dal.ca/igg.htm.) Acessed on September 30, 2009.

78. Norrby-Teglund, A., Ihendyane, N., Kansal R., Basma, H., Kotb, M., Andersson, J., and Hammarström, L. (2000), Relative neutralizing activity in polyspecific IgM, IgA, and IgG preparations against Group A streptococcal superantigens, *Clin. Infect. Dis.*, 31, 1175–1182.

79. Oppenheim, J.J., Biragyn, A., Kwak, L.W., and Yang, D. (2003), Role of antimicrobial peptides such as defensins in innate and adaptive immunity, *Ann. Rheumatol. Dis.*, 62 (Suppl. 2), ii17–ii21.

80. Nelson, K.C., Zhao, M., Schroeder, P.R., Li, N., Wetsel, R.A., Diaz, L.A., and Liu Z. (2006), Role of different pathways of the complement cascade in experimental bullous pemphigoid, *J. Clin. Invest.*, 116, 2892–2900.

81. Akbar, A.N., Vukmanovic-Stejic, M., Taams, L.S., and Macallan, D.C. (2007), The dynamic co-evolution of memory and regulatory CD4+ T cells in the periphery, *Nat. Rev. Immunol.*, 7, 231–237.

82. Lebbink, R.J. and Meyaard, L. (2007), Non-MHC ligands for inhibitory immune receptors: Novel insights and implications for immune regulation, *Mol. Immunol.*, 44, 2153–2164.

83. Figueiredo, C., Horn, P.A., Blasczyk, R., and Seltsam, A. (2007), Regulating MHC expression for cellular therapeutics, *Transfusion*, 47, 18–27.

84. Castriconi, R., Dondero, A., Cantoni, C., Della Chiesa, M., Prato, C., Nanni, M., Fiorini, M., Notarangelo, L., Parolini, S., Moretta, L., Notarangelo, L., Moretta, A., and Bottino, C. (2007), Functional characterization of natural killer cells in type I leukocyte adhesion deficiency, *Blood*, 109, 4873–4881.

85. Stout, R.D. and Suttles, J. (1997), T cells signalling of macrophage function in inflammatory disease, *Front. Biosci.*, 2, 197–206.

86. Prasse, A., Germann, M., Pechkovsky, D.V., Markert, A., Verres, T., Stahl, M., Melchers, I., Luttmann, W., Müller-Quernheim, J., and Zissel, G. (2007), IL-10 producing monocytes differentiate to alternatively activated macrophages and are increased in atopic patients, *J. Allergy Clin. Immunol.*, 119, 464–471.

87. Myung, P.S., Clements, J.L., White, D.W., Malik, Z.A., Cowdery, J.S., Allen, L.H., Harty, J.T., Kusner, D.J., and Koretzky, G.A. (2000), *In vitro* and *in vivo* macrophage function can occur independently of SLP-76, *Int. Immunol.*, 12, 887–897.

88. MacIvor, D.M., Shapiro, S.D., Pham, C.T., Belaaouaj, A., Abraham, S.N., and Ley, T.J. (1999), Normal neutrophil function in cathepsin G-deficient mice, *Blood*, 94, 4282–4293.

89. Orsi, N. (2004), The antimicrobial activity of lactoferrin: current status and perspectives, *Biometals*, 17, 189–196.

90. Chen, C.Y., Gherzi, R., Andersen, J.S., Gaietta, G., Jürchott, K., Royer, H-D., Mann, M., and Karin, M. (2000), Nucleolin and YB-1 are required for JNK- mediated interleukin-2 mRNA stabilization during T-cell activation, *Genes Dev.*, 14, 1236–1248.

91. Allen, S.J., Crown, S.E., and Handel, T.M. (2007), Chemokine: receptor structure, interactions and antagonism, *Annu. Rev. Immunol.*, 25, 787–820.

92. Dumestre-Pérard, C., Doerr, E., Colomb, M.G., and Loos, M. (2007), Involvement of complement pathways in patients with bacterial septicaemia, *Mol. Immunol.*, 44, 1631–1638.

93. Leid, J.G., Wilson, C.J., Shirtliff, M.E., Hassett, D.J., Parsek, M.R., and Jeffers, A.K. (2005), The exopolysaccharide alginate protects *Pseudomonas aeruginosa* biofilm bacteria from IFN-gamma-mediated macrophage killing, *J. Immunol.*, 175, 7512–7518.

94. Wagner, C., Kaksa, A., Müller, W., Denefleh, B., Heppert, V., Wentzensen, A., and Hänsch, G.M. (2004), Polymorphonuclear neutrophils in posttraumatic osteomyletitis: cells recovered from the inflamed site lack chemotactic activity but generate superoxides, *Shock*, 22, 108–115.

95. Quirynen, M., Bollen, C.M.L., Vandekerckhove, B.N.A., Dekeyser, C., Papaioannou, W., and Eyssen, H. (1995), Full- vs. partial-mouth disinfection in the treatment of periodontal infections: short-term clinical and microbiological observations, *J. Dent. Res.*, 74, 1459–1467.

96. Kinniment, S.L., Wimpenny, J.W.T., Adams, D., and Marsh, P.D. (1996), The effect of chlorhexidine on defined, mixed culture oral biofilms grown in a novel model system, *J. Appl. Bacteriol.*, 81, 120–125.

97. Reynolds, M.A., Lavigne, C.K., Minah, G.E., and Suzuki, J.B. (1992), Clinical effects of simultaneous ultrasonic scaling and subgingival scaling and subgingival irrigation with chlorhexidine. Mediating influence of periodontal probing depth, *J. Clin. Periodontol.*, 19, 595–600.

98. Jeong, S.-N., Han, S.-B., Lee, D.-W., and Magnusson, I. (1994), Effects of tetracycline-containing gel and a mixture of tetracycline and citric acid containing gel on non-surgical periodontal therapy, *J. Periodontol.*, 65, 840–847.

99. Jones, D.S., Irwin, C.R., Brown, A.F., Woolfson, A.D., Coulter, W.A., and McClelland, C. (2000), Design, characterisation and preliminary clinical evaluation of a novel mucoadhesive topical formulation containing tetracycline for the treatment of periodontal disease, *J. Cont. Rel.*, 67, 357–368.

100. Medlicott, N.J., Rathbone, M.J., Tucker, I.G., and Holborow, D.W. (1994), Delivery systems for the administration of drugs to the periodontal pocket, *Adv. Drug Deliv. Rev.*, 13, 181–203.

101. Minabe, M., Takeuchi, K., Tamura, T., Hori, T., and Umemoto, T. (1989), Subgingival administration of tetracycline on a collagen film, *J. Periodontol.*, 60, 552–556.

102. Medlicott, N.J., Tucker, I.G., Rathbone, M.J., Holborow, D.W., and Jones, D.S. (1996), Chlorhexidine release from poly(ε-caprolactone) films prepared by solvent evaporation, *Int. J. Pharm.*, 143, 25–35.

103. Agarwal, R.K., Robinson, D.H., Maze, G.I., and Reinhardt, R.A. (1993), *Development* and characterization of tetracycline-poly(lactide/glycolide) films for the treatment of periodontitis, *J. Control. Rel.*, 23, 137–146.

104. Cho, C.S., Ha, J.H., Kim, S.H., Han, S.Y., and Kwon, J.K. (1996), Tetracycline release from bioerodible hydrogels based on semiinterpenetrating polymer networks composed of poly(ε-caprolactone) and poly(ethylene glycol) macromer *in vitro*, *J. Appl.Poly. Sci.*, 60, 161–467.

105. Radvar, M., Pourtaghi, N., and Kinane, D.F. (1996), Comparison of 3 periodontal local antibiotic therapies in persistent periodontal pockets, *J. Periodontol.*, 67, 860–865.

106. Jones, D.S., Woolfson, A.D., and Brown, A.F. (1997), Textural analysis and flow rheometry of novel, bioadhesive antimicrobial oral gels, *Pharm. Res.*, 14, 450–457.

107. Jones, D.S., Woolfson, A.D., and Brown, A.F. (1998), Viscoelastic properties of bioadhesive, chlorhexidine-containing semi-solids for topical application to the oropharynx, *Pharm. Res.*, 15, 1131–1136.

108. Jones, D.S., Irwin, C.R., Woolfson, A.D., Djokic, J., and Adams, V. (1999), Physicochemical characterization and preliminary in vivo efficacy of bioadhesive, semisolid formulations containing flurbiprofen for the treatment of gingivitis, *J. Pharm. Sci.*, 88, 592–598.

109. Jones, D.S., Brown, A.F., and Woolfson, A.D. (2001), Rheological characterization of bioadhesive, antimicrobial, semisolids designed for the treatment of periodontal diseases: transient and dynamic viscoelastic and continuous shear analysis, *J. Pharm. Sci.*, 90, 1978–1990.

110. Jones, D.S., Lawlor, M.S., and Woolfson, A.D. (2003), Rheological and mucoadhesive characterization of polymeric systems composed of poly(methylvinylether-*co*-maleic anhydride) and poly(vinylpyrrolidone), designed as platforms for topical drug delivery, *J. Pharm. Sci.*, 92, 995–1007.

111. Jones, D.S., Lawlor, M.S., and Woolfson, A.D. (2004), Formulation and characterization of tetracycline-containing bioadhesive polymer networks designed for the treatment of periodontal disease, *Curr. Drug Del.*, 1, 17–25.

112. Rosan, B. and Lamont, R.J. (2000), Dental plaque formation, *Microbes. Infect.*, 2, 1599–1607.

113. Allaker, R.P. and Ian Douglas, C.W. (2009), Novel anti-microbial therapies for dental plaque-related diseases, *Int. J. Antimicrob. Agents*, 33, 8–13.

114. Heasman, P.A. and Seymour, R.A. (1994), Pharmacological control of periodontal disease I. Antiplaque agents, *J. Dent.*, 22, 323–335.

115. Gordon, J.M. and Walker, C.B. (1993), Current status of systemic antibiotic usage in destructive periodontal disease, *J. Periodontal.*, 64, 760–771.

116. Kinane, D.F. (2000), Local antimicrobial therapies in periodontal disease, *Ann. R. Aust. Coll. Dent. Surg.*, 15, 57–60.

117. Soskolne, W.A. (1997), Subgingival delivery of therapeutic agents in the treatment of periodontal diseases, *Crit. Rev. Oral Biol. Med.*, 8, 164–174.

118. Pandit, J.K. (1997), Targeted devices for periodontal diseases, In: Jain, N.K. (Ed.), *Controlled Drug Delivery*, CBS Publishers and distributors: New Delhi, India, pp. 130–146.

119. Goodson, J.M., Holborow, D., Dunn, R.L., Hogan, P., and Dunham, S. (1983), Monolithic tetracycline-containing fibers for controlled delivery to periodontal pockets, *J. Periodontol.*, 54, 575–579.

120. Goodson, J.M., Offenbacher, S., Farr, D.H., and Hogan, P.E. (1985), Periodontal disease treatment by local drug delivery, *J. Periodontol.*, 56, 265–272.

121. Coventry, J. and Newman, H.N. (1982), Experimental use of a slow release device employing chlorhexidine gluconate in areas of acute periodontal inflammation, *J. Clin.Periodontol.*, 9, 129–133.

122. Wan Yusof, W.Z., Newman, H.N., Strahan, J.D., and Coventry, J.F. (1984), Subgingival metronidazole in dialysis tubing and subgingival chlorhexidine irrigation in the control of chronic inflammatory periodontal disease,*J. Clin. Periodontol.*, 11, 166–175.

123. Newman, M.G., Kornman, K.S., and Doherty, F.M. (1994), A 6-month multi- center evaluation of adjunctive tetracycline fiber therapy used in conjunction with scaling and root planing in maintenance patients: clinical results, *J. Periodontol.*, 65, 685–691.

124. Addy, M., Rawle, L., Handley, R., Newman, H.N., and Coventry, J.F. (1982), The development and *in vitro* evaluation of acrylic strips and dialysis tubing for local drug delivery, *J. Periodontol.*, 53, 693–699.

125. Addy, M., Hassan, H., Moran, J., Wade, W., and Newcombe, R. (1988), Use of antimicrobial containing acrylic strips in the treatment of chronic periodontal disease. A three month follow-up study, *J. Periodontol.*, 59, 557–564.

126. Friedman, M. and Golomb, G. (1982), New sustained release dosage form of chlorhexidine for dental use. I. Development and kinetics of release, *J. Periodontal. Res.*, 17, 323–328.

127. Stabholz, A., Soskolne, W.A., Friedman, M., and Sela, M.N. (1991), The use of sustained release delivery of chlorhexidine for the maintenance of periodontal pockets: 2-year clinical trial, *J. Periodontol.*, 62, 429–433.

128. Golomb, G., Friedman, M., Soskolne, A., Stabholz, A., and Sela, M.N. (1984), Sustained release device containing metronidazole for periodontal use, *J. Dent. Res.*, 63, 1149–1153.

129. Elkayam, R., Friedman, M., Stabholz, A., Soskolne, A.W., Sela, M.N., and Golub, L. (1988), Sustained release device containing minocycline for local treatment of periodontal disease, *J. Control. Rel.*, 7, 231–236.

130. Kimura, S., Toda, H., Shimabukuro, Y., Kitamura, M., Fujimoto, N., Miki, Y., and Okada, H. (1991), Topical chemotherapy in human periodontitis using a new controlled-release insert containing ofloxacin. I. Microbiological observation, *J. Periodontal. Res.*, 26, 33–41.

131. Higashi, K., Morisaki, K., Hayashi, S., Kitamura, M., Fujimoto, N., Kimura, S., Ebisu, S., and Okada, H. (1990), Local ofloxacin delivery using a controlled-release insert (PT-01) in the human periodontal pocket, *J. Periodontal. Res.*, 25, 1–5.

132. Deasy, P.B., Collins, A.E., MacCarthy, D.J., and Russell, R.J. (1989), Use of strips containing tetracycline hydrochloride or metronidazole for the treatment of advanced periodontal disease, *J. Pharm. Pharmacol.*, 41, 694–699.

133. Steinberg, D., Friedman, M., Soskolne, A., and Sela, M.N. (1990), A new degradable controlled release device for treatment of periodontal disease: *in vitro* release study, *J. Periodontol.*, 61, 393–398.

134. Norling, T., Lading, P., Engström, S., Larsson, K., Krog, N., and Nissen, S.S. (1992), Formulation of a drug delivery system based on a mixture of monoglycerides and

triglycerides for use in the treatment of periodontal disease, *J. Clin. Periodontol.*, 19 (9 Pt 2), 687–692.

135. Kelly, H.M., Deasy, P.B., Ziaka, E., and Claffey, N. (2004), Formulation and preliminary *in vivo* dog studies of a novel drug delivery system for the treatment of periodontitis, *Int. J. Pharm.*, 274, 167–183.
136. Schwach-Abdellaoni, K., Vivien-Castioni, N., and Gurny, R. (2000), Local delivery of antimicrobial agents for the treatment of periodontal diseases, *Eur. J. Pharm. Biopharm.*, 50, 83–99.
137. Ahuja, A., Ali, J., and Rahman, S. (2006), Biodegradable periodontal intrapocket device containing metronidazole and amoxycillin: formulation and characterisation, *Pharmazie*, 61, 25–29.
138. Zilberman, M. and Elsner, J.J. (2008), Antibiotic-eluting medical devices for various applications, *J. Control. Rel.*, 130, 202–215.
139. Shifrovitch, Y., Binderman, I., Bahar, H., Berdicevsky, I., and Zilberman, M. (2009), Metronidazole-loaded bioabsorbable films as local antibacterial treatment of infected periodontal pockets, *J. Periodontol.*, 80, 330–337.
140. Zilberman, M., Shifrovitch, Y., Aviv, M., and Hershkovitz, M. (2009), Structured drug-eluting bioresorbable films: microstructure and release profile, *J. Biomater. Appl.*, 23, 385–406.
141. Siegel, S.C. and von Fraunhofer, J.A. (2002), The effect of handpiece spray patterns on cutting efficiency, *J. Am. Dent. Assoc.*, 133, 184–188.
142. O'Donnell, M.J., MacCarthy, D., and Coleman, D.C. (2006), Microbiology and cross-infection control, in: Ireland, R., Ed., *Clinical Textbook of Dental Hygiene and Therapy*, Blackwell Munksgaard, Oxford, UK, pp. 181–207.
143. Coleman, D.C., O'Donnell, M.J., Shore, A.C., Swan, J., and Russell, R.J. (2007), The role of manufacturers in reducing biofilms in dental chair waterlines, *J. Dent.*, 35, 701–711.
144. Shearer, B.G. (1996), Biofilm and the dental office, *J. Am. Dent. Assoc.*, 127, 181–189.
145. Wirthlin, M.R., Marshall, G.W., Jr., and Rowland, E.W. (2003), Formation and decontamination of biofilms in dental unit waterlines, *J. Periodontol.*, 74, 1595–1609.
146. Tuttlebee, C.M., O'Donnell, M.J., Keane, C.T., Russell, R.J., Sullivan, D.J., Falkiner, F., and Coleman, D.C. (2002), Effective control of dental chair unit waterline biofilm and marked reduction of bacterial contamination of output water using two peroxide-based disinfectants, *J. Hosp. Infect.*, 52, 192–205.
147. Pankhurst, C.L., Johnson, N.W., and Woods, R.G. (1998), Microbial contamination of dental unit waterlines: the scientific argument, *Int. Dent. J.*, 48, 359–368.
148. Mills, S.E. (2000), The dental unit waterline controversy: defusing the myths, defining the solutions, *J. Am. Dent. Assoc.*, 131, 1427–1441.
149. Pankhurst, C.L., Coulter, W., Philpott-Howard, J.N., Surman-Lee, S., Warburton, F., and Challacombe, S. (2005), Evaluation of the potential risk of occupational asthma in dentists exposed to contaminated dental unit waterlines, *Prim. Dent. Care*, 12, 53–59.
150. Dutil, S., Veillette, M., Mériaux, A., Lazure, L., Barbeau, J., and Duchaine, C. (2007), Aerosolization of mycobacteria and legionellae during dental treatment:

low exposure despite dental unit contamination, *Environ. Microbiol.*, 9, 2836–2843.

151. Anon (1993), Council Directive 93 / 42 / EEC of 14 June 1993 concerning medical devices, *OJEU*, L169, 1–43.
152. O'Donnell, M.J., Tuttlebee, C.M., Falkiner, F.R., and Coleman, D.C. (2005), Bacterial contamination of dental chair units in a modern dental hospital caused by leakage from suction system hoses containing extensive biofilm, *J. Hosp. Infect.*, 59, 348–360.
153. Leggat, P.A. and Kedjarune, U. (2001), Bacterial aerosols in the dental clinic: a review, *Int. Dent. J.*, 51, 39–44.
154. Harrel, S.K. and Molinari, J. (2004), Aerosols and splatter in dentistry: a brief review of the literature and infection control implications, *J. Am. Dent. Assoc.*, 135, 429–437.
155. Göksay, D., Cotuk, A., and Zeybek, Z. (2008), Microbial contamination of dental unit waterlines in Istanbul, Turkey, *Environ. Monit. Assess.*, 147, 265–269.
156. Tanzi, M.L., Capobianco, E., Affanni, P., Pizzi, S., Vitali, P., and Veronesi, L. (2006), *Legionella* spp. in hospital dental facilities, *J. Hosp. Infect.*, 63, 232–234.
157. Pankhurst, C.L. and Coulter, W.A. (2007), Do contaminated dental unit waterlines pose a risk of infection? *J. Dent.*, 35, 712–720.
158. Veronesi, L., Capobianco, E., Affanni, P., Pizzi, S., Vitali, P., and Tanzi, M.L. (2007), *Legionella* contamination in the water system of hospital dental settings, *Acta Biomed.*, 78, 117–122.
159. Atlas, R.M., Williams, J.F., and Huntington, M.K. (1995), *Legionella* contamination of dental-unit waters, *Appl. Environ. Microbiol.*, 61, 1208–1213.
160. Borella, P., Bargellini, A., Marchesi, I., Rovesti, S., Stancanelli, G., Scaltriti, S., Moro, M., Montagna, M.T., Tatò, D., Napoli, C., Triassi, M., Montegrosso, S., Pennino, F., Zotti, C.M., Ditommaso, S., and Giacomuzzi, M. (2008), Prevalence of anti-legionella antibodies among Italian hospital workers, *J. Hosp. Infect.*, 69, 148–155.
161. Putnins, E.E., Di Giovanni, D., and Bhullar, A.S. (2001), Dental unit waterline contamination and its possible implications during periodontal surgery, *J. Periodontol.*, 72, 393–400.
162. Schwartz, D.A. (2001), Does inhalation of endotoxin cause asthma? *Am. J. Respir. Crit. Care Med.*, 163, 305–306.
163. Michel, O., Kips, J., Duchateau, J., Vertongen, F., Robert, L., Collet, H., Pauwels, R., and Sergysels, R. (1996), Severity of asthma is related to endotoxin in house dust, *Am. J. Respir. Crit. Care Med.*, 154, 1641–1646.
164. Berlutti, F., Testarelli, L., Vaia, F., Luca, M.D., and Dolci, G. (2003), Efficacy of anti-retraction devices in preventing bacterial contamination of dental unit water lines, *J. Dent.*, 31, 105–100.
165. Montebugnoli, L., Sambri, V., Cavrini, F., Marangoni, A., Testarelli, L., and Dolci, G. (2004), Detection of DNA from periodontal pathogenic bacteria in biofilm obtained from waterlines in dental units, *New Microbiol.*, 27, 391–397.
166. Petti, S. and Tarsitani, G. (2006), Detection and quantification of dental unit water line contamination by oral streptococci, *Infect. Control Hosp. Epidemiol.*, 27, 504–509.

167. Kohn, W.G., Collins, A.S., Cleveland, J.L., Harte, J.A., Eklund, K.J., and Malvitz, D.M. (2003), Centers for Disease Control and Prevention (CDC), Guidelines for infection control in dental health-care settings-2003, *MMWR Recomm. Rep.*, 52, 1–61.

168. Lancellotti, M., de Oliveira, M.P., and de Ávila, F.A. (2007), Research on *Staphylococcus* spp. in biofilm formation in water pipes and sensibility to antibiotics, *Braz. J. O. Sci.*, 6, 1283–1288.

169. Scheid, R.C., Kim, C.K., Bright, J.S., Whitley, M.S., and Rosen, S. (1982), Reduction of microbes in handpieces by flushing before use, *J. Am. Dent. Assoc.*, 105, 658–660.

170. Rice, E.W., Rich, W.K., Johnson, C.H., and Lye, D.J. (2006), The role of flushing dental water lines for the removal of microbial contaminants, *Public Health Rep.*, 121, 270–274.

171. Furuhashi, M. and Miyamae, T. (1985), Prevention of bacterial contamination of water in dental units, *J. Hosp. Infect.*, 6, 81–88.

172. Pankhurst, C.L. and Philpott-Howard, J.N. (1993), The microbiological quality of water in dental chair units, *J. Hosp. Infect.*, 23, 167–174.

173. Dayoub, M.B., Rusilko, D.J., and Gross, A. (1978), A method of decontamination of ultrasonic scalers and high speed handpieces, *J. Periodontol.*, 49, 261–265.

174. Pankhurst, C.L., Philpott-Howard, J.N., Hewitt, J.H., and Casewell, M.W. (1990), The efficacy of chlorination and filtration in the control and eradication of *Legionella* from dental chair water systems, *J. Hosp. Infect.*, 16, 9–18.

175. Murdoch-Kinch, C.A., Andrews, N.L., Atwan, S., Jude, R., Gleason, M.J., and Molinari, J.A. (1997), Comparison of dental water quality management procedures, *J. Am. Dent. Assoc.*, 128, 1235–1243.

176. Walker, J.T., Bradshaw, D.J., Fulford, M.R., and Marsh, P.D. (2003), Microbiological evaluation of a range of disinfectant products to control mixed-species biofilm contamination in a laboratory model of a dental unit water system, *Appl. Environ. Microbiol.*, 69, 3327–3332.

177. Walker, J.T. and Marsh, P.D. (2004), A review of biofilms and their role in microbial contamination of dental unit water systems (DUWS), *Int. Biodeterior. Biodegrad.*, 54, 87–98.

178. Walker, J.T. and Marsh, P.D. (2007), Microbial biofilm formation in DUWS and their control using disinfectants, *J. Dent.*, 35, 721–730.

179. O'Donnell, M.J., Shore, A.C., Russell, R.J., and Coleman, D.C. (2007), Optimisation of the long-term efficacy of dental chair waterline disinfection by the identification and rectification of factors associated with waterline disinfection failure, *J. Dent.*, 35, 438–451.

180. Coleman, D.C., O'Donnell, M.J., Shore, A.C., and Russell, R.J. (2009), Biofilm problems in dental unit water systems and its practical control, *J. Appl. Microbiol.*, 106, 1424–1437.

181. Taylor-Hardy, T.L., Leonard, R.H., Jr., Mauriello, S.M., and Swift, E.J., Jr. (2001), Effect of dental unit waterline biocides on enamel bond strengths, *Gen. Dent.*, 49, 421–425.

182. Roberts, H.W., Karpay, R.I., and Mills, S.E. (2000), Dental unit waterline antimicrobial agents' effect on dentin bond strength, *J. Am. Den. Assoc.*, 131, 179–183.

183. Fan, P.L., Batchu, H., Chou, H.N., Gasparac, W., Sandrik, J., and Meyer, D.M. (2002), Laboratory evaluation of amalgam separators, *J. Am. Dent. Assoc.*, 133, 577–584.

184. Stone, M.E., Kuehne, J.C., Cohen, M.E., Talbott, J.L., and Scott, J.W. (2006), Effect of iodine on mercury concentrations in dental-unit wastewater, *Dent. Mater.*, 22, 119–124.

185. Hammarback, B., Mills, S., and Johnson, R. (2007), Re: Stone et al. "Effect of iodine on mercury concentrations in dental-unit wastewater", *Dent. Mater.*, 23, 1590–1592.

186. Roberts, H.W., Marek, M., Kuehne, J.C., and Ragain, J.C. (2005), Disinfectants' effect on mercury release from amalgam, *J. Am. Dent. Assoc.*, 136, 915–919.

187. Batchu, H., Chou, H.N., Rakowski, D., and Fan, P.L. (2006), The effect of disinfectants and line cleaners on the release of mercury from amalgam, *J. Am. Dent. Assoc.*, 137, 1419–1425.

188. Marais, J.T. and Brozel, V.S. (1999), Electro-chemically activated water in dental unit water lines, *Br. Dent. J.*, 187, 154–158.

189. Kohno, S., Kawata, T., Kaku, M., Fuita, T., Tsutsui, K., Ohtani, J., Tenjo, K., Motokawa, M., Tohma, Y., Shigekawa, M., Kamata, H., and Tanne, K. (2004), Bactericidal effects of acidic electrolyzed water on the dental unit waterline, *Jpn. J. infect. Dis.*, 57, 52–54.

190. Martin, M.V. and Gallagher, M.A. (2005), An investigation of the efficacy of superoxidised (Optident / Sterilox) water for the disinfection of dental unit water lines, *Br. Dent. J.*, 198, 353–354.

191. Zhang, W., Onyango, O., Lin, Z., Lee, S.S., and Li, Y. (2007), Evaluation of Sterilox for controlling microbial biofilm contamination of dental water, *Compend. Contin. Educ. Dent.*, 28, 586–588.

192. Yabune, T., Imazato, S., and Ebisu, S. (2005), Inhibitory effect of PVDF tubes on biofilm formation in dental unit waterlines, *Dent. Mater.*, 21, 780–786.

193. Sacchetti, R., De Luca, G., and Zanetti, F. (2007), Influence of material and tube size on DUWLs contamination in a pilot plant, *New Microbiol.*, 30, 29–34.

194. Rogers, J., Dowsett, A.B., Dennis, P.J., Lee, J.V., and Keevil, C.W. (1994), Influence of plumbing materials on biofilm formation and growth of *Legionella pneumophila* in potable water systems, *Appl. Environ. Microbiol.*, 60, 1842–1851.

195. Rogers, J., Dowsett, A.B., Dennis, P.J., Lee, J.V., and Keevil, C.W. (1994), Influence of temperature and plumbing material selection on biofilm formation and growth of *Legionella pneumophila* in a model potable water system containing complex microbial flora, *Appl. Environ. Microbiol.*, 60, 1585–1592.

196. O'Donnell, M.J., Shore, A.C., and Coleman, D.C. (2006), A novel automated waterline cleaning system that facilitates effective and consistent control of microbial biofilm contamination of dental chair unit waterlines: a one-year study, *J. Dent.*, 34, 648–661.

197. Szymanska, J. (2006), Bacterial decontamination of DUWL biofilm using Oxygenal 6, *Ann. Agric. Environ. Med.*, 13, 163–167.

198. Bakhir, V.M., Zadorozhny, Y., Leonov, B.I., Panicheva, S.A., and Prilutsky, V.I. (2001), Electrochemical activation: water treatment and production of effective solutions, *Proceeding of the Third International Symposium Electrochemical Activation in Medicine, Agriculture and Industry*, 3–25.

199. Bakhir, V.M. (1997), Electrochemical activation; theory and practice, *Proceedings of the First International Symposium on Electrochemical Activation*, 38–45.

200. Bakhir, V. Bibliography: Vitold Bakhir Electrochemical Systems and Technologies Institute [Internet]. 129515, Moscow, post office box 107, Russia; c2005 [cited 2009 March 6]. (Available at http://www.vbinstitute.org/bibliography/.)

201. Bakhir, V. Electrochemical Activation: Vitold Bakhir Electrochemical Systems and Technologies Institute [Internet]. 129515, Moscow, post office box 107, Russia; c2005 [cited 2009 March 6]. (Available at http://www.vbinstitute.org/terms/.)

202. Bakhir, V.M. and Zadorozhny, Y.G., Electrochemical Cell. A flow-through electrochemical modular cell of FEM-3 type, US Patent No. 5,635,040, filed 11.03.1996, pbl. 3.06.1997.

203. O'Donnell, M.J., Russell, R.J., and Coleman, D.C. (2008), Development of hospital-wide centralised automated waterline biofilm control, *J. Dent. Res.*, 87, Special Issue, Abst. No. 0586.

204. O'Donnell, M.J., Boyle, M., Swan, J., Russell, R.J., and Coleman, D.C. (2009), A centralised, automated dental hospital water quality and biofilm management system using neutral Ecasol™ maintains dental unit waterline output at better than potable quality: A 2-year longitudinal study, *J. Dent.*, 37, 748–762.

## CHAPTER 8

# IMPLICATIONS OF BIOFILM FORMATION IN CHRONIC WOUNDS AND IN CYSTIC FIBROSIS

## 8.1. INTRODUCTION

The skin is regarded as the largest organ of the body and has many different functions. Wounds with tissue loss include burn wounds, wounds caused as a result of trauma, diabetic ulcers, and pressure sores. The regeneration of damaged skin includes complex tissue interactions between cells, extracellular matrix molecules, and soluble mediators in a manner that results in skin reconstruction. The moist, warm, and nutritious environment provided by wounds, together with diminished immune functioning secondary to inadequate wound perfusion, may allow a build-up of physical factors (e.g., devitalized, ischemic, hypoxic, or necrotic tissue and foreign material), all of which provide an ideal environment for bacterial growth [1]. In burns, infection is the major complication after the initial period of shock. It is currently estimated that ~75% of the mortality following burn injuries is related to infections rather than to osmotic shock and hypovolemia [2].

Infection is defined as a homeostatic imbalance between the host tissue and the presence of microorganisms at a concentration that exceeds $10^5$ organisms $g^{-1}$ of tissue or the presence of beta-hemolytic streptococci [3]. It is associated with a large variety of wound occurrences ranging from traumatic skin tears and burns to chronic ulcers and complications following surgery and device implantations. If the wound setting manages to overcome the microorganism

*Biofilm Eradication and Prevention: A Pharmaceutical Approach to Medical Device Infections*, By Tamilvanan Shunmugaperumal

invasion by a sufficient immune response, then the wound should heal via the common four-phased process of coagulation, inflammation, proliferation, and remodeling [4]. If not, the formation of an infection can seriously limit the wound healing process, can interfere with wound closure, and may even lead to bacteremia, sepsis, and multisystem failure. People who suffer from immunosuppressive disorders obviously face a higher risk of infection. However, evidence of increasing bacterial resistance is on the rise, and complications associated with infections are therefore expected to increase in the general population. Concern among healthcare practitioners regarding the risk of wound infection is justifiable not only in terms of increased suffering to the patient, but also in view of its economic burden to society.

The epidemic increase in obese humans worldwide is followed by a similar increase in diabetes and cardiovascular diseases. Such patients are particularly prone to the development of chronic wounds, which become colonized by a number of bacterial species. Gjødsbøl et al. [5] investigated the bacterial profile of chronic venous leg ulcers and found *Staphylococcus aureus* (in 93.5% of the investigated ulcers), *Enterococcus faecalis* (71.7%), *Pseudomonas aeruginosa* (52.2%), CoNS (45.7%), *Proteus* species (41.3%), and anaerobic bacteria (39.1%). In Denmark and in the United States, it has been estimated that 1–2% of the populations, respectively, are experiencing a nonhealing wound [6]. Consequently, this has become a burden to the healthcare systems and the patients experience suffering, lost employment, and reduced quality of life.

Among hospitalized patients, 8–10% are susceptible to infection by opportunistic pathogenic bacteria (e.g., *P. aeruginosa* and *S. aureus*), which are notorious for forming chronic, biofilm-based infections in their hosts. At present, there exists some controversy as to the impact of opportunistic pathogens on wound healing. Although the deep dermal tissues of all chronic wounds harbor multiple species, Gjødsbøl et al. [5] found that more than one-half of the chronic wounds investigated in their study were colonized with *P. aeruginosa*. Furthermore, the *P. aeruginosa*-infected wounds appeared significantly larger in terms of area than wounds that did not contain *P. aeruginosa* [5,7,8]. The presence of *P. aeruginosa* also seems to delay or even prevent the healing process [7]. Even now, the biofilm mode of bacterial growth has not yet been satisfactorily identified in chronic wounds. However, the clinical signs are similar to other biofilm-based infections. With regard to antibiotic therapy, there is a lack of evidence concerning optimal regimes or clinical hallmarks for efficient treatment and additionally, biofilm infections are characterized by resistance to the host defenses [9–11]. If this is also true for infected chronic wounds, it explains why the only proficient treatment has proven to be surgical removal, which, in severe cases involves amputation of the extremities [12].

The main goal in treating various types of wound infections is to decrease the bacterial load in the wound to a level that enables wound healing processes to take place. Conventional systemic delivery of antibiotics entails poor penetration into ischemic and necrotic tissue and can cause systemic toxicity with associated renal and liver complications, which result in a need for hospitalization for monitoring. Alternative local delivery of antibiotics by either topical administration or by a delivery device may enable the maintenance of a high local antibiotic concentration for an extended duration of release without exceeding systemic toxicity. Antibiotics already incorporated in controlled-release devices include vancomycin, tobramycin, cefamandol, cephalothin, carbenicillin, amoxicillin, and gentamicin. Many research activities are undertaken based on antibiotic-eluting medical devices that include antibiotic-loaded bioresorbable films for orthopedic and periodontal applications, and bioresorbable fibers for wound healing applications [13].

## 8.2. CHARACTERISTICS OF CHRONIC WOUNDS

Healing of an acute or surgical wound follows a process that has been divided into four stages: coagulation, inflammation, cell proliferation and repair of the matrix, and epithelialization and remodeling.

Chronic wounds all have unique origins. Similarities in pathogenesis have made it feasible to divide these into various groups (e.g., venous leg ulcers, diabetic foot ulcers, and pressure ulcers). Each group has their principles for treatment based on current knowledge of pathogenesis. Venous leg ulcers are precipitated by malfunction of venous valves causing venous hypertension in the crural veins, increased pressure in capillaries, and edema. Venous pressure exceeding 45 mmHg inevitably leads to development of a leg ulcer. The treatment of the venous leg ulcer is compression, which often heals the ulcer. The diabetic foot ulcer is caused by repetitive load on the neurophatic and often ischemic foot and treatment is off-loading and restoration of circulation. Pressure ulcers are caused by sustained or repetitive load on often vulnerable areas (e.g., the sciatic tuberculum, sacral region, heels, and shoulders) in the immobilized patient. Treatment is pressure relief with offloading mattresses, cushion seats, and ambulation of the patient. Despite relevant and correct treatment, some of these wounds will not heal or heal very slowly. The causes of delayed healing in some patients also include dysfunction in the diabetic fibroblasts [14], immunological defects due to inborn defects or cancer, malnutrition, obesity, drug abuse, alcoholism, and smoking.

All these factors act as brakes on the healing process and deteriorate the wound. Even though different wound types do not share origin or cause, they do share characteristics with respect to increased levels of cytokines, matrix metalloproteases (MMPs), and cellular activity when they have progressed to the chronic state.

Nevertheless, chronic wounds might show individual combinations of causes, each contributing to its nonhealing status that makes it unlikely to find a common cure for all patients. Chronic wounds seem to be stuck in the inflammatory stage characterized by a continuing influx of neutrophils (i.e., PMNs)

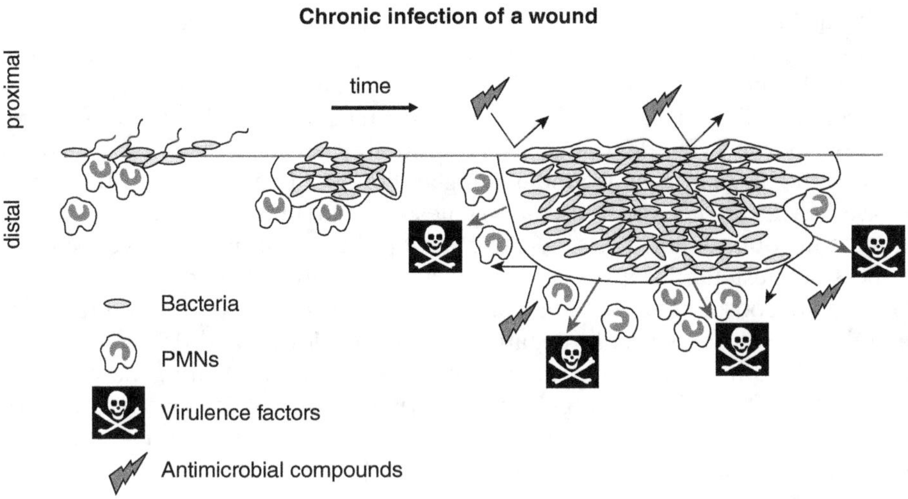


**Figure 8.1.** Development of a biofilm in a chronic infected wound. (See color insert.)

[15] that release cytotoxic enzymes, free oxygen radicals, and inflammatory mediators that cause extensive collateral damage to the host tissue [16]. Additionally, within a chronic wound, healing and destructive processes are out of balance and consequently, by manipulating and counterbalancing these processes, the chronic wound might start to heal.

The stepwise development of a biofilm in a chronic infected wound is shown in Figure 8.1. Small numbers of surviving planktonic bacteria develop over time into microcolonies, with the formation of larger aggregations referred to as biofilms. The biofilm bacteria are encased in a self-produced polymeric matrix. The ability to form biofilms is believed to be one of the main survival strategies of bacteria in a hostile environment. In this state, the bacteria tolerate antimicrobial compounds (e.g., antibiotics, and the action of host cells). Biofilm formation also facilitates the build-up of bacterial cell–cell signaling molecules used in a process termed quorum-sensing (QS). When a certain cell density is reached, the QS system dictates the production of virulence factors, some of which offer a shield against the attended PMNs.

## 8.3. THE HOST RESPONSE

Both cellular and humoral responses take part in the inflammatory process of chronic wounds. Similar to any other infection, PMNs are detected in high amounts in chronic wounds as described by Lobmann et al. [17]. The humoral response involves MMPs, growth factors, and cytokines. In the present context, excess amounts of MMPs are interesting. Matrix metalloproteinases belong to a family of zn-dependent endoproteinases that are involved in the degradation of extracellular matrix (ECM) components. They are produced by several different cells (e.g., fibroblasts, macrophages, eosinophils, but in particular the PMNs). Their production is stimulated by cytokines, growth factors, and cell–cell contact. The MMPs participate in the first phase of the wound-healing process, by removing devitalized tissue, and are therefore believed to play an important role in normal wound healing and remodeling [18]. As for the repair phase, MMPs are necessary for angiogenesis, wound matrix contraction, migration of fibroblasts and keratinocytes, and epithelialization. However, several papers suggest that elevated levels of active MMPs impair wound healing [17,19,20]. Consequently, wound-care products have been developed that aim at relieving the supposedly negative effects of elevated MMPs in order to promote healing [21]. In particular, infections with *P. aeruginosa* show altered levels of MMPs and MMP regulating cytokines [19,22]. Additionally, there are scarcely any reports on antibody development against *P. aeruginosa*, with particular reference to chronic wounds [23,24].

## 8.4. THE BACTERIAL HOST-INFECTION VERSUS COLONIZATION

The term "a colonized wound" implies that a limited number of bacteria are present without affecting the healing process. Many studies of the bacteriology of chronic wounds are likely to be biased because sampling is based on biopsies or swabs. However, initial investigations of Bjarnsholt et al. [25] indicate that the bacteria are assembled in microcolony-based structures found in bacterial biofilms and are far from evenly distributed within the wound (Fig. 8.2). The implications are that cultures from a biopsy or swab are not likely to be representative for the total bacteriological load in the wound. The sample might be lacking the correct information of the colonizing organisms. As can

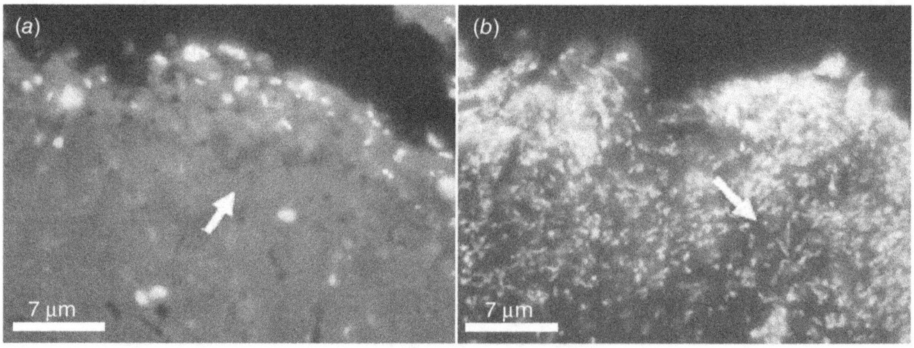


**Figure 8.2.** (*a*) Colonizing planktonic *P. aeruginosa* in a chronic wound; arrows show single cells. (*b*) Shows a larger collection of bacteria in a chronic wound infected with *P. aeruginosa*, with the arrow indicating a microcolony. (See color insert.)

be seen from the pictures, it is clear that the bacteria are not evenly distributed. The microcolony supports the author's hypothesis of biofilms in chronic infected wounds. Parts (*a*) and (*b*) are from the same chronically infected wound collected with the permission of the patient and in accordance with the biomedical project protocol (KA-20051011) approved by the Danish scientific ethical board. The wound material was in phosphate-buffered saline, with 4% imbedded in paraffin. Deparaffinized sections were analyzed by fluorescence *in situ* hybridization (FISH) using peptide nucleic acid (PNA) probes [26]. A mixture of Texas Red-labeled, *P. aeruginosa*-specific PNA probe and fluorescein isothiocyanate (FITC)-labeled universal bacterium PNA was used (AdvanDx Inc., Woburn, MA). Slides were read using a fluorescence microscope equipped with a Texas Red and an FITC filter.

The classic signs of an infected wound are tumor, rubor, dolor, calor, and functio laesa. Other criteria have been proposed [e.g., low transcutaneous oxygen tension ($TcPO_2$), presence of necrotic tissue, foul odor, pain, wound breakdown, or simply lack of healing] [27,28]. This suggests that bacteria might slow down the wound-healing process without sign of invasive growth. An objective, clinically useful endpoint for infection has been proposed: if $<10^5$ bacteria $g^{-1}$ tissue are present, the wound is only colonized, whereas $>10^5\,g^{-1}$ tissue it is infected [29–31]. This is highly debatable, however, because the infective dose of a given bacterium, as well as the host interaction with the surrounding microflora, varies.

When compared with cystic fibrosis (CF) patients who develop a chronic, pulmonary infection, the indifference of the diagnosis for when a wound is "only" colonized seems odd and even dangerous. The CF lower respiratory tract is recurrently colonized before it is infected; however, aggressive antibiotic treatments that delay the onset of chronic infection remarkably prolong the life expectancy of the patients [32]. Cystic fibrosis is an autosomal inherited disease, caused by a defect in the CF Transmembrane Conductance Regulator (CFTR) gene, located on chromosome 7 [33–35]. The major symptoms of a defect in the CFTR appear in the lungs, causing infections that, in most cases, lead to the premature death of the patients. The defect in the CFTR causes a decrease in epithelial chloride secretion and an increase in Na absorption. In the lung, this results in viscous dehydrated mucus that is very difficult to clear mechanically (e.g., by the activity of the mucociliary escalator). The CF patients suffer from acute infections of many different bacteria, the most common being *Haemophilus infuenzae*, *S. aureus*, *Streptococcus pneumonia*, *P. aeruginosa*, the *Burkholderia cepacia* complex, and *Stenotrophomonas maltophilia* [36,37]. The *H. influenzae* and *S. aureus* bacteria predominate early in life, but later *S. aureus* and *P. aeruginosa* become the predominating infectious organisms in the CF lung. Up to 80% of young adults suffering from CF are chronically infected with *P. aeruginosa* [36,38–40]. Several lines of evidence point to the fact that the bacterial infection of the CF lung shows the characteristics of the biofilm formation, which is believed to enable the colonization (or infection) to persist [41–43].

## 8.5. RECENT HYPOTHESIS IN CHRONIC WOUND

There are obvious similarities with respect to the bacterial infection found in CF and chronic wound patients. First, Bjarnsholt et al. [25] proposed that the conditions are kept chronic by the bacterial burden and that one of the main players is *P. aeruginosa*. Similarly, an ongoing transition phase between colonization and infection may exist. Although no indepth investigations have been conducted describing the significance of *P. aeruginosa*, the same authors also propose the presence of this bacterium in the form of a biofilm within the chronic wound. Bacterial persistence including the extreme tolerance to antibiotics and the activities of the immune defense is caused by the ability to form these biofilms and its excretion of damaging virulence factors including production of an efficient PMN shield. In addition, the similarities between CF and chronic wound patients in terms of being compromised hosts are striking. Cystic fibrosis and chronic wound patients both suffer from defects in the primary line of defense. In CF patients, it is the build-up of thickened mucus due to the chronic depletion of the paraciliary liquid layer that hampers the mechanical clearing process. The chronic wound consists primarily of granulation tissue composed of a network of collagen fibers and new capillaries and ECM together with PMNs, macrophages, and fibroblasts. Embedded in this are the above-mentioned microcolonies of bacteria. *Pseudomonas aeruginosa* is rarely present in acute wounds, but their presence is facilitated by antibiotic treatment of an infected or colonized wound. Therefore, the ability of *P. aeruginosa* to form biofilms in combination with the above deficiencies might enable *P. aeruginosa* to evade the host defense system. Because no defects in the cellular defense are present, the PMNs would be expected to eradicate this intruder. Antibiotic treatments are likely to reduce the number of planktonic bacteria in the wound; however, the biofilm-imbedded *P. aeruginosa* survive keeping the wound chronically infected and suppressing the activity of the cellular defense system. Necrotic killing and lysis of PMNs might account for the local high MMP content detected in chronic wounds and the vast amounts of DNA, proteases, and other components from PMNs detected in CF sputa. The implications of sustained PMN lysis are that antimicrobial, as well as tissue-devastating compounds, spill out. Examples of such compounds are myeloperoxidase, elastase [44], and MMP-9 [45], of which chronic wound fluids are particularly rich, in contrast to fluid from human acute wounds [46–48]. Thus, the quora of *P. aeruginosa* have amassed at certain locations in wounds. Such quora are capable of producing the PMN eliminating rhamnolipid, which in turn would reduce the number of functional PMNs at the present locations. This would explain the previously reported impairment of the host cells in chronic infections, which might tip the balance even further away from healing and in a negative feedback loop causes further increase in the production of MMPs from the incoming PMNs.

## 8.6. FUTURE WORK AND PERSPECTIVES

Successful treatment of a nonhealing wound depends on identifying and treating factors that act as brakes on the healing process. Until now, it has not been fully explained why chronic wounds have elevated MMPs or why PMNs exist in vast numbers in the chronic wound. The hypothesis put forward in this chapter could explain this and offer several possibilities to tip the balance in favor of healing. Adaptations from the CF clinic might prove to be resourceful. For many years, it has been known that an early eradication therapy against *P. aeruginosa* and other bacteria prevents the onset of chronic infection [49]. Boosting of the cellular immune system (e.g., using growth factors and MMP inhibitors, as well as local modulation of the invading bacteria by the topical application of Ag ions) might also promote healing. However, if this hypothesis holds, new treatment scenarios are within reach. Finally, a number of studies have proven the concept of QS inhibitors (QSIs) as a relevant antimicrobial treatment [50–52]. Bjarnsholt et al. [9] showed that by blocking the QS communication, freshly isolated PMNs were able to eradicate *in vitro* biofilms of *P. aeruginosa*. In line with this, it recently also has been found that the blocking of the QS system disables killing of the PMNs. In addition, QSI based treatment of mice with pulmonary infection by *P. aeruginosa* resulted in a significantly faster clearing compared with the placebo group [9,52,53]. Jensen et al. [54] analyzed the sputum from *P. aeruginosa*-infected patients and found that extracts caused necrotic lysis of freshly isolated PMNs. This indicates that application of QSI compounds will prevent the buildup of the PMN shield and consequently enable proper antimicrobial activities of the PMNs. Because PMN lysis is likely to be restored to a more moderate level, local tissue destruction is kept at a minimum. Both effects will be expected to tip the balance in favor of bacterial clearance and subsequent healing. Effective QS blockers have been designed, but further clinical tests including their relevance in wound-healing therapy are required.

Furthermore, recent hypothesis assumes permanent or transient defects in the first line of the host defense system enabling the first bacteria to settle down and give rise to populations with a sufficiently high density, organized as biofilms (microcolonies), to trigger QS controlled shields of the second line of defense mechanisms. This hypothesis helps explain why some wounds lack healing potential even though possible immune defects have been reverted. Although the biofilm mode of growth in wounds has to be satisfactorily confirmed (this includes detection of a biofilm matrix and another important hallmark: high-level tolerance to antimicrobial compounds), its presence would explain the inefficiency of antibiotics, and also the proposed biased microbiology of wounds, again disabling the correct antibiotic treatment. It is now generally accepted that all wounds are colonized; however, it is unrealistic to keep the wound sterile. By the use of QSI based drugs, it might be possible to affect the wound to such an extent that the host itself is able to eliminate the infecting bacteria and recreate the normal healing process.

## 8.7. WOUND DRESSINGS

Various wound dressings aim to restore the milieu required for skin regeneration and to protect the wound from environmental threats and penetration of bacteria. Although traditional gauze dressings offer some protection against bacteria, this protection is lost when the outer surface of the dressing becomes moistened by wound exudates or external fluids. Furthermore, traditional gauze dressings exhibit low restriction of moisture evaporation that may lead to dehydration of the wound bed. This situation may lead to adherence of the dressing, particularly as wound fluid production diminishes, thus causing pain and discomfort to the patient during removal.

Most modern dressings are designed according to the well-accepted bilayer structural concept: an upper dense "skin" layer to prevent bacterial penetration and a lower spongy layer designed to adsorb wound exudates and accommodate newly formed tissue. Unfortunately, dressing material absorbed with wound discharges provides conditions that are also favorable for bacterial growth. This has given rise to a new generation of wound dressings with improved curative properties that provide an antimicrobial effect by eluting various germicidal compounds.

## 8.8. WOUND DRESSINGS BASED ON SYNTHETIC POLYMERS

A variety of dressings that contain and release antibiotic agents at the wound surface have been introduced. These dressings are designed to provide controlled release of Ag ions through a slow but sustained release mechanism that helps avoid toxicity yet ensures delivery of a therapeutic dose of Ag ions to the wound [55].

A wide variety of semiocclusive dressing formats [e.g., foams (Contreet® antimicrobial foam, Coloplast), hydrocolloids (Urgotul SSD, Urgo), alginates (Silvercel®, Johnson & Johnson) and hydrofibers® (Aquacel, ConvaTec)] are available. For example, Acticoat® (Smith and Nephew) is a three-ply gauze dressing made of an absorbent rayon polyester core, with upper and lower layers of nanocrystalline Ag coated high-density polyethylene mesh [56]. It is applied wet and is then moistened with water several times daily to allow the release of the Ag ions so as to provide an antimicrobial effect for 3 days. Concerns have been raised by clinicians regarding the safety of the Ag ions included in most of these products. For example, it was found that a young person with 30% mixed depth burns, who received 1 week of local treatment with Acticoat, had shown hepatotoxicity and argyria-like symptoms, and the Ag levels in his plasma and urine were clearly elevated, as well as the liver enzymes, during the treatment period. Therefore the authors raised concern about potential Ag toxicity and suggested that the Ag levels we monitored in the plasma and/or urine during the treatment [57].

In order to address this issue, the Ag in Actisorb® (Johnson & Johnson) is impregnated into an activated charcoal cloth, after which it is encased in a nylon sleeve that does not enable the Ag in the product to be freely released at the wound surface, but nevertheless eradicates bacteria that adsorb onto the activated charcoal component.

Suzuki et al. [58,59] presented a new concept for an antibiotic delivery system that releases gentamicin only in the presence of wounds that are infected by *P. aeruginosa* with a potential for an occlusive dressing. Gentamicin is bound to a poly(vinyl alcohol) (PVA) derivative hydrogel through a specially developed peptide linker cleavable by a proteinase. This allows gentamicin to be released at specific times and locations, namely, when and where *P. aeruginosa* infection occurs. The PVA (linker)–gentamicin demonstrated selective release of gentamicin in *P. aeruginosa*-infected wound fluid, and caused a significant reduction in its growth *in vitro*.

The substantial disadvantage of these temporary dressings is the fact that similarly to textile wound dressings, the necessary change of dressings may be painful and increases the risk of secondary contamination. Bioresorbable dressings successfully address this shortcoming since they can easily be removed from the wound surface once they have fulfilled their role. Film dressings made of lactide–caprolactone copolymers [e.g., Topkin® (Biomet, Europe) and Oprafol® (Lohmann & Rauscher, Germany)] are available for clinical use. Biodegradation of the film occurs via hydrolysis of the copolymer into lactic acid and 6-hydroxycaproic acid. During the hydrolytic process, the pH shifts toward the acid range, with pH values as low as 3.6 measured *in vitro* [60]. Although these two dressings do not contain antibiotic agents, it is claimed that the low pH values induced by the polymer's degradation help reduce germ growth [61] and also promote epithelization [62]. Furthermore, local lactate concentrations can stimulate local collagen synthesis [63]. Film dressings are better suited for small wounds since they lack an absorbing capacity and are impermeable to water vapors and gases, both of which cause accumulation of wound fluids on larger wound surfaces.

Resorbable materials have gained a considerable position in the daily routine of all surgical disciplines. Natural products like catgut and collagen have been used historically. Since the development of synthetic macromolecules >40 years ago, the range of indications for the use of such materials has widened significantly in our daily routine. Suture materials, mesh, tissue pads, clips, screws, and anchors are in use. More recent developments in the field of orthopedic and trauma surgery include screw-plate systems, wound dressing materials and films for the prevention of adhesions and ossifications. This appears to be the beginning of an era of new materials as these implants not only fulfil a temporary biomechanical role, but in theory also can release a controlled amount of biologically active substances at a set timeframe. Also they are potential carriers for transplants on a cellular level. This aspect will be of importance in the orthopedic field, where resorbable films only play a minor role so far.

At the moment there are six resorbable or biodegradable films or foils on the market that are used or can be used in the field of orthopedic and trauma surgery. These are foils and films made of the following materials: carboxymethyl cellulose and hyaloronic acid (Seprafilm®), oxidized regenerated cellulose (Interceed®), polydioxanone (PDS®), and copolymers of lactic acid and caprolactone (Topkin®, Oprafol®, Mesofol®). Main indications for their use are wound dressing, especially after split skin graft and thermal wounds, prophylaxis of adhesions, and prevention of the formation of synostoses and heterotopic ossifications. The results of clinical trials are promising and the increasing number of publications in the last 5 years in this field is an expression of increasing demand of these materials. However, it could also be an expression of the growing interest in drug delivery techniques, as well as in tissue engineering that are possible with these materials.

Katti et al. [64] report the development of a biodegradable poly(D,L-lactide-*co*-glycolide) (PDLGA) nanofiber-based antibiotic delivery system. This system can serve as a biodegradable gauze dressing and an alternative to film dressings. This type of dressing is composed of continuous fibers that form a nonwoven fiber mesh by means of an electrospinning process. Briefly, the process of electrospinning involves use of a polymer solution that is contained in a syringe and held at the end of the needle by its surface tension. Charge is induced on the solution by an external electric field to overcome the surface tension and form a charged jet of solution. As this jet travels through air, it experiences instabilities and follows a spiral path. Evaporation of the solution leaves behind a charged polymer fiber that is collected on a grounded metal screen. Incorporation of antibiotics in the process of electrospinning requires solubility of the incorporated drug in the solvents used in the process [a mixture of dimethylformamide (dmf) and tetrahydrofuran (thf)]. Cefazolin was chosen as complying with this requirement. The effects of orifice diameter, applied voltage as well as polymer and drug solution concentrations were investigated and it was demonstrated that drug-loaded fibers can be electrospun. Although they had larger diameters than the unloaded fibers, they were still in the nanometer range. Antibiotic release from these fibers has not been reported to date.

## 8.9. WOUND DRESSINGS BASED ON NATURAL POLYMERS

Only a handful of natural materials have been investigated for wound dressing applications as either main or additional components to the dressing structure that are able to impact the local wound environment beyond moisture management and to elicit a cellular response. Collagen is the main structural protein of the ECM, and was one of the first natural materials to be utilized for skin reconstruction and dressing applications. Collagen-based products have been available commercially for over a decade. They come in a variety of set-ups ranging from gels, pastes, and powders to more elaborate sheets, sponges, and composite structures. Collagen's limitations as a wound dressing ingredient are mainly due to its rapid biodegradation by collagenase and its susceptibility to bacterial invasion [65–68]. Drug-eluting collagen sponges have been found useful in both partial- and full-thickness burn wounds. Collatamp® (Innocoll GmbH, Germany), Syntacoll (AG, Switzerland), Sulmycin®-Implant (Schering-Plough, USA), and Septocoll® (Biomet Merck, Germany) are several such products that have been found to accelerate both granulation tissue formation and epithelialization, as well as to protect the recovering tissue from potential infection or reinfection by eluting gentamicin. *In vivo*, drug is released by a combination of diffusion and natural enzymatic breakdown of the collagen matrix [69]. A comprehensive clinical study of gentamicin–collagen sponges demonstrated their ability to induce high local concentrations of gentamicin (up to 9000 $\mu g\,mL^{-1}$) at the wound site for at least 72 h while serum levels remained well below the established toxicity threshold of 10–12 $\mu g\,mL^{-1}$ [70].

Simple collagen sponge entrapment systems are characterized by high drug release upon the wetting of the sponge, typically within 1–2 h of application. Sripriya et al. [71] suggested improving the release profile of such systems by using succinylated collagen that can create ionic bonds with the cationic antibiotic ciprofloxacin so as to restrain its diffusion. It is claimed that in this way ciprofloxacin release corresponds to the nature of the wound in line with the amount of wound exudates absorbed in the sponge. Effective *in vitro* release from their system was found to last for 5 days, and was proven successful in controlling infection in rats. Other studies have aimed to better control drug release or improve wound healing properties by combining collagen with other synthetic or natural biodegradable elements. Prabu et al. [72] focused on achieving a more sustained release of the antimicrobial agent and describe a dressing made from a mixture of collagen and poly(caprolactone) loaded with gentamicin and amikacin, whereas Shanmugasundaram et al. [73] chose to impregnate collagen with alginate microspheres loaded with the antibacterial agent silver sulfadiazine (AgSD).

Other studies that focused on improving wound healing capabilities tried to incorporate tobramycin, ciprofloxacin [74], and AgSD [75] into collagen–hyaluronan-based dressings. The two latter studies do not show conclusive evidence of improved healing properties compared to their control. However, hyaluronan (HA), a structure-stabilizing component of the ECM, is thought to play a role in several aspects of the healing process with HA based dressings, and exhibits promising results in the management of chronic wounds (e.g., venous leg ulcers) [76,77].

A wide range of studies describe the employment of the polysaccharides chitin and chitosan as structural materials analogous to collagen for wound dressings. Both materials offer good wound protection and also have been found to encourage wound healing without excessive granulation tissue and scar formation [78]. Chitosan also has been documented as displaying considerable intrinsic antibacterial activity against a broad spectrum of bacteria [79]. Several attempts to improve the dressing's antibacterial capabilities by incorporating various agents (e.g., AgSD [80], chlorhexidine diacetate [81], and minocycline hydrochloride [82]) have been reported.

## 8.10. COMPOSITE FIBER STRUCTURES LOADED WITH ANTIBACTERIAL DRUGS FOR WOUND HEALING APPLICATIONS

Drug-eluting fibers can be used for various biomedical applications. Few controlled-release fiber systems based on polymers have been investigated to date. The two basic types of drug-loaded fibers that have been reported are monolithic fibers in which the drug is dissolved or dispersed throughout the polymer fiber, and hollow reservoir fibers in which the drug is added to the internal section of the fiber. The advantages of drug-loaded fibers include ease of fabrication, high surface area for controlled release, and localized delivery of bioactive agents to their target. Disadvantages of monolithic and reservoir fibers include poor mechanical properties due to drug incorporation and limitations in drug loading. Furthermore, many drugs and all proteins do not tolerate melt processing and organic solvents.

In very recent studies, Zilberman and co-workers [83–86] presented a new concept of core–shell fiber structures that successfully meets these challenges. These composite fibers combine a dense polyglyconate core fiber and a drug-loaded porous PDLGA shell structure (i.e., the antibacterial drug gentamicin is located in a separate compartment (a "shell") around the "core" fiber]. The shell is prepared using freeze drying of inverted emulsions with mild processing conditions. These unique fibers are designed to be used as basic elements of bioresorbable burn and ulcer dressings. Their investigation focused on the effects of the emulsion's composition (formulation) on the shell microstructure, on the drug release profile from the fibers and on the resulting bacterial inhibition. The freeze-drying technique is unique in being able to preserve the liquid structure in solids (Fig. 8.3).

Albumin was found to be the most effective surfactant for stabilizing the inverted emulsions. As a surfactant, it is located at the interface between the aqueous and the organic phase, reduces the interfacial tension between the two phases, and therefore significantly decreases the pore size. It also enables a high encapsulation efficiency and a relatively low burst release (33%) followed by a moderate release profile that enabled release of most of the loaded gentamicin within 2 weeks. This behavior probably results from albumin's ability to bind the gentamicin through specific interactions. The ability of albumin to bind drugs is well known [87]. Albumin can interact with acidic or basic drugs via van der Waals dispersion forces, hydrogen bonds, and ionic interactions. Based on these results, Zilberman and co-workers [83–86]

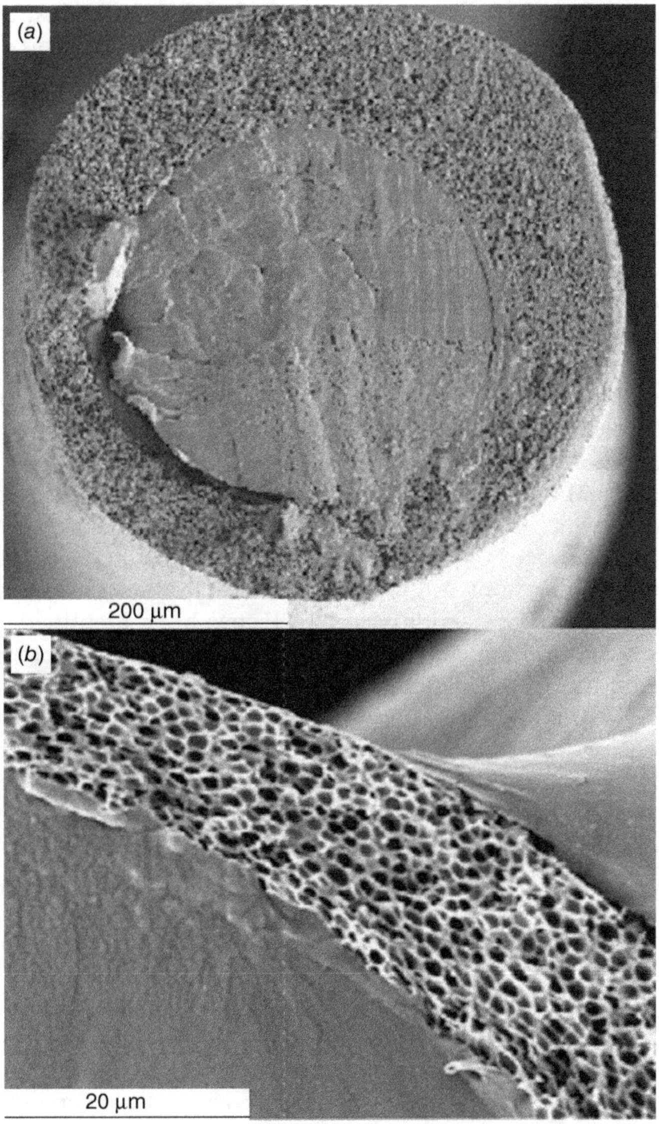


**Figure 8.3.** Scanning electron microscopy (SEM) photographs of a gentamicin-loaded composite fiber demonstrating the concept of core–shell fiber structures. (Reproduced with permission from Zilberman and Elsner *J. Control Rel.*, 130, 202–215, 2008 [13].)

chose albumin as the preferred surfactant in their systems, and most of the study focused on samples that were stabilized with albumin. In order to monitor the effectiveness of various concentrations of the antibiotic released from the fibers in terms of the residual bacteria compared with the initial number of bacteria, *in vitro* microbiological experiments were performed. It was found that these new fiber structures were effective against the relevant bacterial strains (*S. aureus*, *P. aeruginosa*, and *S. epidermidis*) and can be used as basic elements of bioresorbable drug-eluting wound dressings.

## 8.11. CYSTIC FIBROSIS

Cystic fibrosis, a chronic disease of the lower respiratory system, is the most common inherited disease. In this condition, the normal mucociliary clearance system that cleanses the bronchopulmonary epithelium of inhaled particles depends on an upward directional flow of a mucus layer on the tips of cilia that move freely in the underlying watery layer. In CF, there is a net deficiency of water, which hinders the upward flow of the mucus layer. Decreased secretion and increased absorption of electrolytes lead to dehydration and thickening of secretions covering the respiratory epithelium [88].

According to May et al. [89], 70% of patients with CF are defective in the CFTR protein, which results in altered secretions in the secretory epithelia. The hyperviscous mucus that is produced is thought to increase the incidence of bacterial lung infections in CF patients. According to Govan and Deretic [90], the CF gene, which encodes the CFTR, has been identified. The CFTR functions as a chloride ion channel protein. Chloride ion transport is severely impaired when the CFTR is defective in CF patients. *Staphylococcus aureus* is usually the first pulmonary isolate from these patients [89]. It can normally be controlled by antibiotics. *Staphylococcus aureus* and *H. influenzae* infections usually predispose the CF affected lung to colonization with *P. aeruginosa*. *Burkholderia cepacia* has also been shown to infect the lungs of CF patients with lethal consequences, but it has never attained the 80% colonization rate of *P. aeruginosa* [90].

The exact mechanism of *P. aeruginosa* colonization of the lungs of patients with CF is not known. There is evidence that enhanced pseudomonal receptors on the respiratory epithelia may be responsible; impaired mucociliary clearance is another possibility [90]. During initial colonization, the organisms are nonmucoid. Persistence of the organism in the lungs of patients with CF ultimately will result in a mucoid phenotype [88]. There is no clear interval between the initial colonization by *P. aeruginosa* and conversion to mucoid forms; it may take several months to years. The variable timing of the emergence indicates that this is caused by random mutations, followed by selection of mucoid strains in the lungs of patients with CF [90].

This mucoid phenotype was first observed by Lam et al. [42] in postmortem specimens of infected lung tissue and bronchoscopy material from infected patients. The mucoid material was shown to be a polysaccharide material, later identified as alginate. The conditions that trigger the conversion to the mucoid phenotype have been investigated. Hoyle et al. [91] demonstrated, using a chemostat and modified Robbins device, that mucoid exopolysaccharide was transiently produced following adherence of *P. aeruginosa*. May et al. [89]

noted that several *in vitro* conditions (e.g., nutrient limitation, the addition of surfactants, and suboptimal levels of antibiotics) may result in mucoidy. Mucoidy is even elicited by addition of ethanol to the medium, indicating that this phenotype may be a response to dehydration. Mathee et al. [92] showed that biofilms of *P. aeruginosa* challenged with either activated human peripheral blood PMNs or hydrogen peroxide ($H_2O_2$) (a product released in low levels by PMNs) yielded ~0.1% mucoid colonies, while unchallenged biofilms produced none. Alginate was overproduced by all the mucoid colonies. They hypothesized that activated PMNs and the release of toxic products (e.g., $H_2O_2$) could play a role in the generation of mucoid organisms during the inflammatory response.

The sputum from the lungs of patients with CF is usually filled with large numbers of PMNs, and the inflammatory defense mechanisms in the lungs of patients with CF against mucoid *P. aeruginosa* are usually dominated by PMNs and antibodies [92]. In contrast to *P. aeruginosa*, *B. cepacia* does not generally produce alginate-like compounds, though some investigators have reported the production of other exopolysaccharides. Mucoid colonial morphology in *B. cepacia* is rare in both environmental and clinical strains. The presence of biofilms or microcolonies of *Burkholderia* has not been reported for patients colonized solely by this organism [90].

A question posed by a number of investigators is why mucoid *P. aeruginosa* infections are so recalcitrant and resistant to immune system clearance. Koch and Hoiby [88] stated that the biofilm mode of growth protects the organisms from antimicrobial agents and host defenses. The alginate layer of mucoid strains appears to prevent antibody coatings and blocks the immunological determinants required for opsonic phagocytosis [89]. Mucoid strains are apparently more resistant to nonopsonic phagocytosis than are nonmucoid strains [89]. There is evidence that the alginate may promote adherence of the mucoid strains to epithelial cells in the pulmonary tract, thereby inhibiting clearance. *In vivo* experiments with infected rats confirmed this; mucoid *P. aeruginosa* strains were less rapidly removed from the pulmonary tract than were nonmucoid strains [89].

Another mechanism for persistence and survival was proposed by Cochrane et al. [93]. With the use of rats that had been artificially infected with agar beads containing *P. aeruginosa*, they found that the bacteria within these beads produced elevated levels of high-molecular-weight iron–regulated membrane proteins that can function as receptors for iron–siderophore complexes. These molecules aid in the scavenging of low levels of Fe from the bloodstream. A host defense mechanism against pathogenic organisms is to restrict available Fe in order to limit this essential bacterial nutrient. By producing Fe scavenging compounds, the organisms are better able to survive in the host.

Anwar et al. [94] also suggested that biofilm age was a critical factor in *P. aeruginosa* survival. In their experimental system, older biofilm cells of this organism were less susceptible to either whole blood or serum than were either younger biofilms or planktonic organisms.

The possibilities for successful treatment of CF may ultimately hinge on early antimicrobial treatment to prevent or delay chronic infection with *P. aeruginosa*. Koch and Hoiby [88] noted that early treatment with oral ciprofloxacin and inhaled colistin could postpone chronic infection with *P. aeruginosa* for several years. They also suggested that a vaccine against this organism might be effective in preventing initial colonization of the lungs of patients with CF.

## 8.12. *P. AERUGINOSA* BIOFILMS AND THE IMMUNE RESPONSE IN CF

Chronic *P. aeruginosa* lung infection in CF is characterized by (1) the mucoid phenotype producing abundance of alginate *in vitro* and in the patients, (2) microcolonies surrounded by alginate in sputum and in postmortem investigations and the fact that the bacteria stay on the surface of the airways as an endobronchiolitis without spreading to the blood or to other organs, (3) high levels of antibodies against alginate, and (4) resistance against the patients' defense mechanisms and against antibiotic treatment (surveyed by Pedersen [95]).

The noninflammatory defense of the lungs consists of the primary noninflammatory nonspecific defense mechanisms (e.g., the mucociliary escalator, coughing, alveolar macrophages, defensins and surfactant), and the secondary noninflammatory specific defense mechanisms [e.g., secretary IgA (s-IgA)]. The action of these defense mechanisms is silent and very efficient in normal persons and their activity does not give rise to any symptoms. No tissue damage is mediated through these defense mechanisms. Congenital defects of the primary noninflammatory defense mechanisms comprise CF (mucus) [88] and ciliary dyskinesia syndrome (cilia). These defects lead to secondary acute or chronic bacterial infections and recruitment of the inflammatory defense mechanisms of the lungs (e.g., IgG and PMNs). The activity of the inflammatory defense mechanisms may lead to successful killing of the offending pathogens, but in addition gives rise to local and systemic symptoms of inflammation (e.g., fever, tissue damage, and impaired function). If the infection is not eradicated (persistent or chronic infection) then immunopathologic tissue lesions occur (e.g., immune complex mediated tissue damage) (Table 8.1) [96–109]. Once the respiratory tract infections persist, most of the viscosity of sputum is due to DNA from the PMNs [110] as a consequence of the chronic inflammatory response.

*Pseudomonas aeruginosa* is an environmental species, which is found especially in freshwater and soil contaminated with human or animal waste. It is rarely found in the stools of healthy people and then only in small numbers [111]. The most prevalent and severe chronic lung infection in CF patients is caused by mucoid *P. aeruginosa* [95], which has become endemic in CF patients worldwide. The chronic *P. aeruginosa* infection in CF patients is an

**TABLE 8.1. The Pathogenesis of Chronic *P. aeruginosa* Lung Infection in CF: Immune Complex Mediated Tissue Damage**

| Stage of Infection | Mechanisms and Pathogenesis | References | Clinical Signs |
|---|---|---|---|
| Acquisition | Cross-infection, dentist's equipment, whirl pools concomitant virus infection | 96,97 | None<br>Acute exacerbation |
| Attachment | Pili, haemagglutinin, exoenzyme S, alginate, flagella | 98–100 | None |
| Initial persistent colonization | Bacterial toxins: elastase, alkaline protease, exotoxin A and exoenzyme S, phospholipase, lipase etc. | 101,102 | Minimal |
| Chronic infection | Persistence: microcolonies embedded in alginate. PMN pseudomonas mismatch. Tissue damage, immune complexes, PMN elastase, cytokines | 92,103–109 | Chronic suppurative lung inflammation, progressive loss of lung function |

endobronchiolitis caused by bacteria producing a biofilm, and this is probably the reason why they become resistant to antibiotics and to the defense mechanisms of the body (Table 8.1).

*Pseudomonas aeruginosa* produces many toxins and other virulence factors with potential negative effects on the lungs of CF patients [95]. Some of these toxins are thought to play a role during establishment of the initial persistent colonization of the CF resispiratory tract, notably elastase and alkaline protease, which have been shown to interfere with the nonspecific (phagocytes) and immunologically specific (T cells, NK cells, immunoglobulins) defense mechanisms [95,112]. Later, during the infection, the significance of the action of the toxins becomes doubtful, because specific antibodies are produced by the CF patients: For example, free elastase and alkaline protease can only be detected in bronchial secretions during the first few months of the infection before neutralizing antibodies develop [113].

The most characteristic feature of the persistent *P. aeruginosa* infection is the production of mucoid alginate and the formation of microcolonies in the lungs of the patients (Table 8.1). Alginate is an unbranched, linear heteropolysaccharide and it consists of polymannuronic–polyguluronic acid; it is the only antigen that is clinically correlated to poor prognosis in CF patients [114]. The biochemistry and genetics of alginate biosynthesis is to a large extent known

[90]. It has been widely speculated that it is the dehydrated and high osmolarity environment of the CF respiratory tract that contribute to the emergence of mucoid *P. aeruginosa* [115]. The initial colonization of CF patients is, however, accompanied by an inflammatory response [101], and CF patients have increased oxygen free radicals originating from PMNs during lung infection [116]. Likewise, abnormally high concentrations of $H_2O_2$ are present in exhaled air during acute pulmonary exacerbations [116]. Mathee et al. [92] provided experimental evidence that the oxygen radicals produced by the inflammatory response (PMNs) can induce mutations in the *mucA* gene, leading to the phenotypic change from non-alginate-producing to alginate-producing phenotypes of *P. aeruginosa*. Similar mutations were frequently found in mucoid variants from CF patients [92]. The significance of the inflammatory reaction for conversion of *P. aeruginosa* to the mucoid phenotype may also explain why mucoid strains are also sometimes found in non-CF patients suffering from chronic infections in the lungs and other regions of the body [117–119]. The median concentration of the mucoid exopolysaccharide (alginate) in sputum from CF patients is $35.5\,\mu g\,mL^{-1}$ [120]. Mucoid strains are also found in other chronically colonized patients [117–119], but such strains are characteristically very frequent in CF. *Pseudomonas aeruginosa* growing in alginate biofilms is highly resistant to antibiotics probably due to slow growth, penetration barriers, reduced oxygen concentrations in the base of the biofilm and β-lactamase production [121,122], and they are protected against phagocytes and complement [94,123]. In most patients, non-mucoid strains initiate the infection, and the transition to the mucoid variant correlates with the development of a pronounced antibody response against virtually all antigens and toxins of *P. aeruginosa*. The occurrence of mucoid variants also correlates with a poor prognosis [114]. Generally, avidity maturation of the antibody response during the course of the chronic infection, as shown by increase in the binding affinity to the specific antigen, does not occur, the exception being antibodies to chromosomal-lactamase [124]. Complement deposition on the surface of mucoid microcolonies may be deficient in CF patients [125], favoring the survival of such colonies, although the embedded bacteria are unusual in several other aspects. They are often serum sensitive [126], lack the LPS side chain [127], are nonmotile, and express Fe regulated outer-membrane proteins, indicating that the bacteria grow under Fe restricted conditions in the CF lungs. In addition, they sometimes become auxotrophic [128].

Lipopolysaccharide (LPS) has been found to be the major antigen component of immune complexes in sputum of CF patients with chronic *P. aeruginosa* infection [129]. This unique adaptability to the environment in the CF lung is also reflected in the high frequency of development of antibiotic resistance during chemotherapy. The most remarkable host response to the infection is the pronounced antibody response, which continues to increase for several years and is correlated to poor prognosis. As mentioned, these antibodies are eventually directed against most, if not all, antigens of *P. aeruginosa* including alginate, and belong to all classes and subclasses of immunoglobulins. The correlation between the antibody response and poor prognosis has been shown to be due to immune complex mediated chronic inflammation in the lungs of CF patients, including complement activation and high local production of proinflammatory cytokines (e.g., TNF, IL-1, IL-6, IL-8 and IRAP), whereas the concentration of the anti-inflammatory cytokine IL-10 is low [130,131]. This inflammatory reaction is dominated by PMNs, and released leucocyte proteases, myeloperoxidase, and oxygen radicals are the main mechanisms of lung tissue damage (Table 8.1) [103,104,132–134]. The predominance of PMNs is also found in animal models of the chronic *P. aeruginosa* lung infection [135]. This funding is in accordance with the paradigm that the immune response to microorganisms is dichotomic [i.e., dominated by either type 1 helper cells (Th1) (monocytes, gamma-interferon, cell-mediated immunity) or Th2 cells (PMNs, IL-4, antibody-mediated immunity)]. It seems therefore likely that the immune complex mediated PMN dominated inflammation in CF lungs represents a Th2 response to chronic *P. aeruginosa* infection [136–139]. Acyl-homoserine lactone QS of *P. aeruginosa* has been shown to stimulate production of cytokines (e.g., IL-8 in respiratory cells), which attracts PMNs [140] and to downregulate the production of IL-12 in lymphocytes [141,142], which stimulates a Th1 response. This funding may also have an important influence on the inflammation and Th1/Th2 balance of the immune response in CF patients.

## 8.13. ANTIBIOTIC THERAPY OF *P. AERUGINOSA* LUNG INFECTION

It is possible to prevent or at least delay the onset of chronic *P. aeruginosa* infection by early aggressive therapy of the intermittent colonization with oral ciprofloxacin in combination with colistin inhalation [49]. The results showed prevention of chronic *P. aeruginosa* infection in 80% of the patients in the treated group compared to untreated controls [143] and a changed epidemiology of the infection, where fewer young patients become chronically infected [143]. In the Danish CF center "maintenance chemotherapy" = chronic suppressive chemotherapy [144] is used for treatment of chronic *P. aeruginosa* infection. The principle of this treatment regime is to suppress the number and activity of the *P. aeruginosa* bacteria in the lungs of CF patients using tobramycin in combination with a β-lactam antibiotic. This principle has proved superior to "on demand" treatment of acute exacerbations of the chronic infection in the Danish CF center, and >90% of the patients survive for at least 10 years after onset of the chronic infection. This is in contrast to earlier periods when "on demand" treatment resulted in survival of only 50% for 5 years [144–146]. The principles of this chronic suppressive treatment are based upon the observation that the lung function improves during antibiotic treatment and this effect is still detectable 1–2 months after completion of the treatment. This principle is therefore to restore lung function repeatedly by regular 2-week courses of intensive intravenous treatment every 3 months in the CF center. The treatment is intensified in patients with unstable clinical condition by adding daily inhalations of colistin between the courses of intravenous antibiotics and sometimes also by giving oral ciprofloxacin during these intervals. Maintenance treatment with inhaled antibiotics or with ciprofloxacin every 3 months is also efficient [147–149]. The mechanism of action of the antibiotics in the chronic infection caused by biofilm-growing *P. aeruginosa* is not entirely clear. Although the biofilm mode of growth is the characteristic feature of the infection, planktonic bacteria susceptible to antibiotics also occur plentifully in the lungs. In addition, *in vitro* studies have shown that the number of biofilm-growing bacteria can be reduced to 20% by high doses of combinations of antibiotics (piperacillin + tobramycin) [150] and that ciprofloxacin is significantly more efficient than, for example, tobramycin in the treatment of *P. aeruginosa* biofilms [151]. Furthermore, sub-MIC concentrations of antibiotics have been shown to suppress the production of exoproducts (e.g., proteases and phospholipase C), and alginate of *P. aeruginosa* and colistin binds to LPS of *P. aeruginosa* [152–162]. According to these results, therefore, the decrease in the CFU of planktonic bacteria, and to some degree of biofilm bacteria, as well as the inhibition of exoproducts, which are considered to be virulence factors, will reduce the antigenic load in the lungs and therefore possibly the concentration of immune complexes. Accordingly, inflammatory parameters (WBCs, acute-phase proteins), lung function, and well-being improve during antibiotic therapy [149,163–167]. Reduction in bacterial density of *P. aeruginosa* correlated significantly with the improvement of lung function (FVC, FEV1, FEF 25–75%) [163]. Chronic *P. aeruginosa* infection in diffuse panbronchiolitis in Japan is caused by mucoid strains of these bacteria growing as a biofilm that is virtually impossible to eradicate by means of antibiotics [158]. Chronic suppressive therapy by means of long-term daily erythromycin is reported to significantly reduce symptoms and inflammatory parameters. It also increases the 10-year survival from 12 > 90% [168]. Similar results have been obtained with the new macrolides and with fluoroquinolones. In CF patients with chronic *P. aeruginosa* infection, a similar effect has been reported in an uncontrolled study [169]. The efficacy of macrolides in spite of their lack of bacteriostatic or bactericidal effect against *P. aeruginosa* has been studied *in vitro* and in animal models. It seems that it is due to a sub-MIC effect that inhibits the production of proteins (e.g., the exoproteases of *P. aeruginosa* and interference with the biofilm matrix [161] and an anti-inflammatory activity [170]).

## 8.14. AEROSOL DELIVERY TO THE LUNG

Aerosolization offers an attractive approach to deliver antimicrobials directly into the respiratory tract for treatment and prophylaxis of pulmonary infections [171,172]. Aerosolized solutions of aminoglycosides, particularly tobramycin, are used in patients with CF, where high endobronchial concentrations are achieved that may overcome bacterial resistance, as defined by standard susceptibility testing protocols [149,173,174]. Other applications of aerosol technology include aerosolized antibiotics in mechanically ventilated patients [175]. In this instance, an efficient method was achieved that delivers the aerosol beyond the endotracheal tube and drug levels in pulmonary secretions were several orders of magnitude higher than those following intravenous therapy. A number of questions have to be addressed, notably whether the drug can withstand aerosol generation (e.g., the high-frequency ultrasound used in some nebulizers). The resultant aerosol properties will depend not only on the physical process used to nebulize the drug, but also on the intrinsic device characteristics and its performance with a particular drug [176].

## 8.15. IRON AND CHELATED IRON AS ANTIBIOFILM DRUGS FOR CF

Recently, Fe has arisen as a point of focus in the biofilm literature. Singh et al. [177] discovered that treating *P. aeruginosa* with lactoferrin, a ferric iron-chelating protein, prevented the bacteria from forming biofilms. It was later discovered that treatment with supraphysiological Fe concentrations also prevented biofilm formation in *P. aeruginosa* and caused detachment and clearance of preformed biofilms in flow-chamber experiments [178]. This effect of high Fe concentrations inhibiting biofilm formation was also recently observed in a different strain of *P. aeruginosa* [179]. These authors show that high levels of Fe suppress the release of DNA, an important structural component of biofilms. As DNA release in *P. aeruginosa* biofilms is (at least partially) controlled through the pqs operon, it was suggested that Fe exerts its antibiofilm effect through repression of DNA release via the pqs operon [179]. Thus, it appears that *P. aeruginosa* biofilm formation is operative across a narrow range of Fe concentrations (~1–100 $\mu mol\,L^{-1}$), above and below which the organism can grow only in a planktonic state (Fig. 8.4). Biofilm sensitivity to Fe has also been demonstrated in *S. aureus* [180] and *Streptococcus* spp. [181,182].

Because biofilms play an important role in lung infections in CF patients, efficient inhibitors of *P. aeruginosa* biofilm formation hold considerable

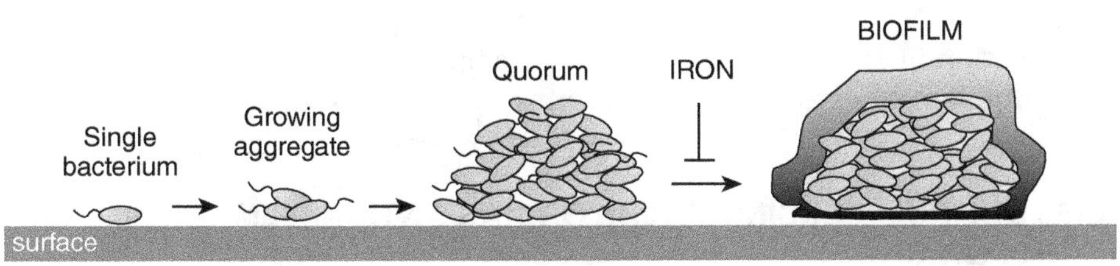


**Figure 8.4.** Schematic of the bacterial-biofilm formation process and its inhibition by high concentrations of Fe. The biofilm is depicted as a cut-away image. (See color insert.)

promise as therapeutic agents. Furthermore, the human lung has a unique potential for selective delivery of antibiofilm drugs directly to the site of a biofilm infection. Indeed, nebulized pharmaceuticals are used by virtually all CF patients who undergo medical treatments. These include Pulmozyme™, a recombinant form of human deoxyribonuclease I (rhDNase I) that thins mucosal secretions [183,184], and TOBI™, an inhaled form of the antibiotic tobramycin to treat *P. aeruginosa* infections [185–187]. These treatments allow for comparatively massive concentrations of antibiotic or other pharmacological agent to be administered to the lungs, concentrations that would be impractical or even deleterious if given systemically. Because of this ability, even therapeutic agents of only moderate potency stand an excellent chance of achieving clinically relevant concentrations through direct delivery to the lungs by nebulization. For example, a standard 300-mg dose of TOBI™ produces a peak sputum concentration in the lungs of ~1200 $\mu g\,mL^{-1}$ and a peak serum concentration of 0.9 $\mu g\,mL^{-1}$, while a serum concentration of only 12 $\mu g\,mL^{-1}$ can cause serious complications owing to cochlear and renal toxicity [185].

It is well-known that free Fe is indeed acutely toxic in humans and thus unsuitable for therapeutic use in CF. Intrigued by a putative role for elevated Fe concentrations in the treatment of CF, Musk and Hergenrother [188] evaluated the antibiofilm properties of Fe (chelated by a number of commercially available and in some cases clinically utilized Fe chelators) against *P. aeruginosa* PA14 in microtiter plate tests. In addition, they have probed the viability of using these chelated Fe forms as nebulized drugs for the treatment of CF by examining their particle size distribution profiles in an Andersen cascade impactor model. The most potent chelated Fe sources were then evaluated for antibiofilm activity in a battery of clinical *P. aeruginosa* strains isolated from the sputum of CF patients. Iron(III) acetohydroxamate and Fe(III) picolinate were both effective in disrupting biofilm formation with moderate potencies in these tests. All the chelated Fe forms tested showed superb distributive properties in an *in vitro* Andersen cascade impactor model for drug distribution in the human lung, suggesting that any of these compounds could be readily delivered directly to the CF lung via nebulization. Both Fe(III) acetohydroxamate and Fe(III) picolinate were also effective at inhibiting the formation of biofilms in a majority of clinical isolates taken from the sputum of CF patients. Taken as a whole, these data serve both to bolster the growing base of literature, which showing that elevated Fe concentrations cause biofilm perturbation in *P. aeruginosa* and suggest continued examination of chelated Fe sources as putative antibiofilm treatments in the CF lung.

## 8.16. QUORUM-SENSING INHIBITORS AS ANTIBIOFILM DRUG FOR CF

*Burkholderia cepacia* complex (Bcc) strains are opportunistic pathogens causing life-threatening infections in CF patients. *Burkholderia cepacia* complex strains are resistant to many antimicrobial agents and commonly produce biofilms *in vitro* and *in vivo*. This contributes to their virulence and makes Burkholderia infections difficult to treat. Although all Bcc species are found in CF patients, overall, *B. multivorans* and *B. cenocepacia* predominate [189,190]. Recently, the QS system of *Burkholderia* spp. has been found to affect their biofilm-forming ability, making it an attractive target for antimicrobial therapy. Brackman et al. [191] evaluated the antibiofilm effect of several known QS inhibitors (Table 8.2). Cinnamaldehyde [192], resveratrol [193], L-canavanine [194], 4-nitropyridine N-oxide, *p*-benzoquinon and indole [51], azithromycin [195], ceftazidime hydrate and tobramycin [196], farnesol [197], (–)-epigallocatechin gallate and (+)-catechin hydrate

**TABLE 8.2. Quorum-Sensing Inhibitor Compounds Tested by Brackman et al.**[a]

| Compound Name | Concentration |
|---|---|
| 4CABA | 50 μM[b] |
| 6CABA | 50 μM[c] |
| 6FABA | 50 μM[d] |
| Azithromycin | 2 μM |
| Baicalein | 1 μM |
| Bacalin hydrate | 100 μM |
| *p*-Benzoquinon | 100 μM |
| L-Canavanine | 20 μM |
| (+)-Catechin | 1000 μM |
| Ceftazidime hydrate | 1 μM |
| Cinnamaldehyde | 250 μM |
| Compound 1 | 500 μM[e] |
| Compound 3 | 500 μM[f] |
| Curcumin | 500 μM |
| (–)-Epigallocatechin gallate | 0.4 μM |
| Esculetin | 500 μM |
| Esculin hydrate | 500 μM |
| Farnesol | 2500 μM |
| Indole | 312 μM |
| 4-Nitropyridine *N*-oxide | 8 μM |
| Resveratrol | 25 μM |
| Tobramycin | 2 μM |

[a]See Ref [19].
[b]4CABA-2-amino-4-chlorobenzoic acid.
[c]6CABA-2-amino-6-chlorobenzoic acid.
[d]6FABA-2-amino-6-fluorobenzoic acid.
[e]Compound 1—*N*′3-(2-thienylcarbonyl)-4-bromo-1,5-dimethyl-1*H*-pyrazole-3-carbohydrazide.
[f]Compound 3—*N*′-(6-*tert*-butyl-2,3-dihydro-2-methylpyridazin-4-yl)-5-chlorothiophene-2-carbohydrazide.

[198], 2-amino-4-chlorobenzoic acid (4CABA), 2-amino-6-chlorobenzoic acid (6CABA) and 2-amino-6-fluorobenzoic acid (6FABA) [199], curcumin [200], baicalein, baicalin hydrate, and esculin hydrate [201] were purchased from Sigma-Aldrich (Bornem, Belgium). Similarly, Esculetin [201] and *N*′3-(2-thienylcarbonyl)-4-bromo-1,5-dimethyl-1*H*-pyrazole-3-carbohydrazide (compound 1) [202] were purchased from Acros Organics (Geel, Belgium). *N*′-(6-*tert*-butyl-2,3-dihydro-2-methylpyridazin-4-yl)-5-chlorothiophene-2-carbohydrazide (compound 3) was synthesized as previously described [202]. All compounds were diluted in 0.5% demethyl sulfoxide (DMSO). The effect of these QS inhibitors on *Burkholderia* spp. biofilm formation was examined using crystal violet, resazurin, and SYTO9 staining, confocal laser scanning microscopy, as well as plating. When used at subinhibitory concentrations, several compounds interfered with biofilm formation by *Burkholderia* spp. on Si disks. Overall, the authors suggest that the QS inhibitors affect later stages of biofilm formation on, and detachment from, Si disks. Whether these compounds, alone or in combination with conventional antimicrobial agents, will ever be useful as antibiofilm agents remains to be determined in future studies.

## REFERENCES


1. Jones, S.A., Bowler, P.G., Walker, M., and Parsons, D. (2004), Controlling wound bioburden with a novel silver-containing Hydrofiber dressing, *Wound Repair Regen.*, 12, 288–294.
2. Revathi, G., Puri, J., and Jain, P.K. (1998), Bacteriology of burns, *Burns*, 24, 347–349.
3. Sussman, C. and Bates-Jensen, B.M. (2001), *Wound Care: A Collaborative Practice Manual for Physical Therapists and Nurses*, 2nd ed., Aspen Publishers, Gaithersburg, MD, pp. 712, xxxii.
4. Campton-Johnston, S. and Wilson, J. (2001), Infected wound management: advanced technologies, moisture-retentive dressings, and die-hard methods, *Crit. Care Nurs. Q.*, 24, 64–77.
5. Gjødsbøl, K., Christensen, J.J., Karlsmark, T., Jørgensen, B., Klein, B.M., and Krogfelt, K.A. (2006), Multiple bacterial species reside in chronic wounds: a longitudinal study, *Int. Wound. J.*, 3, 225–231.
6. Gottrup, F. (2004), A specialized wound-healing center concept: importance of a multidisciplinary department structure and surgical treatment facilities in the treatment of chronic wounds, *Am. J. Surg.*, 187, 38S–43S.
7. Madsen, S.M., Westh, H., Danielsen, L., and Rosdahl, V.T. (1996), Bacterial colonization and healing of venous leg ulcers, *APMIS*, 104, 895–899.
8. Halbert, A.R., Stacey, M.C., Rohr, J.B., and Jopp-McKay, A. (1992), The effect of bacterial colonization on venous ulcer healing, *Australas J. Dermatol.*, 33, 75–80.

9. Bjarnsholt, T., Jensen, P.O., Burmolle, M., Hentzer, M., Haagensen, J.A., Hougen, H.P., Calum, H., Madsen, K.G., Moser, C., Molin, S., Hoiby, N., and Givskov, M. (2005), *Pseudomonas aeruginosa* tolerance to tobramycin, hydrogen peroxide and polymorphonuclear leukocytes is quorum-sensing dependent, *Microbiology*, 151 (Pt. 2), 373–383.

10. Vuong, C., Voyich, J.M., Fischer, E.R., Braughton, K.R., Whitney, A.R., DeLeo, F.R., and Otto, M. (2004), Polysaccharide intercellular adhesin (PIA) protects *Staphylococcus epidermidis* against major components of the human innate immune system, *Cell Microbiol.*, 6, 269–275.

11. Anderson, G.G., Palermo, J.J., Schilling, J.D., Roth, R., Heuser, J., and Hultgren, S.J. (2003), Intracellular bacterial biofilm-like pods in urinary tract infections, *Science*, 301, 105–107.

12. Wieman, T.J. (2005), Principles of management: the diabetic foot, *Am. J. Surg.*, 190, 295–299.

13. Zilberman, M. and Elsner, J.J. (2008), Antibiotic-eluting medical devices for various applications, *J. Control. Rel.*, 130, 202–215.

14. Lerman, O.Z., Galiano, R.D., Armour, M., Levine, J.P., and Gurtner, G.C. (2003), Cellular dysfunction in the diabetic fibroblast: impairment in migration, vascular endothelial growth factor production, and response to hypoxia, *Am. J. Pathol.*, 162, 303–312.

15. Falanga, V. (2000), Classifications for wound bed preparation and stimulation of chronic wounds, *Wound Repair. Regen.*, 8, 347–352.

16. Schultz, G.S., Sibbald, R.G., Falanga, V., Ayello, E.A., Dowsett, C., Harding, K., Romanelli, M., Stacey, M.C., Teot, L., and Vanscheidt, W. (2003), Wound bed preparation: a systematic approach to wound management, *Wound Repair. Regen.*, 11 (Suppl. 1), S1–S28.

17. Lobmann, R., Schultz, G., and Lehnert, H. (2005), Proteases and the diabetic foot syndrome: mechanisms and therapeutic implications, *Diabetes Care*, 28, 461–471.

18. Yager, D.R. and Nwomeh, B.C. (1999), The proteolytic environment of chronic wounds, *Wound Repair Regen.*, 7, 433–441.

19. de, B.S., Polette, M., Zahm, J.M., Hinnrasky, J., Kileztky, C., Bajolet, O., Klossek, J.M., Filloux, A., Lazdunski, A., and Puchelle, E. (2000), *Pseudomonas aeruginosa* virulence factors delay airway epithelial wound repair by altering the actin cytoskeleton and inducing overactivation of epithelial matrix metalloproteinase-2, *Lab. Invest.*, 80, 209–219.

20. Ikema, K., Matsumoto, K., Inomata, Y., Komohara, Y., Miyajima, S., Takeya, M., and Tanihara, H. (2006), Induction of matrix metalloproteinases (MMPs) and tissue inhibitors of MMPs correlates with outcome of acute experimental pseudomonal keratitis, *Exp. Eye Res.*, 83, 1396–1404.

21. Lobmann, R., Zemlin, C., Motzkau, M., Reschke, K., and Lehnert, H. (2006), Expression of matrix metalloproteinases and growth factors in diabetic foot wounds treated with a protease absorbent dressing, *J. Diabetes Complications*, 20, 329–335.

22. Dong, Z., Ghabrial, M., Katar, M., Fridman, R., and Berk, R.S. (2000), Membrane-type matrix metalloproteinases in mice intracorneally infected with *Pseudomonas aeruginosa*, *Invest. Ophthalmol. Vis. Sci.*, 41, 4189–4194.

23. Danielsen, L., Westh, H., Balselv, E., Rosdahl, V.T., and Doring, G. (1996), *Pseudomonas aeruginosa* exotoxin A antibodies in rapidly deteriorating chronic leg ulcers, *Lancet*, 347, 265.
24. Danielsen, L., Balslev, E., Doring, G., Hoiby, N., Madsen, S.M., Agren, M., Thomsen, H.K., Fos, H.H., and Westh, H. (1998), Ulcer bed infection. Report of a case of enlarging venous leg ulcer colonized by *Pseudomonas aeruginosa*, *APMIS*, 106, 721–726.
25. Bjarnsholt, T., Kirketerp-Møller, K., Østrup Jensen, P., Madsen, K.G., Phipps, R., Krogfelt, K., Med, N.H.D., and Techn, M.G.D. (2008), Why chronic wounds will not heal: a novel hypothesis, *Wound Rep. Reg.*, 16, 2–10.
26. Stender, P.N.A. (2003), FISH: an intelligent stain for rapid diagnosis of infectious diseases, *Expert Rev. Mol. Diagn.*, 3, 649–655.
27. Gardner, S.E., Frantz, R.A., and Doebbeling, B.N. (2001), The validity of the clinical signs and symptoms used to identify localized chronic wound infection, *Wound Repair Regen.*, 9, 178–186.
28. Edwards, R. and Harding, K.G. (2004), Bacteria and wound healing, *Curr. Opin. Infect. Dis.*, 17, 91–96.
29. Raahave, D., Friis-Moller, A., Bjerre-Jepsen, K., Thiis-Knudsen, J., and Rasmussen, L.B. (1986), The infective dose of aerobic and anaerobic bacteria in postoperative wound sepsis, *Arch. Surg.*, 121, 924–929.
30. Robson, M.C., Lea, C.E., Dalton, J.B., and Heggers, J.P. (1968), Quantitative bacteriology and delayed wound closure, *Surg. Forum.*, 19, 501–502.
31. Bendy, R.H. Jr., Nuccio, P.A., Wolfe, E., Collins, B., Tamburro, C., Glass, W., and Martin, C.M. (1964), Relationship of quantitative wound bacterial counts to healing of decubiti: effect of topical Gentamicin, *Antimicrobial Agents Chemother.* (Bethesda), 10, 147–155.
32. Hoiby, N., Frederiksen, B., and Pressler, T. (2005), Eradication of early *Pseudomonas aeruginosa* infection, *J. Cyst. Fibros.*, 4 (Suppl. 2), 49–54.
33. Kerem, B., Rommens, J.M., Buchanan, J.A., Markiewicz, D., Cox, T.K., Chakravarti, A., Buchwald, M., and Tsui, L.C. (1989), Identification of the cystic fibrosis gene: genetic analysis, *Science*, 245, 1073–1080.
34. Riordan, J.R., Rommens, J.M., Kerem, B., Alon, N., Rozmahel, R., Grzelczak, Z., Zielenski, J., Lok, S., Plavsic, N., and Chou, J.L. (1989), Identification of the cystic fibrosis gene: cloning and characterization of complementary DNA, *Science*, 245, 1066–1073.
35. Rommens, J.M., Iannuzzi, M.C., Kerem, B., Drumm, M.L., Melmer, G., Dean, M., Rozmahel, R., Cole, J.L., Kennedy, D., and Hidaka, N. (1989), Identification of the cystic fibrosis gene: chromosome walking and jumping, *Science*, 245, 1059–1065.
36. Bauernfeind, A., Bertele, R.M., Harms, K., Horl, G., Jungwirth, R., Petermuller, C., Przyklenk, B., and Weisslein-Pfister, C. (1987), Qualitative and quantitative microbiological analysis of sputa of 102 patients with cystic fibrosis, *Infection*, 15, 270–277.
37. Saiman, L. (2004), Microbiology of early CF lung disease, *Paediatr. Respir. Rev.*, 5 (Suppl. A), S367–S369.
38. Gilligan, P.H. (1991), Microbiology of airway disease in patients with cystic fibrosis, *Clin. Microbiol. Rev.*, 4, 35–51.

39. Koch, C. (2002), Early infection and progression of cystic fibrosis lung disease, *Pediatr. Pulmonol.*, 34, 232–236.

40. Koch, C. and Hoiby, N. (2000), Diagnosis and treatment of cystic fibrosis, *Respiration*, 67, 239–247.

41. Hoiby, N. (1977), *Pseudomonas aeruginosa* infection in cystic fibrosis, Ph.D Thesis, University of Copenhagen, Denmark.

42. Lam, J., Chan, R., Lam, K., and Costerton, J.W. (1980), Production of mucoid microcolonies by *Pseudomonas aeruginosa* within infected lungs in cystic fibrosis, *Infect. Immun.*, 28, 546–556.

43. Baltimore, R.S., Christie, C.D., and Smith, G.J. (1989), Immunohistopathologic localization of *Pseudomonas aeruginosa* in lungs from patients with cystic fibrosis. Implications for the pathogenesis of progressive lung deterioration, *Am. Rev. Respir. Dis.*, 140, 1650–1661.

44. Trengove, N.J., Stacey, M.C., MacAuley, S., Bennett, N., Gibson, J., Burslem, F., Murphy, G., and Schultz, G. (1999), Analysis of the acute and chronic wound environments: the role of proteases and their inhibitors, *Wound Repair Regen.*, 7, 442–452.

45. Mirastschijski, U., Impola, U., Jahkola, T., Karlsmark, T., AGren, M.S., and Saarialho-Kere, U. (2002), Ectopic localization of matrix metalloproteinase-9 in chronic cutaneous wounds, *Hum. Pathol.*, 33, 355–364.

46. Rogers, A.A., Burnett, S., Moore, J.C., and Shakespeare, P.G. (1995), Involvement of proteolytic enzymes-plasminogen activators and matrix metalloproteinases-in the pathophysiology of pressure ulcers, *Wound Repair Regen.*, 3, 273–283.

47. Wysocki, A.B., Staiano-Coico, L., and Grinnell, F. (1993), Wound fluid from chronic leg ulcers contains elevated levels of metalloproteinases MMP-2 and MMP-9, *J. Invest. Dermatol.*, 101, 64–68.

48. Yager, D.R., Zhang, L.Y., Liang, H.X., Diegelmann, R.F., and Cohen, I.K. (1996), Wound fluids from human pressure ulcers contain elevated matrix metalloproteinase levels and activity compared to surgical wound fluids, *J. Invest. Dermatol.*, 107, 743–748.

49. Valerius, N.H., Koch, C., and Hoiby, N. (1991), Prevention of chronic *Pseudomonas aeruginosa* colonisation in cystic fibrosis by early treatment, *Lancet*, 338, 725–726.

50. Bjarnsholt, T., Jensen, P.O., Rasmussen, T.B., Christophersen, L., Calum, H., Hentzer, M., Hougen, H.P., Rygaard, J., Moser, C., Eberl, L., Hoiby, N., and Givskov, M. (2005), Garlic blocks quorum sensing and promotes rapid clearing of pulmonary *Pseudomonas aeruginosa* infections, *Microbiology*, 151 (Pt. 12), 3873–3880.

51. Rasmussen, T.B., Bjarnsholt, T., Skindersoe, M.E., Hentzer, M., Kristoffersen, P., Kote, M., Eberl, L., Nielsen, J., and Givskov, M. (2005), Screening for quorum sensing inhibitors using a novel genetic system-the QSI selector, *J. Bacteriol.*, 187, 1799–1814.

52. Rasmussen, T.B., Skindersoe, M.E., Bjarnsholt, T., Christensen, K.B., Andersen, J.B., Ostenfeld-Larsen, T., Hentzer, M., and Givskov, M. (2005), Idendity and effects of quorum sensing inhibitors produced by *Penicillum* species, *Microbiology*, 151, 1325–1340.

53. Hentzer, M., Wu, H., Andersen, J.B., Riedel, K., Rasmussen, T.B., Bagge, N., Kumar, N., Schembri, M.A., Song, Z., Kristoffersen, P., Manefield, M., Costerton, J.W., Molin, S., Eberl, L., Steinberg, P., Kjelleberg, S., Hoiby, N., and Givskov, M. (2003), Attenuation of *Pseudomonas aeruginosa* virulence by quorum sensing inhibitors, *EMBO J.*, 22, 3803–3815.
54. Jensen, P.O., Bjarnsholt, T., Rasmussen, T.B., Calum, H., Moser, C., Pressler, T., Givskov, M., and Hoiby, N. (2006), Rapid necrotic killing of PMNs is caused by quorum sensing controlled production of rhamnolipid by *Pseudomonas aeruginosa*, *Microbiology*, 3, 225–231.
55. Heggers, J., Goodheart, R.E., Washington, J., McCoy, L., Carino, E., Dang, T., Edgar, P., Maness, C., and Chinkes, D. (2005), Therapeutic efficacy of three silver dressings in an infected animal model, *J. Burn Care Rehabil.*, 26, 53–56.
56. Fraser, J.F., Bodman, J., Sturgess, R., Faoagali, J., and Kimble, R.M. (2004), An *in vitro* study of the anti-microbial efficacy of a 1% silver sulphadiazine and 0.2% chlorhexidine digluconate cream, 1% silver sulphadiazine cream and a silver coated dressing, *Burns*, 30, 35–41.
57. Trop, M. (2006), Silver-coated dressing acticoat caused raised liver enzymes and argyria-like symptoms in burn patient, *J. Trauma*, 61, 1024.
58. Suzuki, Y., Tanihara, M., Nishimura, Y., Suzuki, K., Kakimaru, Y., and Shimizu, Y. (1998), A new drug delivery system with controlled release of antibiotic only in the presence of infection, *J. Biomed. Mater. Res.*, 42, 112–116.
59. Suzuki, Y., Tanihara, M., Nishimura, Y., Suzuki, K., Kakimaru, Y., and Shimizu, Y. (1997), A novel wound dressing with an antibiotic delivery system stimulated by microbial infection, *Asaio J.*, 43, M854–M857.
60. Jürgens, C., Schulz, A.P., Porté, T., Faschingbauer, M., and Seide, K. (2006), Biodegradable films in trauma and orthopedic surgery, *Europ. J. Trauma*, 2, 160–171.
61. Varghese, M.C., Balin, A.K., Carter, D.M., and Caldwell, D. (1986), Local environment of chronic wounds under synthetic dressings, *Arch. Dermatol.*, 122, 52–57.
62. Eisinger, M., Lee, J.S., Hefton, J.M., Darzynkiewicz, Z., Chiao, J.W., and de Harven, E. (1979), Human epidermal cell cultures: growth and differentiation in the absence of differentiation in the absence of dermal components or medium supplements, *Proc. Natl. Acad. Sci. USA*, 76, 5340–5344.
63. Hutchinson, F.G. and Furr, B.J. (1985), Biodegradable polymers for the sustained release of peptides, *Biochem. Soc. Trans.*, 13, 520–523.
64. Katti, D.S., Robinson, K.W., Ko, F.K., and Laurencin, C.T. (2004), Bioresorbable nanofiber-based systems for wound healing and drug delivery: optimization of fabrication parameters, *J. Biomed. Mater. Res. B, Appl. Biomater.*, 70, 286–296.
65. Pruitt, B.A. and Levine, N.S. (1984), Characteristics and uses of biologic dressings and skin substitutes, *Arch. Surg.*, 119, 312–322.
66. Cairns, B.A., deSerres, S., Peterson, H.D., and Meyer, A.A. (1993), Skin replacements. The biotechnological quest for optimal wound closure, *Arch. Surg.*, 128, 1246–1252.
67. Maruguchi, T., Maruguchi, Y., Suzuki, S., Matsuda, K., Toda, K., and Isshiki, N. (1994), A new skin equivalent: keratinocytes proliferated and differentiated on collagen sponge containing fibroblasts, *Plast. Reconstr. Surg.*, 93, 537–544, discussion 545–546.

68. Trafny, E.A., Kowalska, K., and Grzybowski, J. (1998), Adhesion of *Pseudomonas aeruginosa* to collagen biomaterials: effect of amikacin and ciprofloxacin on the colonization and survival of the adherent organisms, *J. Biomed. Mater. Res.*, 41, 593–599.

69. Radu, F.A., Bause, M., Knabner, P., Lee, G.W., and Friess, W.C. (2002), Modeling of drug release from collagen matrices, *J. Pharm. Sci.*, 91, 964–972.

70. Ruszczak, Z. and Friess, W. (2003), Collagen as a carrier for on-site delivery of antibacterial drugs, *Adv. Drug Deliv. Rev.*, 55, 1679–1698.

71. Sripriya, R., Kumar, M.S., and Sehgal, P.K. (2004), Improved collagen bilayer dressing for the controlled release of drugs, *J. Biomed. Mater. Res. B, Appl. Biomater.*, 70, 389–396.

72. Prabu, P., Dharmaraj, N., Aryal, S., Lee, B.M., Ramesh, V., and Kim, H.Y. (2006), Preparation and drug release activity of scaffolds containing collagen and poly(caprolactone), *J. Biomed. Mater. Res. A*, 79, 153–158.

73. Shanmugasundaram, N., Sundaraseelan, J., Uma, S., Selvaraj, D., and Babu, M. (2006), Design and delivery of silver sulfadiazine from alginate microspheres-impregnated collagen scaffold, *J. Biomed. Mater. Res. B, Appl. Biomater.*, 77, 378–388.

74. Park, S.N., Kim, J.K., and Suh, H. (2004), Evaluation of antibiotic-loaded collagen-hyaluronic acid matrix as a skin substitute, *Biomaterials*, 25, 3689–3698.

75. Lee, J.E., Park, J.C., Lee, K.H., Oh, S.H., Kim, J.G., and Suh, H. (2002), An infection-preventing bilayered collagen membrane containing antibiotic-loaded hyaluronan microparticles: physical and biological properties, *Artif. Organs*, 26, 636–646.

76. Taddeucci, P., Pianigiani, E., Colletta, V., Torasso, F., Andreassi, L., and Andreassi, A. (2004), An evaluation of Hyalofill-F plus compression bandaging in the treatment of chronic venous ulcers, *J. Wound Care*, 13, 202–204.

77. Colletta, V., Dioguardi, D., Di Lonardo, A., Maggio, G., and Torasso, F. (2003), A trial to assess the efficacy and tolerability of Hyalofill-F in nonhealing venous leg ulcers, *J. Wound Care*, 12, 357–360.

78. Chung, L.Y. Schmidt, R.J., Hamlyn, P.F., Sagar, B.F., Andrews, A.M., and Turner, T.D. (1994), Biocompatibility of potential wound management products: fungal mycelia as a source of chitin/chitosan and their effect on the proliferation of human F1000 fibroblasts in culture, *J. Biomed. Mater. Res.*, 28, 463–469.

79. Muzzarelli, R., Tarsi, R., Filippini, O., Giovanetti, E., Biagini, G., and Varaldo, P.E. (1990), Antimicrobial properties of *N*-carboxybutyl chitosan, *Antimicrob. Agents Chemother.*, 34, 2019–2023.

80. Mi, F.L., Wu, Y.B., Shyu, S.S., Schoung, J.Y., Huang, Y.B., Tsai, Y.H., and Hao, J.Y. (2002), Control of wound infections using a bilayer chitosan wound dressing with sustainable antibiotic delivery, *J. Biomed. Mater. Res.*, 59, 438–449.

81. Rossi, S., Marciello, M., Sandri, G., Ferrari, F., Bonferoni, M.C., Papetti, A., Caramella, C., Dacarro, C., and Grisoli, P. (2007), Wound dressings based on chitosans and hyaluronic acid for the release of chlorhexidine diacetate in skin ulcer therapy, *Pharm. Dev. Technol.*, 12, 415–422.

82. Aoyagi, S., Onishi, H., and Machida, Y. (2007), Novel chitosan wound dressing loaded with minocycline for the treatment of severe burn wounds, *Int. J. Pharm.*, 330, 138–145.

83. Zilberman, M., Shifrovitch, Y., Aviv, M., and Hershkovitz, M. (2009), Structured drug-eluting bioresorbable films: microstructure and release profile, *J. Biomater. Appl.*, 23, 385–406.
84. Zilberman, M., Golerkansky, E., Elsner, J.J., and Berdicevsky, I. (2009), Gentamicin-eluting bioresorbable composite fibers for wound healing applications, *J. Biomed. Mater. Res. A*, 89, 654–666.
85. Zilberman, M. and Malka, A. (2009), Drug controlled release from structured bioresorbable films used in medical devices—a mathematical model, *J. Biomed. Mater. Res. B Appl. Biomater.*, 89, 155–164.
86. Elsner, J.J. and Zilberman, M. (2009), Antibiotic-eluting bioresorbable composite fibers for wound healing applications: microstructure, drug delivery and mechanical properties, *Acta Biomater.*, 5, 2872–2883.
87. Foye, W.O., Lemke, L., and Williams, D.A. (1998), Principles of Medicinal Chemistry, 3rd ed., Vol. 33, Lea & Febiger, Baltimore, MD, pp. 995.
88. Koch, C. and Hoiby, N. (1993), Pathogenesis of cystic fibrosis, *Lancet*, 341, 1065–1069.
89. May, T.B., Shinabarger, D., Maharaj, R., Kato, J., Chu, L., DeVault, J.D., Roychoudhury, S., Zielinski, N.A., Berry, A., Rothmel, R.K., Misra, T.K., and Chakrabarty, A.M. (1991), Alginate synthesis by *Pseudomonas aeruginosa*: a key pathogenic factor in chronic pulmonary infections of cystic fibrosis patients, *Clin. Microbiol. Rev.*, 4, 191–206.
90. Govan, J.R.W. and Deretic, V. (1996), Microbial pathogenesis in cystic fibrosis: mucoid *Pseudomonas aeruginosa* and *Burkholderia cepacia*, *Microbiol. Rev.*, 60, 539–574.
91. Hoyle, B.D., Williams, L.J., and Costerton, J.W. (1993), Production of mucoid exopolysaccharide during development of *Pseudomonas aeruginosa* biofilms, *Infect. Immun.*, 61, 777–780.
92. Mathee, K., Ciofu, O., Sternberg, C., Lindum, P.W., Campbell, J.I.A., Jensen, P., Givskov, M., Ohman, D.E., Molin, S., Hoiby, N., and Kharazmi, A. (1999), Mucoid conversion of *Pseudomonas aeruginosa* by hydrogen peroxide: a mechanism for virulence activation in the cystic fibrosis lung, *Microbiology*, 145, 1349–1357.
93. Cochrane, D.M.G., Brown, M.R.W., Anwar, H., Weller, P.H., Lam, K., and Costerton, J.W. (1988), Antibody response to *Pseudomonas aeruginosa* surface protein antigens in a rat model of chronic lung infection, *J. Med. Microbiol.*, 27, 255–261.
94. Anwar, H., Strap, J.L., and Costerton, J.W. (1992), Susceptibility of biofilm cells of *Pseudomonas aeruginosa* to bactericidal actions of whole blood and serum, *FEMS Microbiol. Lett.*, 92, 235–242.
95. Pedersen, S.S. (1992), Lung infection with alginate-producing, mucoid *Pseudomonas aeruginosa* in cystic fibrosis, *APMIS* 100 (Suppl. 28), 5–79.
96. Høiby, N., and Pedersen, S.S. (1989), Estimated risk of cross-infection with *Pseudomonas aeruginosa* in Danish cystic fibrosis patients, *Acta Paediatr. Scand.*, 78, 395–404.
97. Johansen, H.K. and Hoiby, N. (1992), Seasonal onset of initial colonisation and chronic infection with *Pseudomonas aeruginosa* in patients with cystic fibrosis in Denmark, *Thorax*, 47, 109–111.
98. Baker, N.R. and Svanborg-Edén, C. (1989), Role of alginate in the adherence of *Pseudomonas aeruginosa*, in: Høiby, N., Pedersen, S.S., Shand, G.H., Döring, G.,

and Holder, I.A. Eds., *Pseudomonas aeruginosa infection*, Karger, Basel, pp. 72–79.

99. Baker, N.R., Minor, V., Deal, C., Shahrabadi, M.S., Simpson, D.A., and Woods, D.E. (1991), *Pseudomonas aeruginosa* exoenzyme-S is an adhesin, *Infect. Immun.*, 59, 2859–2863.
100. Feldman, M., Bryan, R., Rajan, S., Scheffler, L., Brunnert, S., Tang, H., and Prince, A. (1998), Role of flagella in pathogenesis of *Pseudomonas aeruginosa* pulmonary infection, *Infect. Immun.*, 66, 43–51.
101. Elborn, J.S., Cordon, S.M., and Shale, D.J. (1993), Host inflammatory responses to first isolation of *Pseudomonas aeruginosa* from sputum in cystic fibrosis, *Pediatr. Pulmonol.*, 15, 287–291.
102. Kharazmi, A. (1991), Mechanisms involved in the evasion of the host defence by *Pseudomonas aeruginosa, Immunol. Lett.*, 30, 201–206.
103. Goldstein, W. and Döring, G. (1986), Lysosomal enzymes from polymorphonuclear leukocytes and proteinase inhibitors in patients with cystic fibrosis, *Am. Rev. Respir. Dis.*, 134, 49–56.
104. Ammitzbøll, T., Pedersen, S.S., Espersen, F., and Schiøler, H. (1988), Excretion of urinary collagen metabolites correlates to severity of pulmonary disease in cystic fibrosis, *Acta Paediatr. Scand.*, 77, 842–846.
105. Høiby, N. and Koch, C. (1990), *Pseudomonas aeruginosa* infection in cystic fibrosis and its management, *Thorax*, 45, 881–884.
106. Tosi, M.F., Zakem, H., and Berger, M. (1990), Neutrophil elastase cleaves C3Bi on opsonized *Pseudomonas* as well as Cr1 on neutrophils to create a functionally important opsonin receptor mismatch, *J. Clin. Invest.*, 86, 300–308.
107. Bruce, M.C., Poncz, L., Klinger, J.D., Stern, R.C., Tomashefski, J.F., and Dearborn, D.G. (1985), Biochemical and pathologic evidence for proteolytic destruction of lung connective tissue in cystic fibrosis, *Am. Rev. Respir. Dis.*, 132, 529–535.
108. Suter, S. (1989), The imbalance between granulocyte neutral proteases and antiproteases in bronchial secretions from patients with cystic fibrosis, in: Høiby, N., Pedersen, S.S., Shand, G.H., Döring, G., and Holder, I.A., Eds., *Pseudomonas aeruginosa infection*, Karger, Basel, pp. 158–168.
109. McElvaney, N., Nakamura, H., Birrer, P., Hebert, C., Wong, W., Alphonso, M., Baker, J.B., Catalano, M.A., and Crystal, R.G. (1992), Modulation of airway inflammation in cystic fibrosis. In vivo suppression of interleukin-8 levels on the respiratory epithelial surface by aerosolization of recombinant secretory leukoprotease inhibitor, *J. Clin. Invest.*, 89, 1296–1301.
110. Shah, P.L., Scott, S.F., Knight, R.A., Marriott, C., Ranasinha, C., and Hodson, M.E. (1996), *In vivo* effects of recombinant human DNase I on sputum in patients with cystic fibrosis, *Thorax*, 51, 119–125.
111. Speert, D.P., Campbell, M.E., Davidson, A.G.F., and Wong, L.T.K. (1993), *Pseudomonas aeruginosa* colonization of the gastrointestinal tract in patients with cystic fibrosis, *J. Infect. Dis.*, 167, 226–229.
112. Kharazmi, A. (1989), Interactions of *Pseudomonas aeruginosa* proteases with the cells of the immune system, in: Høiby, N., Pedersen, S.S., Shand, G.H., Döring, G., and Holder, I.A., Eds., *Pseudomonas aeruginosa Infection*, Karger, Basel, pp. 42–49.

113. Döring, G., Buhl, V., Høiby, N., Schiøtz, P.O., and Botzenhart, K. (1984), Detection of proteases of *Pseudomonas aeruginosa* in immune complexes isolated from sputum of cystic fibrosis patients, *Acta Pathol. Microbiol. Scand. Sect. C*, 92, 307–312.

114. Pedersen, S.S., Høiby, N., Espersen, F., and Koch, C. (1992), Role of alginate in infection with mucoid *Pseudomonas aeruginosa* in cystic fibrosis, *Thorax*, 47, 6–13.

115. DeVault, J., Kimbara, K., and Chakrabarty, A. (1990), Pulmonary dehydration and infection in cystic fibrosis. Evidence that ethanol activates gene expression and induction of mucoidy in *Pseudomonas aeruginosa*, *Mol. Microbiol.*, 4, 737–745.

116. Jobsis, Q., Raatgeep, H.C., Schellekens, S.L., Kroesbergen, A., Hop, W.C.J., and de Jongste, J.C. (2000), Hydrogen peroxide and nitric oxide in exhaled air of children with cystic fibrosis during antibiotic treatment, *Eur. Resp. J.*, 16, 95–100.

117. Høiby, N. (1974), *Pseudomonas aeruginosa* infection in cystic fibrosis. Relationship between mucoid strains of *Pseudomonas aeruginosa* and the humoral immune response, *Acta Pathol. Microbiol. Scand. Sect. B*, 82, 551–558.

118. Høiby, N. (1975), Prevalence of mucoid strains of *Pseudomonas aeruginosa* in bacteriological specimens from patients with cystic fibrosis and patients with other diseases, *Acta Pathol. Microbiol. Scand. Sect. B*, 83, 549–552.

119. Høiby, N. (1977), Antibodies against *Pseudomonas aeruginosa* in sera from normal persons and from patients colonized with mucoid or non-mucoid *Pseudomonas aeruginosa*: Results obtained by means of crossed immunoelectrophoresis, *Acta Pathol. Microbiol. Scand., Sect. C*, 85, 142–148.

120. Pedersen, S.S., Kharazmi, A., Espersen, F., and Hoiby, N. (1990), *Pseudomonas aeruginosa* alginate in cystic fibrosis sputum and the inflammatory response, *Infect. Immun.*, 58, 3363–3368.

121. Brown, M.R.W., Collier, P.J., and Gilbert, P. (1990), Influence of growth rate on susceptibility to antimicrobial agents-modification of the cell envelope and batch and continuous culture studies, *Antimicrob. Agents Chemother.*, 34, 1623–1628.

122. Giwercman, B., Jensen, E.T., Hoiby, N., Kharazmi, A., and Costerton, J.W. (1991), Induction of beta-lactamase production in *Pseudomonas aeruginosa* biofilm, *Antimicrob. Agents Chemother.*, 35, 1008–1010.

123. Jensen, E.T., Kharazmi, A., Lam, K., Costerton, J.W., and Høiby, N. (1990), Human polymorphonuclear leukocyte response to *Pseudomonas aeruginosa* grown in biofilm, *Infect. Immun.*, 58, 2383–2385.

124. Ciofu, O., Petersen, T.D., Jensen, P., and Hoiby, N. (1999), Avidity of anti-*P. aeruginosa* antibodies during chronic infection in patients with cystic fibrosis, *Thorax*, 54, 141–144.

125. Pier, G.B., Grout, M., and Desjardins, D. (1991), Complement deposition by antibodies to *Pseudomonas aeruginosa* mucoid exopolysaccharide (MEP) and by non-MEP specific opsonins, *J. Immunol.*, 147, 1869–1876.

126. Ojeniyi, B., Høiby, N., and Rosdal, V.T. (1991), Prevalence and persistence of polyagglutinable *Pseudomonas aeruginosa* in cystic fibrosis patients, *APMIS*, 99, 187–195.

127. Ojeniyi, B., Baek, L., and Høiby, N. (1985), Polyagglutinability due to loss of 0-antigenic determinants in *Pseudomonas aeruginosa* strains isolated from cystic fibrosis patients, *Acta Pathol. Microbiol. Scand. Sect. B*, 93, 7–13.

128. Shand, G.H., Pedersen, S.S., Brown, M.R.W., and Høiby, N. (1991), Serum antibodies to *Pseudomonas aeruginosa* outer membrane proteins and iron-regulated membrane proteins at different stages of chronic cystic fibrosis lung infection, *J. Med. Microbiol.*, 34, 203–212.

129. Kronborg, G., Shand, G.H., Fomsgaard, A., and Høiby, N. (1992), Lipopolysaccharide is present in immune complexes isolated from sputum in patients with cystic fibrosis and chronic *Pseudomonas aeruginosa* lung infection, *APMIS*, 100, 175–180.

130. Kronborg, G., Hansen, M., Svenson, M., Fomsgaard, A., Høiby, N., and Bendtzen, K. (1993), Cytokines in sputum and serum from patients with cystic fibrosis and chronic *Pseudomonas aeruginosa* infection as markers of destructive inflammation in the lungs, *Pediatr. Pulmonol.*, 15, 292–297.

131. Konstan, M.W. and Berger, M. (1997), Current understanding of the inflammatory process in cystic fibrosis onset and etiology, *Pediatr. Pulmonol.*, 24, 137–142.

132. Schiøtz, P.O., Clemmensen, I., and Høiby, N. (1980), Immunoglobulins and albumin in sputum from patients with cystic fibrosis. A study of protein stability and presence of proteases, *Acta Pathol. Microbiol. Scand., Sect. C*, 88, 275–280.

133. Høiby, N., Döring, G., and Schiøtz, P.O. (1986), The role of immune complexes in the pathogenesis of bacterial infections, *Ann. Rev. Microbiol.*, 40, 29–53.

134. Konstan, M.W., Hilliard, K.A., Norvell, T.M., and Berger, M. (1995), Bronchoalveolar lavage findings in cystic fibrosis patients with stable, clinically mild lung disease suggest ongoing infection and inflammation (Vol. 150, p. 448, 1994), *Am. J. Respir. Crit. Care Med.*, 151, 260.

135. Johansen, H.K., Hougen, H.P., Rygaard, J., and Hoiby, N. (1996), Interferon-gamma (IFN-gamma) treatment decreases the inflammatory response in chronic *Pseudomonas aeruginosa* pneumonia in rats, *Clin. Exp. Immunol.*, 103, 212–218.

136. Moser, C., Johansen, H.K., Song, Z.J., Hougen, H.P., Rygaard, J., and Hoiby, N. (1997), Chronic *Pseudomonas aeruginosa* lung infection is more severe in Th-2 responding BALB/c mice compared to Th-1 responding C3H/HeN mice, *APMIS*, 105, 838–842.

137. Høiby, N., Johansen, H.K., Moser, C., Song, Z., Ciofu, O., and Kharazmi, A. (2001), *Pseudomonas aeruginosa* and the *in vitro* and *in vivo* biofilm mode of growth, *Microbes Infect.*, 3, 23–35.

138. Johansen, H.K. (1996), Potential of preventing *Pseudomonas aeruginosa* lung infections in cystic fibrosis patients: Experimental studies in animals, *APMIS*, 104 (Suppl. 63), 5–42.

139. Moser, C., Kjaergaard, S., Pressler, T., Kharazmi, A., Koch, C., and Høiby, N. (2000), The immune response to chronic *Pseudomonas aeruginosa* lung infection in cystic fibrosis patients is predominantly of the Th2 type, *APMIS*, 108, 329–335.

140. Dimango, E., Zar, H.J., Bryan, R., and Prince, A. (1995), Diverse *Pseudomonas aeruginosa* gene products stimulate respiratory epithelial cells to produce interleukin-8, *J. Clin. Invest.*, 96, 2204–2210.

141. Telford, G., Wheeler, D., Williams, P., Tomkins, P.T., Appleby, P., Sewell, H., Stewart, G.S., Bycroft, B.W., and Pritchard, D.I. (1998), The *Pseudomonas aeruginosa* quorum-sensing signal molecule *N*-(3-oxododecanoyl)-1-homoserine lactone has immunomodulatory activity, *Infect. Immun.*, 66, 36–42.

142. Kievit, T.D. and Iglewski, B. (2000), Bacterial quorum sensing in pathogenic relationships, *Infect. Immun.*, 68, 4839–4849.
143. Frederiksen, B., Koch, C., and Høiby, N. (1999), Changing epidemiology of *Pseudomonas aeruginosa* infection in Danish cystic fibrosis patients, 1974–1995, *Paediatr. Pulmonol.*, 28, 159–166.
144. Szaff, M., Høiby, N., and Flensborg, E.W. (1983), Frequent antibiotic therapy improves survival of cystic fibrosis patients with chronic *Pseudomonas aeruginosa* infection, *Acta Paediat. Scand.*, 72, 651–657.
145. Pedersen, S.S., Jensen, T., Høiby, N., Koch, C., and Flensborg, E.W. (1987), Management of *Pseudomonas aeruginosa* lung infection in Danish cystic fibrosis patients, *Acta Paediat. Scand.*, 76, 955–961.
146. Frederiksen, B., Lanng, S., Koch, C., and Høiby, N. (1996), Improved survival in the Danish cystic fibrosis centre results of aggressive treatment, *Paediatr. Pulmonol.*, 21, 153–158.
147. Sheldon, C.D., Assoufi, B.K., and Hodson, M.E. (1993), Regular 3 monthly oral ciprofloxacin in adult cystic fibrosis patients infected with *Pseudomonas aeruginosa*, *Resp. Med.*, 87, 587–593.
148. Ramsey, B.W., Dorkin, H.L., Eisenberg, J.D., Gibson, R.L., Harwood, I.R., Kravitz, R.M., Schidlow, D.V., Wilmott, R.W., Astley, S.J., McBurnie, M.A., Wentz, K., and Smith, A.L. (1993), Efficacy of aerosolized tobramycin in patients with cystic fibrosis, *N. Engl. J. Med.*, 328, 1740–1746.
149. Ramsey, B.W., Pepe, M.S., Quan, J.M., Otto, K.L., Montgomery, A.B., Williams Warren, J., Vasiljev-K, M., Borowitz, D., Bowman, C.M., Marshall, B.C., Marshall, S., and Smith, A.L. (1999), Intermittent administration of inhaled tobramycin in patients with cystic fibrosis, *N. Engl. J. Med.*, 340, 23–30.
150. Anwar, H., Strap, J.L., Chen, K., and Costerton, J.W. (1992), Dynamic interactions of biofilms of mucoid *Pseudomonas aeruginosa* with tobramycin and piperacillin, *Antimicrob. Agents Chemother.*, 36, 1208–1214.
151. Preston, C.A.K., Khoury, A.E., Reid, G., Bruce, A.W., and Costerton, J.W. (1996), *Pseudomonas aeruginosa* biofilms are more susceptible to ciprofloxacin than to tobramycin, *Int. J. Antimicrob. Agents*, 7, 251–256.
152. Zhanel, G.G., Kim, S.O., Davidson, R.J., Hoban, D.J., and Nicolle, L.E. (1993), Effect of subinhibitory concentrations of ciprofloxacin and gentamicin on the adherence of *Pseudomonas aeruginosa* to Vero cells and voided uroepithelial cells, *Chemotherapy*, 39, 105–111.
153. Hostacka, A. and Majtan, V. (1993), Alterations in *Pseudomonas* aeruginosa exoproducts by sub-MICs of some antibiotics, *Folia Microbiol.*, 38, 349–352.
154. Grimwood, K., To, M., Rabin, H.R., and Woods, D.E. (1989), Inhibition of *Pseudomonas aeruginosa* exoenzyme expression by subinhibitory antibiotic concentrations, *Antimicrob. Agents Chemother.*, 33, 41–47.
155. Grimwood, K., To, M., Rabin, H.R., and Woods, D.E. (1989), Subinhibitory antibiotics reduce *Pseudomonas aeruginosa* tissue injury in the rat lung model, *J. Antimicrob. Chemother.*, 24, 937–945.
156. Levatte, M.A., Woods, D.E., Shahrabadi, M.S., Semple, R., and Sokol, P.A. (1990), Subinhibitory concentrations of tetracycline inhibit surface expression of the *Pseudomonas aeruginosa* ferripyochelin binding protein *in vivo*, *J. Antimicrob. Chemother.*, 26, 215–225.

157. Sakata, K., Yajima, H., Tanaka, K., Sakamoto, Y., Yamamoto, K., Yoshida, A., and Dohi, Y. (1993), Erythromycin inhibits the production of elastase by *Pseudomonas aeruginosa* without affecting its proliferation *in vitro*, *Am. Rev. Respir. Dis.*, 148, 1061–1065.

158. Kobayashi, H., Ohgaki, N., and Takeda, H. (1993), Therapeutic possibilities for diffuse panbronchiolitis, *Int. J. Antimicrob. Agents*, 3 (Suppl.), 81–86.

159. Yasuda, H., Ajiki, Y., Koga, T., Kawada, H., and Yokota, T. (1993), Interaction between biofilms formed by *Pseudomonas aeruginosa* and clarithromycin, *Antimicrob. Agents Chemother.*, 37, 1749–1755.

160. Molinari, G., Guzman, C.A., Pesce, A., and Schito, G.C. (1993), Inhibition of *Pseudomonas aeruginosa* virulence factors by subinhibitory concentrations of azithromycin and other macrolides antibiotics, *J. Antimicrob. Chemother.*, 31, 681–688.

161. Kobayashi, H. (1995), Airway biofilm disease: its clinical manifestation and therapeutic possibilities of macrolides, *J. Infect. Chemother.*, 1, 1–15.

162. Kondoh, K., Hashiba, M., and Baba, S. (1996), Inhibitory activity of clarithromycin on biofilm synthesis with *Pseudomonas aeruginosa*, *Acta Oto-Laryngol.*, 525 (Suppl.), 56–60.

163. Regelmann, W.E., Elliott, G.R., Warwick, W.J., and Clawson, C.C. (1990), Reduction of sputum *Pseudomonas aeruginosa* density by antibiotics improves lung function in cystic fibrosis more than do bronchodilators and chest physiotherapy alone, *Am. Rev. Resp. Dis.*, 141, 914–921.

164. Pedersen, S.S., Pressler, T., Jensen, T., Rosdal, V.T., Bentzon, M.W., Høiby, N., and Koch, C. (1987), Combined imipenem/cilistatin and tobramycin therapy of multiresistant *Pseudomonas aeruginosa* in cystic fibrosis, *J. Antimicrob. Chemother.*, 19, 101–107.

165. Rayner, R.J., Wiseman, M.S., Cordon, S.M., Norman, D., Hiller, E.J., and Shale, D.J. (1991), Inflammatory markers in cystic fibrosis, *Resp. Med.*, 85, 139–145.

166. Valetta, E.A., Rigo, A., Bonazzi, L., Zanolla, L., and Mastella, G. (1992), Modification of some markers of inflammation during treatment for acute respiratory exacerbations in cystic fibrosis, *Acta Paediatr.*, 81, 227–230.

167. Orenstein, D.M., Pattishall, E.N., Nixon, P.A., Ross, E.A., and Kaplan, R.M. (1990), Quality of well-being before and after antibiotic treatment of pulmonary exacerbation in patients with cystic fibrosis, *Chest*, 98, 1081–1084.

168. Tanimoto, H. (1991), A review of the recent progress in treatment of patients with diffuse panbronchiolitis associated with *Pseudomonas aeruginosa* infection in Japan, in: Homma, J.Y., Yanimoto, H., Holder, I.A., Høiby, N., and Döring, G., Eds., *Pseudomonas aeruginosa in Humans*, Karger, Basel, pp. 94–98.

169. Jaffe, A., Francis, J., Rosenthal, M., and Bush, A. (1998), Long-term azithromycin may improve lung function in children with cystic fibrosis, *Lancet*, 351, 420.

170. Yanagihara, K., Tomono, K., Sawai, T., Hirakata, Y., Kadota, J., Koga, H., Tashiro, T., and Kohno, S. (1997), Effect of clarithromycin on lymphocytes in chronic respiratory *Pseudomonas aeruginosa* infection, *Am. J. Respir. Crit. Care Med.*, 155, 337–342.

171. O'Riordan, T.G. (2004), Inhaled antimicrobial therapy: from cystic fibrosis to the flu, *Respir. Care*, 45, 275–276.

172. Klepser, M.E. (2004), Role of nebulized antibiotics for the treatment of respiratory infections, *Curr. Opin. Infect. Dis.*, 17, 109–112.

173. Smith, A.L. (2002), Inhaled antibiotic therapy: what drug? what dose? what regimen? what formulation? *J. Cyst. Fibros.*, 1, S189–S193.

174. Conway, S.P., Brownlee, K.G., Denton, M., and Peckham, D.G. (2003), Antibiotic treatment of multidrug-resistant organisms in cystic fibrosis, *Am. J. Respir. Medicine*, 2, 321–332.

175. Smaldone, G.C. (2004), Aerosolized antibiotics in mechanically ventilated patients, *Respir. Care*, 49, 635–639.

176. Diot, P., Dequin, P.F., Rivoire, B., Gagnadoux, F., Faurisson, F., Diot, E., Boissinot, E., La Pape, A., Palmer, L., and Lemarie, E. (2001), Aerosols and infectious diseases, *J. Aerosol Med.*, 14, 55–64.

177. Singh, P.K., Parsek, M.R., Greenberg, E.P., and Welsh, M.J. (2002), A component of innate immunity prevents bacterial biofilm development, *Nature (London)*, 417, 552–555.

178. Musk, D.J., Banko, D.A., and Hergenrother, P.J. (2005), Iron salts perturb biofilm formation and disrupt existing biofilms of *Pseudomonas aeruginosa*, *Chem. Biol.*, 12, 789–796.

179. Yang, L., Barken, K.B., Skindersoe, M.E., Christensen, A.B., Givskov, M., and Tolker-Nielsen, T. (2007), Effects of iron on DNA release and biofilm development by *Pseudomonas aeruginosa*, *Microbiology*, 153, 1318–1328.

180. Johnson, M., Cockayne, A., Williams, P.H., and Morrissey, J.A. (2005), Iron-responsive regulation of biofilm formation in *Staphylococcus aureus* involves fur-dependent and fur-independent mechanisms, *J. Bacteriol.*, 187, 8211–8215.

181. Berlutti, F., Ajello, M., Bosso, P., Morea, C., Petrucca, A., Antonini, G., and Valenti, P. (2004), Both lactoferrin and iron influence aggregation and biofilm formation in *Streptococcus mutans*, *Biometals*, 17, 271–278.

182. Martinhon, C.C., Italiani Fde, M., Padilha Pde, M., Bijella, M.F., Delbem, A.C., and Buzalaf, M.A. (2006), Effect of iron on bovine enamel and on the composition of the dental biofilm formed "*in situ*", *Arch. Oral Biol.*, 51, 471–475.

183. Shak, S. (1995), Aerosolized recombinant human DNase I for the treatment of cystic fibrosis, *Chest*, 107, 65S–70S.

184. Suri, R. (2005), The use of human deoxyribonuclease (rhDNase) in the management of cystic fibrosis, *BioDrugs*, 19, 135–144.

185. Geller, D.E., Pitlick, W.H., Nardella, P.A., Tracewell, W.G., and Ramsey, B.W. (2002), Pharmacokinetics and bioavailability of aerosolized tobramycin in cystic fibrosis, *Chest*, 122, 219–226.

186. Cheer, S.M., Waugh, J., and Noble, S. (2003), Inhaled tobramycin (TOBI): a review of its use in the management of *Pseudomonas aeruginosa* infections in patients with cystic fibrosis, *Drugs*, 63, 2501–2520.

187. Murphy, T.D., Anbar, R.D., Lester, L.A., Nasr, S.Z., Nickerson, B., VanDevanter, D.R., and Colin, A.A. (2004), Treatment with tobramycin solution for inhalation reduces hospitalizations in young CF subjects with mild lung disease, *Pediatr. Pulmonol.*, 38, 314–320.

188. Musk, D.J., Jr. and Hergenrother, P.J. (2008), Chelated iron sources are inhibitors of *Pseudomonas aeruginosa* biofilms and distribute efficiently in an *in vitro* model of drug delivery to the human lung, [Comments in *J. Appl. Microbiol.* 2009, Mar;106(3),1058.], *J. Appl. Microbiol.*, 105, 380–388.

189. LiPuma, J.J., Spilker, T., Gill, L.H., Campbel 3rd, P.W., Liu, L., and Mahenthiralingam, E. (2001), Disproportionate distribution of *Burkholderia cepacia* complex species and transmissibility markers in cystic fibrosis, *Am. J. Respir. Crit. Care Med.*, 164, 92–96.

190. Speert, D.P., Henry, D., Vandamme, P., Corey, M., and Mahenthiralingam, E. (2002), Epidemiology of *Burkholderia cepacia* complex in patients with cystic fibrosis, Canada, *Emerg. Infect. Dis.*, 8, 181–187.

191. Brackman, G., Hillaert, U., Van Calenbergh, S., Nelis, H.J., and Coenye, T. (2009), Use of quorum sensing inhibitors to interfere with biofilm formation and development in *Burkholderia multivorans* and *Burkholderia cenocepacia*, *Res. Microbiol.*, 160, 144–151.

192. Niu, C., Alfre, S., and Gilbert, E.S. (2006), Subinhibitory concentrations of cinnamaldehyde interfere with quorum sensing, *Lett. Appl. Microbiol.*, 43, 489–494.

193. Wang, W.B., Lai, H.C., Hsueh, P.R., Chiou, R.Y., Lin, S.B., and Liaw, S.J. (2006), Inhibition of swarming and virulence factor expression in *Proteus mirabilis* by resveratrol, *J. Med. Microbiol.*, 55, 1313–1321.

194. Keshavan, N.D., Chowdhary, P.K., Haines, D.C., and Gonzalez, J.E. (2005), L-canavanine made by *Medicago sativa* interferes with quorum sensing in *Sinorhizobium meliloti*, *J. Bacteriol.*, 187, 8427–8436.

195. Tateda, K., Cote, R., Pechere, J.C., Kohler, T., Yamaguchi, K., and Van Delden, C. (2001), Azithromycin inhibits quorum sensing in *Pseudomonas aeruginosa*, *Antimicrob. Agents Chemother.*, 6, 1930–1933.

196. Garske, L.A., Beatson, S.A., Leech, A.J., Walsh, S.L., and Bell, S.C. (2004), Subinhibitory concentrations of ceftazidime and tobramycin reduce the quorum sensing signals of *Pseudomonas aeruginosa*, *Pathology*, 36, 571–575.

197. Cugini, C., Calfee, M.W., Farrow 3rd, J.M., Morales, D.K., Pesci, E.C., and Hogan, D.A. (2007), Farnesol, a common sesquiterpene, inhibits PQS production in *Pseudomonas aeruginosa*, *Mol. Microbiol.*, 65, 896–906.

198. Huber, B., Eberl, L., Feucht, W., and Polster, J. (2003), Influence of polyphenols on bacterial biofilm formation and quorum-sensing, *Z. Naturforsch.*, [C] 58(11–12), 879–884.

199. Lesic, B., Lepine, F., Deziel, E., Zhang, J., Zhang, Q., Padfield, K., Castonguay, M.H., Milot, S., Stachel, S., Tzika, A.A., Tompkins, R.G., and Rahme, L.G. (2007), Inhibitors of pathogen intercellular signals as selective anti-infective compounds, *PLoS Pathog.*, 3, 1229–1239.

200. Rudrappa, T. and Bais, H.P. (2008), Curcumin, a known phenolic from *Curcuma longa*, attenuates the virulence of *Pseudomonas aeruginosa* PAO1 in whole plant and animal pathogenicity models, *J. Agric. Food Chem.*, 56, 1955–1962.

201. Zeng, Z., Qian, L., Cao, L., Tan, H., Huang, Y., Xue, X., Shen, Y., and Zhou, S. (2008), Virtual screening for novel quorum sensing inhibitors to eradicate biofilm formation of *Pseudomonas aeruginosa*, *Appl. Microbiol. Biotechnol.*, 79, 119–126.

202. Riedel, K., Köthe, M., Kramer, B., Saeb, W., Gotschlich, A., Ammendola, A., and Eberl, L. (2006), Computer-aided design of agents that inhibit the *cep* quorum-sensing system of *Burkholderia cenocepacia*, *Antimicrob. Agents Chemother.*, 50, 318–323.

# PART III

# DRUG DELIVERY CARRIERS TO ERADICATE BIOFILM FORMATION ON MEDICAL DEVICES

# CHAPTER 9

# STRATEGIES FOR PREVENTION OF DEVICE-RELATED NOSOCOMIAL INFECTIONS

## 9.1. INTRODUCTION

In the previous and present decades, a number of strategies have been or are being developed for the prevention and/or eradication of biofilm formation over implanted or inserted medical devices. Even some of the novel approaches developed at laboratory levels is really interesting, and the obtained results are encouraging from the medical and social points of view. Part III is designed to describe briefly the possible prevention strategies to eradicate the biofilm community from forming over medical devices.

## 9.2. STRATEGIES FOR PREVENTION OF DEVICE-RELATED NOSOCOMIAL INFECTIONS

Whenever an infection of an indwelling or implanted foreign body is suspected, a general decision has to be addressed: whether to remove the foreign body and/or whether to initiate calculated antimicrobial treatment (Fig. 9.1). Answering the following key questions relevant to the clinical situation of the patient may help the physician to manage these infections adequately based on a rationale approach.

*Biofilm Eradication and Prevention: A Pharmaceutical Approach to Medical Device Infections*,
By Tamilvanan Shunmugaperumal

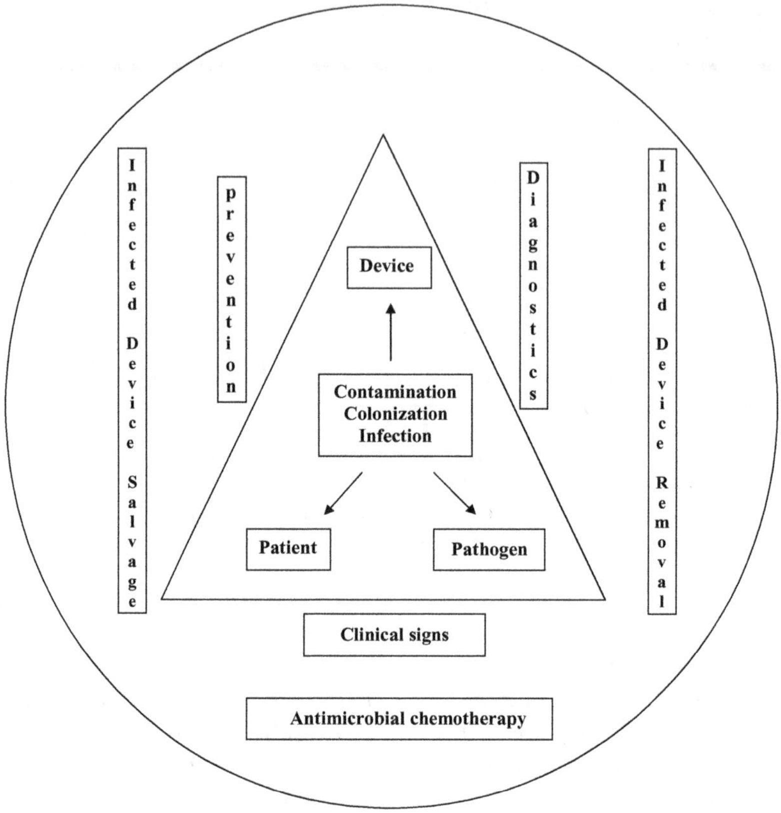


**Figure 9.1.** Complex interaction of host (patient), pathogen, and device to be considered for a decision on either removal or salvage of the infected device.

1. Is a foreign body-related infection (FBRI) a plausible explanation for the patient's signs (e.g., fever, skin inflammation at the exit site, soft tissue inflammation along the tunnel of an implanted catheter, septic thrombophlebitis)?
2. Are there any risk factors predisposing for FBRI (e.g. neutropenia, malignant hematological disorders, acquired immunodeficiency syndrome (AIDS), type of catheter)?
3. In which clinical situation is the patient (e.g., sepsis, pregnancy, premature infant)?
4. In light of a possible necessity to remove the foreign body, how important is the medical device for the patient regarding: (a) the survival of the patient [cardiac devices or "highly needed" catheters, e.g., tunneled Broviac–Hickman-type catheters or totally implantable venous access devices (i.e., ports) for intravenous administration of vital medications and parenteral nutrition]; (b) prosthetic therapy (e.g., prosthetic joints,

lens); (c) optimal intravenous application of fluids, medications, and blood products (e.g., all kinds of vascular prostheses; hemodialysis shunts); and (d) cosmetic and reconstructive surgery?
5. Which diagnostic methods should be applied to confirm the diagnosis?
6. Is calculated antimicrobial therapy necessary and, if so, which antibacterials should be given?

Several comprehensive reviews on the clinical management of infections due to an increasing palette of medical devices have been published focusing on different aspects concerning the removal of the infected device, antimicrobial therapy, and on additional procedures to detect and prevent complications associated with FBRIs [1–17].

## 9.3. REMOVAL OF THE DEVICE

The optimal treatment of a FBRI is the removal of the infected device when possible and its replacement, if still needed. This is the therapy of choice, especially for easy-to-change devices (e.g., short-term peripheral catheters) [1,2].

Regardless of the type of device, removal of implanted devices is recommended when the patient shows signs of severe sepsis, septic phlebitis, and septic shock. Furthermore, catheters should be removed in patients with bacteraemia persisting >48–72 h. In addition, presence of local skin or soft tissue infections (e.g., tunnel infection, gross purulence at the exit site), metastatic complications (e.g., endocarditis, osteomyelitis, septic thrombosis), and/or relapse of infection after antibacterial therapy has been discontinued should lead to removal of the device. In addition, local debridement at the exit site of a medical device should be considered if a subcutaneous abscess or extensive tunnelitis is present. The removal of the device is regularly necessary if the microorganisms that are isolated are known to be difficult to eradicate or to be high-virulence nonfermenter, mycobacteria, and yeasts [18–20].

Studies have shown that long-term tunneled catheters (mainly hemodialysis catheters) may be exchanged successfully with guidewire in patients with uncomplicated catheter-related bloodstream infections (CRBSI) and no signs of exit, tunnel tract, or pocket infection [21,22].

## 9.4. SALVAGE OF THE DEVICE AND TREATMENT WITH ANTIMICROBIAL AGENTS

Removing the infected medical device is not always possible, easy to perform, and/or without risk. Therefore, salvage of the device is sometimes the preferred option. In particular, FBRIs associated with long-term or permanent catheters (e.g., Hickman-type catheter or Port-a-Cath) are frequently treated successfully "through the line" [23–25].

To reduce the incidence of intravascular CRBSI, specific guidelines comprising both technological and nontechnological strategies for prevention have been established [26]. Quality assurance and continuing education, type of catheter material, choice of the catheter insertion site, hand hygiene, and aseptic techniques are aspects of particular interest [27]. Table 9.1 provides the general recommendations for the prevention of intravascular device-related (IVDR) bloodstream infections (BSIs). Additional strategies (e.g., skin antisepsis, catheter site dressing regimens, catheter securement devices, in-line filters, antimicrobial or antiseptic impregnated catheters and cuffs, systemic antibiotic prophylaxis, antibiotic or antiseptic ointments, antibiotic lock prophylaxis, and anticoagulants) are routinely employed for reducing–preventing device-related nosocomial infections [26,27]. Another concept for

**TABLE 9.1. General Recommendations for the Prevention of IVDRBSIs[a]**

| Recommendation | Strength of Evidence[a] |
|---|---|
| *General Measures* | |
| Educate all healthcare workers involved with vascular access regarding indications for use, proper insertion technique, and maintenance of IVDs | IA |
| Surveillance | |
| Routinely monitor institutional rates of IVDR BSI | IA |
| Determine rates of CVC related BSI, using standardized definitions and denominators, expressed per 1000 CVC[c] days$^{-1}$ | IB |
| *At Insertion* | |
| Use aseptic technique | |
| Wash hands before insertion or manipulation of any IVD | IA |
| Wear cleantion of or sterile gloves during insertion or manipula noncentral IVD | IC |
| Use maximal barrier precautions (mask, cap, long-sleeved sterile gown, sterile gloves, and sterile sheet drape) during insertion of CVCs | IA |
| Use dedicated intravenous-device teams strongly recommended | IA |
| Use cutaneous antisepsis (chlorhexidine is preferred; however, an iodophor (e.g., 10% povidone-iodine, tincture of iodine, or 70% alcohol) are also acceptable) | IA |
| Use of sterile gauze or a sterile semipermeable polyurethane film dressing | IA |
| Use of systemic antibiotics at insertion strongly discouraged | IA |

**TABLE 9.1.** ***Continued***

| Recommendation | Strength of Evidence[a] |
|---|---|
| *Maintenance* | |
| Remove IVDs as soon as their use is no longer essential | IA |
| Monitor the IVD site on regular basis: ideally, daily | IB |
| Change dressing of CVC insertion site at least weekly | II |
| Use of topical antibiotic ointments not recommended | IA |
| Perform systemic anticoagulation with low-dose warfarin (1 mg daily) for patients with long-term IVDs and no contraindication | IA |
| Replace PIVCs every 96 h | IA |
| Replace administration sets every 96 h, unless lipid-containing admixture or blood products given, in which case administration sets should be replaced every 24 h | IA |
| *Technology* | |
| Consider use of chlorhexidine-impregnated sponge dressing for adolescent and adult patients who have noncuffed CVCs or arterial catheters expected to remain in place for ≥4 days | NR |
| If, after consistent application of basic infection-control precautions, the institutional rate of IVDR BSI is still high for short-term CVCs (i.e., ≥3.3 BSIs/1000 IVD days), consider the use of a CVC coated with an anti-infective agent (i.e., chlorhexidine–silver sulfadiazine or minocycline-rifampin) | IB |
| For individual patients with long-term IVDs in place who have had recurrent IVDR BSIs, despite consistent application of infection-control practices, consider the use of a prophylactic antibiotic lock solution (i.e., heparin with vancomycin [25 $\mu$g mL$^{-1}$] with or without ciprofloxacin [2 $\mu$g mL$^{-1}$] | II |

[a]*Note.* Adapted from the Healthcare Infection Control Practices Advisory Committee (HICPAC) draft guideline for the prevention of intravascular catheter-related infections [25].

IVD, iv device; PIVC, peripheral iv catheter.

[b]Adapted from the Centers for Disease Control/HICPAC system for weighting recommendations based on the quality of scientific evidence. IA, strongly recommended for implementation and strongly supported by well-designed experimental, clinical, or epidemiological studies; IB, strongly recommended for implementation and supported by some experimental, clinical, or epidemiological studies and a strong theoretical rationale; II, suggested for implementation and supported by suggestive clinical or epidemiological studies or theoretical rationale; NR, no recommendation for or against use at this time.

[c]Central venous catheter = CVC.

the prevention of device-related infections involves the impregnation–coating of devices with various substances (e.g., antibacterials, antiseptics, and/or metals) [26,28–35]. Further strategies include minimizing the length of time of catheterization and using catheters provided with a surgically implanted cuff [36]. Furthermore, all steps in the pathogenesis of biofilm formation may represent targets against which prevention strategies may be directed. For example, enzymes involved in bacterial cell wall synthesis could provide novel targets for the development of antibiofilm agents. The work conducted at Kane Biotech (Winnipeg, Canada), which has led to the development of an antibiofilm composition comprising an *N*-acetyl-*d*-glucosamine-1-phosphate acetyltransferase (GlmU) inhibitor and protamine sulfate, a cationic polypeptide. This composition demonstrated antimicrobial efficacy against a range of microbes and represents a licensing opportunity. Since biofilm formation represents a problem that extends past the urinary tract, such technology is likely to have wide-ranging relevance in infectious diseases including, for example, vascular cannula infections, a serious problem in the intensive care unit (ICU) setting [37].

## 9.5. STANDARDIZATION OF ASEPTIC CARE

In the following sections, some of the most important strategies for prevention of catheter-related infections are summarized, including those most recently developed. Quality assurance and continuing education are aspects of particular interest. Several studies have shown that the risk for intravascular device-associated BSIs declines following standardiation of aseptic care [26,38–40]. While insertion and maintenance of intravascular catheters by inexperienced staff (as well as nursing staff reductions) might increase the risk for catheter colonization and CRBI, specialized "IV (intravenous) teams" have shown effectiveness in reducing the incidence of infections and associated complication and costs [40–42].

## 9.6. CHOICE OF CATHETER INSERTION SITE

The density of local skin flora and, thus, also the site of catheter insertion, influences the subsequent risk for CRI [43–45]. In adult patients, a subclavian site is preferred for infection control purposes, although other factors (e.g., the potential for mechanical complications or risk for subclavian vein stenosis) should be considered when deciding where to place the catheter [46–48]. Consideration of comfort, security, and maintenance of asepsis, as well as patient-specific factors (e.g., anatomic deformity and bleeding diathesis), relative risk (RR) of mechanical complications, the availability of bedside ultrasound, and the risk for infection should guide site selection [45].

In addition, phlebitis has long been recognized as a risk for infection. Lower extremity insertion sites are associated with a higher risk for phlebitis than are upper extremity sites (for adults), and hand veins have a lower risk for infection than do veins on the wrist or upper arm [49].

## 9.7. HAND HYGIENE AND ASEPTIC TECHNIQUE

The most important and simple strategy to reduce the rate of FBRIs is the attention of an adequate hand hygiene and aseptic technique [50–52]. While for short peripheral catheters good hand hygiene before catheter insertion or maintenance combined with proper aseptic technique during catheter manipulation is of major importance, the level of barrier precautions needed to prevent infection during insertion of CVCs should be more stringent. That is, maximal sterile barrier precautions are necessary to reduce the incidence of CRBI in patients with CVCs.

Good hand hygiene comprises the use of either a waterless, alcohol-based product or an antibacterial soap and water with adequate rinsing [51]. Maximal sterile barrier precautions should be achieved through the use of a cap, mask, sterile gown, sterile gloves, and a large sterile drape [52,53]. For the insertion of peripheral venous catheters, a new pair of disposable nonsterile gloves can be used in conjunction with a "no-touch" technique, thus, appropriate aseptic technique does not necessarily require sterile gloves [26].

A review of data regarding hand washing and antisepsis in healthcare settings and recommendations to promote improved hand hygiene practices and reduce transmission of pathogenic microorganisms to patients and personnel in healthcare settings is given in the *Guideline for Hand Hygiene in Health-Care Settings* by Boyce and Pittet [54].

## 9.8. SKIN ANTISEPSIS AND CATHETER SITE DRESSING REGIMENS

Currently, it was shown that most CRBIs with short-term percutaneously inserted, noncuffed CVCs were extraluminally acquired and derived from the cutaneous microflora. It was concluded that strategies achieving successful suppression of cutaneous colonization can substantially reduce the risk of CRBI with short-term CVCs [55]. In the past, a number of different commercially available products for cleansing arterial catheter and CVC insertion sites have been studied [56–59]. Preparation of central venous and arterial sites with 2% aqueous chlorhexidine gluconate lowered BSI rates compared with site preparation with 10% povidone iodine or 70% alcohol [58]. In another prospective, randomized study of adults, a tincture of 0.5% chlorhexidine was shown to be as or less effective in preventing CRBI or CVC colonization than 10% povidone iodine [57]. In contrast, in a study comprising neonates, 0.5% chlorhexidine reduced peripheral intravenous colonization compared with povidone iodine [59].

Different dressing regimens have also been compared. In the largest controlled trial of dressing regimens on 2000 peripheral catheters, the rate of colonization among catheters dressed with transparent dressings (5.7%) was shown to be comparable with that of those dressed with gauze (4.6%) [60]. No clinically substantial differences in either the incidences of catheter site colonization or phlebitis were observed. In a meta-analysis assessing studies that compared the risk for CRBIs for groups using transparent dressings versus groups using gauze dressing, the risk was found not to differ between the groups [61]. A chlorhexidine-impregnated sponge placed over the site of short-term arterial and CVCs reduced the risk for catheter colonization and CRBI [62]. Concerning catheter securement devices, a study, which compared a sutureless device with suture for the securement of peripherally inserted central catheters, revealed that CRBI was reduced in the group of patients who received the sutureless device [63].

## 9.9. CATHETER MATERIAL AND IN-LINE FILTERS

The type of catheter material used is also of importance regarding the risk for subsequent infections. For example, several studies showed that Teflon® or polyurethane catheters are associated with fewer infectious complications than catheters made of poly(vinyl chloride) (PVC) or polyethylene [60,64]. Steel needles have the same rate of infectious complications as do Teflon catheters. However, their use is frequently complicated by infiltration of IV fluids into the subcutaneous tissues [65] (see the following chapters for prevention by material modification or by incorporation of antimicrobial agents).

The routine use of IV in-line filters on infusion lines has been controversial for many years and is still under debate [66,67]. So far, no data have been published that support the efficacy of in-line filters in preventing infections associated with intravascular catheters and infusion systems. However, they reduce the incidence of infusion-related phlebitis [67]. While these filters may reduce the risk for infection from contaminated infusate or proximal contamination (i.e., introduced proximal to the filter) or may reduce the risk for phlebitis in patients who require high doses of medication or in those in whom infusion-related phlebitis has already occurred, no strong recommendation can be made in favor of using in-line filters because infusate-related BSI are rare and in-line filters might become blocked, especially with certain solutions (e.g., dextran and lipids).

Once a biofilm has formed on an implanted medical device it is difficult to treat such infections because of significantly decreased levels of susceptibility of antimicrobial agents (some 10–1000 times less) and lower levels of phagocytosis relative to the levels of resistance or tolerance and phagocytosis for their planktonic counterparts [68]. Thus, supraphysiological concentrations of antibacterial agents may be required to eliminate the microorganisms embedded in biofilms [69]. As shown in a number of experimental FBRIs, the pharmacokinetic parameters are modified and do not correspond to the efficacy of antibacterial treatment *in vivo* when a foreign body is implanted. These changes are obvious if mouse model-based results of *Staphylococcus aureus* caused intraabdominal abscess surrounding intraperitoneally placed Si catheter treated by meticillin and gentamicin are analyzed [70]. Whereas both agents showed strong effects *in vitro* in time-kill studies on bacteria colonizing catheters taken out of infected mice and on catheters contaminated *in vitro*, only poor results were observed *in vivo*, despite high local concentrations (> minimum inhibitory concentration (MIC) for at least 72 h] of meticillin and high peak concentrations of gentamicin (>13 μg mL$^{-1}$). The failure was not caused by development of antibacterial resistance or influenced by protein concentration, pH, or local presence of inhibitors of antibacterials in the pus.

Of importance, antibacterials administered in subinhibitory concentrations may influence the mechanisms of adherence and slime production, especially in staphylococci (e.g., leading to higher polysaccharide intracellular adhesin production or to increased expression of fibronectin-binding proteins) [71–73].

The special conditions surrounding a foreign body have guided the search for alternative applications of antibacterials (e.g., lipid-based sustained release formulations). Roehrborn et al. [74] described the use of such biodegradable, locally injectable formulation of amikacin in a mouse model in which Teflon (the use of trade names is for product identification purpose only and does not imply endorsement) tubes were subcutaneously implanted and challenged by inoculation of *S. aureus.* Whereas treatment with local or systemic free amikacin had no effect, the number of infected foreign bodies was reduced from 86 to 25% ($p = 0.02$) following treatment with encapsulated amikacin formulation, and log CFU (colony forming units) per gram of tissue was significantly decreased from 4.8 ± 0.9 to 1.3 ± 0.6.

Typically, initial treatment of catheter-related bacteraemia is administration of systemic antibacterials. Additionally, when a catheter-related infection is documented and a specific pathogen is identified, "antibiotic-look" therapy should be considered if salvage of the catheter is necessary. Note that recommendations from the treatment of medical device associated infections are based almost exclusively on observational studies, animal models, case reports, and expert opinion rather than on the results of appropriate clinical trials.

## 9.10. USE OF LOCK SOLUTIONS FOR INTRALUMINAL THERAPY ("ANTIBIOTIC-LOCK" TECHNIQUE)

A technique of filling and closing a catheter lumen with a lock solution may prevent or cure catheter-related infections, as active ingredients can be maintained directly with the internal surface of the device for prolonged periods of time (hour to days). Thus, to circumvent the need for catheter withdrawal, Messing et al. [75] were the first to describe the intraluminal application of antibacterial agents, referred to as antibiotic-lock technique. Avoiding systemic adverse effects, this method allows the delivery of a high concentration of antibacterials (or, rarely, disinfectants) in the catheter in order to decontaminate the intraluminal surface of the catheter *in situ*. In an analysis of 14 open-label trials of standard parenteral therapy for the treatment of CRBI and the salvage of tunneled catheters, a salvage in 342 (66.5%) of 514 episodes was documented [12].

Currently, the antibiotic-lock technique is recommended for the treatment of uncomplicated catheter-related bacteraemia by several medical societies (e.g., the Infectious Diseases Society of America, the Society of Critical Care Medicine, and the Society for Healthcare Epidemiology of America) [12]. However, several parameters of intraluminal antibacterial therapy are not clearly defined (e.g., the duration of the antibiotic-lock therapy is not established). In most studies, this technique was administered for 7–14 days. Furthermore, the usefulness of different types of antibacterial agents, their optimum concentration, and the necessity of simultaneous systemic treatment remain to be defined [76]. Glycopeptides, aminoglycosides, and ciprofloxacin have been shown to be suitable agents [20,75,77,78].

Some studies used the antibiotic-lock technique in conjunction with the administration of systemic antibacterials and/or thrombolytic–anticoagulant agents [77,79–81]. However, bacteria (e.g., staphylococci) may survive and grow in heparin locked catheters [82]. The drawback of using lock solutions containing antibacterials used for systemic therapy is that it may lead to the emergence of antibacterial resistance. In particular, the prophylactic and therapeutic long-term application of vancomycin could be of high risk for the development of staphylococcal subpopulations with reduced susceptibility against glycopeptides as a result of the existence of more or less "occult" device-related infection sites [83].

To meet concerns regarding a selection of highly resistant bacteria and an insufficient clearance of the device, the antimicrobial activity of alternative agents (e.g., catheter lock solutions) were investigated. Taurolidine, known as a nontoxic substance with antiadherence properties, was shown to be active against a broad range of bacteria, as well as fungi [84,85]. The findings of Shah et al. [86] evaluating taurolidine–citrate (Neutrolin™, Biolink Crop., Norwell, MA) for its antimicrobial and biofilm eradication activity in a catheter model suggested that this lock solution is a promising combination agent for the prevention and treatment of intravascular catheter-related infections [86]. Alternatively, the ethanol (alcohol)-lock technique was introduced for the treatment of BSIs in patients with tunneled central venous catheters (CVCs) and proven to be a safely used, well tolerated, and effective way to treat central venous line infections [87]. However, further studies are needed to ascertain whether ethanol or taurolidine locks might be equal or superior to the antibiotic lock technique. Since its effect does not depend on sensitivity to antibacterial agents, this approach may be of particular value for infections with multiresistant microorganisms. Furthermore, highly antibacterial-resistant microorganisms will not be selected by the use of disinfectants or other alternative agents and, in principle, its use could reduce the consumption of broad-spectrum antibacterials, especially vancomycin.

## 9.11. RECOMMENDATIONS FOR CALCULATED ANTIMICROBIAL THERAPY

Because of the high risk of complications, CVC related and surgically implanted venous access infections should be treated with parenteral drugs, using high doses and short courses (~7–10 days), irrespective of the removal of the device [88,89]. Antimicrobial therapy for the time period prior to a microbiological diagnosis should be initiated on a calculated basis considering the spectrum of expected pathogens and their local–regional resistance situation. However, treatment should be de-escalated to narrow-spectrum drugs on the basis of susceptibility tests as soon as test results are available.

Considering that staphylococci (especially CoNS, e.g., *Staphylococcus epidermidis* and *Staphylococcus haemolyticus*) are by far the most frequent pathogens isolated in FBRIs, calculated antimicrobial therapy should include the administration of a glycopeptide (especially vancomycin) with an aminoglycoside (e.g., gentamicin) or rifampicin because a significant percentage of staphylococci recovered from hospitalized patients are meticillin resistant [12,17,90]. In critically ill patients, coverage against Gram-negative bacteria, including *Pseudomonas aeruginosa,* and even fungi may be considered until definitive data from microbiological diagnostics are available.

## 9.12. RECOMMENDATIONS FOR AETIOLOGICALLY GUIDED ANTIMICROBIAL THERAPY

Aetiologically guided antimicrobial therapy should be initiated as soon as possible on the basis of appropriate microbiological diagnostics. Choice and duration of this therapy depends mainly on the isolated causative microorganisms, the resistance pattern, and the presence of complications, especially deep-seated soft-tissue infections.

### 9.12.1. Coagulase-Negative Staphylococci

Implant infections due to coagulase-negative staphylococci (CoNS) remain a therapeutic challenge since they frequently result in failure of conservative therapy and often require withdrawal of the foreign body. Although cure rates are not affected by removal, investigations on the impact of CVC removal on the recurrence of catheter-related CoNS bacteraemia have shown that there is a 20% chance of recurrence of bacteraemia when the CVC is not removed [91,92]. In contrast, the risk is significantly reduced to 3% if the catheter is removed [92]. This risk is especially high if the catheter stays in place for >3 weeks after bacteraemia.

Most CoNS isolates causing FBRI are meticillin resistant as a result of the possession of the *mecA* gene. Consequently, these isolates are resistant to all β-lactam antibacterials. Thus, most CoNS infections require treatment with glycopeptides, in particular vancomycin. In addition, teicoplanin has the potential for use as an alternative in the treatment of infections due to CoNS [9]. Notably, glycopeptides are poorly bactericidal against staphylococci. If an isolate is susceptible, replacement of vancomycin by a semisynthetic penicillin is advisable. Superior rapid action of rifampicin compared with vancomycin was noted in a mouse model of intraperitoneally implanted preformed bacterial biofilm catheter segments [84]. While simultaneous use of antibacterials of the cell wall-active class (including vancomycin) and rifampicin was shown to act synergistically, other antibacterials (including aminoglycosides) antagonized rifampicin activity [93]. However, combination of antibacterials is not generally recommended for CRBI due to CoNS [12]. Recently, two oxazolidinones (linezolid and eperezolid) were shown to achieve eradication of *S. epidermidis* biofilms more rapidly than vancomycin and gentamicin in an *in vitro* model using polyurethane coupons in a modified Robbins device [93].

The duration of parenteral therapy may be quite short (5–7 days) when treating uncomplicated FBBRI due to CoNS if the catheter is removed. If an intraluminal infection is suspected and an intravenous catheter or a surgically implanted device is retained, systemic antibacterial therapy and antibiotic-lock therapy for 10–14 days are recommended [12,77,94,95]. Note that persistent or relapsing fever and other signs of treatment failure are clear indications for removal of the device [12].

The widespread use of vancomycin for the treatment of FBRIs is of concern because of the emergence of vancomycin-resistant enterococci and of staphylococci with reduced sensitivity to glycopeptides (vancomycin–glycopeptide intermediate *S. aureus*). Moreover, the most recent recovery of true vancomycin-resistant *S. aureus* strains underscores the need of control regarding the use of vancomycin in healthcare settings [96].

### 9.12.2. *Staphylococcus aureus*

The FBRIs caused by *S. aureus* infections are dreaded because of possible accompaniment by serious infectious complications (severe sepsis, septic thrombosis, and/or several deep-seated infections, e.g., endocarditis, osteomyelitis, and other metastatic infections). Thus, it is generally accepted that the colonized foreign body, especially in the case of nontunneled CVC, must be removed [12,97,98]. Tunneled CVCs should be removed if there is evidence of exit-site infection as well as tunnel or pocket infections [99,100]. Only in selected cases of uncomplicated infections may tunneled CVCs or medical devices be retained and treated with appropriate systemic antibacterial therapy accompanied by antibiotic-lock therapy (for details see Section 9.12) [12,101,102].

Since metastatic infections may occur in the course of *S. aureus* infections, it is clinically important to rule out at least their most devastating consequence (i.e., acute endocarditis). Transesophageal echocardiography, which has been shown to be a highly sensitive method to diagnose endocarditis, should be performed in each patient with *S. aureus* BSI unless contraindications are present [103,104]. Clinical symptoms of bone infections should lead to scintigrapic and radiographic examinations [8]. A scoring system that was published based on the presence or absence of four risk factors (community acquisition, skin examination findings suggesting acute systemic infection, persistent fever at 72 h, and positive follow-up blood culture results at 48–96 h) accurately identified complicated *S. aureus* bacteraemia [105].

In contrast to CoNS, most experts favor parenteral treatment for CRBI caused by *S. aureus* with a minimum duration of 10–14 days of parenteral antibacterials [106,107]. Some authors recommended a subsequent additional treatment with oral antistaphylococcal antibacterials over a period of 1–2 weeks [108]. If persisting bacteraemia or complications (e.g., prolonged fever, metastatic or deep-seated infection) are occurring, much longer periods (4–6 weeks for endocarditis, 6–8 weeks for osteomyelitis) of parenteral antistaphylococcal therapy are recommended [12,23]. The first choice for treatment of CRBIs caused by *S. aureus* should be the parenteral application of β-lactam antibacterials (penicillinase-resistant penicillins, e.g. flucloxacillin and oxacillin) when the isolate is susceptible [12]. First-generation cephalosporins (e.g., cefazolin), may be used for patients with a penicillin allergy without anaphylaxis or angio-oedema [12]. For patients with a serious allergy to β-lactams and for those infected with methicillin-resistant *S. aureus* (MRSA), vancomycin is the drug of choice [12,109]. However, vancomycin has higher failure rates than have penicillinase-resistant penicillins and some complications are difficult to treat with glycopeptide monotherapy for pharmacological reasons [110,111]. In the case of MRSA, lincosamide antibacterials (clindamycin) and newer fluoroquinolones, as well as combinations with rifampicin, fusidic acid, cotrimoxazole, and fosfomycin, may be included into the therapeutic regimen if isolates are sensitive [110]. New antimicrobials (e.g., the oxazolidinones, streptogramins, and newer glycopeptides) exhibit high activity against MRSA (and other multiresistant Gram-positive pathogens), but resistance to some of these agents has already occurred. In a recent study encompassing children with hospital-acquired pneumonia or bacteraemia due to multiresistant Gram-positive bacteria, linezolid was well tolerated. No significant difference was detected in clinical cure rates in the clinically evaluable population between the linezolid and vancomycin groups for patients with catheter-related bacteraemia [112]. However, the potential of these alternative agents for the treatment of CRBIs should be analyzed in further trials.

Several animal models of FBRIs were developed in order to investigate the effects of antibacterial treatment [113–117] (see Table 9.2) [118–122]. In one study, Chuard et al. [114] showed that two- or three-drug combinations [e.g., fleroxacin and rifampicin (and vancomycin)], respectively, were highly

**TABLE 9.2. Experimental Animal Models of Foreign-Body Infection to Study the Effects of Treatment With Antibacterials**

| Foreign-Body Model | Animal Used | Causative Pathogen | Antibacterial Agent Used | Reference |
|---|---|---|---|---|
| Subcutaneous tissue cages | Rat | *S. aureus* | Vancomycin, fleroxacin, rifampicin | Chuard et al. 1991 [114] |
| Intraperitoneal catheter segments | Mouse | *S. epidermidis* | Rifampicin, vancomycin | Gagnon et al. 1992 [118] |
| Intraperitoneal silicone catheter | Mouse | *S. aureus* | Methicillin, gentamicin | Espersen et al. 1994 [70] |
| Subcutaneous tissue cages | Guinea pig | *S. aureus* (MRSA) | Teicoplanin, rifampicin | Schaad et al. 1994 [116] |
| Subcutaneous tissue cages | Rat | *S. aureus* (MSSA,[a] MRSA) | Imipenem, oxacillin, vancomycin | Schaad et al. 1994 [117] |
| Subcutaneous tissue cages | Guinea pig | *S. aureus* (MRSA) | Sparfloxacin, temafloxacin, ciprofloxacin, vancomycin | Cagni et al. 1995 [115] |
| Subcutaneous Teflon tubes | Mouse | *S. aureus* | Amikacin (lipid-based, slow-release) | Roehrborn et al. 1995 [74] |
| Subcutaneous catheter | Rat | *S. epidermidis* | Teicoplanin, rifampicin | Van Wijngaerden et al. 1999 [119] |
| Central venous catheter | Rat | *Enterococcus faecium* (VRE)[b] | Oritavancin | Rupp et al. 2001 [120] |
| Subcutaneous tissue cages | Guinea pig | *S. aureus* (MRSA) | Levofloxacin, alatrofloxacin, vancomycin | Vaudaux et al. 2002 [113] |
| Subcutaneous Teflon® catheter | Mouse | *S. aureus* (bioluminescent mutant) | Linezolid | Kuklin et al. 2003 [121] |
| Subcutaneous tissue cages | Rat | *S. aureus* | Daptomycin, vancomycin | Vaudaux et al. 2003 [122] |

[a]Meticillin-susceptible *S. aureus* = MSSA.
[b]Vancomycin-resistant enterococci = VRE.

effective and superior to single drugs in treating chronic staphylococcal FBRIs. Applying different fluoroquinolones, partly in comparison with vancomycin, in two different experimental models (rat and guinea pig), it was shown that the newer fluoroquinolones (temafloxacin and sparfloxacin), were significantly more active than ciprofloxacin for the prophylaxis or treatment of FBRIs caused by a fluoroquinolone-susceptible MRSA strain. As with temafloxacin and sparfloxacin, vancomycin was also significantly more active than ciprofloxacin in decreasing the viable counts of MRSA in tissue cage fluids in the rat model [115]. A further comparison of fluoroquinolones with vancomycin for treatment of experimental FBRI by MRSA showed levofloxacin was significantly more active than vancomycin in decreasing the viable counts of MRSA [113]. A second-generation glycopeptide, oritavancin (LY 333328), was shown to be effective against *S. aureus* in a rat CVC infection model [123]. The therapeutic activity of daptomycin was compared with that of vancomycin in a rat model of subcutaneously implanted tissue cages chronically infected with *S. aureus* [122]. The authors concluded that a low-dose regimen of daptomycin was at least equivalent to vancomycin; however, three of four cages implanted in daptomycin-treated rats yielded subpopulations with reduced susceptibility to daptomycin.

### 9.12.3. Gram-Positive Rods (Including Rapidly Growing Mycobacteria)

The majority of IV line infections caused by *Corynebacterium* spp. and *Bacillus* spp. require catheter withdrawal. Vancomycin has been widely used to treat infections caused by these bacteria, although treatment should be de-escalated based on the results of susceptibility testing. Catheter removal is essential for successful treatment of CVC related infections due to rapidly growing mycobacteria of the *Mycobacterium fortuitum* complex [124]. Since these mycobacteria exhibit variable, species-specific susceptibility to traditional antimycobacterial drugs and other antibacterials (including cefoxitin, imipenem–cilastatin, aminoglycosides, tetracyclines, macrolides, and co-trimoxazole) trimethoprim–sulfamethoxazole therapy should be based on culture and susceptibility results [125].

### 9.12.4. Gram-Negative Rod

Gram-negative rods are commonly associated with contaminated infusate and are usually found to be the cause of BSIs in immunocompromised patients with indwelling devices. Controlled studies regarding withdrawal of the infected device or the choice of optimal antibacterial agents and the duration of therapy are missing. However, patients with catheter-related infections due to Gram-negative rods should have the catheter removed, if possible, and should receive appropriate antibacterial therapy. Patients with devices that cannot be removed should be treated for 2 weeks with systemic and antibiotic-lock therapy provided that the Gram-negative bacteraemia is not associated with organ dysfunction, hypoperfusion, or hypotension [12,126,127]. In cases of catheter-related bacteraemia with nonfermenter species other than *P. aeruginosa*, *B. cepacia*, *Acinetobacter baumannii*, and *Stenotrophomonas maltophilia*, some reports have demonstrated that catheter withdrawal reduces the rate of treatment failure and improves survival [128]. Approximately 10–14 days of parenteral therapy is recommended when treating CRBIs caused by Gram-negative rods. However, a longer duration (4–6 weeks) of antibacterial therapy should be performed if prolonged bacteraemia occurs despite catheter removal [129].

### 9.12.5. Yeasts

Since several *Candida* species readily form biofilms, they are frequently isolated from patients with FBRIs [130]. *Candida albicans* represents the predominant and most virulent species. However, the importance of infections caused by non-*albicans Candida* spp. and other unusual yeasts (e.g., *Malassezia* spp., *Rhodotorula* spp., *Hansenula anomala*) has emerged over the last decade [131]. Notably, current routine methods for yeast identification may be insufficient to identify isolates of lipohilic *Malassezia* spp., which have been found to be associated to low, but not negligible, extent with infections of CVCs for parenteral nutrition-bearing lipid emulsions [132]. In particular, infections due to *C. parapsilosis* have been shown to correlate strongly with the presence of an intravascular device and the use of total parenteral nutrition due to the slime-forming ability of this species [133].

In the case of CRBIs due to yeasts, removal of all existing intravascular catheters is desirable, if feasible [134,135]. Following isolation of *C. parapsilosis* and *C. glabrata* in blood, initial management must include withdrawal of the catheter [136–138]. The evidence for these recommendations is strongest in the non-neutropenic patient population [139]. In neutropenic patients it is difficult to determine whether the gut or a catheter may act as the primary source of fungaemia. Management of Candida infection by catheter removal alone is not sufficient because of an increased risk of disseminated and/or metastatic fungal infections [140,141]. Thus, it is recommended to treat catheter-related *Candida* infections with appropriate antifungal agents for a minimum duration of 2 or 3 weeks after the last positive blood culture [15]. Infections due to *Malassezia* spp. should include discontinuation of IV lipids [142].

Since its introduction to the pharmaceutical market in the 1950s, amphotericin B has been the gold standard antifungal agent for life-threatening invasive fungal infections. However, its use is considerably hampered by the high rate of toxicity, which has led to the development of lipid-based formulations of amphotericin B with their superior safety profiles. These lipid formulations can be considered as suitable replacements for amphotericn B for primary therapy for many invasive fungal infections [143]. In general, *C. albicans* is susceptible to all antifungal agents. However, its potential to develop azole resistance has been documented. In a randomized trial in patients without neutropenia and major immunodeficiency, high-dose fluconazole appeared to be effective as amphotericin B, with less toxicity [144]. In contrast, some *Candida* spp. other than *C. albicans* are characterized by decreased susceptibility against azoles. Thus, knowledge of the species is increasingly important for the choice of the specific antifungal treatment and, especially in the setting of infections due to non-*albicans Candida* spp., susceptibility testing by standardized methods is most helpful. Whereas *C. krusei* and *C. glabrata* are intrinsically–innately more resistant to fluconazole, *C. Tropicalis*, *C. gullermondii*, and *C. dubliniensis* are generally susceptible to azoles, but fluconazole may be less active against these yeasts. In patients infected by these yeasts or in institutions where isolates of these *Candida* spp. are more frequent, the prescription of amphotericin B or the administration of higher doses of fluconazole should be the preferred treatment until the susceptibility data are available [136]. Note, the azole-sensitive species *C. lusitaniae* has innately higher MICs to amphotericin B.

The first of the second-generation triazole agents to receive regulatory approval is voriconazole, which has shown an expanded *in vitro* activity against a wide variety of yeasts and moulds. In addition, caspofungin, a new echinocandin antifungal agent with broad-spectrum activity against *Candida* and *Aspergillus* spp., was shown to be highly active against *Candida* isolates exhibiting high-level resistance to fluconazole and itraconazole [145]. In a recent study designed to compare the efficacy of caspofungin with that of amphotericin B, caspofungin was shown to be at least as effective as amphotericin B for the treatment of invasive candidiasis and, more specifically, candidaemia [146]. Regarding *C. glabrata*, *C. krusei*, and *C. albicans*, voriconazole and caspofungin appear to have enhanced activity. However, the clinical relevance of these findings should be studied in treatment trials [145,147,148].

Therapy of patients with FBRIs due to *Candida* spp. should be accompanied by ophthalmoscopic examination to rule out metastatic endophthalmitis. Remember that candidal endocarditis has also been observed following FBRIs.

## 9.13. USE OF ALTERNATIVE SUBSTANCES AND APPROACHES

Considering the extremely robust defense mechanisms of biofilms, designing novel therapeutics may seem like a daunting task. However, some have accepted this challenge and in the process have devised some clever and creative solutions as shown below.

With the emergence of antibacterial-resistant staphylococci, the antibacterial enzyme lysostaphin has, in the past few years, gained renewed interest as an antistaphylococcal therapeutic agent [149,150]. This glycylglycine endopeptidase is specifically capable of cleaving the cross-linking pentaglycine bridges in the cell wall of staphylococci, making it highly active against both actively growing and quiescent bacteria. With a $MIC_{90}$ of 0.001–0.064 $\mu g\,mL^{-1}$, lysostaphin kills planktonic *S. aureus* within minutes and is also effective against *S. epidermidis* at higher concentrations ($MIC_{90}$: 12.5–64 μg mL$^{-1}$) [151]. With the use of biofilm plate assays, Wu et al. [151] demonstrated that lysostaphin is also effective against sessile staphylococci associated with biofilms. Once established, staphylococcal biofilms are very difficult to disrupt. Therefore, the fact that lysostaphin is specifically able to disrupt the extracellular matrix of *S. aureus* biofilms *in vitro* on plastic and glass surfaces (confirmed by scanning electron microscopy, SEM) has to be regarded as a major progress in the management of FBRIs. Lactoferrin, another constituent of human secretions, blocks biofilm development by *P. aeruginosa* [152]. Various other enzymes have been studied for the removal and disinfection of bacterial biofilms. However, they are hampered by the fact that these procedures require two or more compounds. One enzyme is for removal of the adherent bacteria in the biofilms and another agent has antibacterial activity [153].

To address this issue, a variety of chemicals have been shown to be active against bacteria in biofilms. A combination of streptokinase and streptodornase has been shown by Nemoto et al. [154,155] to be active against *S. aureus* and *P. aeruginosa* biofilms. Similarly, oxidoreductases were bactericidal against biofilms and a complex mixture of polysaccharide-hydrolyzing enzymes removed bacterial biofilm [153]. Hatch and Schiller [156] showed that alginate lyase permitted increased diffusion of aminoglycosides through alginate in *P. aeruginosa*. Yasuda et al. [157] studied interactions between clarithromycin and biofilms formed by *S. epidermidis* using a clarithromycin-resistant strain and showed that treatment with a relatively low concentration of clarithromycin resulted in eradication of the "slime-like structure" and in a decrease in the amount of hexose. Allicin, which is derived from garlic, is a sulfur-containing compound formed in small quantities from the enzymatic action of allinase on alliin. Allicin has been shown to be active *in vitro* against *S. epidermidis* and *C. albicans*, and diminishes *S. epidermidis* and *C. albicans* biofilm formation [158,159].

### 9.13.1. Combination Therapy with Rifampin

*Staphylococcus epidermidis* and *S. aureus* are often susceptible to rifampin, although emergence of rifampin resistance can be problematic. Use of combination therapy generally avoids this pitfall. Gagnon et al. [160] determined the effect of combinations of 13 different antimicrobics with rifampin against *S. epidermidis* biofilms *in vitro*. Synergy with rifampin was observed with cloxacillin, cephalothin, cefazolin, cefamandole, vancomycin, ciprofloxacin, tetracycline, and amikacin. Whereas tobramycin, erythromycin, clindamycin, and fusidic acid did not influence the outcome, gentamicin unexpectedly showed antagonism with rifampin.

In continuous-flow biofilm cultures using a medium mimicking cystic fibrosis (CF) bronchial secretions, *P. aeruginosa* was not eradicated from biofilms after 1 week of treatment with high concentrations of ceftazidime and gentamicin, to which the strains were susceptible by conventional testing [161]. Addition of rifampin, however, which had little activity against the strains as measured by minimum inhibitory concentrations, led to nonstrain-specific elimination of bacteria from biofilms [161].

The comparative activities of vancomycin, clindamycin, novobiocin, and minocycline, alone or in combination with rifampin, were tested in an *in vitro* model of colonization using the modified Robbins device with antibiotic impregnated cement filling the lumen of catheter segments [162]. The combination of minocycline and rifampin was the most active in preventing bacterial colonization of biofilm-producing strains of *S. epidermidis* and *S. aureus* to the catheter surfaces [162]. A similar trend was observed when the inhibitory activities of polyurethane catheters coated with minocycline and rifampin were compared with the inhibitory activities of catheters coated with other antimicrobial agents [162]. The inhibitory activities of catheters coated with minocycline and rifampin against *S. epidermidis*, *S. aureus*, and *Enterococcus faecalis* strains, were significantly better than those of catheters coated with vancomycin [162]. The inhibitory activities of catheters coated with minocycline and rifampin against Gram-negative bacilli and *C. albicans* were comparable to those of catheters coated with ceftazidime and amphotericin B, respectively [162].

Rifampin penetrates biofilms formed by *S. epidermidis*, but does not kill biofilm *S. epidermidis* [163]. The combination of sparfloxacin or vancomycin with amikacin or rifampin show activity against *S. epidermidis* biofilms on catheters [164,165]. Peck et al. [166] showed that the combination of erythromycin or rifampin and vancomycin was more active than vancomycin alone against *S. epidermidis* biofilms formed on polyurethane sheets.

### 9.13.2. Ultrasound Enhancement of Antimicrobial Transport

Ultrasound, defined as acoustic energy or sound waves with frequencies >20 kHz, is commonly used to remove bacterial cells from the surface of foreign bodies, especially if applied as high-intensity ultrasound (>10 W cm$^{-2}$) [167]. This intensity is known to lyse bacterial and eukaryotic cells on surfaces and in suspension. The application of low-frequency ultrasound to enhance the activity of vancomycin against implanted *S. epidermidis* biofilms was examined using polyethylene disks covered with a biofilm of *S. epidermidis* and implanted subcutaneously in rabbits on both sides of their spine [168]. Carmen et al. [168] reported that *S. epidermidis* biofilms responded favorably to combinations of ultrasound and vancomycin at 48 h of insonation. In addition, pulsed ultrasound enhances the killing of *Escherichia coli* biofilms by aminoglycosides in a rabbit model with subcutaneously implanted polyethylene disks [169,170]. These authors applied low-frequency (28.48 kHz) and low-power density (300 mW cm$^{-2}$) ultrasound treatment for 24 h with and without systemic administration of gentamicin. Whereas exposure to ultrasound alone caused no considerable difference in bacterial viability, in the presence of gentamicin, there was a substantial reduction in bacterial viability (the bioacoustic effect). Here, the ultrasound significantly reduced bacterial viability below that of nontreated biofilms without damage to the skin [169,170]. However, Pitt and Ross [171] found that low-frequency ultrasound (70 kHz) of low acoustic intensity ($<2\,W\,cm^{-2}$) increased the growth rate of the cells compared with growth without ultrasound when *S. epidermidis*, *P. aeruginosa*, and *E.coli* cells adhered to and grew on a polyethylene surface.

### 9.13.3. Electrical Enhancement of Antimicrobial Activity

Direct electric current already has been safely used in humans for fracture healing. In a recent review by Del Pozo et al. [172], they have shown that electrical currents may be potentially applied even in the human setting either alone (the electricidal effect) or combined with antimicrobial agents (the bioelectric effect). Both approaches, however, need more *in vitro* studies as well as studies in experimental models before they can be translated to clinical practice.

As we know, the development of biofilm-related infections begins with the adhesion of the microorganisms to the biomaterial surface, mediated by the van der Waals forces, acid–base interactions, and electrostatic forces [173]. Electrostatic forces between bacteria and surfaces are generally repulsive, since almost all biomaterials are negatively charged, as are bacteria [174]. It has been proposed that these repulsive forces can be enhanced by the application of electrical current, which provokes the surface detachment of bacterial biofilms [175–177].

The antibacterial activity of electrical current previously has been demonstrated against *S. aureus* and *S. epidermidis* in agar [178–180]; the normal flora on human skin [181]; *E. coli*, *Proteus* species, and *Klebsiella pneumoniae* in synthetic urine [182]; *E. coli*, *Staphylococcus aureus*, and *Bacillus subtilis* in water [183,184]; and *E. coli* in salt solutions [185]. The mechanism of the antibacterial activity of electrical current has been suggested to result from toxic substances (e.g., $H_2O_2$, oxidizing radicals, and chlorine molecules) produced as a result of electrolysis [178], the oxidation of enzymes and coenzymes, membrane damage leading to the leakage of essential cytoplasmic constituents [186], and/or a decreased bacterial respiratory rate [183]. Recently, using an *in vitro* model, it was demonstrated that the low-intensity electrical current (i.e., 20, 200, and 2000 μA) substantially reduced the numbers of viable bacteria in staphylococcal or *Pseudomonas* biofilm over a prolonged exposure time period of up to 7 days [187].

Electrochemical approaches are now proposed both as a means to prevent biofilm formation and also to enhance the activity of antimicrobials against established biofilms, (i.e., by the bioelectric effect). For example, an iontophoretic approach has been taken where a low-current power source is used to drive the release of antimicrobial Ag ions from the device [188]. The bioelectric effect refers to the concurrent application of antibiotics (tobramycin) and a weak electric field to kill bacteria. With application of direct current (dc) electric fields between 1.5 and 20 V cm$^{-2}$ (current densities of ~15 × 10$^{-6}$ to 2.1 × 10$^{-3}$ A cm$^{-2}$), the concentrations of antibiotics needed to be effective against biofilm bacteria fell from ~5000 times to 4 times greater than those necessary for planktonic bacteria in the absence of electricity [189]. Electrolytic generation of oxygen may be partly responsible for this bioelectric effect, at least with the aminoglycoside antibiotic tobramycin against *P. aeruginosa* biofilm [190]. Enhanced antibiotic activity also has been shown using pulsed electromagnetic fields (PEMF). Here, PEMF enhanced the activity of gentamicin against 5-day biofilms of *S. epidermidis*, although there was no significant effect when gentamicin was replaced with vancomycin [191]. A research group led by Prof. Patel of Mayo clinic, Minnesota, has designed a model that permitted them to study the interaction between the biofilm itself, the electric field, and the antimicrobial agents. An eight-channel current generator–controller and eight chambers delivering a continuous flow of fresh media with or without antimicrobial agents and / or electric current (20, 200, or 2000 mA) via graphite or stainless steel electrodes to biofilm-coated Teflon coupons was used. This technology was used to extensively assess whether the *in vitro* enhancement of killing of biofilm-associated *P. aeruginosa* and *S. epidermidis* by electric current plus aminoglycoside, quinolone, and tetracycline antimicrobial agents generalizes to antimicrobial agents representing a variety of antimicrobial classes (cephalosporin, oxazolidinone, sulfonamide, macrolide, cyclic lipopeptide, and ansamycin antimicrobial agents) and to MRSA. However, the results of these experiments indicate that the enhancement of the activity of antimicrobial agents against biofilm organisms by electric current is not a generalizable phenomenon across microorganisms and antimicrobial agents [192].

Direct currents (DC) currents have been used clinically to drive chemotherapeutic molecules into solid tumors [193], and antibiotic molecules into the inner-ear and other tissues [194]. The obvious human application of the bioelectric effect could be in the management of infections associated with orthopedic hardware. In addition to systemic delivery of antimicrobics, local delivery of antimicrobics (e.g., in poly(methyl methacrylate, PMMA) also deserves further study concerning the bioelectric effect. Furthermore, implantable devices could be assessed to produce effective electric fields to enhance the perioperative use of antimicrobials to kill developing bacterial biofilms, thereby preventing device-related infections.

Electrode composition may have an impact on bioelectric effect. Stainless steel electrodes have been most commonly studied [195,196], but C, Pt, and Au electrodes also have been used [182,197–199]. Using an *in vitro* model, del Pozo et al. [199] demonstrated that electrode composition plays a role in the observed *in vitro* bioelectric effect. These authors studied the *in vitro* enhancement of bactericidal activity of rifampicin by electric current against MRSA biofilms using two different electrode materials. Rifampicin combined with electrical current (2000 μA) delivered by stainless steel electrodes demonstrated a 3.5–4.4 log reduction of MRSA biofilms. However, a lesser effect

(2.1–2.4 log reduction) was observed when electrical current was delivered by graphite electrodes.

Issues concerning electric current mediated tissue toxicity, delivery systems for electric current, and electrode geometry would need to be addressed before it could be applied in a human setting. Ideally, if the bioelectric effect is applied to human infections, the electric current should be delivered in a noninvasive (e.g., transcutaneous) or minimally invasive (e.g., subcutaneous) fashion. Attaching wires directly to the surfaces of foreign bodies is not ideal since the wires themselves may be a conduit for microorganisms. According to some authors, the bioelectric effect requires a current flow, not just an electric field. Stewart et al. [190] reported that when electrodes were placed outside the treatment chamber to create essentially the same electric field, but with zero current, the electrical enhancement of killing was completely eliminated. However, Pickering et al. [191] investigated the *in vitro* effect of a pulsed electromagnetic field on the efficacy of tobramycin and vancomycin against *S. epidermidis* on the tips of stainless steel pegs. As described in their study, exposure to a pulsed electromagnetic field increased the activity of gentamicin, but not vancomycin, against *S. epidermidis* biofilms.

### 9.13.4. Photodynamic Approaches to Biofilm Treatment

Photodynamic approaches have been proposed as a means to overcome resistance and to break down biofilms [200]. Photosensitizing drugs produce reactive oxygen species that are difficult for the microorganism to defend against. Necessarily, applications are limited to those sites where light can reach. For this reason, the focus has been on pathogens associated with the skin and with the oral cavity. For example, biofilms of the oral pathogen *Actinomyces viscosus* have been exposed to laser light at 666 nm in the presence of methylene blue. Confocal microscopy revealed that a single photomechanical wave increased the penetration of methylene blue by 75% and enhanced the photodestruction of the biofilm [201]. Similar findings have been obtained using multispecies biofilms of oral bacteria irradiated with light from a helium–neon laser in the presence of toluidine blue, where >95% of biofilm bacteria were killed [202]. Comparisons of growth phase and extracellular slime on the photodynamic inactivation of *S. epidermidis* and *S. aureus* indicated that slime production and stationary phase, both characteristics of biofilm infections, were obstacles to this therapy. However, use of polylysine-based cationic photosensitizers may overcome some of the growth-phase effects [203].

### 9.13.5. Studies Investigating the Use of Novel Catheter Lock Solutions for the Eradication of Biofilms

Other less conventional approaches, using agents that are not classified as antimicrobial agents, have been evaluated using the lock approach. For example, tetrasodium EDTA (ethylenediaminetetraacetic acid) or disodium

EDTA used alone or in combination with minocycline have been used effectively against bacterial and fungal biofilms. The EDTA has antimicrobial properties [204,205] against bacteria and fungi, and may also destabilize the biofilm structure [206]. Percival et al. [207] and Kite et al. [208] showed that 40 mg $mL^{-1}$ of tetrasodium EDTA could eradicate biofilms in an *in vitro* model and on explanted hemodialysis catheters, respectively. In the *in vitro* model system, the treatment eradicated biofilms after a 21-h dwell time against biofilms of *S. epidermidis*, *P. aeruginosa*, *K. pneumoniae*, and *E. coli* grown for 48-h; after a 25-h dwell time, biofilms of MRSA and *C. albicans* were also eradicated. Raad et al. [209] tested 18-h biofilms of *S. epidermidis*, *S. aureus*, and *C. albicans* against combinations of minocycline and disodium EDTA and found that 0.1 mg $mL^{-1}$ minocycline plus 30 mg $mL^{-1}$ EDTA significantly reduced (but did not eradicate) biofilms of each organism. Biofilms on explanted catheter tips were also substantially reduced (10-fold or more) by a combination of 3 mg $mL^{-1}$ minocycline and 30 mg $mL^{-1}$ EDTA. This treatment approach was also effective in the treatment of CRBSI in three different patient studies, as evidenced by remission of symptoms and negative catheter tip cultures [204]. A novel lock treatment containing taurolidine (2 *H*-1,2,4-thiadiazine-4,4′-methylenebis(tetrahydro-1,1,1′-tetroxide) eradicated 72-h biofilms of *S. aureus*, *S. epidermidis*, and *E. faecalis* in an *in vitro* model when they were exposed to 5000 U $mL^{-1}$ for 24 h [86]. Taurolidine is a derivative of the amino acid taurine, which inhibits and kills a broad range of microorganisms. Its proposed mechanism of action is based on the interaction of methylol derivatives with components of the bacterial cell wall, resulting in cell damage [210]. Metcalf et al. [211] instilled 70% ethanol in a Hickman catheter and combined this treatment with IV amoxicillin to resolve an *E. coli* bloodstream infection in a patient. The catheter was locked with ethanol between total parenteral nutrition infusions for a period of 3 days, and remained free of infection for >3 years, when the study was completed. Assuming patient compatibility, the next step would be to evaluate these treatments in animal models and in patient studies, using guidelines that have been suggested for the traditional antimicrobial lock treatments.

### 9.13.6. Bacteriophage

Phages are commonly defined as viruses that infect bacteria and carry a single copy of genetic material containing the necessary information needed to reproduce inside a host within a protein or lipoprotein coat [212]. Bacteriophages are estimated to be the most abundant life form on the planet, with a total species count believed to be in the range of $1 \times 10^8$ [213]. Attachment to the host cell is receptor mediated, and specificity of these receptors precludes whether or not the phage can infect at the bacterial-strain level or exhibit more broad-spectrum infection properties. Once infection by a phage has initiated, two life cycles are most commonly observed [214] (Fig. 9.2). They are referred to as being lytic (in which the phage hijacks the hosts' machinery and the

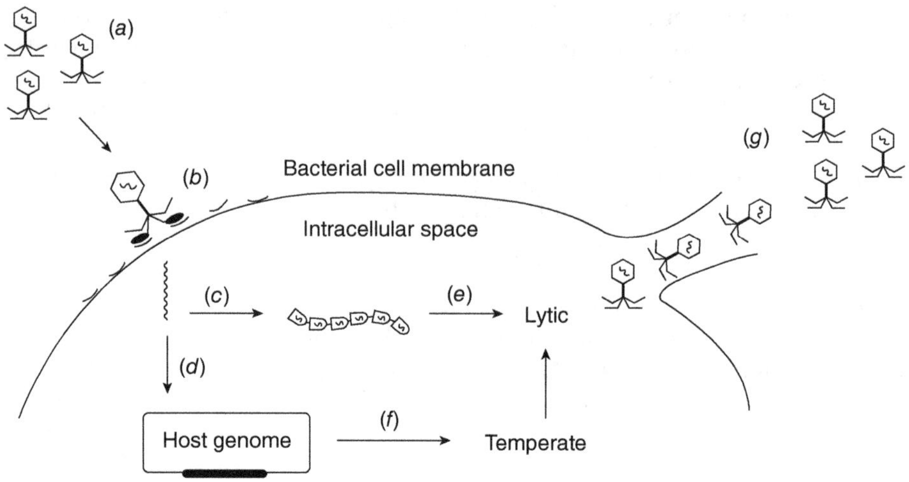


**Figure 9.2.** General life cycle of bacteriophages. (*a*) Phage outside the bacterial cell (*b*) attach to the membrane through specific receptors. The phage genome is injected into the host cell where it can either (*c*) hijack the host machinery to produce progeny or (*d*) integrate itself into the hosts' genetic material. The phage may then proceed through (*e*) the lytic cycle in which the phage kills the host cell through (*f*) release of its progeny, or (*g*) the temperate cycle in which the phage lays dormant within the host and can become lytic at anytime.

bacteria is killed when cell lysis occurs in the process of releasing progeny) or temperate (in which the phage may incorporate itself into the host genome and lay dormant within the bacteria). Whereas the implementation of phage therapies in human medicine dates back to the early 20th century [215], it was quickly overlooked with the discovery of broad-spectrum antibiotics [216]. With the emergence of multiple-drug-resistant bacteria and a paucity of newly discovered antibiotic treatments, renewed interest in this area of late has come from both academic and industrial research. Attempts have been made to employ phage infections of resistant bacteria as a means to circumvent the problem of drug resistance. This finding is significant because the coevolution of phages and bacteria is a perpetual struggle, with both mutating in concert in hopes of one gaining the upper hand for continued existence. Harnessing the power of phage-mediated infections of bacteria that form biofilms has led to a few promising results; ultimately this begs the question if these therapies represent a viable avenue for biofilm remediation efforts.

One of the most unique characteristics of phages is their ability to produce depolymerases and other surface enzymes that degrade bacterial polysaccharides. These enzymes demonstrate great specificity and have been observed to elicit activity against a number of Gram-negative bacteria [217–219]. As previously discussed, biofilms secrete an extracellular polymeric substances (EPS), which essentially acts as the glue that holds the biofilm together. Additionally, note that the EPS acts as a defense mechanism by providing the bacteria a barrier from microbicides or other entities that would cause harm. It is believed that phage treatment of a biofilm through this mechanism might expose a small portion of host bacterial cells for infection. Furthermore, due to the fluid architecture of the biofilm, the phage might seek transport to distal sites of biofilm by transporting through channels normally employed for the distribution of planktonic bacteria and nutrients. In one particular study, it was reported that the EPS of *E. coli* biofilms did not provide resistance to infection with the phage T4 [220]. Another study noted that phages labeled with fluorescent and chromogenic probes attached to or associated with *E. coli* biofilm matrices; this further demonstrates that the phage was not inhibited by the EPS [221]. Phages have also been observed to diffuse through the EPS of *P. aeruginosa* [222] and *Lactococcus lactis* [223].

There have been other reports that involved the engineering of the *E. coli* specific phage T7 to express the protein Dispersin B (DspB) intracellularly during infection so that DspB would be released into the extracellular space upon cell lysis [224]. The DspB had been previously documented to promote enzymatic degradation of an EPS polysaccharide, and thus it results in reduced biofilm mass when applied exogenously to the biofilms of several different species of bacteria [225]. This is similar to reports of the use of other enzymes (e.g., DNAase) to break down the bacterial EPS, as DNA is one of its major components [153,226]. The group reported that the engineered phage was able to succeed in dispersing established biofilms in comparison to a number of control phages. Furthermore, as a proof-of-concept work, it was postulated that this approach could be employed in the design and application of other phages to help target a range of medically relevant biofilms. A more recent report in this area documented that other engineered bacteriophages were successful in targeting the SOS gene network in *E. coli* and this resulted in the enhanced killing of bacterial cells with quinolines by several orders of magnitude *in vitro* and significantly increased survival times of infected mice *in vivo* [227]. A key observation made in the course of this study was that the engineered phage was also capable of reducing the number of persister cells in bacterial populations that had already been exposed to antibiotics as well as displaying an increased efficacy against biofilm bacteria.

Device-associated infections are of major concern due to the relative inability of the medical community to treat established biofilm infections on many of these substrates. Phage technology has shown promise in reducing the formation of catheter-associated biofilms formed by bacteria (e.g., *S. epidermidis*). One particular experiment focused on hydrogel-coated catheters impregnated with phage 456. This formulation was found to be successful in inhibiting the formation of *S. epidermidis* biofilms under a number of conditions [228]. However, the effect of the phage on established biofilms was not investigated. There are a few examples of clinical uses of phage therapies that provide groundwork for the implementation of phages in the control of biofilms. One approach that has been utilized in treatment is the phage cocktail

PhageBioDerm®, a biodegradable polymer wound dressing composed of ciprofloxacin, benzocaine, chymotrypsin, bicarbonate, and six lytic phages. These phages have shown activity against *S. aureus*, *P. aeruginosa*, *E. coli*, *Streptococcus* species, and *Proteus* species. Another phage cocktail under evaluation is WPP-201, a blend of eight lytic phages that possess the same spectrum of activity as PhageBioDerm®.

Despite the advantages of bacteriophages as controls over biofilms, there are a number of potential drawbacks. Mutations of the proteins that serve as receptor sites on the bacterial cell surface for phages can confer phage resistance to the bacteria. For use as a viable therapy in human treatment, phage immunogenicity is a concern [229]. The immune system recognizes the phage as a foreign substance and therefore triggers an antigenic response. Serum studies have indicated that repeated exposure to phage infection triggers increases in antibody titers [212,230]. Additionally, once cell lysis occurs, the bacterial cells' contents spill into the surrounding environments with the ability to act as toxins. This event can lead to a number of biological consequences including inflammation and endotoxic shock [212]. Finally, the problem posed by biofilms originating from more than one bacteria are of concern [231]. Although phage specificity is high, this could necessitate the use of multiple phages in a cocktail so that complete infection of the biofilm community is obtained.

Bacteriophage can also be instilled into catheters as a lock treatment to eradicate biofilms. Doolittle et al. [220] reported that Phage T4 significantly reduced biofilms of *E. coli* in an *in vitro* model system. Biofilms were grown for 28 h in a Modified Robbins Device (MRD) prior to the addition of phage. Viable biofilm cell counts were reduced by 6 logs within 5 h of treatment. Phage numbers increased initially during the first 5 h, then decreased as the number of surviving biofilm cells diminished. Hanlon et al. [222] investigated the effect of Phage F116 on biofilms of *P. aeruginosa* in microtiter plates and showed that intact biofilms were more tolerant to phage attack than suspended biofilm cells. They also found that an increase in biofilm age did not appear to significantly decrease susceptibility, as has been observed during the treatment of biofilms with antimicrobial agents. Phage treatment was effective on biofilms grown for 20 days prior to treatment. Other published studies also have demonstrated the efficacy of phage lock treatments against biofilms of different bacteria [232,233]. For this treatment approach to work, organisms isolated from the colonized device would need to be screened against a bank of phages to determine the specific phage strain with greatest lytic ability. This strain could be grown to a high titer then instilled into the indwelling catheter as a lock treatment.

### 9.13.7. Quorum-Sensing Inhibitors

Bacteria are social organisms capable of interacting with each other and their surroundings. Particularly well described is the ability to coordinate gene

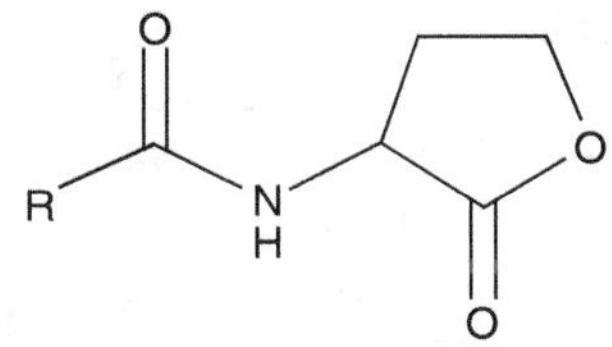

**Figure 9.3.** Basic structure of the acylated homoserine lactones commonly used as signal molcecules by many Gram-negative bacteria. R = $C_{12}$.

expression in accordance with population density, a process termed quorum-sensing (QS) [234]. In Gram-negative bacteria, this is achieved by production and reception of diffusible signal molecules in the form of acylhomoserine lactones (AHLs) (Fig. 9.3). The signal molecules are produced by an AHL synthase encoded by homologues of the AHL synthase gene *luxI*, which was first identified in *Vibrio fischeri* [235]. At low-population densities, *luxI* is constitutively expressed at a low, basal level. Hence, the AHLs accumulate in the surroundings. The LuxR family of receptor–response regulator proteins perceives the AHLs. At a certain threshold concentration of AHL, the signal molecule forms a complex with the receptor protein, which becomes activated. The activated receptor–signal complex in turn forms dimers or multimers with other activated LuxR–AHL complexes. These dimers or multimers function as transcriptional regulators controlling expression of QS regulated target genes. The QS paradigm states that transcription of QS target genes is activated at a certain population density, which is proportional to the AHL concentration, known as the "quorum size": The number of bacteria required to activate the QS system [234,236–238]. In *P. aeruginosa*, however, research has shown that each individual QS regulated gene possesses its own specific quorum size. There is not a single population density at which all QS genes are activated; rather, different genes are activated at different population densities [239–241].

One area of intense interest is the development of inhibitors of bacterial QS [242,243]. Quorum-sensing systems are a vital component in community behavior and biofilm formation for a wide range of diverse bacteria, and treatment with QS inhibitors could lead to a severe abrogation of biofilm formation. Many large screening projects are currently underway to identify such inhibitors. Numerous chemical libraries of both natural and synthetic origin have been screened and several QS inhibitory compounds have been identified. These endeavors have led to the discovery of three types of molecules:

1. Those that block production of the QS signal.
2. Enzymes or other factors that degrade the signal.
3. Signal analogues that disrupt QS by blocking binding of the true signal, thus preventing activation of the receptor [242,243].

***9.13.7.1. Blockage of AHL Production.*** To date, the least investigated strategy to interfere with QS is blockage of AHL production. Although a few substrate analogues, including holo-ACP, L/D-*S*-adenosylhomocysteine, sinefungin, and butyryl-*S*-adenosylmethionine (butyryl-SAM), have been found to be able to block AHL production *in vitro* [238]. None of them have been tested on bacteria *in vivo*. How these analogues of the AHL building blocks, SAM and acyl-ACP, which are also used in central amino acid and fatty acid catabolism, would affect other cellular functions is presently unknown.

***9.13.7.2. Inactivation of Signal Molecules.*** Another strategy is inactivation or complete degradation of the generated signal molecules. This can be achieved by different methods: chemical degradation, enzymic destruction, and metabolism of the AHL. A simple way to achieve inactivation of the AHL signal molecules is by increasing the pH to >7; this causes lactonolysis (ring opening) of the AHL [244]. A number of higher organisms employ this strategy in defence against invading QS bacteria. Plants that are infected with the tissue-macerating plant pathogen *Erwinia carotovora* will, as a first response at the site of attack, actively increase pH [245]. This alkalinization will in turn prevent expression of QS controlled genes and virulence factors. Several factors influence the kinetics of ring opening. A temperature increase accelerates ring opening, but this effect is counteracted by the length of the side chain, which decreases the rate of lactonolysis. These characteristics suggest that in order to be active under physiological conditions, an AHL signal molecule must possess a side-chain length of at least four carbons [244,245]. To date, no bacteria producing AHLs with side chains shorter than four carbon atoms have been identified.

Lactonolysis of AHLs can also be accomplished by enzymic activity. Members of the genus *Bacillus*, including *B. cereus*, *B. mycoides*, and *B. thuringiensis*, produce an enzyme, AiiA, specific for degradation of AHLs [246–249]. The activity of these enzymes lowers the amount of bioactive AHL signal molecules by catalyzing the ring-opening reaction. Within 2 h, up to 20-mM 3-oxo-C6 HSL (homoserine lactone) can be completely inactivated by a suspension culture producing the enzyme. When *Er. carotovora* is transformed with a plasmid carrying the *aiiA* gene, its virulence against potatoes and eggplants is attenuated. In addition, when the plant-colonizing bacterium *Pseudomonas fluorescens* was transformed with the *aiiA* gene, it was able to prevent soft rot in potatoes caused by *Er. carotovora* and crown gall disease in tomatoes caused by *Agrobacterium tumefaciens*. Furthermore, expression of *aiiA* in transgenic tobacco plants made them much less vulnerable to infection by *Er. carotovora* compared to their wild-type counterparts [247,250,251]. This finding indeed indicates that enzymic degradation of AHLs would be useful as a means of biocontrol. Production of AHL lactonases is not limited to *Bacillus* species. Several bacteria including *P. aeruginosa* PAI-A, *Arthrobacter* sp., *K. pneumoniae*, *Ag. tumefaciens*, and *Rhodococcus* sp., have been found to produce AiiA homologues [252–255]. Other bacteria (e.g., *Comamonas* sp.)

have been found to degrade AHLs [252]. It seems likely that production of AHL degrading enzymes constitutes a non-antibiotic-based strategy employed by some bacteria in competition with AHL producers. Certainly, this class of enzymes has obvious commercial applications, especially in the food manufacturing sector. Unfortunately, there is a drawback to the lactonolysis reaction: It is reversible at acidic pH. A ring-opened AHL molecule spontaneously undergoes ring formation if the environment is not alkaline, regardless of the method by which it was opened (chemical or enzymic) [256]. One way to prevent this could be by chemically modifying the ring-opened AHL (e.g., by mild nucleophilic substitution or reduction of the carboxylic acid), thus preventing reconversion to the ring form. Blocking QS in the environment may have the unintentional effect of interfering with beneficial bacteria. *Pseudomonas chlororaphis* controls production of an antibiotic with QS. Under normal circumstances, this bacterium and its antibiotic can be used to control tomato vascular wilt caused by *Fusarium oxysporum*. In an experiment where the bacterium was cocultured with an AiiA producing bacterium, the biocontrol activity was lost, rendering the plants susceptible to infection [251].

The lactone ring is not the only chemical target point of the AHL molecules. The oxidized AHL signal molecules (e.g., 3-oxo-C12 HSL) can react with oxidized halogen compounds (e.g., hypobromous and hypochlorous acids). Again, nature has developed this into a defence strategy against invading bacteria. The marine alga *Laminaria digitata* produces and secretes oxidized halogen compounds that interfere with QS controlled gene expression of colonizing bacteria [257].

A different, enzyme-based method to inactivate the signal molecules is simply to metabolize the AHLs. Both *Variovorax paradoxus* and *P. aeruginosa* PAI-A are able to proliferate with AHLs as a sole source of energy, carbon and nitrogen. The bacteria produce an amino acylase that cleaves the peptide bond of the signal molecule. The side chain is used as a carbon source, the nitrogen from the amide bond is made available as ammonium via the action of lactonases, and the ring part is used as an energy donor [255,258].

Interestingly, differentiated human airway epithelial cells have been found to be specific with respect to breakdown of AHL molecules. The cells were able to inactivate 3-oxo-C12 HSL and C6 HSL, but were unable to exert an effect on 3-oxo-C6 HSL and C4 HSL, indicating that both side-chain length and oxidation state are important for this kind of inactivation [259,260].

These examples demonstrate that inactivation of QS signal molecules occurs in natural environments as a functional protective strategy adopted by plants, bacteria, and mammals against pathogens.

***9.13.7.3. Interference with the Signal Receptor.*** A third approach to interfere with bacterial QS is to prevent the signal from being perceived by the bacteria, by either blockage or destruction of the receptor protein (the LuxR homologue). Several synthetic and natural QS inhibitors were already identified and some of the specific examples are only discussed further as below.

Identification of signal analogues has been a particularly productive endeavor. Many eukaryotes, as a microbial defense mechanism, produce secondary metabolites and other compounds that interfere with QS and other bacterial processes [261]. The marine alga *Delisea pulchra*, for example, secretes a class of molecules called furanones [261]. This organism produces the molecules in central vesicle gland cells, from which they are secreted to the surface of the fronds in amounts of $100\,\text{ng}\,\text{cm}^{-2}$ [262]. Here they prevent bacterial colonization, and thereby macrofouling, by interfering with QS controlled motility. Furanones are structurally quite similar to the acylhomoserine lactone class of QS signals, and thus disrupt community behavior of bacteria that utilize this class of autoinducers [242,243]. The effects of furanones on bacteria and biofilms are many and varied. Treatment of *Serratia liquifaciens* cultures with furanone abrogated swarming motility by inhibiting expression of the QS regulated gene *swrA*, involved in production of the swarming surfactant serrawettin W2 [263]. Furanone also inhibited QS regulated virulence of *Vibrio harvey* and *P. aeruginosa* [264,265]. Furanone compounds penetrated *P. aeruginosa* microcolonies, affected biofilm architecture, and enhanced bacterial detachment from established biofilms. A furanone derivative could even inhibit the growth, swarming, and biofilm formation of the Gram-positive microorganism *B. subtilis* [266,267]. Thus, by interfering with cell–cell communication, furanones can perturb a number of functions of a wide range of different bacteria. The different effects on these several bacterial species most likely relates to differences in regulatory circuitry activated by QS in these microorganisms. Still, it is clear that furanone compounds inhibit community behaviors.

Several other inhibitors of bacterial QS have also been discovered. Screens of *Penicillium* extracts revealed two molecules (patulin and penicillic acid) that inhibited QS regulation in *P. aeruginosa* [268]. Patulin also exhibited efficacy as a treatment for *P. aeruginosa* pulmonary infection in a mouse model. Intriguingly, this study found a synergistic effect on *in vitro* biofilm clearance when patulin and tobramycin were used in combination [268]. Synergy has also been observed between RNAIII inhibiting peptide (RIP) and a number of different antibiotics during clearance of device-related *S. epidermidis* infections *in vivo* [269]. A modified version of a heptapeptide (RIP) isolated from cultures of *Staphylococcus xylosus*, prevented phosphorylation of target of RNAIII activating protein (TRAP), which under normal circumstances would activate the *agr* regulatory system of *Staphylococcus* species [269,270]. This hindrance resulted in decreased adherence and biofilm formation of both *S. aureus* and *S. epidermidis* on a variety of abiotic materials, as well as mammalian cells, in culture. Taken together, these studies point to a profound effect of natural compounds on bacterial QS. Especially when considering antibiotic synergy, QS inhibitor molecules have shown great potential for treatment of bacterial biofilms. Furthermore, mutant TRAP strains of *S. aureus* (simulating cells that had been inhibited with RIP) also produced significantly less biofilm in flow cells and in membrane colony biofilm systems

[271]. These authors also found that biofilm formation by several *S. aureus* and *S. epidermidis* strains on Dacron grafts implanted into rats was significantly reduced when these grafts were first soaked in a 20-mg $L^{-2}$ RIP solution. This treatment was combined with parenteral administration of RIP. These results suggest that RIP might also be capable of eradicating biofilms on CVCs.

Antibiofilm modulators identified from natural products represent one of the last approaches in the discovery of biologically active agents against biofilms. Salicylic acid [272,273], cinnamaldehyde [274], and extracts from both garlic [159,275] and cranberries [276–279], all have shown various degrees of antibiofilm properties against a number of bacteria in various studies.

If we turn to plants, crown vetch, carrot, soybean, water lily, tomato, pea seedlings (*Pisum sativum*), habanero (chilli), and garlic have been found to produce compounds capable of interfering with bacterial QS [280,281]. Crude extracts of garlic have been shown to specifically inhibit QS gene expression in *P. aeruginosa* [280]. By using an *in vitro* model, it was demonstrated that *P. aeruginosa* PAO1 biofilms grown in the presence of garlic extract were substantially more susceptible to tobramycin treatment than were untreated or garlic extract-only-treated biofilms [280]. When examined in detail, garlic extract proved to contain a minimum of three different QS inhibitors, one of which has been identified to be a cyclic disulfur compound [280,282]. This QSI exerts a strong antagonistic effect on LuxR based QS but, interestingly, has no effect against *P. aeruginosa* QS [280]. Bjarnsholt et al. [275] showed that PMNs were activated in the presence of biofilms of *P. aeruginosa* grown for 3 days in media containing 2% garlic extract, resulting in extensive grazing and phagocytosis of the biofilm. The QS deficient *P. aeruginosa* mutants also exhibited polymorphonuclear (PMN) activation and phagocytosis, supporting the role of garlic extract as a QS inhibitor. Studies in an animal model also showed that treatment could stimulate the immune response and clear the introduced bacteria. It is still unclear from this work how established biofilms of this organism would respond to this treatment.

In summary, although it has been recognized that QS inhibitors could provide a viable approach for the control of clinically relevant bacteria [283], they are not "magic bullets". However, combinatory chemotherapy with both antibiotics and antipathogenic treatment that includes a synergistic effect with the host innate immune system could form the basis of a possible future treatment scenario for chronic infections caused by bacteria that regulate pathogenicity by means of QS.

### 9.13.8. Non-Quorum-Sensing Inhibitors

Additional antibiofilm molecules have been discovered that appear to affect bacterial mechanisms other than QS. Another molecule that interferes with *S. aureus* biofilm formation is farnesol, produced by *Candida albicans* [284]. Farnesol compromised membrane integrity of *S. aureus* biofilm bacteria and acted synergistically in reducing the minimum inhibitory concentration of gentamicin for both methicillin-sensitive and -resistant *S. aureus*. In a separate study, Ren et al. [285] screened thousands of natural plant extracts and discovered that ursolic acid disrupts biofilms formed by *E. coli*, *P. aeruginosa*, and *V. harvey*. It was demonstrated that QS was not involved in this effect. While the exact mechanism of inhibition remained elusive, microarray profiling implicated motility, heat shock, cysteine synthesis, and sulfur metabolism, as affected by ursolic acid treatment. Finally, subinhibitory concentrations of the macrolide antibiotic clarithromycin inhibited twitching motility of *P. aeruginosa* [286]. While macrolides have generally not exhibited activity against *Pseudomonas*, clarithromycin treatment altered *P. aeruginosa* biofilm architecture, raising the possibility of utilizing macrolides in combination with other antibiotics for biofilm eradication.

### 9.13.9. Extracellular Signal (Molecules) Responsible for Biofilm Dispersion

The search for an extracellular signal responsible for biofilm dispersion has uncovered a range of factors that have been shown to stimulate biofilm disruption. In 2000, Chen and Stewart [287] reported that reactive chemicals (e.g., NaCl, $CaCl_2$, hypochlorite, monochloramine, and concentrated urea), chelating agents, surfactants (e.g., sodium dodecyl sulfate, Tween 20, and Triton X-100), and lysozyme, as well as a number of antimicrobial agents, when added to mixed biofilms of *P. aeruginosa* and *K. pneumoniae*, resulted in the removal of >25% of protein from the surface, indicating cell release from the biofilms. Sauer et al. [288] showed that a sudden increase in the concentration of organic carbon causes bacteria to disaggregate from a biofilm. Thormann et al. [289] reported that a rapid reduction in oxygen could induce biofilm dispersion after cessation of flow in an oxygen-limited growth medium. Other studies showed that starvation may be a trigger for dispersion [290], that a prophage in *P. aeruginosa* may mediate cell death providing a vehicle for cell-cluster disaggregation [291], and that nitric oxide may play a role in the biofilm dispersion process [292]. Finally, the chelator ethylenediaminetetraacetic acid (EDTA) has been shown to induce killing and dispersion in *P. aeruginosa* biofilms [293]. Although the mechanism of dispersion induction is unknown in these cases, a common thread throughout these studies is that they induce major perturbations of cellular metabolism and likely also activate stress regulons, which may be involved in biofilm dispersion.

The identification of a cell–cell communication molecule responsible for biofilm dispersion has been the focus of a number of researchers over the past decade. Recently, indole has been shown to act as an intercellular messenger, inhibiting biofilm formation in *E. coli*, but enhancing biofilm formation in *P. aeruginosa* [294,295]. To date, however, indole has not been shown to activate a dispersion response in existing biofilms. Rice et al. [296] described a limited role for *N*-butanoyl-L-homoserine lactone in modulating detachment, or sloughing, of *Serratia marcescens*; however, the role of QS molecules in biofilm dispersion remains controversial. Dow et al. [297] characterized a substituted fatty acid messenger (*cis*-11-methyl-2-dodecenoic acid) called diffusible signal factor (DSF), recovered from *Xanthomonas campestris* and shown it to be responsible for virulence, as well as induction of the release of *endo*-β-1,4-mannanase. Intriguingly, DSF was shown to be able to disaggregate cell flocs formed in broth culture by *X. campestris*, although no activity against extracellular xanthan was detected [297].

Very recently, Davies and Marques [298] demonstrated that an unsaturated fatty acid (*cis*-2-decenoic acid) produced by *P. aeruginosa* both in batch and biofilm cultures, is responsible for inducing a dispersion response (in a cross-kingdom manner) in biofilms formed by a range of Gram-negative bacteria, including *P. aeruginosa*, and by Gram-positive bacteria. Furthermore, *cis*-2-decenoic acid was also capable of inducing dispersion in biofilms of *C. albicans*, indicating that this molecule has cross-kingdom functional activity. By dispersing established biofilms from bacteria and fungi (e.g., *E. coli*, *S. auerus*, and *C. albicans*), the authors demonstrate that the broad-spectrum activity of *cis*-2-decenoic acid might result from an evolutionary point that is similarly shared between these organisms.

The discovery of a signaling molecule responsible for biofilm dispersion has important implications for the exogenous induction of the transition of biofilm bacteria to a planktonic state. The unusual resistance of biofilm bacteria to treatment with antimicrobial agents and the persistence and chronic nature of biofilm infections could potentially be reversed if, in treatment, biofilm bacteria could be forced to transition to a planktonic phenotype. The application of a dispersion inducer prior to, or in combination with, treatment by antimicrobial agents provides a novel mechanism for enhancing the activity of these treatments through the disruption of existing biofilms. In situations where microbicides are unwanted or unnecessary, dispersion induction could be used as an alternative to toxic compounds or reactive chemicals.

### 9.13.10. Enzymes That Degrade the Biofilm EPS

The biofilm structural matrix, termed EPS, is composed of polymers, primarily polysaccharide in nature. Alginate, the EPS of *P. aeruginosa*, can be enzymatically depolymerized by alginate lyases. Hatch and Shiller [156] showed that alginate retarded the diffusion of aminoglycosides and inhibited their antimicrobial activity. However, addition of alginate lyase allowed greater penetration of gentamicin and tobramycin through alginate and greater activity of these agents against *P. aeruginosa*. This would suggest that alginate lyase might enhance the effectiveness of antimicrobial agents in the treatment of biofilms. In a subsequent study, Alkawash et al. [299] treated biofilms of two different mucoid *P. aeruginosa* strains in an *in vitro* model with $64\,\mu g\,mL^{-1}$ gentamicin with and without $20\,U\,mL^{-1}$ alginate lyase and found that the enzyme significantly improved the efficacy of the antimicrobial agent. After 120 h, the combination treatment had eradicated the biofilms of each strain, whereas between

6 and 7 log CFU $mL^{-1}$ of the biofilm organisms that were treated with gentamicin alone still survived after this exposure period. Enzymes targeting other polymers comprising the bacterial EPS might also be effective in this regard.

### 9.13.11. Delivery on Demand: An Infection-Responsive System

An interesting approach to antibiotic delivery has been the development of systems that are responsive to microbial infection. The rationale here is to develop systems that release antibiotics only during infection, recognizing that prophylactic or prolonged use of antibiotics can favor selection of resistant variants, and also lead to renal and liver toxicity with some agents (e.g., the aminoglycosides). One such approach has been based on the observation that wound fluid from *S. aureus* infection showed high levels of thrombin-like activity. An insoluble polymer–drug conjugate was prepared in which gentamicin was bound to poly(vinyl alcohol) (PVA) through a thrombin-sensitive peptide linker [300]. The conjugate released gentamicin when it was incubated together with thrombin and leucine aminopeptidase, but not with either component alone. Gentamicin was also released upon incubation with *S. aureus* wound fluid and the conjugate reduced the bacterial number in an animal model of infection.

## 9.14. NEW DIRECTIONS IN MEDICAL BIOFILM CONTROL

Traditional treatment of microbial infections is based on compounds that kill or inhibit growth of the microbe. One major concern with this approach is the frequent development of resistance to antibiotics [301]. As stated above, biofilm communities tend to be significantly less responsive to antibiotics and antimicrobial stressors than planktonic organisms of the same species. A further complication is that the spread of antibiotic resistance genes borne on plasmid DNA (pDNA), within and between species, is greatly exacerbated in biofilm communities [302,303]. As a consequence to this increase in resistance, researchers have turned to a number of alternatives to synthetic antibiotics, including bacteriophage [215] and bacteriophage lytic enzymes [304], probiotics [305,306], and human antimicrobial peptides (defensins, cathelicidins, and histatins) [307]. The success of these alternatives awaits much development and optimization. Unfortunately, most of these alternatives are still based upon some mechanism of killing or terminating the target bacteria; an approach some feel preordains the development of resistance in bacteria.

### 9.14.1. Antipathogenic Drugs

Recently, it has been proposed to develop substances that specifically inhibit bacterial virulence. Such "antipathogenic" drugs, in contrast to antibacterial drugs, do not kill bacteria or stop their growth, and are assumed not to lead

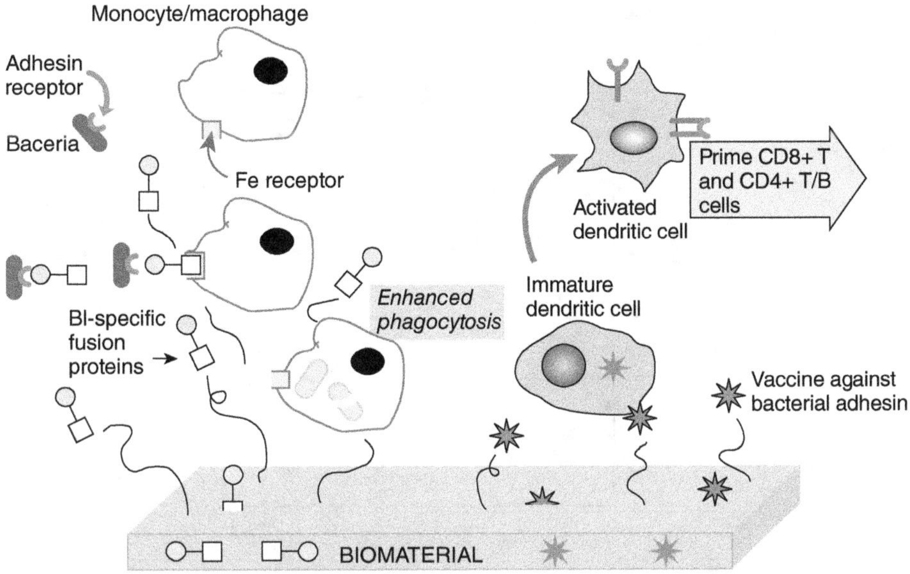


**Figure 9.4.** Hypothetical biomaterial engineered to enhance short- and long-term infection immune response. (Adapted from Bryers [314].) (See color insert.)

to the development of resistant strains. A very elegant approach comprises the inhibition of regulatory systems that govern the expression of a series of bacterial virulence factors [e.g., antiadhesion therapy (passive antibody therapy [308,309], and synthetic peptide vaccine and antibody therapy [310]), inhibiting or negating cell–cell signaling [311], negating biofilm formation by disrupting Fe metabolism [312], and up-regulation of biofilm detachment promoters (rhamnolipids)] [313].

In this section, as shown by Bryers [314], examples of three such "antibiofilm" strategies are presented. These three strategies could be deployed from a biomedical device coating or implant (Fig. 9.4) as a defense aimed at negating biofilm formation; defenses based on (1) disrupting bacterial Fe metabolism, (2) enhancing macrophage phagocytosis, and (3) "self-vaccinating" biomaterials.

### 9.14.2. Iron Metabolism Interference

Iron is critical for bacterial growth and the function of key metabolic enzymes [315–317] and sequestration of Fe is an early evolutionary strategy of host defense [318]. Gallium has many features similar to $Fe^{3+}$, including a nearly identical ionic radius, and biologic systems are often unable to distinguish Ga from $Fe^{3+}$. Unlike $Fe^{3+}$, Ga cannot be reduced to the divalent state and sequential reduction–oxidation is critical for Fe to function in many enzymes. Thus, placing Ga, rather than Fe, in such enzymes renders them nonfunctional

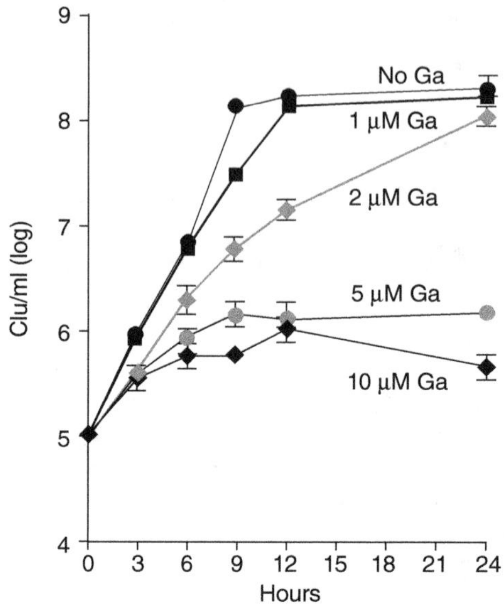


**Figure 9.5.** Effect of Ga on *P. aeruginosa* suspended growth. $Ga(NO_3)_3$ inhibits *P. aeruginosa* growth in a concentration-dependent manner. Experiments were performed in biofilm medium at 37 °C, and data are the mean of four experiments; error bars indicate SEM.

[319,320]. Importantly, Ga binds to the siderophores of *Pseudomonas* sp. [319,321] and is taken up by other bacteria including *S. aureus*, *S. epidermidis*, *E. coli*, *Enterococcus faecalis*, and *S. typhimurium* [322,323].

Kaneko et al. [312] recently reported (Fig. 9.5) that concentrations 1–2 μM $Ga(NO_3)_3$ did not effect the growth rate or extent of *P. aeruginosa*. Concentrations of Ga >2 μM decreased *P. aeruginosa* growth rate in a dose-dependent manner. These results led the Singh group to investigate the effects of Ga on biofilm formation. For these studies, they used a Ga concentration (1 μM) that did not impair the growth of suspended *P. aeruginosa*, since they were interested specifically in *antibiofilm* effects of Ga. (In a therapeutic application, both growth inhibitory and antibiofilm actions would be desirable.) Gallium effects were studied using a green fluorescent protein (GFP) expressing *P. aeruginosa* strain. In subinhibitory concentrations of Ga, *P. aeruginosa* attached to the growth surface, but biofilm formation was completely inhibited. Biofilm formation by mucoid and non-mucoid *P. aeruginosa* isolates from CF patients was also blocked by 1 μM $Ga(NO_3)_3$ [312].

To determine if Ga applications can kill *P. aeruginosa* biofilms, Kaneko et al. [312] reports growing biofilms for 3 days (with no Ga present) and then switching to a medium containing Ga for 48 h. Bacterial viability was assayed using a live–dead stain. While most commercial antibiotic agents show markedly less activity against biofilms than against planktonic organisms, *P. aerugi-*

*nosa* bacteria within mature biofilms were killed by concentrations of Ga similar to those that killed planktonic cells [312].

### 9.14.3. Enhancing Phagocytosis

Another alternative antibiofilm defense is one that seeks to enhance macrophage phagocytosis of bacteria by developing artificial "opsonins". Opsonization is the process where microorganisms are coated with host-produced molecules (immunoglobulins, complement factors), which in turn facilitates their binding to specific receptor molecules present on phagocytes [neutrophils, macrophage, dendritic cells (DCs)].

The IgG antibodies bind to their antigens on the surface of bacteria through coupling of the variable binding sites in the Fab region of the antibody, leaving the Fc region exposed. Phagocytes possess Fc gamma receptors and therefore can bind to the coated bacteria and internalize them. Complement fragment (C3b) also specifically binds to surface proteins or polysaccharides on microorganisms, thus allowing binding to C3b receptors on the phagocytes. As described earlier, bacteria evolved numerous ways to circumvent these natural opsonins and thus avoid elimination. One strategy bacteria evolved to avoid phagocytosis is to avoid opsonization by IgG and complement. Therefore, research has initiated development of "artificial opsonins", designed to uniquely bind to both target bacteria and macrophage, thus enhancing phagocytosis. There are several reports of enhancing phagocytosis employing either (1) fusion proteins that couple recognition moieties of both bacteria and macrophage or (2) synthetically derived "opsonins".

Whitesides' group [324] describes the application of a bifunctional polyacrylamide containing both vancomycin and fluorescein groups, which recognized the surfaces of different species of Gram-positive bacteria (*S. aureus*, *S. pneumoniae*, and *E. faecalis*). Vancomycin groups recognize bacterial cell-wall component peptides terminated in D–Ala–D–Ala. Fluorescein groups allowed the imaging of bound opsonin plus they are recognized by antifluorescein Mab, which promoted binding to macrophage. Flow cytometry revealed that bispecific polymer-labeled *S. aureus* and *S. pneumoniae* were opsonized by antifluorescein Mabs 20-fold more than were untreated bacteria and promoted subsequent phagocytosis of the *S. aureus* bacteria by cultured J774 macrophage-like cells approximately twofold more efficiently than in control groups.

The Taylor group, in a series of elegant papers, reports the use of several different bispecific fusion proteins (BiFPs) that enhanced phagocytosis by macrophage of various pathogens, including: *E. coli* [325], *P. aeruginosa* [326], and *S. aureus* [327]. In all cases, BiFPs consisted of (1) a molecule that recognizes a surface marker on the pathogen that was chemically coupled with (2) a Mab that is specific to the complement receptor 1 (CR1) present on primate erythrocytes. In *in vitro* and *in vivo* studies, these works from the Taylor groups have shown that BiFPs promote binding of the target pathogen first to circulating erythrocytes, which then enhances macrophage phagocytosis of the

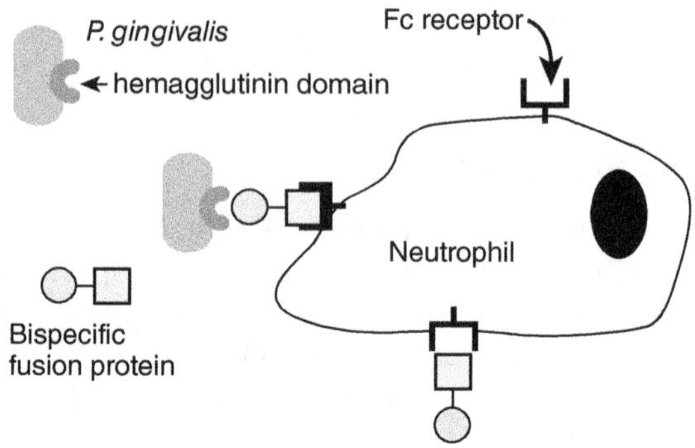


**Figure 9.6.** Use of bispecific fusion proteins to opsonize pathogenic bacteria and enhance phagocytosis. Fc receptor is an antibody possessing its binding specificity known as the Fc (fragment, crystallizable) region. (See color insert.)

bacteria. The BiFP mediated phagocytosis apparently did not harm the erythrocyte, as verified in both *in vitro* and *in vivo* experiments [325].

Kobayashi et al. [328] reports improved *in vivo* and *in vitro* phagocytosis of a periodontal pathogen (*P. gingivalis*) using a BiFP composed of two monoclonal antibody fragments; one against (1) the hemagglutinin domain of *P. gingivalis* (anti-r130k-HMGD antibody) and (2) the PMN leukocyte *Fc*$\alpha$RI (CD89) receptor (FcR) (Fig. 9.6). The Kobayashi work selectively targeted Fc receptors that were dominant on PMNs collected from gingival crevicular fluid of chronic periodontitis patients versus Fc receptors dominant on peripheral blood PMNs. Data show that PMNs exhibited a higher capacity to phagocytose and kill *P. gingivalis* opsonized with the BiFP targeting *P. gingivalis* r130k-HMGD to leukocyte Fc RI as compared to opsonizing the bacterial with only the anti-r130k-HMGD antibody.

### 9.14.4. "Self-Vaccinating" Biomaterials

The goal of any vaccine is to produce a long-term protective immune response against a pathogen. For most bacteria, initial attachment to a eukaryotic cell surface or ligand-coated biomedical device leads to biofilm formation, then upregulation of virulence factors leading to infection. Both an innate and an induced antibody response could prevent attachment and abrogate colonization. The ideal antigens to promote both levels of immune response would be the very surface proteins (bacterial "adhesins") that mediate specific bacterial cell adhesion to ligands present on host tissue or device surfaces [329]. Given that bacterial specific adhesion can trigger expression of many virulence factors leading to acute and chronic inflammation, a vaccine approach that blocks bacterial adhesion may have multiple advantages. Recent research has greatly expanded the molecular details of the specific adhesin: ligand couples employed by Gram-positive bacteria (*S. epidermidis*, *S. aureus*, Group A and

B *Streptococcus*, *E. faecalis*), Gram-negative bacteria (*P. aeruginosa*, *Klebsiella pneumoniae*, *P. gingivalis*, *E. coli*), and yeast (*Candida* spp.) to colonize and infect living tissue and implanted biomedical devices [329]. Thus, the concept of biomaterials designed to engineer infection immunity, targeting specific adhesins, could be applied to numerous situations.

There are essentially three ways to activate a dendritic cell to present antigen with the subsequent immune response being dependent on the form of the antigen (Fig. 9.7): direct antigen presentation (e.g., whole attenuated pathogen, isolated antigen molecule, recombinant protein), DNA vaccine (i.e., pDNA encoding for the antigen protein), and mRNA vaccine (i.e., RNA encoding for the antigen protein). *Exogenously* acquired protein antigens are processed by late endosomes into MHC-II complexes that initiate CD+4 T cell activation, and then activate B-cell antibody secretion. *Endogenously* acquired antigen (e.g., viral infection; DNA/mRNA vaccine) is tagged with ubiquitin, then partially degraded in the proteosome, transported next to the endoplasmic recticulum, and presented on the DCs surface as a MHC-I complex, thus activating an immediate (but short-lived) cytotoxic CD8+ T cell response. In contrast to protein vaccines, DNA or RNA based vaccines can provide the ability to *potentially* generate both a strong cytotoxic T cell and humoral response.

Upon internalization of a pDNA vaccine carrier by dendritic cells, the carrier must escape endosomal entrapment or be degraded, the carrier must release the pDNA into the cytoplasm, and then the pDNA must be incorporated into the dendritic nucleus for expression. Subsequently expressed antigen can be processed as above into MHC-I complexes to initiate priming of CD8+ T-cell or the antigen can be secreted. The secreted antigen can be taken up exogenously by the same or other DCs and presented by the MHC-II pathway to CD4+ T cells, which can secrete soluble cytokine signals to T or B cells to induce antibody secretion. In dendritic cells, in a process known as "cross-priming", endogenously expressed antigen could be routed to the MHC-II complex.

The DNA vaccines have certain advantages over protein antigens including: (1) DNA can serve as a natural adjuvant by including unmethylated CpG [cytosine and guanine separated by a phosphate (-C-phosphate-G-)] motifs; (2) by coding for multiple gene expression, DNA vaccine can also induce costimulatory molecules; (3) pDNA can target expression to certain cellular locations, thus fine tuning the immune response; and (4) pDNAs allow the possibility of multiple antigen expression. Disadvantages of DNA vaccines include (1) incorporated CpG motifs can lead to overstimulation and toxicity [330,331]; (2) the required large amounts of highly characterized pDNA require processing with antibiotics and antibiotic-resistance markers; (3) pDNA contain sequences meant to control gene expression (e.g., promoters, polyA signals, introns) that may deregulate gene expression after integration into the genome; and (4) genome-incorporated pDNA can result in uncontrolled duration and strength of antigen expression.

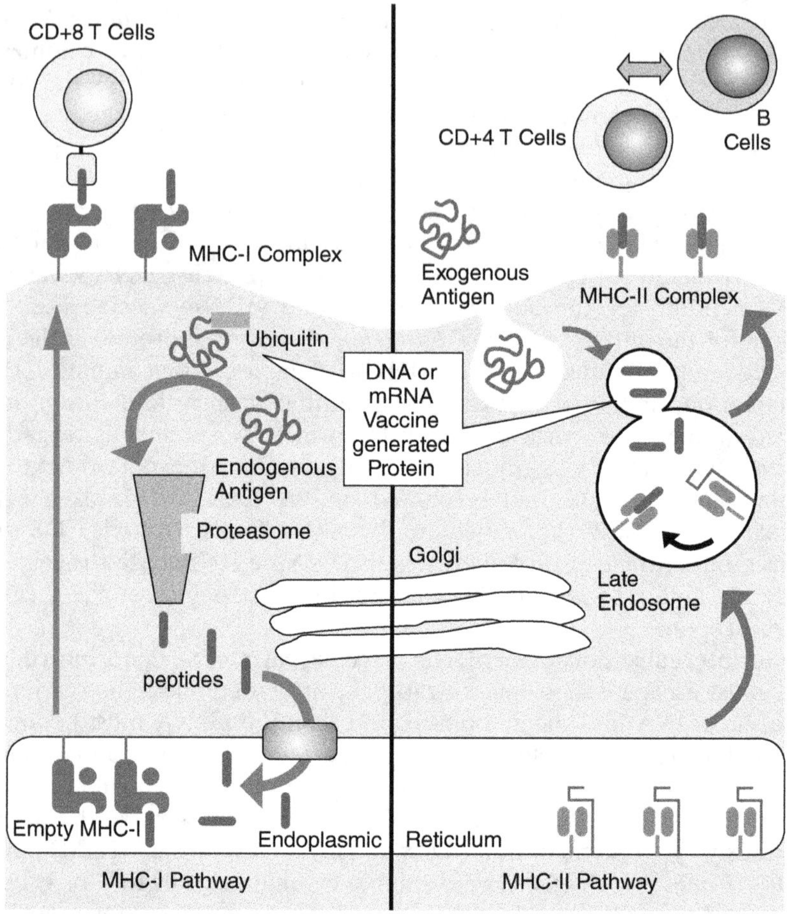


**Figure 9.7.** Antigen presentation and pathways of vaccine response. Plasmid DNA or messenger ribonucleic acid (mRNA) is taken up by dendritic cells for intracellular expression of antigen. Antigen can be secreted (not shown) and subsequently taken up by another DC as an exogenous antigen. Antigen expressed intracellularly by a dendritic cell or taken up through cross-priming is presented by MHC-I to CD8+ T-cells (cytotoxic leukocytes; CTLs). Antigen taken in exogenously or directed by DNA or mRNA trafficking signals are processed by the MHC-II pathway and presented to CD4+ TH cells, which can subsequently secrete: soluble cytokine signals (e.g., IL-12) back to the dendritic cell, proliferative signals (e.g., IL-2 and IFN-γ) to Tc cells, or signals directed toward B-cells (e.g., IL-4) to induce B-cell proliferation and antibody secretion. (See color insert.)

The RNA vaccines offer the same advantages as DNA vaccines versus protein antigens, but RNA vaccines have several added benefits versus DNA vaccines [332]. Due to RNAs smaller size, a larger amount can be delivered per carrier, thus they generally are more efficient. Since RNA does not require nuclear incorporation, expression of antigen in transfected cells occurs much faster than with pDNA. Potentially dangerous side effects are reduced since eukaryotic promoters needed for pDNA are not present in RNA constructs. The mRNAs are relatively easy to manufacture in high amounts, to purify to homogeneity, and to characterize. Finally, RNA does not persist in the organism, and RNA is not incorporated into the genome.

One main drawback to both DNA and RNA vaccines is that the efficiency of naked oligonucleotide transfection is very low (cf. viral systems) due to lack of protection from systemic nucleases, inability to migrate through cell membranes, or entrapment and degradation within endosomes. Consequently, nonviral gene constructs typically require some type of polymer delivery system. The reader is directed to a number of excellent reviews on the subject of polymer gene delivery published by Dang and Leong [333], Gao et al. [334], Keegan and Saltzman [335], Little and Langer [336], and Pack et al. [337].

Researchers are currently developing a platform of polymer constructs that would release condensed RNA vaccines meant to transfect dendritic cells arriving upon implantation of a biomedical device. This polymer construct will release nanoparticles of condensed DNA or mRNA vaccine that will target a selected adhesion protein employed by the microorganism to initiate colonization. Such targets could be the fibronectin binding receptors on *S. aureus* or *S.epidermidis* used to bind to surface-immobilized fibronectin; cell surface Arg-specific (RgpA) and Lys-specific (Kgp) proteinases (designated the RgpA–Kgp complex) used by the oral pathogen, *P. gingivalis*, the major cause of chronic periodontitis; Group B *Streptococcus* to immobilized fibronectin; *Steptococcus mutans* and *S. sangiuns* to mucin-coated dental devices; *P. aeruginosa* binding to a G4 glycolipid on cornea and contact lenses; or *E. coli* FimH binding to mannose-coated catheters.

One distinct advantage of nucleotide transfection of dendritic cells (vs direct antigen protein) is that one can modify the DNA or mRNA with targeting signals to better control MHC Classes I and II presentation. For example, incorporation of ubiquitin mRNA with the target antigen will result in an enhanced formation of peptides for MHC Class I presentation [338–340], whereas targeting sequences from the invariant chain (Ii) or lysosome-associated membrane protein (Lamp1) will lead to presentation of the antigen in the context of both MHC Classes I and II, thus providing antigen-specific help [341–344]. Furthermore, researchers are currently developing mRNA vaccines anti to the *S. epidermidis* fibronectin binding receptors used to colonize fibronectin-coated cardiovascular devises. The *S. epidermidis* upregulate these fibronectin binding receptors once exposed to serum at high shear stresses, which distinguishes infecting strains from benign *S. epidermidis* skin flora.

One might ask: Why transfect dendritic cells from the medical device rather than simply inoculate the patient? Transfection rates of DCs cells within the vicinity of a surgically implanted scaffold have been shown by Babensee's group [345] to greatly exceed transfection rates observed by injecting the same amount of antigen within small microparticles. They suggested that "danger signals" associated with the surgical implantation of the larger scaffolds due to tissue injury attracted far greater numbers of localized DCs, leading to an enhanced immune response. Further, Babensee's group has shown that staged transfections (initial and a series of "boosters") greatly improved transfections efficiency.

## 9.15. EVALUATING BIOFILM ERADICATING STRATEGIES

A systematic approach to assess biofilm eradication treatment strategies might include the following suggestions:

1. Develop an *in vitro* model that reasonably simulates the indwelling catheter biofilm with respect to substratum, properties of the growth medium, biofilm age and cell density, the presence of serum proteins, and that uses bloodstream isolates of clinically relevant organisms. Murga et al. [346] and Curtin and Donlan [228] can be consulted for examples of *in vitro* model systems for growing and testing biofilms on indwelling medical devices. The Drip Flow Reactor [228] can be modified and used for growing and testing biofilms on the lumens of central venous catheters. Other screening approaches incorporating the MEBC Device [69] or the CDC Biofilm Reactor [347] can provide higher throughput testing, but under less relevant conditions.
2. Validate results obtained in the *in vitro* model under more rigorous conditions by using either explanted biofilms, as done by Kite et al. [208] and/or an animal model. Treatments that appear effective in *in vitro* models often do not show the same level of effectiveness in animal models, due to such complicating factors as the response of the host immune system and presence of serum proteins.
3. Ascertain that the treatment can be tolerated by the patient and is compatible with the normal-use regimen of the device.
4. Assure that catheter biofilms are recovered and quantified when conducting clinical studies to evaluate the treatment. Resolution of patient symptoms may not predict eradication of the biofilm from the catheter. Biofilm recovery and detection methods should also be validated [348].

## 9.16. CONCLUDING REMARK

As our population ages, there will be an increase in the number of people experiencing hospitalization and receiving short- or long-term biomedical implants. As engineered biomaterials and tissue regenerative medicine advance, an increasing portion of the population will receive one or multiple biomedical devices, ranging from disposable contact lenses, dental implants, orthopedic implants, and vascular grafts to tissue engineered livers, small diameter vascular grafts that promote stem cells differentiation into endothelial cells, and polymer transfection systems that deliver micro-RNA knockout therapy to control chronic inflammation. The classic view of the host response to biomaterials divides the response into several overlapping phases: blood–material interactions, acute inflammation, chronic inflammation, foreign-body reaction, and fibrous encapsulation. The current healthcare approach to clean and sterilize has done little to prevent an epidemic in nosocomial infections. Biomaterials technologies employing disinfectant rinses, tethered, or release antibiotics have also done little to reduce this epidemic and may have contributed to the raise of antibiotic-resistant bacteria. Thus, the judicious use of novel drug delivery carriers is an alternative, but effective, approach to eradicate biofilm consortia from forming over biomedical devices. The following sections and chapters are devoted to describing the applications of various pharmaceutical approaches involving novel drug-delivery carriers to control biofilm-related nosocomial infections.

## 9.17. NOVEL DRUG DELIVERY CARRIERS

Considering the increasing use of relatively invasive medical and surgical procedures to salvage normal functions of vital organs, the material properties of medical devices have received much attention. Nevertheless, alteration of the foreign-body material surface may lead to a change in specific and non-specific interactions with microorganisms and, thus, to a reduced microbial adherence [349]. Medical devices made out of a material that would be anti-adhesive or at least colonization resistant would be the most suitable candidates to avoid colonization and subsequent infection [349]. A number of elements in the process of biofilm formation have been studied as targets for novel drug-delivery technologies. These include surface modification of devices to reduce bacterial attachment and biofilm development, as well as incorporation of antimicrobials, again to prevent colonization. In addition, dental plaque and oral hygiene is another common therapeutic target. There is now a considerable body of work using carrier systems (especially, lipidic and polymeric based) to target antibiotics against intracellular infections. Other technologies not specifically focused on biofilms include aerosolized delivery of antibiotics to the lung and formulation into liposome- and polymer-based vehicles. Precisely, the potential of novel drug delivery carriers to eradicate biofilms from device-related nosocomial infections are considered in three main categories:

(1) Prevention of colonization and biofilm formation (antibiofilm approach).
(2) Accumulation at the biofilm surface or interface.
(3) Drug penetration into the biofilm (intracellular infection).

The method by which a drug is delivered can have a significant effect on its efficacy. Some drugs have an optimum concentration range within which maximum benefit is derived, and concentrations above or below this range can be toxic or produce no therapeutic benefit at all. On the other hand, the very slow progress in the efficacy of the treatment of severe diseases has suggested a growing need for a multidisciplinary approach to the delivery of therapeutics to targets in tissues. From this, new ideas on controlling the pharmacokinetics, pharmacodynamics, nonspecific toxicity, immunogenicity, biorecognition, and efficacy of drugs, were generated. These new strategies, often called novel drug delivery carriers (NDDC), are based on interdisciplinary approaches that combine polymer science, pharmaceutics, bioconjugate chemistry, and molecular biology.

Otherwise inaccessible internal (lung, liver, kidney heart, brain, etc.) and easily accessible external (eye, nose, ear, penis, vagina, anus, etc.) organs of the human body always consist of several different types of physiological barriers. The majority of these barriers block or prevent the entry of any foreign material including drug and NDDC into both the internal and external organs of the human body. Therefore, it becomes necessary for a successful NDDC to overcome several different types of barriers that originate from the complexity of the human organism [350]. Indeed, specific molecular responses are required for each barrier from the NDDC, and thus demand the integration of diverse molecular and supramolecular responsive designs within a single drug delivery structure. On the other hand, it has been estimated that anywhere from 40 to as much as 70% of all new chemical entities (NCE) entering drug development programs possess insufficient aqueous solubility to allow consistent gastrointestinal absorption of a magnitude sufficient to ensure therapeutic efficacy [351]. Hence, the NDDC should have the potential to overcome the major problems of currently available drugs or NCE, which include not only poor aqueous solubility, but also toxic side effects and lack of selectivity for the diseased tissue.

Indeed, depending on the immediate requirements, the NDDC should simultaneously carry on its surface various moieties capable of functioning in a certain orchestrated order for demonstrating sequentially the following properties [352]: (1) circulate for a long time in the blood or, more generally, stay for a long time in the body; (2) specifically target the site of the disease through different mechanisms, like enhanced permeability and retention effect (EPR) and ligand-mediated recognition; (3) respond to a local stimuli characteristic of the pathological site (e.g., abnormal pH values or temperature or respond to externally applied stimuli (e.g., heat, magnetic field, or ultrasound), by, for example, releasing an entrapped drug or facilitating the contact between drug-loaded nanocarriers and target cells; (4) provide an enhanced intracellular delivery of an entrapped drug in case the drug is expected to exert its action inside the cell; and (5) afford real-time information about the carrier (and drug) biodistribution and target accumulation, as well as about the outcome of the therapy due to the presence within the structure of the carrier of a certain reporter moiety.

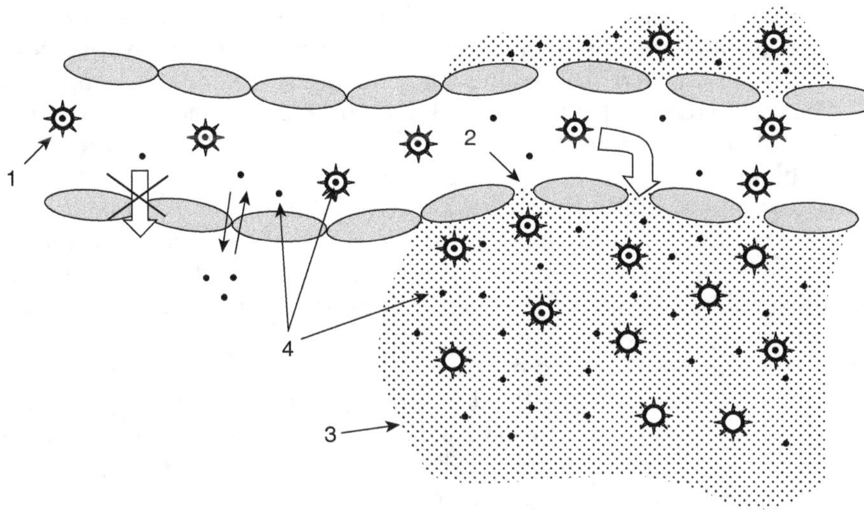


**Figure 9.8.** The EPR effect. *Key*: Long-circulating drug carriers (1) penetrate through the leaky pathological vasculature (2) into the tumor interstitium (3) and degrade there, releasing a free drug (4), and creating its high local concentration.

To address all of the above-mentioned issues, NDDC have initially been designed to take advantage of the enhanced vascular permeability present at disease sites (Fig. 9.8). The NDDC can easily extravasate at these sites, in contrast to nontarget tissues. This, in combination with the decreased clearance and enhanced blood residence time of a NDDC associated drug, will actually promote the drug concentrations in the diseased tissues, increasing the therapeutic efficacy of the incorporated drug molecules. If the NDDC is so designed to possess particle sizes in the submicron or nanometric level, then the use of nanometric NDDC results in a reduced volume of distribution for the entrapped drug, entailing further diminished extravasation in nontarget tissues, with resultant reduction of toxic side effects. The selectivity of NDDC can be even further enhanced by including targeting ligands that allow for recognition of specific markers expressed at the diseased site. All of the above said modifications are being made on the nanometric NDDC in order to allow drug penetration into the biofilm for the treatment of intracellular infection.

## 9.18. LIPID- AND POLYMER-BASED DRUG DELIVERY CARRIERS

Efforts and attempts are continuing to control–eradicate biofilms by novel antimicrobial agents either alone or in combination NDDC. With the failure of conventional means to achieve adequate therapeutic levels at the infectious sites of biofilm localization, either due to the ecological niche of the sites or the bacterial resistance toward the already existing therapeutic strategies, novel drug-delivery strategies are receiving considerable attention in recent years. More particularly, for the eradication of microbial biofilm on device-related nosocomial infections, lipid- and polymer-based drug-delivery carriers are widely investigated in conjunction with other antibiofilm approaches like electrical [188–191], ultrasound [168,170,171,353], and photodynamic [200–203]-mediated enhancement of antimicrobial activity or transport through biofilms.

Uni- and multilamellar liposomes are covered in this book as lipid-based systems. The selected examples for polymer-based drug-delivery carriers include implantable matrices, microparticles, fibrous scaffolds, micells and thermoreversible gels, and surface modified polymeric materials having antimicrobial–antiseptic–Ag coatings onto them.

Also, most of these selected drug-delivery carriers are prepared from biodegradable polymers like poly(lactide) (PLA) and its copolymers with glycolide (PLGA). Some of the delivery systems (e.g., micells and thermoreversible gels) are obtained from poloxamer 188 and poloxamer 407 (also known as Pluronics® or Lutrols®). The poloxamers are a well-studied series of commercially available, nonionic, triblock copolymers with a central block composed of the relatively hydrophobic poly(propylene oxide) flanked on both sides by blocks of the relatively hydrophilic poly(ethylene oxide) and possess an impressive safety profile. They are the U.S. Food and Drug Administration (FDA) approved selectively for pharmaceutical and medical applications, including parenteral administration [354–356]. Another concept for the prevention of device-related infections involves the impregnation–coating of devices with various substances (e.g., antibacterials, antiseptics, and/or metals). Surface-modified polymeric devices with impregnation–coating of various substances (e.g., antibacterials, antiseptics, and/or metals) are also covered.

## REFERENCES


1. Sitges-Serra, A. and Girvent, M. (1999), Catheter-related bloodstream infections, *World J. Surg.*, 6, 589–595.
2. Bouza, E., Burilllo, A., and Munoz, P. (2002), Catheter-related infections: diagnosis and intravascular treatment, *Clin. Microbiol. Infect.*, 5, 265–274.
3. Hodge, D. and Puntis, J.W. (2002), Diagnosis, prevention, and management of catheter-related bloodstream infection during long term parenteral nutrition, *Arch. Dis. Child Fetal. Neonatal Ed.*, 87, F21–F24.
4. Mermel, L.A., Farr, B.M., Sherertz, R.J., Raad, I.I., O'Grady, N., Harris, J.S., and Craven, D.E., Infectious Diseases Society of America; American College of Critical Care Medicine; Society for Healthcare Epidemiology of America. (2001), Guidelines for the management of intravascular catheter-related infections, *Infect. Control Hosp. Epidemiol.*, 22, 222–242.
5. Karchmer, A.W. and Longworth, D. (2003), Infections of intracardiac devices, *Cardiol. Clin.*, 21, 253–271.
6. Berns, J.S. (2003), Infection with antimicrobial-resistant microorganisms in dialysis patients, *Semin. Dial.*, 16, 30–37.

7. Carratalà, J. (2001), Role of antibiotic prophylaxis for the prevention of intravascular catheter-related infection, *Clin. Microbiol. Infect.*, 7 Suppl. 4, 83–90.
8. Fätkenheuer, G., Cornely, O., and Seifert, H. (2002), Clinical management of catheter-related infections, *Clin. Microbiol. Infect.*, 8, 545–550.
9. Graninger, W., Assadian, O., Lagler, H., and Ramharter, M. (2002), The role of glycopeptides in the treatment of intravascular catheter-related infections, *Clin. Microbiol. Infect.*, 8, 310–315.
10. Karchmer, A.W. and Longworth, D.L. (2002), Infections of intracardiac devices, *Infect. Dis. Clin. North Am.*, 16, 477–505, xii.
11. Kovalik, E.C. and Schwab, S.J. (2002), Treatment approaches for infected hemodialysis vascular catheters, *Curr. Opin. Nephrol. Hypertens.*, 11, 593–596.
12. Mermel, L.A., Farr, B.M., Sheretz, R.J., Raad, I.I., O'Grady, N., Harris, J.S., and Craven, D.E., Infectious Diseases Society of America; American College of Critical Care Medicine; Society for Healthcare Epidemiology of America. (2001), Guidelines for the management of intravascular catheter-related infections, *Clin. Infect. Dis.*, 32, 1249–1272.
13. Nichols, R.L. and Raad, I.I. (1999), Management of bacterial complications in critically ill patients: surgical wound and catheter-related infections, *Diagn. Microbiol. Infect. Dis.*, 33, 121–130.
14. Oppenheim, B.A. (2000), Optimal management of central venous catheter-related infections: what is the evidence? *J. Infect.*, 40, 26–30.
15. Paiva, J.A. and Pereira, J.M. (2002), Treatment of the afebrile patient after catheter withdrawal: drugs and duration, *Clin. Microbiol. Infect.*, 8, 290–294.
16. Raad, I. (1998), Intravascular-catheter-related infections, *Lancet*, 351, 893–898.
17. Rodriguez-Bano, J. (2002), Selection of empiric therapy in patients with catheter-related infections, *Clin. Microbiol. Infect.*, 8, 275–281.
18. Seifert, H., Strate, A., and Pulverer, G. (1995), Nosocomial bacteraemia due to *Acinetobacter baumannii*: clinical features, epidemiology, and predictors of mortality, *Medicine (Baltimore)*, 74, 340–349.
19. Elting, L.S. and Bodey, G.P. (1990), Septicemia due to *Xanthomonas* species and non-aeruginosa *Pseudomonas* species: increasing incidence of catheter-related infections, *Medicine (Baltimore)*, 69, 296–306.
20. Seifert, H., Strate, A., Schulze, A., and Pulverer, G. (1993), Vascular catheter-related bloodstream infection due to *Acinetobacter johnsonii* (formerly *Acinetobacter calcoaceticus var. lwoffi*): report of 13 cases, *Clin. Infect. Dis.*, 17, 632–636.
21. Benoit, J.L., Carandang, G., Sitrin, M., and Arnow, P.M. (1995), Intraluminal antibiotic treatment of central venous catheter infection in patients receiving parenteral nutrition at home, *Clin. Infect. Dis.*, 21, 1286–1288.
22. Shaffer, D. (1995), Catheter-related sepsis complicating long-term, tunneled central venous dialysis catheters: management by guidewire exchange, *Am. J. Kidney Dis.*, 25, 593–596.
23. Samore, M.H. and Burke, J.P. (2000), Infections of long intravenous lines: new developments and controversies, *Curr. Clin. Top Infect. Dis.*, 20, 256–270.
24. Schuman, E.S., Winters, V., Gross, G.F., and Hayes, J.F. (1985), Management of Hickman catheter sepsis, *Am. J. Surg.*, 149, 627–628.

25. Press, O.W., Ramsey, P.G., Larson, E.B., Fefer, A., and Hickman, R.O. (1984), Hickman catheter infections in patients with malignancies, *Medicine (Baltimore)*, 63, 189–200.
26. O'Grady, N.P., Alexander, M., Dellinger, E.P., Gerberding, J.L., Heard, S.O., Maki, D.G., Masur, H., McCormick, R.D., Mermel, L.A., Pearson, M.L., Raad, I.I., Randolph, A., and Weinstein, R.A. (2002), Guidelines for the prevention of intravascular catheter-related infections, Centers for Disease Control and Prevention, *MMWR Recomm. Rep.*, 51(RR-10), 1–29.
27. von Eiff, C., Jansen, B., Kohnen, W., and Becker, K. (2005), Infections associated with medical devices: pathogenesis, management and prophylaxis, *Drugs*, 65, 179–214.
28. Stickler, D.J., Morris, E.A., and Hughes, G. (2002), Strategies for the control of catheter encrustation, *Int. J. Antimicrob. Agents*, 19, 499–506.
29. Stickler, D., Morris, N., Moreno, M-C., and Sabbuba, N. (1998), Studies on the formation of crystalline bacterial biofilms on urethral catheters, *Eur. J. Clin. Microbiol. Infect. Dis.*, 17, 649–652.
30. Domenico, P., Baldassarri, L., Schoch, P.E., Kaehler, K., Sasatsu, M., and Cunha, B.A. (2001), Activities of bismuth thiols against staphylococci and staphylococcal biofilms, *Antimicrob. Agents Chemother.*, 45, 1417–1421.
31. Ahearn, D.G., Grace, D.T., Jennings, M.J., Borazjani, R.N., Boles, K.J., Rose, L.J., Simmons, R.B., and Ahanotu, E.N. (2000), Effects of hydrogel/silver coatings on in vitro adhesion to catheters of bacteria associated with urinary tract infection, *Curr. Microbiol.*, 41, 120–125.
32. Darouiche, R.O., Raad, I.I., Heard, S.O., Thornby, J.I., Wenker, O.C., Gabrielli, A., Berg, J., Khardori, N., Hanna, H., Hachem, R., Harris, R.L., and Mayhall, G. (1999), A comparison of two antimicrobial-impregnated central venous catheters, *New Engl. J. Med.*, 340, 1–8.
33. Johnson, J.R., Delavari, P., and Azar, M. (1999), Activities of a nitrofurazone containing urinary catheter and a silver hydrogel catheter against multidrug resistant bacteria characteristic of catheter-associated urinary tract infection, *Antimicrob. Agents Chemother.*, 43, 2990–2995.
34. Stickler, D.J. (1996), Bacterial biofilms and the encrustation of urethral catheters, *Biofouling*, 94, 293–305.
35. Sobho, F., Khoury, A.E., Zamboni, A.C., Davidson, D., and Mittelman, M.W. (1995), Effects of ciprofloxacin and protamine sulfate combinations against catheter-associated *Pseudomonas aeruginosa* biofilms, *Antimicrob. Agents Chemother.*, 39, 1281–1286.
36. Flowers, R.H., Schwenzer, K.J., Kopel, R.F., Fish, M.J., Tucker, S.I., and Farr, B.M. (1989), Efficacy of an attachable subcutaneous cuff for the prevention of intravascular catheter-related infection, A randomized controlled trial, *JAMA*, 261, 878–883.
37. Burton, E., Gawande, P.V., Yakandawala, N., LoVetri, K., Zhanel, G.G., Romeo, T., Friesen, A.D., and Madhyastha, S. (2006), Antibiofilm activity of GlmU enzyme inhibitors against catheter-associated uropathogens, *Antimicrob. Agents Chemother.*, 50, 1835–1840.
38. Rosenthal, V.D., Guzman, S., Pezzotto, S.M., and Crnich, C.J. (2003), Effect of an infection control program using education and performance feedback on rats of

intravascular device-associated blood-stream infections in intensive care units in Argentina, *Am .J. Infect. Control.*, 31, 405–409.

39. Sheretz, R.J., Ely, E.W., Westbrook, D.M., Gledhill, K.S., Streed, S.A., Kiger, B., Flynn, L., Hayes, S., Strong, S., Cruz, J., Bowton, D.L., Hulgan, T., and Haponik, E.F. (2000), Education of physicians-in-training can decrease the risk for vascular catheter infection, *Ann. Intern. Med.*, 132, 641–648.
40. Eggimann, P., Harbarth, S., Constantin, M.N., Touveneau, S., Chevrolet, J.C., and Pittet, D. (2000), Impact of a prevention strategy targeted at vascular-access care on incidence of infections acquired in intensive care, *Lancet*, 355, 1864–1868.
41. Soifer, N.E., Borzak, S., Edlin, B.R., and Weinstein, R.A. (1998), Prevention of peripheral venous catheter complications with an intravenous therapy team: a randomized controlled trial, *Arch. Intern. Med.*, 158, 473–477.
42. Fridkin, S.K., Pear, S.M., Williamson, T.H., Galgiani, J.N., and Jarvis, W.R. (1996), The role of understaffing central venous catheter-associated blood stream infections, *Infect. Control Hosp. Epidemiol.*, 17, 150–158.
43. Bertone, S.A., Fisher, M.C., and Mortensen, J.E. (1994), Quantitative skin cultures at potential catheter sites in neonates, *Infect. Control Hosp. Epidemiol.*, 15, 315–318.
44. Goetz, A.M., Wagener, M.M., Miller, J.M., and Muder, R.R. (1998), Risk of infection due to central venous catheters: effect of site of placement and catheter type, *Infect. Control Hosp. Epidemiol.*, 19, 842–845.
45. Randolph, A.G., Cook, D.J., Gonzales, C.A., and Pribble, C.G. (1996), Ultrasound guidance for placement of central venous catheters: a meta-analysis of the literature, *Crit. Care Med.*, 24, 2053–2058.
46. Merrer, J., De Jonghe, B., Golliot, F., Lefrant, J.Y., Raffy, B., Barre, E., Rigaud, J.P., Casciani, D., Misset, B., Bosquet, C., Outin, H., Brun-Buisson, C., and Nitenberg, G. (2001), French Catheter Study Group in Intensive Care, Complications of femoral and subclavian venous catheterization in critically ill patients: a randomized controlled trial, *JAMA*, 286, 700–707.
47. Heard, S.O., Wagle, M., Vijayakumar, E., McLean, S., Brueggemann, A., Napolitano, L.M., Edwards, L.P., O'Connell, F.M., Puyana, J.C., and Doern, G.V. (1998), Influence of triple-lumen central venous catheters coated with chlorhexidine and silver sulfadiazine on the incidence of catheter-related bacteraemia, *Arch. Intern. Med.*, 158, 81–87.
48. Richet, H., Hubert, B., Nitemberg, G., Andremont, A., Buu-Hoi, A., Ourbak, P., Galicier, C., Veron, M., Boisivon, A., Bouvier, A.M., Ricome, J.C., Wolff, M.A., Pean, Y., Beradigrassias, L., Bourdain, J.L., Hautefort, B., Laaban, J.P., and Tillant, D. (1990), Prospective multicenter study of vascular-related complications and risk factors for positive central-catheter cultures in intensive care unit patients, *J. Clin. Microbiol.*, 28, 2520–2525.
49. Maki, D. and Mermel, L., (1998), Infections due to infusion therapy, in: Bennett, J.V., Brachman, P.S., Eds., Hospital infections, 4th ed., Lippincott-Raven, Philadelphia, pp. 689–724.
50. MacDonald, A., Dinah, F., Mackenzie, D., and Wilson, A. (2004), Performance feedback of hand hygiene, using alcohol gel as the skin decontaminant, reduces the number of inpatients newly affected by MRSA and antibiotic costs, *J. Hosp. Infect.*, 56, 56–63.

51. Pittet, D., Hugonnet, S., Harbarth, S., Mourouga, P., Sauvan, V., Touveneau, S., and Perneger, T.V. (2000), Effectiveness of a hospital-wide programme to improve compliance with hand hygiene: Infection Control Programme, *Lancet*, 356, 1307–1312.
52. Raad, I.I., Hohn, D.C., Gilbreath, B.J., Suleiman, N., Hill, L.A., Bruso, P.A., Marts, K., Mansfield, P.F., and Bodey, G.P. (1994), Prevention of central venous catheter-related infections by using maximal sterile barrier precautions during insertion, *Infect. Control Hosp. Epidemiol.*, 15 (4 Pt 1), 231–238.
53. Galway, R., Harrod, M.E., Crisp, J., Donnellan, R., Hardy, J., Harvey, A., Maurice, L., Petty, S., and Senner, A. (2003), Central venous access and hand washing: variability in policies and practices, *Paediatr. Nurs.*, 15, 14–18.
54. Boyce, J.M. and Pittet, D. (2002), Guideline for Hand Hygiene in Health-Care Settings: recommendations of the Healthcare Infection Control Practices Advisory Committee and the HICPAC/SHEA/APIC/IDSA Hand Hygiene Task Force. Society for Healthcare Epidemiology of America/Association for Professionals in Infection Control/Infectious Diseases Society of America, *MMWR Recomm. Rep.*, 51 (RR-16), 1–45.
55. Safdar, N. and Maki, D.G. (2004), The Pathogenesis of catheter-related blood stream infection with noncuffed short-term central venous catheters, *Intensive Care Med.*, 30, 62–67.
56. Clemence, M.A., Walker, D., and Farr, B.M. (1995), Central venous catheter practices: results of a survey, *Am. J. Infect. Control.*, 23, 5–12.
57. Humar, A., Ostromecki, A., Direnfeld, J., Marshall, J.C., Lazar, N., Houston, P.C., Boiteau, P., and Conly, J.M. (2000), Prospective randomized trial of 10% povidone-iodine versus 0.5% tincture of chlorhexidine as cutaneous antisepsis for prevention of central venous catheter infection, *Clin. Infect. Dis.*, 31, 1001–1007.
58. Maki, D.G., Ringer, M., and Alvarado, C.J. (1991), Prospective randomized trial of povidone-iodine, alcohol, and chlorhexidine for prevention of infection associated with central venous and arterial catheters, *Lancet*, 338, 339–343.
59. Garland, J.S., Buck, R.K., Maloney, P., Durkin, D.M., Toth-Lloyd, S., Duffy, M., Szocik, P., McAuliffe, T.L., and Goldmann, D. (1995), Comparison of 10% povidone-iodine and 0.5% chlorhexidine gluconate for the prevention of peripheral intravenous catheter colonization in neonates: a prospective trial, *Pediatr. Infect. Dis. J.*, 14, 510–516.
60. Maki, D.G. and Ringer, M. (1987), Evaluation of dressing regimens for prevention of infection with peripheral intravenous catheters: gauze, a transparent polyurethane dressing, and an iodophor transparent dressing, *JAMA*, 258, 2396–2403.
61. Hoffmann, K.K., Weber, D.J., Samsa, G.P., and Rutala, W.A. (1992), Transparent polyurethane film as an intravenous catheter dressing: a meta-analysis of the infection risks, *JAMA*, 267, 2072–2076.
62. Maki, D.G., Mermel, L.A., and Kluger, D.M. (2000), The efficacy of a chlorhexidine-impregnated sponge (biopatch) for the prevention of intravascular catheter-related infection: a prospective, randomized, controlled, multicenter trial [abstract no.1430], *40th Interscience Conference on Antimicrobial Agents and Chemotherapy of the American society for Microbiology*, Sept. 17–20; Toronto, Canada.
63. Yamamoto, A.J., Solomon, J.A., Soulen, M.C., Tang, J., Parkinson, K., Lin, R., and Schears, G.J. (2002), Sutureless securement device reduces complications of peripherally inserted central venous catheters, *J. Vasc. Interv. Radiol.*, 13, 77–81.

64. Maki, D.G. and Ringer, M. (1991), Risk factors for infusion-related phlebitis with small peripheral venous catheters: a randomized controlled trial, *Ann. Intern. Med.*, 114, 845–854.

65. Tully, J.L., Friedland, G.H., Baldini, L.M., and Goldmann, D.A. (1981), Complications of intravenous therapy with steel needles and Teflon catheters: a comparative study, *Am. J. Med.*, 70, 702–706.

66. Ball, P.A. (2003), Intravenous in-line filters: filtering the evidence, *Curr. Opin. Clin. Nutr. Metab. Care*, 6, 319–325.

67. Maddox, R.R., John, Jr, J.F., Brown, L.L., and Smith, C.E. (1983), Effect of inline filtration on post-infusion phlebitis, *Clin. Pharm.*, 2, 58–61.

68. Allison, D.G., McBain, A.J., and Gilbert, P. (2000), Microbial biofilms: problems of control, in: Allison, D.G., Gilbert, P., Lappin-Scott, H. et al., Eds., *Community Structure and Cooperation in Biofilms*, Society for General Microbiology Press, Reading, UK, pp. 309–327.

69. Ceri, H., Olson, M.E., Stremick, C., Read, R.R., Morck, D., and Buret, A. (1999), The Calgary Biofilm Device: new technology for rapid determination of antibiotic susceptibilities of bacterial biofilms, *J. Clin. Microbiol.*, 37, 1771–1776.

70. Espersen, F., Frimodt-Moller, N., Corneliussen, L., Riber, U., Rosdahl, V.T., and Skinhøj, P. (1994), Effect of treatment with methicillin and gentamicin in a new experimental mouse model of foreign body infection, *Antimicrob. Agents Chemother.*, 38, 2047–2053.

71. Rachid, S., Ohlsen, K., Witte, W., Hacker, J., and Ziebuhr, W. (2000), Effect of subinhibitory antibiotic concentrations on polysaccharide intercellular adhesin expression in biofilm-forming *Staphylococcus epidermidis*, *Antimicrob. Agents Chemother.*, 44, 3357–3363.

72. Wilcox, M.H., Finch, R.G., Smith, D.G., Williams, P., and Denyer, S.P. (1991), Effects of carbon dioxide and sub-lethal levels of antibiotics on adherence of coagulase-negative staphylococci to polystyrene and silicone rubber, *J. Antimicrob. Chemother.*, 27, 577–587.

73. Bisognano, C., Vaudaux, P.E., Lew, D.P., Ng, E.Y., and Hooper, D.C. (1997), Increased expression of fibronectin-binding proteins by fluoroquinolone-resistant *Staphylococcus aureus* exposed to subinhibitory levels of ciprofloxacin, *Antimicrob. Agents Chemother.*, 41, 906–913.

74. Roehrborn, A.A., Hansbrough, J.F., Gualdoni, B., and Kim, S. (1995), Lipid- based slow-release formulation of amikacin sulfate reduces foreign body-associated infections in mice, *Antimicrob. Agents Chemother.*, 39, 1752–1755.

75. Missing, B., Peitra-Cohen, S., Debure, A., Beliah, M., and Bernier, J.J. (1988), Antibiotic-lock technique; a new approach to optimal therapy for catheter-related sepsis in home-parenteral nutrition patients, *JPEN J. Parenter. Enteral. Nutr.*, 12, 185–189.

76. Kentos, A., Struelens, M.J., and Thys, J.P. (1996), Antibiotic-lock technique for the treatment of central venous catheter infections, *Clin. Infect. Dis.*, 23, 418–419.

77. Ruiz-Valverde, M.P., Barbera, J.R., Segarra, A., Capdevila, J.A., Evangelista, A., and Piera, L. (1997), Successful treatment of catheter-related sepsis and extraluminal catheter thrombosis with vancomycin and fraxiparin without catheter removal, *Nephron*, 75, 354–355.

78. Capdevila, J.A., Segarra, A., Planes, A.M., Ramírez-Arellano, M., Pahissa, A., Piera, L., and Martínez-Vázquez, J.M. (1993), Successful treatment of hemodialysis catheter-related sepsis without catheter removal, *Nephrol. Dial. Transplant.*, 8, 231–234.
79. Krishnasami, Z., Carlton, D., Bimbo, L., Taylor, M.E., Balkovetz, D.F. Barker, J., and Allon, M. (2002), Management of hemodialysis catheter-related bacteraemia with an adjunctive antibiotic lock solution, *Kidney Int.*, 61, 1136–1142.
80. La Quaglia, M.P., Caldwell, C., Lucas, A., Corbally, M., Heller, G., Steinherz, L., Brown, A.E., Groeger, J., and Exelby, P.R. (1994), A prospective randomized double-blind trial of bolus urokinase in the treatment of established Hickman catheter sepsis in children, *J. Pediatr. Surg.*, 29, 742–745.
81. Jones, G.R., Konsler, G.K., and Dunaway, R.P. (1996), Urokinase in the treatment of bacteraemia and candidaemia in patients with right atrial catheters, *Am. J. Infect. Control.*, 24, 160–166.
82. Capdevila, J.A., Gavalda, J., Fortea, J., López, P., Martin, M.T., Gomis, X., and Pahissa, A. (2001), Lack of antimicrobial activity of sodium heparin for treating experimental catheter related infection due to *Staphylococcus aureus* using the antibiotic-lock technique, *Clin. Microbiol. Infect.*, 7, 206–212.
83. Sieradzki, K., Roberts, R.B., Serur, D., Hargrave, J., and Tomasz, A. (1999), Heterogeneously vancomycin-resistant *Staphylococcus epidermidis* strain causing recurrent peritonitis in a dialysis patient during vancomycin therapy, *J. Clin. Microbiol.*, 37, 39–44.
84. Traub, W.H., Leonhard, B., and Bauer, D. (1993), Taurolidine: *in vitro* activity against multiple-antibiotic-resistant, nosocomially significant clinical isolates of *Staphylococcus aureus*, *Enterococcus faecium*, and diverse Enterobacteriaceae, *Chemotherapy*, 39, 322–330.
85. Jones, D.S., Gorman, S.P., McCafferty, D.F., and Woolfson, A.D. (1991), The effects of three non-antibiotic, antimicrobial agents on the surface hydrophobicity of certain microorganisms evaluated by different methods, *J. Appl. Bacteriol.*, 71, 218–227.
86. Shah, C.B., Mittelman, M.W., Costerton, J.W., Parenteau, S., Pelak, M., Arsenault, R., and Mermel, L.A. (2002), Antimicrobial activity of a novel catheter lock solution, *Antimicrob. Agents Chemother.*, 46, 1674–1679.
87. Dannenberg, C., Bierbach, U., Rothe, A., Beer, J., and Körholz, D. (2003), Ethanol-lock technique in the treatment of bloodstream infections in pediatric oncology patients with broviac catheter, *J. Pediatr. Hematol. Oncol.*, 25, 616–621.
88. Arnow, P.M., Quimosing, E.M., and Beach, M. (1993), Consequences of intravascular catheter sepsis, *Clin. Infect. Dis.*, 16, 778–784.
89. Rotstein, C., Brock, L., and Roberts, R.S. (1995), The incidence of first Hickman catheter-related infection and predictors of catheter removal in cancer patients, *Infect. Control Hosp. Epidemiol.*, 16, 451–458.
90. Perdreau-Remington, F., Stefanik, D., Peters, G., Ludwig, C., Rütt, J., Wenzel, R., and Pulverer, G. (1996), A four-year prospective study on microbial ecology of explanted prosthetic hips in 52 patients with "aseptic" prosthetic joint loosening, *Eur. J. Clin. Microbiol. Infect. Dis.*, 15, 160–165.
91. Flynn, P.M., Shenep, J.L., Stokes, D.C., and Barrett, F.F., (1987), *In situ* management of confirmed central venous catheter-related bacteraemia, *Pediatr. Infect. Dis. J.*, 6, 729–734.

92. Raad, I., Davis, S., Khan, A., Tarrand, J., Elting, L., and Bodey, G.P. (1992), Impact of central venous catheter removal on the recurrence of catheter-related coagulase-negative staphylococcal bacteremia, *Infect. Control Hosp. Epidemiol.*, 13, 215–221.

93. Gagnon, R.F., Richards, G.K., and Wiesenfeld, L. (1991), *Staphylococcus epidermidis* biofilms: unexpected outcome of double and triple antibiotic combination with rifampin, *ASAIO Trans.*, 37, M158–M160.

94. Boyce, J.M., Mermel, L.A., Zervos, M.J., Rice, L.B., Potter-Bynoe, G., Giorgio, C., and Medeiros, A.A. (1995), Controlling vancomycin-resistant enterococci, *Infect. Control Hosp. Epidemiol.*, 16 (11), 634–637. [Erratum in: Infect Control Hosp Epidemiol 1996 Apr; 17(4), 211].

95. Gaillard, J.L., Merlino, R., Pajot, N., Goulet, O., Fauchere, J.L., Ricour, C., and Veron, M. (1990), Conventional and nonconventional modes of vancomycin administration to decontaminate the internal surface of catheters colonized with coagulase-negative staphylococci, *JPEN J. Parenter. Enteral. Nutr.*, 14, 593–597.

96. *Staphylococcus aureus* resistant to vancomycin: United States, (2002), *MMWR Morb. Mortal. Wkly. Rep.*, 51, 565–567.

97. Fowler, Jr, V.G., Sanders, L.L., Sexton, D.J., Kong, L., Marr, K.A., Gopal, A.K., Gottlieb, G., McClelland, R.S., and Corey, G.R. (1998), Outcome of *Staphylococcus aureus* bacteraemia according to compliance with recommendations of infectious diseases specialists: experience with 244 patients, *Clin. Infect. Dis.*, 27, 478–486.

98. Libman, H. and Arbeit, R.D. (1984), Complications associated with *Staphylococcus aureus* bacteraemia, *Arch. Intern. Med.*, 144, 541–545.

99. Benezra, D., Kiehn, T.E., Gold, J.W., Brown, A.E., Turnbull, A.D., and Armstrong, D. (1988), Prospective study of infections in indwelling central venous catheters using quantitative blood cultures, *Am. J. Med.*, 85, 495–498.

100. Dugdale, D.C. and Ramsey, P.G. (1990), *Staphylococcus aureus* bacteraemia in patients with Hickman catheters, *Am. J. Med.*, 89, 137–141.

101. Williams, N., Carlson, G.L., Scott, N.A., and Irving, M.H. (1994), Incidence and management of catheter-related sepsis in patients receiving home parenteral nutrition, *Br. J. Surg.*, 81, 392–394.

102. Rubin, L.G., Shih, S., Shende, A., Karayalcin, G., and Lanzkowsky, P. (1999), Cure of implantable venous port-associated bloodstream infections in pediatric hematology-oncology patients without catheter removal, *Clin. Infect. Dis.*, 29, 102–105.

103. Fowler, Jr, V.G., Li, J., Corey, G.R., Boley, J., Marr, K.A., Gopal, A.K., Kong, L.K., Gottlieb, G., Donovan, C.L., Sexton, D.J., and Ryan, T. (1997), Role of echocardiography in evaluation of patients with *Staphylococcus aureus* bacteraemia: experience in 103 patients, *J. Am. Coll. Cardiol.*, 30, 1072–1078.

104. Rosen, A.B., Fowler, Jr, V.G., Corey, G.R., Downs, S.M., Biddle, A.K., Li, J., and Jollis, J.G. (1999), Cost-effectiveness of transesophageal echocardiography to determine the duration of therapy for intravascular catheter-associated *Staphylococcus aureus* bacteraemia, *Ann. Intern. Med.*, 130, 810–820.

105. Fowler, Jr, V.G., Olsen, M.K., Corey, G.R., Woods, C.W., Cabell, C.H., Reller, L.B., Cheng, A.C., Dudley, T., and Oddone, E.Z. (2003), Clinical identifiers of complicated *Staphylococcus aureus* bacteraemia, *Arch. Intern. Med.*, 163, 2066–2072.

106. Raad, I.I. and Sabbagh, M.F. (1992), Optimal duration of therapy for catheter—related *Staphylococcus aureus* bacteraemia: a study of 55 cases and review, *Clin. Infect. Dis.*, 14, 75–82.

107. Malanoski, G.J., Samore, M.H., Pefanis, A., and Karchmer, A.W. (1995), *Staphylococcus aureus* catheter-associated bacteraemia: minimal effective therapy and unusual infectious complications associated with arterial sheath catheters, *Arch. Intern. Med.*, 155, 1161–1166.

108. Rahal, J.J. (1989), Preventing second-generation complications due to *Staphylococcus aureus*, *Arch. Intern. Med.*, 149, 503–504.

109. Hospital Infection Control Practices Advisory Committee (HICPAC), Recommendations for preventing the spread of vancomycin resistance, (1995), *Infect. Control Hosp. Epidemiol.*, 16, 105–113.

110. Peters, G. and Becker, K. (1996), Epidemiology, control and treatment of methicllin-resistant *Staphylococcus aureus*, *Drugs*, 52 Suppl. 2, 50–54.

111. Levine, D.P. and Fromm, B.S. (1991), Slow response to vancomycin or vancomycin plus rifampin in methicillin-resistant *Staphylococcus aureus* endocarditis, *Ann. Intern. Med.*, 115, 674–680.

112. Jantausch, B.A., Deville, J., Adler, S., Morfin, M.R., Lopez, P., Edge-Padbury, B., Naberhuis-Stehouwer, S., and Bruss, J.B. (2003), Linezolid for the treatment of children with bacteraemia or nosocomial pneumonia caused by resistant gram-positive bacterial pathogens, *Pediatr. Infect. Dis. J.*, 22 (9 Suppl.), S164–S171.

113. Vaudaux, P., Francois, P., Bisognano, C., Schrenzel, J., and Lew, D.P. (2002), Comparison of Levofloxacin, alatrofloxacin, and vancomycin for prophylaxis and treatment of experimental foreign-body-associated infection by methicillin-resistant *Staphylococcus aureus*, *Antimicrob. Agents Chemother.*, 46, 1503–1509.

114. Chuard, C., Herrmann, M., Vaudaux, P., Waldvogel, F.A., and Lew, D.P. (1991), Successful therapy of experimental chronic foreign-body infection due to methicillin resistant *Staphylococcus aureus* by antimicrobial combinations, *Antimicrob. Agents Chemother.*, 35, 2611–2616.

115. Cagni, A., Chuard, C., Vaudaux, P.E., Schrenzel, J., and Lew, D.P. (1995), Comparison of sparfloxacin temafloxacin, and ciprofloxacin for prophylaxis and treatment of experimental foreign-body infection by methicillin-resistant *Staphylococcus aureus*, *Antimicrob. Agents Chemother.*, 39 (8), 1655–1660.

116. Schaad, H.J., Chuard, C., Vaudaux, P., Rohner, P., Waldvogel, F.A., and Lew, D.P. (1994), Comparative efficacies of imipenem, oxacillin and vancomycin for therapy of chronic foreign body infection due to methicllin-susceptible and- resistant *Staphylococcus aureus*, *J. Antimicrob. Chemother.*, 33, 1191–1200.

117. Schaad, H.J., Chuard, C., Vaudaux, P., Waldvogel, F.A., and Lew, D.P. (1994), Teicoplanin alone or combined with rifampin compared with vancomycin for prophylaxis and treatment *Staphylococcus aureus*, *Antimicrob. Agents Chemother.*, 33, 1703–1710.

118. Gagnon, R.F., Richards, G.K., and Subang, R. (1992), Experimental *Staphylococcus epidermidis* implant infection in the mouse; Kinetics of rifampin and vancomycin action, *ASAIO J.*, 38 (3), M596–M599.

119. Van Wijngaerden, E., Peetermans, W.E., Vandersmissen, J., Van Lierde, S., Bobbaers, H., and Van Eldere, J. (1999), Foreign body infection: a new rat model for prophylaxis and treatment, *J. Antimicrob. Chemother.*, 44, 669–674.

120. Rupp, M.E., Fey, P.D., and Longo, G.E. (2001), Effect of LY333328 against vancomycin-resistant *Enterococcus faecium* in a rat central venous catheter-associated infection model, *J. Antimicrob. Chemother.*, 47, 705–707.

121. Kuklin, N.A., Pancari, G.D., Tobery, T.W., Cope, L., Jackson, J., Gill, C., Overbye, K., Francis, K.P., Yu, J., Montgomery, D., Anderson, A.S., McClements, W., and Jansen, K.U. (2003), Real-time monitoring of bacterial infection *in vivo*: development of bioluminescent staphylococcal foreign-body and deep-thigh-wound mouse infection models, *Antimicrob. Agents Chemother.*, 47, 2740–2748.

122. Vaudaux, P., Francois, P., Bisognano, C., Li, D., Lew, D.P., and Schrenzel, J. (2003), Comparative efficacy of daptomycin and vancomycin in the therapy of experimental foreign body infection due to *Staphylococcus aureus*, *J. Antimicrob. Chemother.*, 52, 89–95.

123. Rupp, W.E. and Ulphani, J. (1998), Efficacy of LY333328 in a rat model of *Staphylococcus aureus* central venous catheter-associated infection [abstract no. F111], *Programs and Abstracts of the Thirty-Eighth Interscience Conference on Antimicrobial Agents and Chemotherapy, San Diego, CA, 1998*, American Society for Microbiology, Washington, DC, p. 260.

124. Swanson, D.S. (1998), Central venous catheter-related infections due to nontuberculous Mycobacterium species, *Pediatr. Infect. Dis. J.*, 17 (12), 1163–1164.

125. Guay, D.R. (1996), Nontuberculous mycobacterial infections, *Ann. Pharmacother.*, 30, 819–830.

126. Ishida, H., Ishida, Y., Kurosaka, Y., Otani, T., Sato, K., and Kobayashi, H. (1998), *In vitro* and *in vivo* activities of levofloxacin against biofilm-producing *Pseudomonas aeruginosa*, *Antimicrob. Agents Chemother.*, 42, 1641–1645.

127. Ashby, M.J., Neale, J.E., Knott, S.J., and Critchley, I.A. (1994), Effect of antibiotics on non-growing planktonic cells and biofilms of *Escherichia coli*, *J. Antimicrob. Chemother.*, 33, 443–452.

128. Gill, M.V., Ly, H., Mueenuddin, M., Schoch, P.E., and Cunha, B.A. (1997), Intravenous line infection due to *Ochrobactrum anthropi* (CDC Group Vd) in a normal host, *Heart Lung*, 26, 335–336.

129. Seifert, H. (1997), Catheter-related infections due to gram-negative bacilli, in: Seifert, H., Jansen, B., and Farr, B.M., Eds., *Catheter-related infections*, Marcel Dekker, New York, pp 111–138.

130. Douglas, L.J. (2003), *Candida* biofilms and their role in infection, *Trends Microbiol.*, 11, 30–36.

131. Hazen, K.C. (1995), New and emerging yeast pathogens, *Clin. Microbiol. Rev.*, 8, 462–478.

132. Sizun, J., Karangwa, A., Giroux, J.D., Masure, O., Simitzis, A.M., Alix, D., and De Parscau, L. (1994), *Malassezia furfur*-related colonization and infection of central venous catheters; a prospective study in a pediatric intensive care unit, *Intensive Care Med.*, 20, 496–499.

133. Anaissie, E.J., Rex, J.H., Uzun, Ö., and Vartivarian, S. (1998), Predictors of adverse outcome in cancer patients with candidemia, *Am. J. Med.*, 104, 238–245.

134. Pappas, P.G., Rex, J.H., Sobel, J.D., Filler, S.G., Dismukes, W.E., Walsh, T.J., and Edwards, J.E., Infectious Diseases Society of America. (2004), Guidelines for treatment of candidiasis, *Clin. Infect. Dis.*, 38, 161–189.

135. Karlowicz, M.G., Hashimoto, L.N., Kelly, Jr, R.E., and Buescher, E.S. (2000), Should central venous catheters be removed as soon as candidemia is detected in neonates? *Pediatrics*, 106, E63.

136. Munoz, P., Burillo, A., and Bouza, E. (2000), Criteria used when initiating antifungal therapy against *Candida* spp. in the intensive care unit, *Int. J. Antimicrob. Agents*, 15, 83–90.

137. Rex, J.H., Walsh, T.J., Sobel, J.D., Filler, S.G., Pappas, P.G., Dismukes, W.E., and Edwards, J.E. (2000), Practice guidelines for the treatment of candidiasis: Infectious Diseases Society of America, *Clin. Infect. Dis.*, 30, 662–678.

138. Kibbler, C.C., Seaton, S., Barnes, R.A., Gransden, W.R., Holliman, R.E., Johnson, E.M., Perry, J.D., Sullivan, D.J., and Wilson, J.A. (2003), Management and outcome of bloodstream infections due to Candida species in England and Wales, *J. Hosp. Infect.*, 54, 18–24.

139. Nguyen, M.H., Peacock, Jr, J.E., Tanner, D.C., Morris, A.J., Nguyen, M.L., Snydman, D.R., Wagener, M.M., and Yu, V.L. (1995), Therapeutic approaches in patients with candidemia: evaluation in a multicenter, prospective, observational study, *Arch. Intern. Med.*, 155, 2429–2435.

140. Lecciones, J.A., Lee, J.W., Navarro, E.E., Witebsky, F.G., Marshall, D., Steinberg, S.M., Pizzo, P.A., and Walsh, T.J. (1992), Vascular catheter-associated fungemia in patients with cancer: analysis of 155 episodes, *Clin. Infect. Dis.*, 14, 875–883.

141. Rose, H.D. (1978), Venous catheter-associated candidemia, *Am. J. Med. Sci.*, 275, 265–269.

142. Barber, G.R., Brown, A.E., Kiehn, T.E., Edwards, F.F., and Armstrong, D. (1993), Catheter-related *Malassezia furfur* fungemia in immunocompromised patients, *Am. J. Med.*, 95, 365–370.

143. Ostrosky-Zeichner, L., Marr, K.A., Rex, J.H., and Cohen, S.H. (2003), Amphotericin B: time for a new "gold standard", *Clin. Infect. Dis.*, 37, 415–425.

144. Rex, J.H., Bennett, J.E., Sugar, A.M., Pappas, P.G., van der Horst, C.M., Edwards, J.E., Washburn, R.G., Scheld, W.M., Karchmer, A.W., Dine, A.P., Levenstein, M.J., Webb, C.D., for The Candidemia Study Group and the National Institute of Allergy and Infectious Diseases Mycoses Study Group. (1994), A randomized trial comparing fluconazole with amphotericin B for the treatment of candidemia in patients without neutropenia: Candidemia Study Group and the National Institute, *N. Engl. J. Med.*, 331, 1325–1330.

145. Diekema, D.J., Messer, S.A., Hollis, R.J., Jones, R.N., and Pfaller, M.A. (2003), Activities of caspofungin, itraconazole. posaconazole, ravuconazole, voriconazole, and amphotericin B against 448 recent clinical isolates o filamentous fungi, *J. Clin. Microbiol.*, 41, 3623–3626.

146. Mora-Duarte, J., Betts, R., Rotstein, C., Colombo, A.L., Thompson-Moya, L., Smietana, J., Lupinacci, R., Sable, C., Kartsonis, N., and Perfect, J., Caspofungin Invasive Candidiasis Study Group. (2002), Comparison of caspofungin and amphotericin B for invasive candidiasis, *N. Engl. J. Med.*, 347, 2020–2029.

147. Kauffman, C.A. and Zarins, L.T. (1998), *In vitro* activity of voriconazole against *Candida* species, *Diagn. Microbiol. Infect. Dis.*, 31, 297–300.

148. Silling, G., Fegeler, W., Roos, N., Essink, M., and Büchner, T. (1999), Early empiric antifungal therapy of infections in neutropenic patients comparing fluconazole with amphotericin B/flucytosine, *Mucoses*, 42 Suppl. 2, 101–104.

149. Patron, R.L., Climo, M.W., Goldstein, B.P., and Archer, G.L. (1999), Lysostaphin treatment of experimental aortic value endocarditis caused by a *Staphylococcus aureus* isolate with reduce susceptibility to vancomycin, *Antimicrob. Agents Chemother.*, 43, 1754–1755.

150. von Eiff, C., Kokai-Kun, J.F., Becker, K., and Peters, G. (2003), *In vitro* activity of recombinant lysostaphin against *Staphylococcus aureus* isolates from anterior nares and blood, *Antimicrob. Agents Chemother.*, 47, 3613–3615.

151. Wu, J.A., Kusuma, C., Mond, J.J., and Kokai-Kun, J.F. (2003), Lysostaphin disrupts *Staphylococcus aureus* and *Staphylococcus epidermidis* biofilms on artificial surfaces, *Antimicrob. Agents Chemother.*, 47, 3407–3414.

152. Singh, P.K., Parsek, M.R., Greenberg, E.P., and Welsh, M.J. (2002), A component of innate immunity prevents bacterial biofilm development, *Nature (London)*, 417, 552–555.

153. Johansen, C., Falholt, P., and Gram, L. (1997), Enzymatic removal and disinfection of bacterial biofilms, *Appl. Environ. Microbiol.*, 63, 3724–3728.

154. Nemoto, K., Hirota, K., Murakami, K., Taniguti, K., Murata, H., Viducic, D., and Miyake, Y. (2003), Effect of Varidase (streptodornase) on biofilm formed by *Pseudomonas aeruginosa*, *Chemotherapy*, 49, 121–125.

155. Nemoto, K., Hirota, K., Ono, T., Murakami, K., Murakami, K., Nagao, D., and Miyake, Y. (2000), Effect of Varidase (streptokinase) on biofilm formed by *Staphylococcus aureus*, *Chemotherapy*, 46, 111–115.

156. Hatch, R.A. and Schiller, N.L. (1998), Alginate lyase promotes diffusion of aminoglycosides through the extracellular polysaccharide of mucoid *Pseudomonas aeruginosa*, *Antimicrob. Agents Chemother.*, 42, 974–977.

157. Yasuda, H., Ajiki, Y., Koga, T., and Yokota, T. (1994), Interaction between clarithromycin and biofilms formed by *Staphylococcus epidermidis*, *Antimicrob. Agents Chemother.*, 38, 138–141.

158. Perez-Giraldo, C., Cruz-Villalon, G., Sanchez-Silos, R., Martínez-Rubio, R., Blanco, M.T., and Gómez-García, A.C. (2003), *In vitro* activity of allicin against *Staphylococcus epidermidis* and influence of subinhibitory concentrations on biofilm formation, *J. Appl. Microbiol.*, 95, 709–711.

159. Shuford, J., Steckelberg, J., and Patel, R. (2005), Effects of fresh garlic extract on *Candida albicans* biofilms, *Antimicrob. Agents Chemother.*, 49, 473,

160. Gagnon, R.F., Richards, G.K., and Kostiner, G.B. (1994), Time-kill efficacy of antibiotics in combination with rifampin against *Staphylococcus epidermidis* biofilms, *Adv. Perit. Dial.*, 10, 189–192.

161. Ghani, M. and Soothill, J.S. (1997), Ceftazidime, gentamicin, and rifampicin, in combination, kill biofilms of mucoid *Pseudomonas aeruginosa*, *Can. J. Microbiol.*, 43, 999–1004.

162. Raad, I., Darouiche, R., Hachem, R., Sacilowski, M., and Bodey, G.P. (1995), Antibiotics and prevention of microbial colonization of catheters, *Antimicrob. Agents Chemother.*, 39, 2397–2400.

163. Zheng, Z. and Stewart, P.S. (2002), Penetration of rifampin through *Staphylococcus epidermidis* biofilms, *Antimicrob. Agents Chemother.*, 46, 900–903.

164. Pascual, A., Garcia, I., Ramirez de Arellano, E., and Perea, E.J. (1995), Activity of sparfloxacin on *Staphylococcus epidermidis* attached to plastic catheters, *J. Antimicrob. Chemother.*, 36, 425–430.

165. Pascual, A., Ramirez de Arellano, E., and Perea, E.J. (1994), Activity of glycopeptides in combination with amikacin or rifampin against *Staphylococcus epidermidis* biofilms on plastic catheters, *Eur. J. Clin. Microbiol. Infect. Dis.*, 13, 515–517.

166. Peck, K.R., Kim, S.W., Jung, S.I., Kim, Y.S., Oh, W.S., Lee, J.Y., Jin, J.H., Kim, S., Song, J.H., and Kobayashi, H. (2003), Antimicrobials as potential adjunctive agents in the treatment of biofilm infection with *Staphylococcus epidermidis*, *Chemotherapy*, 49, 189–193.

167. Dewhurst, E., Rawson, D.M., and Steele, G.C. (1986), The use of a model system to compare the efficiency of ultrasound and agitation in the recovery of *Bacillus subtilis* spores from polymer surfaces, *J. Appl. Bacteriol.*, 61, 357–363.

168. Carmen, J.C., Roeder, B.L., Nelson, J.L., Beckstead, B.L., Runyan, C.M., Schaalje, G.B., Robison, R.A., and Pitt, W.G. (2004), Ultrasonically enhanced vancomycin activity against *Staphylococcus epidermidis* biofilms *in vivo*, *J. Biomater. Appl.*, 18, 237–245.

169. Rediske, A.M., Roeder, B.L., Brown, M.K., Nelson, J.L., Robison, R.L., Draper, D.O., Schaalje, G.B., Robison, R.A., and Pitt, W.G. (1999), Ultrasonic enhancement of antibiotic action on *Escherichia coli* biofilms: An *in vivo* model, *Antimicrob. Agents Chemother.*, 43, 1211–1214.

170. Rediske, A.M., Roeder, B.L., Nelson, J.L., Robison, R.L., Schaalje, G.B., Robison, R.A., and Pitt, W.G. (2000), Pulsed ultrasound enhances the killing of *Escherichia coli* biofilms by aminoglycoside antibiotics *in vivo*, *Antimicrob. Agents Chemother.*, 44, 771–772.

171. Pitt, W.G. and Ross, S.A. (2003), Ultrasound increases the rate of bacterial cell growth, *Biotechnol. Prog.*, 19, 1038–1044.

172. del Pozo, J.L., Rouse, M.S., and Patel, R. (2008), Bioelectric effect and bacterial biofilms. A systematic review, *Int. J. Artif. Organs*, 31, 786–795.

173. Rijnaarts, H.M., Norde, W., Bouwer, E.J., Lyklema, J., and Zehnder, A.J.B. (1993), Bacterial adhesion under static and dynamic conditions, *Appl. Environ. Microbiol.*, 59, 3255–3265.

174. Jucker, B.A., Harms, H., and Zehnder, A.J. (1996), Adhesion of the positively charged bacterium *Stenotrophomonas* (*Xanthomonas*) *maltophilia* 70401 to glass and Teflon, *J. Bacteriol.*, 178, 5472–5479.

175. Poortinga, A.T., Bos, R., and Busscher, H.J. (2000), Controlled electrophoretic deposition of bacteria to surfaces for the design of biofilms, *Biotechnol. Bioeng.*, 67, 117–120.

176. Ueshima, M., Tanaka, S., Nakamura, S., and Yamashita, K. (2002), Manipulation of bacterial adhesion and proliferation by surface charges of electrically polarized hydroxyapatite, *J. Biomed. Mater. Res.*, 60, 578–584.

177. van der Borden, A.J., van der Mei, H.C., and Busscher, H.J. (2005), Electric block current induced detachment from surgical stainless steel and decreased viability of *Staphylococcus epidermidis*, *Biomaterials*, 26, 6731–6735.

178. Liu, W.K., Brown, M.R., and Elliott, T.S. (1997), Mechanisms of the bactericidal activity of low amperage electric current (DC), *J. Antimicrob. Chemother.*, 39, 687–695.

179. Pareilleux, A. and Sicard, N. (1970), Lethal effects of electric current on *Escherichia coli*, *Appl. Microbiol.*, 19, 421–424.

180. Barranco, S.D., Spadaro, J.A., Berger, T.J., and Becker, R.O. (1974), *In vitro* effect of weak direct current on *Staphylococcus aureus*, *Clin. Orthop. Relat. Res.*, 100, 250–255.
181. Bolton, L., Foleno, B., Means, B., and Petrucelli, S. (1980), Direct-current bactericidal effect on intact skin, *Antimicrob. Agents Chemother.*, 18, 137–141.
182. Davis, C.P., Wagle, N., Anderson, M.D., and Warren, M.M. (1991), Bacterial and fungal killing by iontophoresis with long-lived electrodes, *Antimicrob. Agents Chemother.*, 35, 2131–2134.
183. Matsunaga, T., Nakasono, S., and Masuda, S. (1992), Electrochemical sterilization of bacteria absorbed on granular activated carbon, *FEMS Microbiol. Lett.*, 72, 255–259.
184. Matsunaga, T., Nakasono, S., Takamuku, T., Burgess, J.G., Nakamura, N., and Sode, K. (1992b), Disinfection of drinking water by using a novel electrochemical reactor employing carbon-cloth electrodes, *Appl. Environ. Microbiol.*, 58, 686–689.
185. Rajnicek, A.M. (1993), Bacterial galvanotropism: mechanisms and applications, *Sci. Prog.*, 77(Pt 1–2), 139–151.
186. Shimada, K. and Shimahara, K. (1985), Leakage of cellular contents and morphological changes in resting *Escherichia coli* B cells exposed to an alternating changes. *Agric. Biol. Chem.*, 49, 3605–3607.
187. del Pozo, J.L., Rouse, M.S., Mandrekar, J.N., Steckelberg, J.M., and Patel, R. (2009), The electricidal effect: reduction of Staphylococcus and pseudomonas biofilms by prolonged exposure to low-intensity electrical current, *Antimicrob. Agents Chemother.*, 53, 41–45.
188. Raad, I., Hachem, R., Zermeno, A., Stephens, L.C., and Bodey, G.P. (1996), Silver iontophoretic catheter: a prototype of a long-term antiinfective vascular access device, *J. Infect. Dis.*, 173, 495–498.
189. Costerton, J.W., Ellis, B., Lam, K., Johnson, F., and Khoury, A.E. (1994), Mechanism of electrical enhancement of efficacy of antibiotics in killing biofilm bacteria, *Antimicrob. Agents Chemother.*, 38, 2803–2809.
190. Stewart, P.S., Wattanakaroon, W., Goodrum, L., Fortun, S.M., and McLeod, B.R. (1999), Electrolytic generation of oxygen partially explains electrical enhancement of tobramycin efficacy against *Pseudomonas aeruginosa* biofilm, *Antimicrob. Agents Chemother.*, 43, 292–296.
191. Pickering, S.A.W., Bayston, R., and Scanlin, T.F. (2003), Electromagnetic augmentation of antibiotic efficacy in infection of orthopaedic implants, *J. Bone Jt. Surg.*, BR 85B, 588–593.
192. del Pozo, J.L., Rouse, M.S., Mandrekar, J.N., Sampedro, M.F., Steckelberg, J.M., and Patel, R. (2009), Effect of electrical current on the activities of antimicrobial agents against *Pseudomonas aeruginosa*, *Staphylococcus aureus*, and *Staphylococcus epidermidis* biofilms, *Antimicrob. Agents Chemother.*, 53, 35–40.
193. Sersa, G. and Miclavcic, D. (1990), Inhibition of SA-1 tumor growth in mice by human leukocyte interferon alpha combined with low-level direct current, *Mol. Biother.*, 2, 165–168.
194. Belehradek, J., Jr, Orlowski, S., Ramirez, L.H., Pron, G., Poddevin, B., and Mir, L.M. (1994), Electropermeabilization of cells in tissues assessed by the qualitative and quantitative electroloading of bleomycin, *Biochim Biophys Acta.*, 1190, 155–163.

195. Wattanakaroon, W. and Stewart, P.S. (2000), Electrical enhancement of *Streptococcus gordonii* biofilm killing by gentamicin, *Arch. Oral. Biol.*, 45, 167–171.

196. Caubet, R., Pedarros-Caubet, F., Chu, M., Freye, E., de Belém Rodrigues, M., Moreau, J.M., and Ellison, W.J. (2004), A radio frequency electric current enhances antibiotic efficacy against bacterial biofilms, *Antimicrob. Agents Chemother.*, 48, 4662–4664.

197. Khoury, A.E., Lam, K., Ellis, B., and Costerton, J.W. (1992), Prevention and control of bacterial infections associated with medical devices, *ASAIO J.*, 38, M174–M178.

198. del Pozo, J.L., Rouse, M.S., Nguyen, G., Steckelberg, J.M., and Patel, R. (2007), The effect of electrical current on antimicrobial activity against methicillin resistant *Staphylococcus aureus* (MRSA) biofilms, 47th Annual Interscience Conference on Antimicrobial Agents and Chemotherapy (ICAAC), September, Chicago IL.

199. del Pozo, J.L., Rouse, M.S., Fernandez-Sampedro, M., Steckelberg, J.M., and Patel, R. (2007), Electrode composition and electrical enhancement of rifampin activity against methicillin resistant *Staphylococcus aureus* (MRSA) biofilms, 47th Annual Interscience Conference on Antimicrobial Agents and Chemotherapy (ICAAC), September, Chicago IL.

200. Wainwright, M. and Crossley, K.B. (2004), Photosensitising agents—circumventing resistance and breaking down biofilms: a review, *Int. Biodeterior. Biodegrad.*, 53, 119–126.

201. Soukos, N.S., Socransky, S.S., Mulholland, S.E., Lee, S., and Doukas, A.G. (2000), Photomechanical drug delivery into bacterial biofilms, *Pharm. Res.*, 17, 405–409.

202. O'Neill, J.F., Hope, C.K., and Wilson, M. (2002), Oral bacteria in multispecies biofilms can be killed by red light in the presence of toluidine blue, *Lasers Surg. Med.*, 31, 86–90.

203. Gad, F., Zahra, T., Hasan, T., and Hamblin, M.R. (2004), Effects of growth phase and extracellular slime on photodynamic inactivation of gram-positive pathogenic bacteria, *Antimicrob. Agents Chemother.*, 48, 2173–2178.

204. Raad, I., Buzaid, A., Rhyne, J., Hachem, R., Darouiche, R., Safar, H., Albitar, M., and Sherertz, R.J. (1997), Minocycline and ethylenediaminetetraacetate for the prevention of recurrent vascular catheter infections, *Clin. Infect. Dis.*, 25, 149–151.

205. Root, J.L., McIntyre, O.R., Jacobs, N.J., and Daghlian, C.P. (1988), Inhibitory effect of disodium EDTA upon the growth of Staphylococcus epidermidis *in vitro*: relation to infection prophylaxis of Hickman catheters, *Antimicrob. Agents Chemother.*, 32, 1627–1631.

206. Turakbia, M.H., Cooksey, K.E., and Characklis, W.G. (1983), Influence of a calcium-specific chelant on biofilm removal, *Appl. Environ. Microbiol.*, 46, 1236–1238.

207. Percival, S.L., Kite, P., Eastwood, K., Murga, R., Carr, J., Arduino, M.J., and Donlan, R.M. (2005), Tetrasodium EDTA as a novel central venous catheter lock solution against biofilm, *Infect. Control Hosp. Epidemiol.*, 26, 515–519.

208. Kite, P., Eastwood, K., Sugden, S., and Percival, S.L. (2004), Use of *in vivo*-generated biofilms from hemodialysis catheters to test the efficacy of a novel

antimicrobial catheter lock for biofilm eradication *in vitro*, *J. Clin. Microbiol.*, 42, 3073–3076.

209. Raad, I., Chatzinikolaou, I., Chaiban, G., Hanna, H., Hachem, R., Dvorak, T., Cook, G., and Costerton, W. (2003), *In vitro* and *ex vivo* activities of minocycline and EDTA against microorganisms embedded in biofilm on catheter surfaces, *Antimicrob. Agents Chemother.*, 47, 3580–3585.
210. Torres-Viera, C., Thauvin-Eliopoulos, C., Souli, M., DeGirolami, P., Farris, M.G., Wennersten, C.B., Sofia, R.D., and Eliopoulos, G.M. (2000), Activities of taurolidine in vitro and in experimental enterococcal endocarditis, *Antimicrob. Agents Chemother.*, 44, 1720–1724.
211. Metcalf, S.C.L., Chambers, S.T., and Pithie, A.D. (2004), Use of ethanol locks to prevent recurrent central line sepsis, *J. Infect.*, 49, 20–22.
212. Donlan, R.M. (2009), Preventing biofilms of clinically relevant organisms using bacteriophage, *Trends Microbiol.*, 17, 66–72.
213. Rohwer, F. (2003), Global phage diversity, *Cell*, 113, 141–141.
214. Richards, J.J. and Melander, C. (2009), Controlling bacterial biofilms, *Chembiochem.*, 10, 2287–2294.
215. Sulakvelidze, A., Alavidze, Z., and Morris, J.G. (2001), Bacteriophage therapy, *Antimicrob. Agents Chemother.*, 45, 649–659.
216. Deresinski, S. (2009), Bacteriophage therapy: smaller fleas, *Clin. Infect. Dis.*, 48, 1096–1101.
217. Adams, M.H. and Park, B.H. (1956), An enzyme produced by a phage-host cell system. II. The properties of the polysaccharide depolymerase, *Virology*, 2, 719–736.
218. Kimura, K. and Itoh, Y. (2003), Characterization of poly-gamma-glutamate hydrolase encoded by a bacteriophage genome: possible role in phage infection of Bacillus subtilis encapsulated with poly-gamma-glutamate, *Appl. Environ. Microbiol.*, 69, 2491–2497.
219. Sutherland, I.W. (1967), Phage-induced fucosidases hydrolysing the exopolysaccharide of Klebsiella arogenes type 54 [A3(S1)], *Biochem. J.*, 104, 278–285.
220. Doolittle, M.M., Cooney, J.J., and Caldwell, D.E. (1995), Lytic infection of *Escherichia coli* biofilms by bacteriophage T4, *Can. J. Microbiol.*, 41, 12–18.
221. Doolittle, M.M., Cooney, J.J., and Caldwell, D.E. (1996), Tracing the interaction of bacteriophage with bacterial biofilms using fluorescent and chromogenic probes, *J. Ind. Microbiol.*, 16, 331–341.
222. Hanlon, G.W., Denyer, S.P., Olliff, C.J., and Ibrahim, L.J. (2001), Reduction in exopolysaccharide viscosity as an aid to bacteriophage penetration through *Pseudomonas aeruginosa* biofilms, *Appl. Environ. Microbiol.*, 67, 2746–2753.
223. Deveau, H., Van Calsteren, M.R., and Moineau, S. (2002), Effect of exopolysaccharides on phage-host interactions in *Lactococcus lactis*, *Appl. Environ. Microbiol.*, 68, 4364–4369.
224. Lu, T.K. and Collins, J.J. (2007), Dispersing biofilms with engineered enzymatic bacteriophage, *Proc. Natl. Acad. Sci. USA*, 104, 11197–11202.
225. Itoh, Y., Wang, X., Hinnebusch, B.J., Preston, J.F., and Romeo, T. (2005), Depolymerization of beta-1,6-N-acetyl-D-glucosamine disrupts the integrity of diverse bacterial biofilms, *J. Bacteriol.*, 187, 382–387.

226. Whitchurch, C.B., Tolker-Nielsen, T., Ragas, P.C., and Mattick, J.S. (2002), Extracellular DNA required for bacterial biofilm formation, *Science*, 295, 1487–1487.

227. Lu, T.K. and Collins, J.J. (2009), Engineered bacteriophage targeting gene networks as adjuvants for antibiotic therapy, *Proc. Natl. Acad. Sci. USA*, 106, 4629–4634.

228. Curtin, J.J. and Donlan, R.M. (2006), Using bacteriophages to reduce formation of catheter-associated biofilms by Staphylococcus epidermidis, *Antimicrob. Agents Chemother.*, 50, 1268–1275.

229. Azeredo, J. and Sutherland, I.W. (2008), The use of phages for the removal of infectious biofilms, *Curr. Pharm. Biotechnol.*, 9, 261–266.

230. Jerne, N.K. and Avegno, P. (1956), The development of the phage-inactivating properties of serum during the course of specific immunization of an animal: reversible and irreversible inactivation, *J. Immunol.*, 76, 200–208.

231. Sutherland, I.W., Hughes, K.A., Skillman, L.C., and Tait, K. (2004), The interaction of phage and biofilms, *FEMS Microbiol. Lett.*, 232, 1–6.

232. Hughes, K.A., Sutherland, I.W., and Jones, M.V. (1998), Biofilm susceptibility to bacteriophage attack: the role of phage-borne polysaccharide depolymerase, *Microbiol.*, 144, 3039–3047.

233. Sillankorva, S., Oliviera, R., Vieira, M.J., Sutherland, I.W., and Azeredo, J. (2004), Bacteriophage Phi S1 infection of *Pseudomonas fluorescens* planktonic cells versus biofilms, *Biofouling*, 20, 133–138.

234. Fuqua, W.C., Winans, S.C., and Greenberg, E.P. (1994), Quorum sensing in bacteria: the LuxR-LuxI family of cell density-responsive transcriptional regulators, *J. Bacteriol.*, 176, 269–275.

235. Engebrecht, J. and Silverman, M. (1984), Identification of genes and gene products necessary for bacterial bioluminescence, *Proc. Natl. Acad. Sci. USA.*, 81, 4154–4158.

236. Eberl, L. (1999), N-acyl homoserinelactone-mediated gene regulation in gram-negative bacteria, *Syst. Appl. Microbiol.*, 22, 493–506.

237. Salmond, G.P., Bycroft, B.W., Stewart, G.S., and Williams, P. (1995), The bacterial "enigma": cracking the code of cell-cell communication, *Mol. Microbiol.*, 16, 615–624.

238. Parsek, M.R., Val, D.L., Hanzelka, B.L., Cronan, J.E., and Greenberg, E.P. (1999), Acyl homoserine-lactone quorum-sensing signal generation, *Proc. Natl. Acad. Sci. USA.*, 96, 4360–4365.

239. Hentzer, M., Wu, H., Andersen, J.B., Riedel, K., Rasmussen, T.B., Bagge, N., Kumar, N., Schembri, M.A., Song, Z., Kristoffersen, P., Manefield, M., Costerton, J.W., Molin, S., Eberl, L., Steinberg, P., Kjelleberg, S., Hoiby, N., and Givskov, M. (2003), Attenuation of *Pseudomonas aeruginosa* virulence by quorum sensing inhibitors, *EMBO J.*, 22, 3803–3815.

240. Schuster, M., Urbanowski, M.L., and Greenberg, E.P. (2004), Promoter specificity in *Pseudomonas aeruginosa* quorum sensing revealed by DNA binding of purified LasR, *Proc. Natl. Acad. Sci. USA*, 101, 15833–15839.

241. Wagner, V.E., Bushnell, D., Passador, L., Brooks, A.I., and Iglewski, B.H. (2003), Microarray analysis of *Pseudomonas aeruginosa* quorum-sensing regulons: effects of growth phase and environment, *J. Bacteriol.*, 185, 2080–2095.

242. Rasmussen, T.B. and Givskov, M. (2006), Quorum-sensing inhibitors as anti-pathogenic drugs, *Int. J. Med. Microbiol.*, 296, 149–161.

243. Rasmussen, T.B. and Givskov, M. (2006), Quorum sensing inhibitors: a bargain of effects, *Microbiology*, 152, 895–904.

244. Yates, E.A., Philipp, B., Buckley, C., Atkinson, S., Chhabra, S.R., Sockett, R.E., Goldner, M., Dessaux, Y., Cámara, M., Smith, H., and Williams, P. (2002), N-Acylhomoserine lactones undergo lactonolysis in a pH-, temperature-, and acyl chain length-dependent manner during growth of *Yersinia pseudotuberculosis* and *Pseudomonas aeruginosa*, *Infect. Immun.*, 70, 5635–5646.

245. Byers, J.T., Lucas, C., Salmond, G.P., and Welch, M. (2002), Nonenzymatic turnover of an *Erwinia carotovora* quorum-sensing signaling molecule, *J. Bacteriol.*, 184, 1163–1171.

246. Dong, Y.H., Xu, J.L., Li, X.Z., and Zhang, L.H. (2000), AiiA, an enzyme that inactivates the acylhomoserine lactone quorum-sensing signal and attenuates the virulence of *Erwinia carotovora*, *Proc. Natl. Acad. Sci. USA.*, 97, 3526–3531.

247. Dong, Y.H., Wang, L.H., Xu, J.L., Zhang, H.B., Zhang, X.F., and Zhang, L.H. (2001), Quenching quorum-sensing-dependent bacterial infection by an N-acyl homoserine lactonase, *Nature (London)*, 411, 813–817.

248. Wang, L.H., Weng, L.X., Dong, Y.H., and Zhang, L.H. (2004), Specificity and enzyme kinetics of the quorum-quenching N-acyl homoserine lactone lactonase (AHL-lactonase), *J. Biol. Chem.*, 279, 13645–13651.

249. Lee, S.J., Park, S.Y., Lee, J.J., Yum, D.Y., Koo, B.T., and Lee, J.K. (2002), Genes encoding the N-acyl homoserine lactone-degrading enzyme are widespread in many subspecies of *Bacillus thuringiensis*, *Appl. Environ. Microbiol.*, 68, 3919–3924.

250. Dong, Y.H., Zhang, X.F., Xu, J.L., and Zhang, L.H. (2004), Insecticidal *Bacillus thuringiensis* silences *Erwinia carotovora* virulence by a new form of microbial antagonism, signal interference, *Appl. Environ. Microbiol.*, 70, 954–960.

251. Molina, L., Constantinescu, F., Michel, L., Reimmann, C., Duffy, B., and Defago, G. (2003), Degradation of pathogen quorum-sensing molecules by soil bacteria: a preventive and curative biological control mechanism, *FEMS Microbiol. Ecol.*, 45, 71–81.

252. Uroz, S., D'Angelo-Picard, C., Carlier, A., Elasri, M., Sicot, C., Petit, A., Oger, P., Faure, D., and Dessaux, Y. (2003), Novel bacteria degrading *N*-acylhomoserine lactones and their use as quenchers of quorum-sensing-regulated functions of plant-pathogenic bacteria, *Microbiology*, 149, 1981–1989.

253. Carlier, A., Uroz, S., Smadja, B., Fray, R., Latour, X., Dessaux, Y., and Faure, D. (2003), The Ti plasmid of Agrobacterium tumefaciens harbors an attM-paralogous gene, aiiB, also encoding N-acyl homoserine lactonase activity, *Appl. Environ. Microbiol.*, 69, 4989–4993.

254. Park, S.Y., Lee, S.J., Oh, T.K., Oh, J.W., Koo, B.T., Yum, D.Y., and Lee, J.K. (2003), AhlD, an *N*-acylhomoserine lactonase in *Arthrobacter* sp., and predicted homologues in other bacteria, *Microbiology*, 149, 1541–1550.

255. Huang, J.J., Han, J.I., Zhang, L.H., and Leadbetter, J.R. (2003), Utilization of acyl-homoserine lactone quorum signals for growth by a soil pseudomonad and *Pseudomonas aeruginosa* PAO1, *Appl. Environ. Microbiol.*, 69, 5941–5949.

256. Camara, M., Williams, P., and Hardman, A. (2002), Controlling infection by tuning in and turning down the volume of bacterial small-talk, *Lancet Infect. Dis.*, 2, 667–676.

257. Borchardt, S.A., Allain, E.J., Michels, J.J., Stearns, G.W., Kelly, R.F., and McCoy, W.F. (2001), Reaction of acylated homoserine lactone bacterial signaling molecules with oxidized halogen antimicrobials, *Appl. Environ. Microbiol.*, 67, 3174–3179.

258. Leadbetter, J.R. and Greenberg, E.P. (2000), Metabolism of acylhomoserine lactone quorum-sensing signals by *Variovorax paradoxus*, *J. Bacteriol.*, 182, 6921–6926.

259. Chun, C.K., Ozer, E.A., Welsh, M.J., Zabner, J., and Greenberg, E.P. (2004), Inactivation of a *Pseudomonas aeruginosa* quorum-sensing signal by human airway epithelia, *Proc. Natl. Acad. Sci. USA*, 101, 3587–3590.

260. Hastings, J.W. (2004), Bacterial quorum-sensing signals are inactivated by mammalian cells, *Proc. Natl. Acad. Sci. USA*, 101, 3993–3994.

261. Steinberg, P.D., Schneider, R., and Kjelleberg, S. (1997), Chemical defenses of seaweeds against microbial colonization, *Biodegradation*, 8, 211–220.

262. Maximilien, R., de Nys, R., Holmstrom, C., Gram, L., Givskov, M., Kjelleberg, S., and Steinberg, P. (1998), Chemical mediation of bacterial surface colonisation by secondary metabolites from the red alga *Delisea pulchra*, *Aquat. Microb. Ecol.*, 15, 233–246.

263. Rasmussen, T.B., Manefield, M., Andersen, J.B., Eberl, L., Anthoni, U., Christophersen, C., Steinberg, P., Kjelleberg, S., and Givskov, M. (2000), How *Delisea pulchra* furanones affect quorum sensing and swarming motility in *Serratia liquefaciens* MG1, *Microbiology*, 146, 3237–3244.

264. Hentzer, M., Riedel, K., Rasmussen, T.B., Heydorn, A., Andersen, J.B., Parsek, M.R., Rice, S.A., Eberl, L., Molin, S., Hoiby, N., Kjelleberg, S., and Givskov, M. (2002), Inhibition of quorum sensing in *Pseudomonas aeruginosa* biofilm bacteria by a halogenated furanone compound, *Microbiology*, 148, 87–102.

265. Manefield, M., Harris, L., Rice, S.A., de Nys, R., and Kjelleberg, S. (2000), Inhibition of luminescence and virulence in the black tiger prawn (*Penaeus monodon*) pathogen *Vibrio harveyi* by intercellular signal antagonists, *Appl. Environ. Microbiol.*, 66, 2079–2084.

266. Ren, D., Sims, J.J., and Wood, T.K. (2002), Inhibition of biofilm formation and swarming of *Bacillus subtilis* by (5*Z*)-4-bromo-5-(bromomethylene)-3-butyl-2(5*H*)-furanone, *Lett. Appl. Microbiol.*, 34, 293–299.

267. Ren, D., Bedzyk, L.A., Setlow, P., England, D.F., Kjelleberg, S., Thomas, S.M., Ye, R.W., and Wood, T.K. (2004), Differential gene expression to investigate the effect of (5*Z*)-4-bromo-5-(bromomethylene)-3-butyl-2(5*H*)-furanone on *Bacillus subtilis*, *Appl. Environ. Microbiol.*, 70, 4941–4949.

268. Rasmussen, T.B., Skindersoe, M.E., Bjarnsholt, T., Christensen, K.B., Andersen, J.B., Ostenfeld-Larsen, T., Hentzer, M., and Givskov, M. (2005), Idendity and effects of quorum sensing inhibitors produced by *Penicillum* species, *Microbiology*, 151, 1325–1340.

269. Balaban, N., Giacometti, A., Cirioni, O., Gov, Y., Ghiselli, R., Mocchegiani, F., Viticchi, C., Del Prete, M.S., Saba, V., Scalise, G., and Dell'Acqua, G. (2003), Use of the quorum-sensing inhibitor RNAIII-inhibiting peptide to prevent biofilm

formation *in vivo* by drug-resistant *Staphylococcus epidermidis*, *J. Infect. Dis.*, 187, 625–630.

270. Balaban, N., Gov, Y., Bitler, A., and Boelaert, J.R. (2003), Prevention of *Staphylococcus aureus* biofilm on dialysis catheters and adherence to human cells, *Kidney Int.*, 63, 340–345.

271. Balaban, N., Stoodley, P., Fux, C.A., Wilson, S., Costerton, J.W., and Dell'Acqua, G. (2005), Prevention of staphylococcal biofilm-associated infections by the quorum sensing inhibitor RIP, *Clin. Orthop. Relat. Res.*, 437, 48–54.

272. Bryers, J.D., Jarvis, R.A., Lebo, J., Prudencio, A., Kyriakides, T.R., and Uhrich, K. (2006), Biodegradation of poly(anhydride-esters) into non-steroidal anti-inflammatory drugs and their effect on *Pseudomonas aeruginosa* biofilms *in vitro* and on the foreign-body response *in vivo*, *Biomaterials*, 27, 5039–5048.

273. Rosenberg, L.E., Carbone, A.L., Römling, U., Uhrich, K.E., and Chikindas, M.L. (2008), Salicylic acid-based poly(anhydride esters) for control of biofilm formation in *Salmonella enterica serovar Typhimurium*, *Lett. Appl. Microbiol.*, 46, 593–599.

274. Brackman, G., Defoirdt, T., Miyamoto, C., Bossier, P., Van Calenbergh, S., Nelis, H., and Coenye, T. (2008), Cinnamaldehyde and cinnamaldehyde derivatives reduce virulence in Vibrio spp. by decreasing the DNA-binding activity of the quorum sensing response regulator LuxR, *BMC Microbiol.*, 8, 149.

275. Bjarnsholt, T., Jensen, P.O., Rasmussen, T.B., Christophersen, L., Calum, H., Hentzer, M., Hougen, H.P., Rygaard, J., Moser, C., Eberl, L., Hoiby, N., and Givskov, M. (2005), Garlic blocks quorum sensing and promotes rapid clearing of pulmonary *Pseudomonas aeruginosa* infections, *Microbiology*, 151(Pt. 12), 3873–3880.

276. Duarte, S., Gregoire, S., Singh, A.P., Vorsa, N., Schaich, K., Bowen, W.H., and Koo, H. (2006), Inhibitory effects of cranberry polyphenols on formation and acidogenicity of *Streptococcus mutans* biofilms, *FEMS Microbiol. Lett.*, 257, 50–56.

277. Labrecque, J., Bodet, C., Chandad, F., and Grenier, D. (2006), Effects of a high-molecular-weight cranberry fraction on growth, biofilm formation and adherence of *Porphyromonas gingivalis*, *J. Antimicrob. Chemother.*, 58, 439–443.

278. Yamanaka, A., Kimizuka, R., Kato, T., and Okuda, K. (2004), Inhibitory effects of cranberry juice on attachment of oral streptococci and biofilm formation, *Oral Microbiol. Immunol.*, 19, 150–154.

279. Yamanaka, A., Kouchi, T., Kasai, K., Kato, T., Ishihara, K., and Okuda, K. (2007), Inhibitory effect of cranberry polyphenol on biofilm formation and cysteine proteases of *Porphyromonas gingivalis*, *J. Periodontal Res.*, 42, 589–592.

280. Rasmussen, T.B., Bjarnsholt, T., Skindersoe, M.E., Hentzer, M., Kristoffersen, P., Kote, M., Eberl, L., Nielsen, J., and Givskov, M. (2005), Screening for quorum sensing inhibitors using a novel genetic system-the QSI selector, *J. Bacteriol.*, 187, 1799–1814.

281. Teplitski, M., Robinson, J.B., and Bauer, W.D. (2000), Plants secrete substances that mimic bacterial N-acyl homoserine lactone signal activities and affect population density-dependent behaviors in associated bacteria, *Mol. Plant. Microbe. Interact.*, 13, 637–648.

282. Persson, T., Hansen, T.H., Rasmussen, T.B., Skinderso, M.E., Givskov, M., and Nielsen, J. (2005), Rational design and synthesis of new quorum-sensing inhibitors

derived from acylated homoserine lactones and natural products from garlic, *Org. Biomol. Chem.*, 3, 253–262.

283. Rice, S.A., McDougald, D., Kumar, N., and Kjelleberg, S. (2005), The use of quorum-sensing blockers as therapeutic agents for the control of biofilm-associated infections, *Curr. Opin. Investig. Drugs*, 6, 178–184.
284. Jabra-Rizk, M.A., Meiller, T.F., James, C.E., and Shirtliff, M.E. (2006), Effect of farnesol on *Staphylococcus aureus* biofilm formation and antimicrobial susceptibility, *Antimicrob. Agents Chemother.*, 50, 1463–1469.
285. Ren, D., Zuo, R., Gonzalez Barrios, A.F., Bedzyk, L.A., Eldridge, G.R., Pasmore, M.E., and Wood, T.K. (2005), Differential gene expression for investigation of *Escherichia coli* biofilm inhibition by plant extract ursolic acid, *Appl. Environ. Microbiol.*, 71, 4022–4034.
286. Wozniak, D.J. and Keyser, R. (2004), Effects of subinhibitory concentrations of macrolide antibiotics on *Pseudomonas aeruginosa*, *Chest*, 125, 62S–69S, quiz 69S.
287. Chen, X. and Stewart, P.S. (2000), Biofilm removal caused by chemical treatments, *Water Res.*, 34, 4229–4233.
288. Sauer, K., Camper, A.K., Ehrlich, G.D., Costerton, J.W., and Davies, D.G. (2002), *Pseudomonas aeruginosa* displays multiple phenotypes during development as a biofilm, *J. Bacteriol.*, 184, 1140–1154.
289. Thormann, K.M., Saville, R.M., Shukla, S., and Spormann, A.M. (2005), Induction of rapid detachment in *Shewanella oneidensis* MR-1 biofilms, *J. Bacteriol.*, 187, 1014–1021.
290. Gjermansen, M., Ragas, P., Sternberg, C., Molin, S., and Tolker-Nielsen, T. (2005), Characterization of starvation-induced dispersion in *Pseudomonas putida* biofilms, *Environ. Microbiol.*, 7, 894–993.
291. Webb, J.S., Thompson, L.S., James, S., Charlton, T., Tolker-Nielsen, T., Koch, B., Givskov, M., and Kjelleberg, S. (2003), Cell death in *Pseudomonas aeruginosa* biofilm development, *J. Bacteriol.*, 185, 4585–4592.
292. Barraud, N., Hassett, D.J., Hwang, S-H., Rice, S.A., Kjelleberg, S., and Webb, J.S. (2006), Involvement of nitric oxide in biofilm dispersal of *Pseudomonas aeruginosa*, *J. Bacteriol.*, 188, 7344–7353.
293. Banin, E., Brady, K.M., and Greenberg, E.P. (2006), Chelator-induced dispersal and killing of *Pseudomonas aeruginosa* cells in a biofilm, *Appl. Environ. Microbiol.*, 72, 2064–2069.
294. Lee, J., Jayaraman, A., and Wood, T.K. (2007), Indole is an inter-species biofilm signal mediated by SdiA, *BMC Microbiol.*, 7, 1–15, 42.
295. Lee, J., Bansal, T., Jayaraman, A., Bentley, W.E., and Wood, T.K. (2007), Enterohemorrhagic *Escherichia coli* biofilms are inhibited by 7-hydrozyindole and stimulated by isatin, *Appl. Environ. Microbiol.*, 73, 4100–4109.
296. Rice, S.A., Koh, K.S., Queck, S.Y., Labbate, M., Lam, K.W., and Kjelleberg, S. (2005), Biofilm formation and sloughing in *Serratia marcescens* are controlled by quorum sensing and nutrient cues, *J. Bacteriol.*, 187, 3477–3485.
297. Dow, J.M., Crossman, L., Findlay, K., He, Y.-Q., Feng, J.,-X., and Tang, J.-L. (2003), Biofilm dispersal in *Xanthomonas campestris* is controlled by cell-cell signaling and is required for full virulence to plants, *Proc. Natl. Acad. Sci. USA*, 100, 10995–11000.

298. Davies, D.G. and Marques, C.N., (2009), A fatty acid messenger is responsible for inducing dispersion in microbial biofilms, *J Bacteriol.*, 191, 1393–1403.

299. Alkawash, M.A., Soothill, J.S., and Schiller, N.L. (2006), Alginate lyase enhances antibiotic killing of mucoid Pseudomonas aeruginosa in biofilms, *APMIS.*, 131–138.

300. Tanihara, M., Suzuki, Y., Nishimura, Y., Suzuki, K., Kakimaru, Y., and Fukunisi, Y. (1999), A novel microbial infection-responsive drug release system, *J. Pharm. Sci.*, 88, 510–514.

301. Geddes, A. (2000), Infection in the twenty-first century: Predictions and postulates, *J. Antimicrob. Chemother.*, 46, 873–878.

302. Beaudoin, D., Bryers, J.D., Cunningham, A.B., and Peretti, S.W. (1998), Mobilization of broad host range plasmid from *Pseudomonas putida* to established biofilm of *Bacillus azotoformans*. I, Experiments, *Biotechnol. Bioeng.*, 57, 272–279.

303. Beaudoin, D., Bryers, J.D., Cunningham, A.B., and Peretti, S.W. (1998), Mobilization of broad host range plasmid from *Pseudomonas putida* to established biofilm of *Bacillus azotoformans*. II, Modeling, *Biotechnol. Bioeng.*, 57, 280–286.

304. Fischetti, V.A. (2005), Bacteriophage lytic enzymes: Novel anti-infectives, *Trends Microbiol.*, 13, 491–496.

305. Hong, H.A., Ducle, H., and Cutting, S.M. (2005), The use of bacterial spore formers as probiotics, *FEMS Microbiol. Rev.*, 29, 813–835.

306. Persson, G.R. (2005), Immune responses and vaccination against periodontal infections, *J. Clin. Periodontol.*, 32 (Suppl. 6), 39–53.

307. De Smet, K. and Contreras, R. (2005), Human antimicrobial peptides: Defensins, cathelicidins and histatins, *Biotechnol. Lett.*, 27, 1337–1347.

308. Casadevall, A., Dadachova, E., and Pirofski, L.A. (2004), Passive antibody therapy for infectious diseases, *Nat. Rev. Microbiol.*, 2, 695–703.

309. Martinez, L.R. and Casadevall, A. (2005), Specific antibody can prevent fungal biofilm formation and this effect correlates with protective efficacy, *Infect. Immun.*, 73, 6350–6362.

310. Cachia, P.J. and Hodges, R.S. (2003), Synthetic peptide vaccine and antibody therapeutic development: Prevention and treatment of *Pseudomonas aeruginosa*, *Biopolymers*, 71, 141–168.

311. Otto, M. (2004), Quorum-sensing control in *Staphylococci*—A target for antimicrobial drug therapy? *FEMS Microbiol. Lett.*, 241,135–141.

312. Kaneko, Y., Thoendel, M., Olakanmi, O., Britigan, B.E., and Sink, P.K. (2007), The transition metal gallium disrupts *Pseudomonas aeruginosa* iron metabolism and has antimicrobial and antibiofilm activity, *J. Clin. Invest.*, 117, 877–888.

313. Boles, B.R., Thoendel, M., and Sink, P.K. (2005), Rhamnolipids mediate detachment of *Pseudomonas aeruginosa* from biofilms, *Mol. Microbiol.*, 57, 1210–1223.

314. Bryers, J.D. (2008), Medical biofilms, *Biotechnol Bioeng.*, 100, 1–18.

315. Ankenbauer, R., Sriyosachati, S., and Cox, C.D. (1985), Effects of siderophores on the growth of *Pseudomonas aeruginosa* in human serum and transferrin, *Infect. Immun.*, 49, 132–140.

316. Barclay, R. and Ratledge, C. (1986), Participation of iron on the growth inhibition of pathogenic strains of *Mycobacterium avium* and *M. paratuberculosis* in serum, *Zentralbl. Bakteriol. Mikrobiol. Hyg. [A]*, 262, 189–194.

317. Finkelstein, R.A., Sciortino, C.V., and McIntosh, M.A. (1983), Role of iron in microbe-host interactions, *Rev. Infect. Dis.*, 5 (Suppl.4), S759–S777.

318. Jurado, R.L. (1997), Iron, infections, and anemia of inflammation, *Clin. Infect. Dis.*, 25, 888–895.

319. Narasimhan, J., Antholine, W.E., and Chitambar, C.R. (1992), Effect of gallium on the tyrosyl radical of the iron-dependent M2 subunit of ribonucleotide reductase, *Biochem. Pharmacol.*, 44, 2403–2408.

320. Singh, P.K. (2004), Iron sequestration by human lactoferrin stimulates *P. aeruginosa* surface motility and blocks biofilm formation, *Biometals*, 17, 267–270.

321. Chitambar, C.R. and Narasimhan, J. (1991), Targeting iron-dependent DNA synthesis with gallium and transferrin-gallium, *Pathobiology*, 59, 3–10.

322. Atkinson, R.A., Salah, E.l., Din, A.L., Kieffer, B., Lefevre, J.F., and Abdallah, M.A. (1998), Bacterial iron transport: 1H NMR determination of the three-dimensional structure of the gallium complex of pyoverdin G4R, the peptidic siderophore of *Pseudomonas putida* G4R, *Biochemistry*, 37, 15965–15973.

323. Hubbard, J.A., Lewandowska, K.B., Hughes, M.N., and Poole, R.K. (1986), Effects of iron-limitation of *Escherichia coli* on growth, the respiratory chains and gallium uptake, *Arch. Microbiol.*, 146, 80–86.

324. Krishnamurthy, V.M., Quinton, L.J., Estroffa, L.A., Metalloa, S.J., Issacs, J.M., Mizgerd, J.P., and Whitesides, G.M. (2006), Promotion of opsonization by antibodies and phagocytosis of Gram-positive bacteria by a bifunctional polyacrylamide, *Biomaterials*, 27, 3663–3674.

325. Kuhn, S.E., Nardin, A., Klebba, P.E., and Taylor, R.P. (1998), *Escherichia coli* bound to the primate erythrocyte complement receptor via bispecific monoclonal antibodies are transferred to and phagocytosed by human monocytes in an *in vitro* model, *J. Immunol.*, 160, 5088–5097.

326. Lindorfer, M.A., Nardin, A., Foley, P.L., Solga, M.D., Bankovich, A.J., Martin, E.N., Henderson, A.L., Price, C.W., Gyimesi, E., Wozencraft, C.P., Goldberg, J.B., Sutherland, W.M., and Taylor, R.P. (2001), Targeting of *Pseudomonas aeruginosa* in the bloodstream with bispecific monoclonal antibodies, *J. Immunol.*, 167, 2240–2249.

327. Gyimesi, E., Bankovich, A.J., Schuman, T.A., Goldberg, J.B., Lindorfer, M.A., and Taylor, R.P. (2004), *Staphylococcus aureus* bound to complement receptor 1 on human erythrocytes by bispecific monoclonal antibodies is phagocytosed by acceptor macrophages, *Immunol. Lett.*, 95, 185–192.

328. Kobayashi, T., Takauchi, A., van Spriel, A.B., Vilé, H.A., Hayakawaf, M., Abiko, Y., van de Winkel, J.G.J., and Yoshiea, H. (2004), Targeting of *Porphyromonas gingivalis* with a bispecific antibody directed to Fc$\alpha$RI (CD89) improves *in vitro* clearance by gingival crevicular neutrophils, *Vaccine*, 23, 585–594.

329. Wizemann, T.M., Adamou, J.E., and Langermann, S. (1999), Adhesins as targets for vaccine development, *Emerg. Infect. Dis.*, 5, 395–403.

330. Heikenwalder, M., Polymenidou, M., Junt, T., Sigurdson, C., Wagner, H., Akira, S., Zinkernagel, R., and Aguzzi, A. (2004), Lymphoid follicle destruction and immunosuppression after repeated CpG oligodeoxynucleotide administration, *Nat. Med.*, 10, 187–192.

331. Zhao, H., Hemmi, H., Akira, S., Cheng, S.H., Scheule, R.K., and Yew, N.S. (2004), Contribution of Toll-like receptor 9 signaling to the acute inflammatory response to nonviral vectors, *Mol. Ther.*, 9, 241–248.

332. Gilboa, E. and Vieweg, J. (2004), Cancer immunotherapy with mRNA transfected dendritic cells, *Immunol. Rev.*, 199, 251–263.

333. Dang, J.M. and Leong, K.W. (2006), Natural polymers for gene delivery and tissue engineering, *Adv. Drug Deliv. Rev.*, 58, 487–499.

334. Gao, Y., Gu, W., Chen, L., Xu, Z., and Li, Y. (2007), A multifunctional nano device as non-viral vector for gene delivery: *In vitro* characteristics and transfection, *J. Control. Release*, 118, 381–388.

335. Keegan, M.E. and Saltzman, W.M. (2006), Surface-modified biodegradable microspheres for DNA vaccine delivery, *Methods Mol. Med.*, 127, 107–113.

336. Little, S.R. and Langer, R. (2005), Nonviral delivery of cancer genetic vaccines, *Adv. Biochem. Eng. Biotechnol.*, 99, 93–118.

337. Pack, D.W., Hoffman, A.S., Pun, S., and Stayton, P.S. (2005), Design and development of polymers for gene delivery, *Nat. Rev. Drug Discov.*, 4, 581–593.

338. Tobery, T.W. and Siliciano, R.F. (1997), Targeting of HIV-1 antigens for rapid intracellular degradation enhances cytotoxic T lymphocyte (CTL) recognition and the induction of *de novo* CTL responses *in vivo* after immunization, *J. Exp. Med.*, 185, 909–920.

339. Tobery, T. and Siliciano, R.F. (1999), Cutting edge: Induction of enhanced CTL-dependent protective immunity *in vivo* by N-end rule targeting of a model tumor antigen, *J. Immunol.*, 162, 639–642.

340. Varshavsky, A., Turner, G., Du, F., and Xie, Y. (2000), Felix Hoppe-Seyler Lecture 2000. The ubiquitin system and the N-end rule pathway, *Biol. Chem.*, 381, 779–789.

341. Bonini, C., Lee, S.P., Riddell, S.R., and Greenberg, P.D. (2001), Targeting antigen in mature dendritic cells for simultaneous stimulation of CD4+ and CD8+ T cells, *J. Immunol.*, 166, 5250–5257.

342. Lin, K.Y., Guarnieri, F.G., Staveley-O'Carroll, K.F., Levitsky, H.I., August, J.T., Pardoll, D.M., and Wu, T.C. (1996), Treatment of established tumors with a novel vaccine that enhances major histocompatibility class II presentation of tumor antigen, *Cancer Res.*, 56, 21–26.

343. Thomson, S.A., Burrows, S.R., Misko, I.S., Moss, D.J., Coupar, B.E., and Khanna, R. (1998), Targeting a polyepitope protein incorporating multiple class II-restricted viral epitopes to the secretory/endocytic pathway facilitates immune recognition by CD4+ cytotoxic T lymphocytes: A novel approach to vaccine design, *J. Virol.*, 72, 2246–2252.

344. Wu, T.C., Guarnieri, F.G., Staveley-O'Carroll, K.F., Viscidi, R.P., Levitsky, H.I., Hedrick, L., Cho, K.R., August, J.T., and Pardoll, D.M. (1995), Engineering an intracellular pathway for major histocompatibility complex class II presentation of antigens, *Proc. Natl. Acad. Sci. USA*, 92, 11671–11675.

345. Bennewitz, N.L. and Babensee, J.L. (2005), The effect of the physical form of poly(lactic-*co*-glycolic acid) carriers on the humoral immune response to co-delivered antigen, *Biomaterials*, 26, 2991–2999.

346. Murga, R., Miller, J.M., and Donlan, R.M. (2001), Biofilm formation by gram-negative bacteria on central venous catheter connectors: effect of conditioning films in a laboratory model, *J. Clin. Microbiol.*, 39, 2294–2297.

347. Donlan, R.M., Piede, J.A., Heyes, C.D., Sanii, L., Murga, R., Edmonds, P., El-Sayed, I., and El-Sayed, M.A. (2004), Model system for growing and quantifying

*Streptococcus pneumoniae* biofilms in situ and in real time, *Appl. Environ. Microbiol.*, 70, 4980–4988.

348. Donlan, R.M., Murga, R., Bell, M., Toscano, C.M., Carr, J.H., Novicki, T.J., Zuckerman, C., Corey, L.C., and Miller, J.M. (2001), Protocol for the detection of biofilms on needleless connectors attached to central venous catheters, *J. Clin. Microbiol.*, 39, 750–753.
349. Duran, L.W. (2000), Preventing medical device related infections, *Med. Device Technol.*, 11, 14–17.
350. Sofou, S. (2007), Surface-active liposomes for targeted cancer therapy, *Nanomedicine*, 2, 711–724.
351. Gursoy, R.N. and Benita, S. (2004), Self-emulsifying drug delivery systems (SEDDS) for improved oral delivery of lipophilic drugs, *Biomed. Pharmacother.*, 58, 173–182.
352. Torchilin, V.P. (2007), Nanocarriers, *Pharm. Res.*, 24, 2333–2334.
353. Carmen, J.C., Nelson, J.L., Beckstead, B.L., Runyan, C.M., Robison, R.A., Schaap, A.P., and Pitt, W.G. (2004), Ultrasonic-enhanced gentamicin transport through colony biofilms of *Pseudomonas aeruginosa* and *Escherichia coli*, *J. Infect. Dis.*, 10, 193–199.
354. Available at http:/www.accessdata.fda.gov/scripts/cder/iig/index.cfm. Accessed on 21 June 2009, Inactive Ingredient Search for Approved Drug Products, "Poloxamer" 2003 (7 Nov.), U.S. Food and Drug Administration.
355. Kabanov, A., Batrakova, E., and Alakhov, V. (2002), Pluronic block copolymers as novel polymer therapeutics for drug and gene delivery, *J. Control. Rel.*, 82, 189–212.
356. Nace, V. *Nonionic Surfactants: Polyoxyalkylene Block Copolymers*, Marcel Dekker, New York, 1996.

# CHAPTER 10

# LIPOSOMES AS DRUG DELIVERY CARRIERS TO BIOFILMS

## 10.1. INTRODUCTION

Liposomes are artificial lipid vesicles consisting of one or more lipid bilayers enclosing a similar number of aqueous compartments. There are a number of components present in liposomes, with phospholipid and cholesterol being the main ingredients. This type of phospholipid includes phosphoglycerides and sphingolipids, together with their hydrolysis products. Liposomes can be subcategorized into: (1) small unilamellar vesicles (SUV), 25–70 nm in size, that consist of a single lipid bilayer; (2) large unilamellar vesicles (LUV), 100–400 nm in size, that consist of a single lipid bilayer; and (3) multilamellar vesicles (MLV), 200 nm to several microns, that consist of two or more concentric bilayers. A typical liposome structure is shown in Fig. 10.1.

## 10.2. LIPOSOMES AS DRUG DELIVERY CARRIERS TO BIOFILMS

Liposomes are attractive as drug delivery–targeting vehicles by virtue of their compatibility with biological constituents and the range and extent of pay loads that they can carry. Liposomes have the potential to carry hydrophobic and hydrophilic drugs over long periods of time and also to decrease drug side effects by protecting the environment from direct contact with the drugs. It is illustrated in the liposomal literature that liposomes need to be stable when

*Biofilm Eradication and Prevention: A Pharmaceutical Approach to Medical Device Infections*,
By Tamilvanan Shunmugaperumal

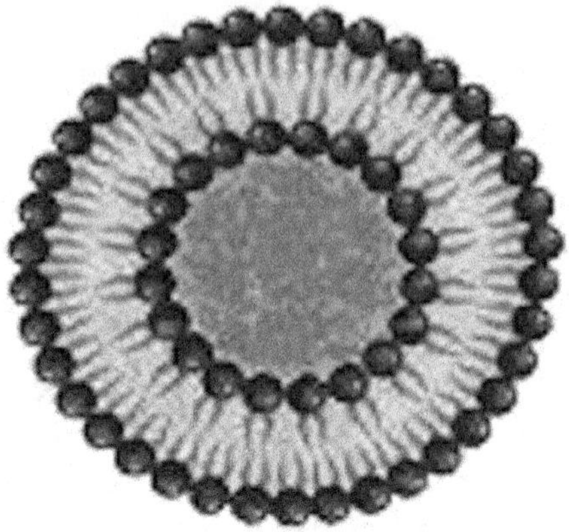

**Figure 10.1.** Liposome structure.

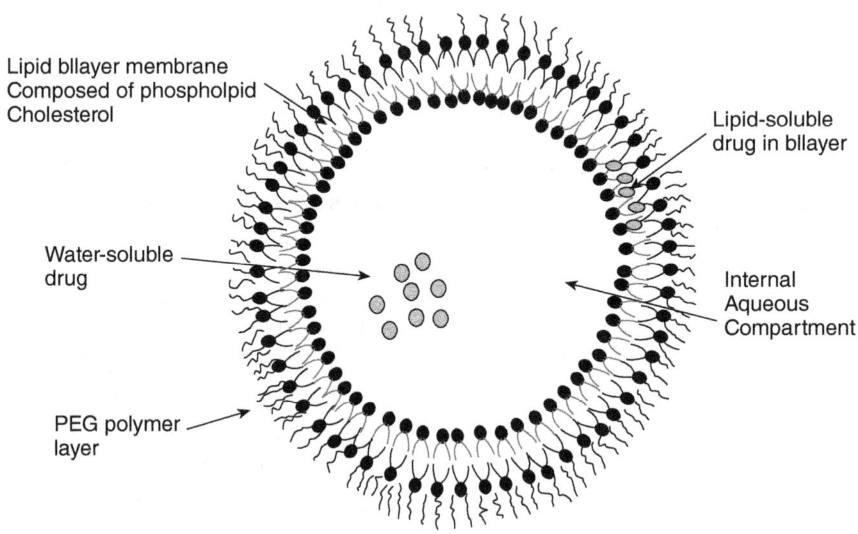


**Figure 10.2.** Structure of unilamellar Stealth liposome.

used as drug delivery tools *in vivo*. There are three forms of liposome stability to consider in relation to drug delivery: chemical, physical, and biological stabilities. Stability can be controlled by manipulating factors (e.g., pH, size distribution, and ionic strength), or by using the alternative method of coating liposomes with inert hydrophilic polymers (Stealth® liposomes) (see Fig. 10.2.). A new approach for achieving chemical and physical liposome stabilization was developed by adsorbing them on solid surfaces, like zinc citrate [1]. When liposomes are adsorbed on solid surfaces, adsorption is irreversible. Liposomes can either disrupt on adsorption, adsorb intact, or a combination of the two processes can occur. However, liposomes adsorbed predominantly intact on solid particulates of zinc citrate.

Apart from the above-described modifications on native liposomes, surface (charge)-modified liposomes (cationic or anionic liposomes), have been recognized as an interesting and promising delivery vehicle for active and passive drug targeting purposes even with or without ligand–antibody attachment onto their surfaces. In the context of antibiotics and treatment of infection, liposomes have been studied for their ability to act against colonizing microorganisms [2], to concentrate agents at biofilm interfaces [3,4], and also to be taken up into cells harboring intracellular pathogens [5–9].

## 10.3. LIPOSOMES TO REDUCE MICROBIAL ADHESION–COLONIZATION ONTO MEDICAL DEVICES

Since many catheter-related infections are due to skin organisms acquired at the time of catheter insertion, anticolonization strategies are worth exploring. There is evidence that the intrinsic properties of a material might be advantageous regarding resistance to infection. Thus, improvement of surface texture, tailoring the protein adsorption characteristics, and improving the antithrombogenicity of a given material would be key factors in the development of innovative, infection resistant materials. However, this goal has to be achieved even after insertion of the devices into the bloodstream and despite the ever-occurring interactions of the device surface with host factors (e.g., proteins and cells).

The surfaces of the medical devices are simply modified with the application of external coating substances onto them. For example, surfaces containing immobilized long-chain N-alklyated polyvinylpyridines and structurally unrelated N-alklyated polyethylenimines were lethal to *Staphylococcus aureus*, *Staphylococcus epidermidis*, *Pseudomonas aeruginosa*, and *Escherichia coli*. The structure–activity analysis revealed that for surfaces to be bactericidal, the immobilized long polymeric chains have to be hydrophobic, but not excessively so, and positively charged [10,11].

Alternatively, there are many instances where plain broad-spectrum antimicrobials (without any carriers) have been incorporated into the device [12–16]. These are then eluted in an attempt to prevent biofilm formation by killing early colonizing bacteria. However, as noted by some authors, such a strategy is not without its problems [17]. Sufficient antibiotic must be incorporated for the "user-lifetime" of the device and such incorporation must not damage the properties of the material (e.g., lubrication, lifetime, and host compatibility). There is also the nagging concern that low levels of antimicrobials could favor acquisition of antibiotic resistant organisms [18]. It is worrysome that in staphylococcal species, which are commonly associated with device-related infections, subinhibitory concentrations of tetracycline and quinupristin–dalfopristin enhanced biofilm development are increased by expression of the intercellular adhesion *ica* locus [19].

Among the other alternative drug-delivery strategies that have been developed for the anticolonization or antibiofilm approach, a liposomal hydrogel that reduces bacterial adhesion to a Si catheter material is most promising. Liposomes containing ciprofloxacin are sequestered within a poly(ethylene glycol)-gelatin hydrogel. Bacterial adhesion was completely inhibited on catheter surfaces throughout a 7-day adhesion assay [2]. Using peritoneum of male Sprague-Dawley rats, a new model of persistent *P. aeruginosa* peritonitis was developed. The ability of liposomal ciprofloxacin hydrogel (LCH)-coated Si versus plain Si to prevent bacterial colonization at optimal conditions was compared [20]. While Plain Si coupons in all tested rats were colonized and peritoneal washings were consistently culture-positive, LCH coupons removed after 7 days from the tested rats were sterile, as were the peritoneal washings, and there was no evidence of peritonitis. This finding indicates that the LCH coated Si resists colonization in this rat model of persistent *P. aeruginosa* peritonitis [20].

Pugach et al. [21] developed an antibiotic liposome (ciprofloxacin-loaded liposome) containing hydrogel for external coating of Si foley catheters (Fig. 10.3) and evaluated its efficacy in a rabbit model. Their goal was to create a catheter that would hinder the development of catheter-associated nosocomial urinary tract infections. They inserted either an untreated, liposomal hydrogel coated or a liposome hydrogel with ciprofloxacin coated 10F silicone foley catheter into New Zealand white rabbits and challenged the system with $5 \times 10^6$ virulent *E. coli* at the urethral meatus twice daily for 3 days. Urine cultures were evaluated twice daily for 7 days. When urine cultures became

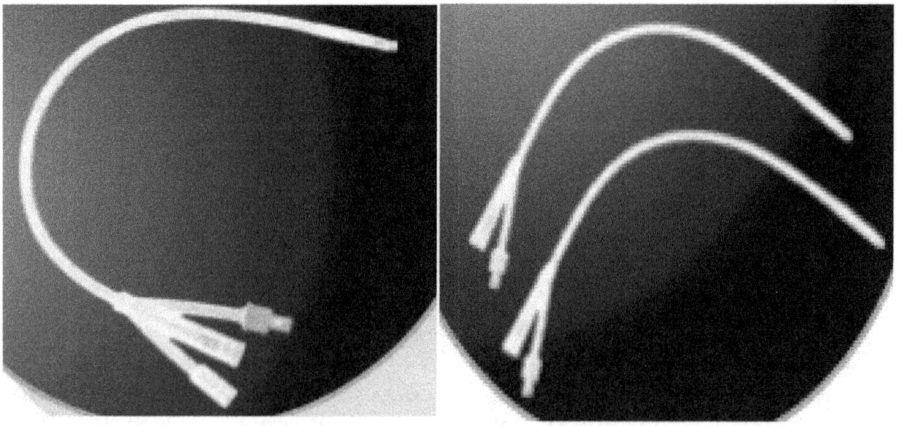

(*a*) (*b*)

**Figure 10.3.** Silicone two-way (*a*) and three-way (*b*) foley balloon catheters.

positive, the rabbits were sacrificed and urine, urethral catheter, and urethral tissue were cultured. The time to bacteriuria detection in 50% of the specimens was double for hydrogel with ciprofloxacin-coated catheters versus untreated and hydrogel-coated catheters. A significant ($p = 0.04$) improvement in average time to positive urine culture from 3.5 to 5.3 days and a 30% decrease in the bacteriuria rate for hydrogel with ciprofloxacin-coated catheters were noted compared to untreated catheters. A significant benefit was realized by coating the extraluminal catheter surface with a ciprofloxacin liposome impregnated hydrogel. Therefore, this procedure will provide a significant clinical advantage, while reducing healthcare costs substantially.

For interested readers' four antimicrobial urinary catheters are currently marketed in the United States. They are coated with a Ag alloy (3 latex or Si base catheters) or nitrofurazone, a nitrofurantoin-like drug (1 Si base catheter). Johnson et al. [22] assessed, through randomized and quasirandomized clinical trials, the currently marketed antimicrobial urinary catheters for preventing catheter-associated urinary tract infection. According to fair-quality evidence, antimicrobial urinary catheters can prevent bacteriuria in hospitalized patients during short-term catheterization, depending on antimicrobial coating and several other variables. A similar type of clinical trial was also conducted and evaluated to find out the efficacy of Si based, Ag ion-impregnated urinary catheters in the prevention of nosocomial urinary tract infections [23]. Unlike previous trials of latex-based, Ag ion-impregnated foley catheters and Si based, Ag impregnated foley catheters were not effective in preventing the nosocomial urinary tract infections. However, this study was affected by differences in the study groups. Prospective trials remain important in assessing the efficacy and cost-effectiveness of new Ag coated products. Additionally, from the above-described two different clinical experiments, an identical clinical trial should also be conducted for antibiotics containing-liposomal hydrogel-coated medical devices in the future.

## 10.4. LIPOSOMES AS DRUG DELIVERY CARRIERS TO BIOFILM INTERFACES

Jones and co-workers [1,3,4,24–30] extensively studied the interaction between liposomes and bacterial biofilms. Confocal laser-scanning microscopy has been used to visualize the adsorption of fluorescently labeled liposomes on immobilized biofilms of the bacterium *S. aureus* [24]. The liposomes were prepared with a wide range of compositions with phosphatidylcholines as the predominant lipids using the extrusion technique. They had weight average diameters of $125 \pm 5$ nm and were prepared with encapsulated carboxyfluorescein. Cationic liposomes were prepared by incorporating dimethyldioctadecylammonium bromide (DDAB) or 3, beta [*N*-(*N*1,*N*1-dimethylammonium ethane)-carbamoyl] cholesterol (DC-chol) and anionic liposomes were prepared by incorporation of phosphatidylinositol (PI). Pegylated cationic liposomes were prepared by incorporation of DDAB and 1,2-dipalmitoylphosphatidylethanolamine-*N*-[polyethylene glycol)-2000]. Confocal laser scanned images showed the preferential adsorption of the fluorescent cationic liposomes at the biofilm–bulk phase interface, which on quantitation gave fluorescent peaks at the interface when scanned perpendicular (*z*-direction) to the biofilm surface (*x*–*y* plane). The biofilm fluorescence enhancement (BFE) at the interface was examined as a function of liposomal lipid concentration and liposome composition. Studies of the extent of pegylation of the cationic liposomes incorporating DDAB, on adsorption at the biofilm-bulk-phase interface were made. The results demonstrated that pegylation inhibited adsorption to the bacterial biofilms as seen by the decline in the peak of fluorescence as the mol% DPPE–PEG-2000 was increased in a range from 0 to 9mol%. The results indicate that confocal laser-scanning microscopy is a useful technique for the study of liposome adsorption to bacterial biofilms and complements the method based on the use of radiolabeled liposomes.

Using cationic liposomes prepared from dimyristoylphosphatidylcholine (DMPC), cholesterol, and DDAB or anionic liposomes substituting DMPC with PI, Robinson et al. [29] noted that each bacterium in the biofilm adsorbed independently and that the extent of adsorption of anionic liposomes was smaller. Interestingly, when targeting mixed biofilms of *Streptococcus sanguis* and *Streptococcus salivarius* by liposomes loaded with the bactericide triclosan, anionic liposomes were most effective against *S. sanguis*, but relatively ineffective against *S. salivarius* [29]. An additional approach has been to load antibacterials into liposomes adsorbed on the surface of zinc citrate particles, used in toothpaste formulations, to produce solid supported vesicles (SSV) containing either triclosan or aqueous-soluble penicillin-G. Anionic liposomes were prepared by incorporation of PI into DMPC liposomes and cationic liposomes were prepared by incorporation of DDAB and cholesterol into DMPC. While zinc citrate is itself antibacterial, it was noted that particles and empty liposomes had an additional or synergistic effect, whereas particles and liposomally encapsulated antimicrobials had an inhibitory effect on each other against *S. oralis* biofilms [1]. Other oral hygiene approaches have included liposomal encapsulation of the enzyme glucose oxidase and horseradish peroxidase, which generate hydrogen peroxide ($H_2O_2$) and oxyacids in the presence of their substrates. They were effective against *S. gordonii* biofilms in a manner dependent on liposome–biofilm and substrate–biofilm incubation times [30].

Another work by Jones [25] described methods for the use of liposomes to deliver bactericides to bacterial biofilms. Anionic liposomes, cationic liposomes, and proteoliposomes with covalently linked lectins or antibodies are designed by the extrusion technique (vesicles by extrusion, VET). The liposomes are prepared from the phospholipid dipalmitoylphosphatidylcholine (DPPC), together with the anionic lipid PI or the cationic amphiphile DDAB, with the reactive lipid DPPE–MBS, the *m*-maleimidobenzoyl-*N*-hydroxysuccinimide (MBS) derivative of dipalmitoylphosphatidylethanolamine (DPPE).

Proteins (lectin or antibody), after derivatization with *N*-succinimidyl-*S*-acetylthioacetate (SATA), can be covalently linked to the surface of the liposomes by reaction with the reactive lipid, DPPE–MBS. The physical and chemical characterization of the liposomes and proteoliposomes by photon correlation spectroscopy (PCS) and protein analysis, to determine the number of chemically linked protein molecules (lectin or antibody) per liposome, are described. The liposomes can be used for carrying oil-soluble bactericides (e.g., Triclosan) or water-soluble antibiotics (e.g., vancomycin or benzylpenicillin) and targeted to immobilized bacterial biofilms of oral- or skin-associated bacteria adsorbed on microtiter plates. Techniques for the preparation of immobilized bacterial biofilms, applicable to a wide range of bacterial suspensions, and for the analysis of the adsorption (targeting) of the liposomes to the bacterial biofilms are given. The mode of delivery and assessment of antibacterial activity of liposome encapsulating bactericides and antibiotics, when targeted to the bacterial biofilms, by use of an automated microtiter plate reader, are illustrated. Specific reference is given to the delivery of the antibiotic benzylpenicillin encapsulated in anionic liposomes to biofilms of *S. aureus*. The methods have potential application for the delivery of oil- or water-soluble bactericidal compounds to a wide range of adsorbed bacteria responsible for infections in implanted devices (e.g., catheters, heart valves, and artificial joints).

## 10.5. LIPOSOMES AS DRUG DELIVERY CARRIERS IN INTRACELLULAR INFECTION

Infectious diseases caused by intracellular bacteria present a significant challenge to antibiotic therapy. Antibiotic treatment of these types of infections has been associated with high failure and/or relapse rates [31,32]. Intracellular pathogens, whether obligate or facultative, can hide, reside, and multiply within the phagocytic cells of the reticuloendothelial system (RES), and by virtue to their intracellular location, are protected from the actions of the immunological defence cells and of antimicrobial agents [31,33–35]. The ineffectiveness of conventional antibiotics against intracellular infections may also be attributable to poor drug penetration, limited drug accumulation in subcellular compartments, and/or drug inactivation by acidity in subcellular compartments [33–35]. These factors may explain why some antibiotics are bactericidal against extracellular bacteria *in vitro*, but are ineffective in killing intracellular forms of the bacteria [33,35,36]. Since this book mainly describes the potential of liposomes for eradicating biofilm consortia on device-related nosocomial infections, the applicability of liposomes on the treatment of biofilm-mediated intracellular infections are not elaborated and only a short outline is presented. Interested readers may find further details in this particular area through review articles [37,38].

Ciprofloxacin, a fluoroquinolone, is a potent and broad-spectrum antibiotic. It has good antibacterial activity against most Gram-negative bacteria and

Gram-positive cocci. Ciprofloxacin has been shown to have a superior ability to penetrate most tissues compared to other antibiotics [39–42], accumulates in macrophages [43] and neutrophils [44], and is bactericidal in a low pH environment [45]. These attributes contribute partly to ciprofloxacin being the drug of choice for the treatment of infectious diseases caused by intracellular pathogens. Furthermore, ciprofloxacin, when orally or intravenously administered, is known to reach such organs as liver, spleen, lungs, and lymph nodes [46], which are important infection sites for intracellular bacteria. However, ciprofloxacin does not preferentially accumulate well at these tissues and may therefore not reach high sustainable therapeutic levels at these sites.

The spontaneous uptake of liposomes by cells of the RES following parenteral administration has been exploited to target antibiotics to those intracellular sites where parasitic bacteria reside, and by virtue of sustained release properties, extend the half-life of the drug in the body. Ciprofloxacin has been incorporated with high efficiency into DSPC–cholesterol liposomes and examined in a mouse model of *Francisella tularensis* [47]. Intravenous injection of liposome-encapsulated ciprofloxacin resulted in increased drug retention in the lungs, liver, and spleen compared with that of the free drug. Aerosolized liposomal ciprofloxacin gave complete protection against a lethal pulmonary infection of *F. tularensis*, whereas free ciprofloxacin was ineffective [47]. Caution should be exercised in extrapolating data, as it is clear that liposomal efficacy is dependent on the infecting organism. Liposome-encapsulated ciprofloxacin, delivered intravenously, has been compared to free drugs in a rat model of *S. pneumoniae* pneumonia and, while serum and lung lavage levels were higher (peak and area under curve), survival rates were similar [48].

An interesting development of the liposomal concept has been the use of pH sensitive liposomes in a murine salmonellosis model [49]. Here, gentamicin encapsulated in liposomes including a pH sensitive lipid fusion between unsaturated phosphatidylethanolamine (PE) and *N*-succinyldioleyl–PE gave 153- and 437-fold greater drug levels in the liver and spleen, respectively, compared with free drug. Overall, liposomal delivery was associated with 10,000-fold greater activity than that of the free drug.

## 10.6. STEALTH® LIPOSOMES

There is an increasing interest in developing injectable liposomes that are not cleared quickly from the circulation when liposomes are designed to reach non-RES tissues in the vascular system, extravascular sites of action, or to act as circulating drug reservoirs. Because, it is well established that when colloidal drug delivery carriers like liposomes are mixed with blood, many plasma proteins, mainly the apolipoproteins, associate with the surface of these carriers. A number of factors have been reported to influence plasma protein–liposome interactions and clearance rates including surface charge, surface coatings, and lipid doses [50]. It has been shown that cationic liposomes exhibit extensive interactions with plasma, resulting in immediate clot formation at charge concentrations higher than 0.5 mmol $mL^{-1}$ [51]. The circulation time for these liposomes was on the order of minutes. These findings were further confirmed for other cationic liposome formulations showing significant serum turbidity and protein binding [52]. These results were expected since the majority of plasma proteins carry a net negative charge at physiological pH. The ability of anionic liposomes to interact with blood proteins depends on the nature of the anionic lipid, mainly the composition of the acyl chain [53]. In addition, it was found that liposomes composed of neutral saturated lipids with acyl chains lengths >16 carbon atoms bound large quantities of blood proteins and were rapidly cleared from the circulation [54]. This phenomenon was attributed to the occurrence of hydrophobic domains at the surface of the vesicles. These vesicle–blood protein interactions also depend on the lipid dose administered. Increased lipid doses result in decreased protein levels on the surface of the liposomes and longer circulation time suggesting the occurrence of a saturable protein-binding mechanism [55]. Finally, the most widely used approach for enhancing the circulation time of liposomes is the inclusion of amphipathic poly(ethyleneglycols), with a typical molecular weight of 2000–5000, in the vesicle bilayers that sterically decrease the adsorption of plasma proteins onto the liposome surfaces [56,57].

Somewhat counterintuitively, stealth approaches have been adopted for delivering antibiotics. Pegylated long-circulating liposomes loaded with gentamicin were superior to free gentamicin in a rat model of *K. pneumoniae* unilateral pneumonia–septicaemic [58]. Studies *in vitro* have also confirmed that pegylation of liposomes reduced their affinity for *S. aureus* biofilms [28]. Here, they found that liposomes prepared from the phospholipids DMPC, dipalmitoyl PC, and distearoyl PC containing DDAB (cationic) or PI (anionic) and variable amounts of dipalmitoylphosphatidylethanolamine bonded to polyet(hylene glycol) (PEG) of molecular mass 2000 (DPPE–PEG-2000), exhibited decreasing electrophoretic mobilities and zeta potentials with increasing DPPE–PEG-2000 incorporation. The adsorption of liposomes to *S. aureus* biofilms followed the Langmuir isotherm and both surface coverage and the magnitude of the Gibbs energy of adsorption decreased with the extent of pegylation [28]. A study by Bakker-Woudenberg et al. [59] in an experimental *K. pneumoniae* pneumonia, the therapeutic potential of ciprofloxacin was significantly improved by encapsulation in PEG coated (pegylated) long-circulating (STEALTH) liposomes. Pegylated liposomal ciprofloxacin in high doses was nontoxic and resulted in relatively high and sustained ciprofloxacin concentrations in blood and tissues. Hence, an increase in the area under the plasma concentration–time curve (AUC). These data correspond to data from animal and clinical studies showing that for fluoroquinolones, the AUC–MIC (minimum inhibitory concentration) ratio is associated with a favorable outcome in serious infections. Clinical failures and the development of resistance are observed for marginally susceptible organisms like *P. aeruginosa* and for which sufficient AUC/MIC ratios cannot be achieved.

In the next study, the therapeutic efficacy of pegylated liposomal ciprofloxacin was investigated in two rat models of *P. aeruginosa* pneumonia [60]. In the acute model, pneumonia developed progressively, resulting in a rapid onset of septicemia and a high mortality rate. Ciprofloxacin twice daily for 7 days was not effective at doses at or below the maximum tolerated dose (MTD). However, pegylated liposomal ciprofloxacin, either at high dosage or given at low dosage in combination with free ciprofloxacin on the first day of treatment, was fully effective (100% survival). Obviously, prolonged concentrations of ciprofloxacin in blood prevented death of the animals due to early-stage septicemia in this acute infection. However, bacterial eradication from the left lung was not effected. In the chronic model, pneumonia was characterized by bacterial persistence in the lung without bacteremia, and no signs of morbidity or mortality were observed. Ciprofloxacin administered for 7 days at the MTD twice daily resulted in killing of >99% of bacteria in the lung and this result can also be achieved with pegylated liposomal ciprofloxacin given once daily, although complete bacterial eradication is never observed [60].

## 10.7. CASE STUDY 1: URINARY TRACT INFECTION

Urinary tract infections (UTIs), the majority of which (80%) are caused by uropathogenic *E. coli* (UPEC) and 25% of all UTIs recur within 6 months. The UTIs occur as a continuum of steps by ascension of UPEC from the perineum through the urethra to the bladder, passing though the ureters to the kidneys. Clinically, the symptoms of cystitis, dysuria, and frequency often precede those of upper-tract disease (e.g., flank pain and chills). The UPEC have also been shown to persist and re-emerge in the bladder despite antibiotic therapy. Superficial facet cells of urinary bladder express integral membrane proteins called uroplakins (UP), which can serve as the receptors for UPEC. Upon entry into the superficial facet cells, UPEC are able to rapidly replicate and form intracellular bacterial communities (IBCs), characterized by a defined differentiation program and enhanced resistance to antibiotics. Their intracellular proliferation results in communal formations with biofilm-like properties: the IBCs. The bacteria thrive tightly enmeshed in the protective matrix of the IBC within the host epithelium. Ultimately, they need to be released and dispersed in order to exit the infected cell and find new naive cells for residence. The dispersion and the existence of the host cell are, therefore, central steps in the UPEC life cycle in the bladder [termed the intracellular bacterial communities (IBC) pathway]. Although the vast majority of UPEC are cleared by host defenses within a few days, small clusters of intracellular bacteria have occasionally been observed to persist for months in an antibiotic-insensitive state. The long-term persistence in the face of antibiotic therapy suggests that these bacteria are within a protected location within bladder epithelial cells. In addition to the superficial facet epithelial cell barrier, invading bacteria also face a chemical barrier: The complex network of proteoglycans–glycosaminoglycans (GAG layer) that is woven into the urothelium and is known to act as an antimicrobial adherence factor. Parsons et al. [61] showed that protamine sulfate (PS), a highly cationic protein (pI ~ 12) can lead to both exfoliation of the superficial facet cell barrier and biochemical inactivation of GAGs. Furthermore, PS treatment also increases urothelial ionic permeability and facilitates bacterial entry. Thus, Hultgren and co-workers [62,63] reasoned that PS could be used as both a chemical exfoliant of infected superficial facet cells and as an adjuvant to facilitate bacterial entry into nonexfoliating transitional cells underlying the superficial facet cell layer. Their findings raise possible therapeutic avenues for the treatment of recurrent UTIs. They also show that inducing epithelial exfoliation by using cationic proteins (e.g., PS) can, in some cases, expel bacteria from their intracellular locations. Therefore, protamine sulfate or other similar cationic compounds could act as potential therapeutic adjuvants in conjunction with antibiotic treatment to induce stripping of the urothelial lining containing cryptic bacterial quiescent intracellular reservoirs (QIRs), thus eliminating a potential source of chronic same-strain recurrent UTI episodes.

It is well known that cationic liposomes have already re-emerged as a promising new vaccine adjuvant technology and these lipid-bilayer vesicles have positive surface charge [64]. It has also been shown that cationic liposomes have the ability to incorporate adequate quantities of a number of antimicrobial agents [65]. Hence, it would be reasonable to speculate that antimicrobial agent-laden cationic liposomes should possibly pave a new way of treating QIRs, in principle but no experimental proof, to eliminate a potential source of chronic same-strain recurrent UTI episodes.

## 10.8. CASE STUDY 2: CHRONIC GRANULOMATOUS DISEASE

Chronic granulomatous disease (CGD) is a genetically determined primary immunodeficiency disease in which phagocytic cells are unable to reduce molecular oxygen and create the reactive oxygen metabolites. Thus they are unable to kill ingested catalase-positive microorganisms [66]. The ingested organisms will remain viable within the phagocytes where they are protected from antibiotics. This leads to recurrent life-threatening bacterial and fungal infections resulting in marked inflammation, abscess, and granuloma formation. Chronic granulomatous disease is now known to be caused by a defect in the nicotinamide adenine dinucleotide phosphate (NADPH) (reduced form) oxidase enzyme of phagocytes (collected on 21-06-2009, available at http://www.emedicine.com/ped/topic1590.htm). Most cases of CGD are transmitted as a mutation on the x-chromosomes and are thus called an x-linked trait or x-linked recessive. A less common mode of inheritance is by autosomal recessive pattern. In this form of inheritance the disease is less severe and tends to occur at an older age [67].

People with CGD are sometimes infected with unique organisms that usually do not cause disease in people with normal immune systems. The microorganisms that can cause disease in CGD patients include *S. aureus*, *E. coli*, *Klebsiella species*, *Aspergillus* species, and *candida* species. Manifestations of CGD include recurrent infections of the lungs, lymph nodes, and skin. Bones, liver, and gastrointestinal tract are less commonly involved. The great majority of the infective episodes are caused by *S. aureus* followed by *Aspergillus* [67,68]. Obstructive lesions of the gastrointestinal and urinary tract occur in CGD, especially in the x-linked form. In CGD, the persistence of viable bacteria within the phagocyte in the colonic mucosa may cause excessive stimulation of the inflammatory process and subsequent mucosa damage [69,70].

Antimicrobial prophylaxis, early and aggressive treatment of infections, and interferon- gamma (IFN-γ) are the cornerstones of current therapy for CGD [71]. Although hematopoietic stem cell transplantation (HSCT) from a human leukocyte antigen (HLA)—compatible donor can cure CGD, this approach is fraught with clinically significant morbidity and a finite risk of death. The HSCT remains a controversial therapeutic modality in this disease, even when stem cells from a matched sibling donor are available. Therefore, only daily prophylaxis of infections with more potent antibacterial and antifungal antibiotics (trimethoprim-sulfamethoxazole or cephalosporin and ketoconazole or itraconazole) in conventional dosage forms is indicated in CGD. On the other hand, patients with established superficial or deep infections (vs those with obstructing granulomas) should receive aggressive intravenous antibiotics for several weeks. Note that it might not be convenient for the CGD patients to take the conventional antibiotic medication for several weeks to contain their infections. Moreover, liposomes incorporating the antimicrobial agents are already available in the market in order to diminish the injection frequencies and possible side effects of the non-liposomal, conventional dosage forms of antimicrobial agent. Therefore, note that significant investigations should be directed for the judicious use of antimicrobial agent-laden liposomes in CGD patients when infections occur. While some debate continues among the scientific community as to whether improved delivery of antimicrobials to the biofilm actually represents a viable approach to eradication, given the altered metabolic state of the microorganisms in the biofilm matrix, the current status of the field is further comprehensively reviewed under polymer-based drug delivery carriers in Chapter 11.

## 10.9. CASE STUDY 3: MENINGOENCEPHALITIS IN IMMUNOCOMPROMISED PATIENTS

*Cryptococcus neoformans* is an encapsulated opportunistic yeast-like fungus that is a relatively frequent cause of meningoencephalitis in immunocompromised patients, especially in individuals with acquired immunodeficiency syndrome (AIDS) or solid organ transplants, and also occasionally causes disease in apparently healthy individuals [72]. The *C. neoformans* capsular polysaccharide is mainly composed of glucuronoxylomannan (GXM), which is a major contributor to its virulence since acapsular strains are not pathogenic [73]. Copious amounts of GXM are released during cryptococcal infection, causing deleterious effects on the host immune response [73,74]. Martinez and Casadevall [75] previously reported that *C. neoformans* GXM release is necessary for adhesion to a solid support and subsequent biofilm formation, which facilitates the evasion of the yeast from host responses [76] and antifungal therapies [77]. The *C. neoformans* forms biofilms on polystyrene plates [75] and medical devices [78–81] after GXM shedding. For example, Walsh et al. [78] reported on *C. neoformans* biofilms in ventriculoatrial shunt catheters. In addition, several reports of *C. neoformans* infection of polytetrafluoroethylene peritoneal dialysis fistula [80] and prosthetic cardiac valves [79] demonstrate the ability of this organism to adhere to medical devices.

Chitosan, a hydrophilic biopolymer industrially obtained by N-deacetylation of crustacean chitin, has antimicrobial activities [82]. This natural compound is inexpensive and nontoxic. Chitosan has been utilized in diverse applications, including as an antimicrobial compound in agriculture, as a potential elicitor of plant defense responses, as a flocculating agent in wastewater treatment, as an additive in the food industry, as a hydrating agent in cosmetics, and as a pharmaceutical agent in biomedicine [82]. The antimicrobial activity of chitosan has been observed against a wide variety of microorganisms including fungi, algae, and bacteria [82]. Based on these applications and its antimicrobial activity, Martinez et al. [83] hypothesized that chitosan could interfere with *C. neoformans* biofilm formation and, by penetrating mature biofilms, bind to yeast cells to deliver direct microbicidal activity. Although considerable work on the effect of chitosan on bacterial biofilms has been done [84–86], no comparable studies have been done with fungal biofilms. Therefore, Martinez et al. [83] exploited the ability of *C. neoformans* to form biofilms *in vitro* on polystyrene microtiter plates to study the susceptibilities of cryptococcal biofilms to chitosan. A semiquantitative measurement of fungal biofilm formation was obtained from the 2,3-bis(2-methoxy-4-nitro-5-sulfophenyl)-5-[(phenylamino) carbonyl]-2*H*-tetrazolium-hydroxide (XTT) reduction assay [87]. Melanization was induced by growing the biofilms on defined minimal medium broth at 30 °C with the addition of 1 mM L-3,4-dihydroxyphenylalanine (L-dopa) for 7 days. Nonmelanized controls were obtained by growing the yeast cells on defined minimal medium broth without L-dopa for 7 days. To evaluate the susceptibilities of fungal biofilms to chitosan, phosphate-buffered saline (PBS) containing different concentrations of chitosan (0, 0.625, 1.25, 2.5, and 5 mg mL$^{-1}$) in 200 μL was added to each well. Mature biofilms and chitosan were mixed for 1 min by use of a microtiter plate reader to ensure a uniform distribution and were incubated at 37 °C for 0.5 and 1 h. After incubation, biofilm metabolic activity was quantified by the XTT reduction assay. The susceptibilities of the mature cryptococcal biofilms to chitosan were determined by comparing the metabolic activities of the biofilms coincubated with chitosan with those of the biofilms grown in PBS.

*Cryptococcus neoformans* has substantial chitosan in its cell wall during vegetative growth and the polymer may be an essential factor for the proper maintenance of cell wall integrity [88]. The results of Martinez et al. [83] show that addition of exogenous chitosan to *C. neoformans* biofilms significantly reduces metabolic activity and prevents the adhesion of the yeast cells to the polystyrene surface (Fig. 10.4). The cell wall is the structure that mediates the cell's interactions with the environment, and might be important in adhesion of fungi to solid surfaces (e.g., indwelling medical devices). Bachmann et al. [89] proposed the use of the cell wall as an attractive target for the development of strategies to combat biofilm-associated infections. Chitosan has anti-adherent activity and prevents *Candida albicans* biofilm development [90]. Other studies have proposed the treatment of medical devices with antifungal agents before they are implanted in patients [89,91]. Chitosan may be a strong

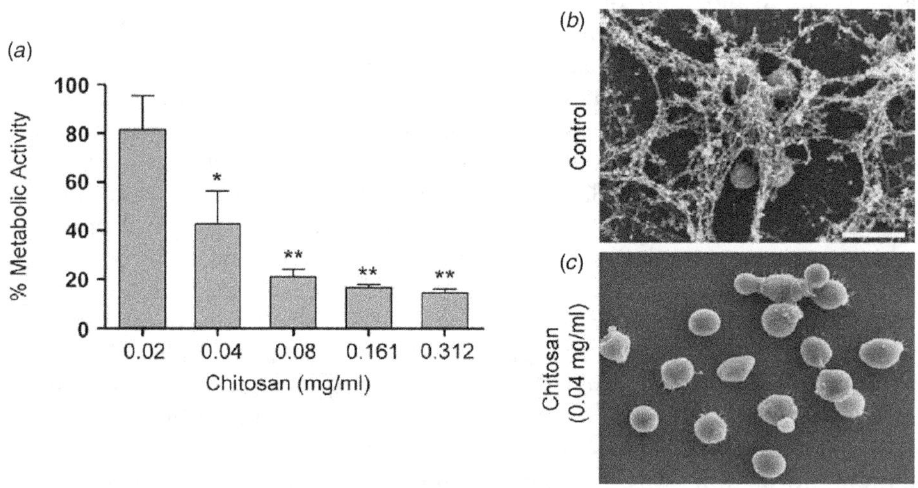


**Figure 10.4.** Chitosan inhibits *C. neoformans* biofilm formation. (*a*) Percent metabolic activity of untreated and chitosan-treated *C. neoformans* strain B3501 biofilms measured by the XTT reduction assay. Yeast cells were coincubated with various concentrations (0.02, 0.04, 0.08, 0.16, and 0.31 $mg\,mL^{-1}$) of chitosan for 48 h; and their biofilm production was compared to that of fungal cells incubated in PBS. Bars are the averages of four XTT measurements, and brackets denote standard deviations. *, $p < 0.05$ and **, $p < 0.001$ in comparing the untreated and chitosan-treating groups. This experiment was done twice, with similar results each time. (*b*) Scanning electron microscopy image of untreated *C. neoformans* B3501 biofilms formed on glass coverslips revealed that cryptococcal cells are internally connected by copious amounts of polysaccharide. (*c*) The SEM image of *C. neoformans* B3501 grown with 0.04 $mg\,mL^{-1}$ showed yeast cells with no exo- or capsular polysaccharide. Scale bar: 5 mm. (Reproduced with permission from Martinez et al. *Biomaterials*, 31, 669–679, 2010 [83].)

candidate for this endeavor, due to its antiadherent and antifungal properties against fungal biofilms. Indeed, Martinez et al. [83] demonstrated that chitosan might be effective against *C. neoformans*, possibly because melanin production did not provide significant resistance to chitosan's antifungal activity. In addition, using immunofluorescence (IF), these authors concluded that chitosan effectively damages melanin from yeast cells making *C. neoformans* cells more accessible to the chitosan's antimicrobial activity. Importantly, the concentrations used in their experiments were not toxic to human endothelial cells, which are the cells most readily exposed to chitosan if applied to a venous or arterial catheter. Chitosan inhibits *C. neoformans* biofilm formation in polystyrene microtiter plates. A recent study showed that chitosan-coated surfaces have antibiofilm properties *in vitro* against certain bacteria and fungi [84]. This phenomenon has been attributed to the ability of cationic chitosan to disrupt negatively charged cell membranes as microbes settle on the surface [82].

The surface charge of untreated and chitosan-treated melanized and non-melanized cryptococcal cells was measured (Table 10.1). Previous studies showed that the polysaccharide capsule and melanin of *C. neoformans* are responsible for the high negative charge of the cells [92]. Chitosan-treated cells were significantly less negative (25.40 ± 0.57) than untreated cells (–21.89 ± 0.31) ($p < 0.001$). Melanization significantly increased the negative charge (–34.75 ± 0.57) of cryptococcal cells when compared to non-melanized cells (–21.89 ± 0.31) ($p < 0.001$). However, treatment with chitosan imparted a high positive charge to melanized (26.15 ± 0.38) and non-melanized (25.40 ± 0.57) *C. neoformans* cells.

The zeta potential analysis of Martinez et al. [83] demonstrates that chitosan has a profound effect on the negative charge of the cryptococcal cellular membrane, which may translate into interference with surface colonization or adhesion and cell–cell interactions during biofilm formation [93]. For example, a net positive charge to the fungal surfaces may keep yeast cells in suspension, preventing biofilm formation [94].

**TABLE 10.1. Zeta Potential of *C. neoformans* B3501 Strain Grown With (Melanized) and Without (Non-melanized) L-DOPA and Untreated or Treated With Chitosan**[a]

| Experiments | Zeta Potentials |
|---|---|
| Control (PBS) | –21.89 ± 0.31 |
| 0.312 mg mL$^{-1}$ chitosan | 25.40 ± 0.57[b] |
| L-DOPA | –34.75 ± 0.57[c] |
| L-DOPA + 0.312 mg mL$^{-1}$ chitosan | 26.15 ± 0.38[b] |

[a]Taken from Martinez et al. [83].

[b]Value significantly greater than the value for control ($p < 0.001$).

[c]Value significantly less than the value for control ($p < 0.001$).

This phenomenon reflects physical stress on the biofilm structure due to permeabilization of the cellular membrane, which allowed increased penetration of chitosan and effective delivery of its antifungal activity [95]. Binding of chitosan with DNA and inhibition of mRNA synthesis occurs through chitosan penetration toward the nuclei of the microorganisms and interference with the synthesis of mRNA and proteins [95]. It is most likely that the interaction between positively charged chitosan molecules and negatively charged microbial cell membranes leads to the leakage of proteinaceous and other intracellular constituents causing cell death [96]. Finally, the findings of Martinez et al. [83] suggested that chitosan might offer a flexible, biocompatible platform for designing coatings to protect surfaces from infection. Hence, chitosan can be potentially developed as an antimicrobial agent for prophylaxis against and/or the treatment of catheter or other medical device related fungal biofilm diseases.

There are plenty of interesting opportunities left to be investigated with cationic substance-loaded nanocarriers. Interestingly, inclusion of cation-producing substances (e.g., stearylamine and chitosan) into the native (plain or negatively charged) nanocarriers make the particles to acquire positive charge onto them. On one hand, the cationization on nanocarriers like liposomes and oil-in-water nanosized emulsions prove their tremendous application for drug absorption enhancement and for "ferrying" compounds across cell membranes. Furthermore, the cationic liposomes and nanosized emulsions provide an interesting opportunity for use as drug delivery vehicles for numerous therapeutics that can range in size from small molecules to macromolecules. Even antifungal drug-loaded liposomes (Ambiosome®) are available in the market for the management of systemic candidiasis that occurred nosocomially during the patient stay in an ICU. On the other hand, the development of chitosan-containing cationic nanocarriers (liposomes or nanosized emulsions) will certainly be of importance for intracellular targeting to eradicate *C. neoformans*, which is one of the causative agents for meningoencephalitis in immunocompromised patients, especially in individuals with AIDS or solid organ transplants. Additionally, chitosan-containing cationic nanocarriers might offer a flexible, biocompatible platform for designing coatings for prophylaxis against and/or the treatment of catheter or other medical device related fungal biofilm diseases.

## REFERENCES

1. Catuogno, C. and Jones, M.N. (2003), The antibacterial properties of solid supported liposomes on *Streptococcus oralis* biofilms, *Int. J. Pharm.*, 257, 125–140.
2. DiTizio, V., Ferguson, G.W., Mittelman, M.W., Khoury, A.E., Bruce, A.W., and DiCosmo, F. (1998), A liposomal hydrogel for the prevention of bacterial adhesion to catheters, *Biomaterials*, 19, 1877–1884.

3. Kim, J.H., Gias, M., and Jones, M.N. (1999), The adsorption of cationic liposomes to *Staphylococcus aureus* biofilms, *Colloids Surf. A Physicochem, Eng. Asp.*, 149, 561–570.

4. Jones, M.N., Song, Y.H., Kaszuba, M., and Reboiras, M.D. (1997), The interaction of phospholipid liposomes with bacteria and their use in the delivery of bactericides, *J. Drug Target.*, 5, 25–34.

5. Pinto-Alphandary, H., Andremont, A., and Couvreur, P. (2000), Targeted delivery of antibiotics using liposomes and nanoparticles: research and implications, *Int. J. Antimicrob. Agents*, 13, 155–168.

6. Nightingale, S.D., Saletan, S.L., Swenson, C.E., Lawrence, A.J., Watson, D.A., Pilkiewicz, F.G., Silverman, E.G., and Cal, S.X. (1993), Liposome encapsulated gentamicin treatment of Mycobacterium avium-Mycobacterium intracellulare complex bacteremia in AIDS patients, *Antimicrob. Agents Chemother.*, 37, 1869–1872.

7. Majumdar, S., Flasher, D., Friend, D.S., Nassos, P., Yajko, D., Hadley, W.K., and Duzgunes, N. (1992), Efficacies of liposome-encapsulated streptomycin and ciprofloxacin against *Mycobacterium avium-Mycobacterium intracellulare* complex infections in human peripheral blood monocyte/macrophages, *Antimicrob. Agents Chemother.*, 36, 2808–2815.

8. Eduardo, L., Bermudez, M., Wu, M., and Young, L.S. (1987), Intracellular killing of *Mycobacterium avium* complex by rifapentine and liposome encapsulated Amikacin, *J. Infect. Dis.*, 156, 510–513.

9. Fountain, M.W., Weiss, S.J., Fountain, A.G., Shen, A., and Lenk, R.P. (1985), Treatment of *Brucella canis* and *Brucella abortus in vitro* and *in vivo* by stable plurilamellar vesicle-encapsulated aminoglycosides, *J. Infect. Dis.*, 152, 529–535.

10. Lin, J., Qiu, S., Lewis, K., and Klibanov, A.M. (2003), Mechanism of bactericidal and fungicidal activities of textiles covalently modified with alkylated polyethylenimine, *Biotechnol. Bioeng.*, 83, 168–172.

11. Lin, J., Qui, S., Lewis, K., and Klibanov, A.M. (2002), Bactericidal properties of flat surfaces and nanoparticles derivatized with alkylated polyethylenimines, *Biotechnol. Prog.*, 18, 1082–1086.

12. Donelli, G., Francolini, I., Piozzi, A., Di Rosa, R., and Marconi, W. (2002), New polymer-antibiotic systems to inhibit bacterial biofilm formation: a suitable approach to prevent central venous catheter-associated infections, *J. Chemother.*, 14, 501–507.

13. Bach, A., Eberhardt, H., Frick, A., Schmidt, H., Bottinger, B.W., and Martin, E. (1999), Efficacy of silver-coating central venous catheters in reducing bacterial colonization, *Crit. Care Med.*, 27, 515–520.

14. Schierholz, J.M., Lucas, L.J., Rump, A., and Pulverer, G. (1998), Efficacy of silver-coated medical devices, *J. Hosp. Infect.*, 40, 257–262.

15. Sheretz, R.J., Carruth, W.A., Hampton, A.A., Byron, M.P., and Solomon, D.D. (1993), Efficacy of antibiotic-coated catheters in preventing subcutaneous *Staphylococcus aureus* infection, *J. Infect. Dis.*, 167 (1993) 98–106.

16. Tebbs, S.E. and Elliott, T.S.J. (1993), A novel antimicrobial central venous catheter impregnated with benzalkonium chloride, *J. Antimicrob. Chemother.*, 31, 261–271.

17. Donelli, G. and Francolini, I. (2001), Efficacy of antiadhesive, antibiotic and antiseptic coatings in preventing catheter-related infections: review, *J. Chemother.*, 13, 595–606.
18. Danese, P.N. (2002), Antibiofilm approaches: prevention of catheter colonization, *Chem. Biol.*, 9, 873–880.
19. Rachid, S., Ohlsen, K., Witte, W., Hacker, J., and Ziebuhr, W. (2000), Effect of subinhibitory antibiotic concentrations on polysaccharide intercellular adhesin expression in biofilm-forming *Staphylococcus epidermidis*, *Antimicrob. Agents Chemother.*, 44, 3357–3363.
20. Finelli, A., Burrows, L.L., DiCosmo, F.A., DiTizio, V., Sinnadurai, S., Oreopoulos, D.G., and Khoury, A.E. (2002), Colonization-resistant antimicrobial-coated peritoneal dialysis catheters: evaluation in a newly developed rat model of persistent *Pseudomonas aeruginosa* peritonitis, *Perit. Dial. Int.*, 22, 27–31.
21. Pugach, J.L., DiTizio, V., Mittelman, M.W., Bruce, A.W., DiCosmo, F., and Khoury, A.E. (1999), Antibiotic hydrogel coated Foley catheters for prevention of urinary tract infection in a rabbit model, *J. Urol.*, 162, 883–887.
22. Johnson, J.R., Kuskowski, M.A., and Wilt, T.J. (2006), Systematic Review: Antimicrobial urinary catheters to prevent catheter-associated urinary tract infection in hospitalized patients, *Ann. Intern. Med.*, 144, 116–126.
23. Srinivasan, A., Karchmer, T., Richards, A., Song, X., and Perl, T.M. (2006), A prospective trial of a novel, silicone-based, silver-coated foley catheter for the prevention of nosocomial urinary tract infections, *Infect. Control. Hosp. Epidemiol.*, 27, 38–43.
24. Ahmed, K., Gribbon, P.N., and Jones, M.N. (2002), The application of confocal microscopy to the study of liposome adsorption onto bacterial biofilms, *J. Liposome Res.*, 12, 285–300.
25. Jones, M.N. (2005), Use of liposomes to deliver bactericides to bacterial biofilms, *Methods Enzymol.*, 391, 211–228.
26. Kim, H.J. and Jones, M.N. (2004), The delivery of benzyl penicillin to *Staphylococcus aureus* biofilms by use of liposomes, *J. Liposome Res.*, 14, 123–139.
27. Ahmed, K. and Jones, M.N. (2003), The effect of shear on the desorption of liposomes adsorbed to bacterial biofilms, *J. Liposome Res.*, 13, 187–197.
28. Ahmed, K., Muiruri, P.W., Jones, G.H., Scott, M.J., and Jones, M.N. (2001), The effect of grafted poly(ethylene glycol) on the electrophoretic properties of phospholipid liposomes and their adsorption to bacterial biofilms, *Colloids Surf A Physicochem. Eng. Asp.*, 194, 287–296.
29. Robinson, A.M., Bannister, M., Creeth, J.E., and Jones, M.N. (2001), The interaction of phospholipid liposomes with mixed bacterial biofilms and their use in the delivery of bactericide, *Colloids Surf A Physicochem. Eng. Asp.*, 186, 43–53.
30. Hill, K.J., Kaszuba, M., Creeth, J.E., and Jones, M.N. (1997), Reactive liposomes encapsulating a glucose oxidase-peroxidase system with antibacterial activity, *Biochim. Biophys. Acta Biomembr.*, 1326, 37–46.
31. Donowitz, G.R. (1994), Tissue directed antibiotics and intracellular parasites: complex interactions of phagocytes, pathogens and drugs, *Clin. Infect. Dis.*, 19, 926–930.
32. al-Orainey, I.O., Bashandi, A.M., and Saeed, E.N. (1990), Failure of ciprofloxacin to eradicate brucellosis in experimental animals, *J. Chemother.*, 2, 380–383.

33. Raoult, M.M. (1996), Optimum treatment of intracellular infection, *Drugs*, 52, 45–59.
34. Holmes, B., Quie, P.G., Windhorst, D.B., Pollara, B., and Good, R.A. (1966), Protection of phagocytized bacteria from the killing action of antibiotics, *Nature (London)*, 11, 1131–1133.
35. Rous, P. and Jones, F.S. (1916), The protection of pathogenic microorganisms by living tissue cells, *J. Exp. Med.*, 23, 601–605.
36. van den Broek, P.J. (1989), Antimicrobial drugs, microorganisms, and phagocytes, *Rev. Infect. Dis.*, 11, 213–245.
37. Salem, I.I., Flasher, D.L., and Düzgüneş, N. (2005), Liposome-encapsulated antibiotics, *Methods Enzymol.*, 391 (Special issue), 261–291.
38. Gupta, C.M. and Haq, W. (2005), Tuftsin-bearing liposomes as antibiotic carriers in treatment of macrophage infections, *Methods Enzymol.*, 391 (Special issue), 291–304.
39. Kuhlmann, J., Schaefer, H.G., and Beermann, D. (1998), Clinical pharmacology, in: Kuhlmann, J., Dalhoff, A., and Zeiler, H.J., Eds., *Quinolone Antibacterials, Handbook of Experimental Pharmacology*, Vol. 127, Springer-Verlag, Berlin, pp. 339–406.
40. Nix, D.E., Goodwin, D., Peloquin, C.A., Rotella, D.L., and Schentag, J.J. (1991), Antibiotic tissue penetration and its relevance: models of tissue penetration and their meaning, *Antimicrob. Agents Chemother.*, 35, 1947–1950.
41. Dalhoff, A. (1989), A review of quinolone pharmacokinetics, in: Fernandes, P.B., Ed., *Telesymposium on Quinolones*, J.R. Prous.Science Publishers, Barcelona, Spain, pp. 277–312.
42. Bergan, T., Dalhoff, A., and Rohwedder, R. (1988), Pharmacokinetics of ciprofloxacin, *Infection*, 16 (Suppl. 1), 3–13.
43. Easmon, C.S. and Crane, J.P. (1985), Uptake of ciprofloxacin by macrophages, *J. Clin. Pathol.*, 38, 442–444.
44. Easmon, C.S. and Crane, J.P. (1985), Uptake of ciprofloxacin by human neutrophils, *J. Antimicrob. Chemother.*, 16, 67–73.
45. Rastogi, N. and Blom Potar, M.C. (1990), Intracellular bactericidal activity of ciprofloxacin and ofloxacin against *Mycobacterium tuberculosis* H37Rv multiplying in the J-775 macrophage cell line, *Zbl. Bakt.*, 273, 195–199.
46. Hanan, M.S., Riad, E.M., and el-Khouly, N.A. (2000), Antibacterial efficacy and pharmacokinetic studies of ciprofloxacin on *Pasteurella multocida* infected rabbits, *Dtsch. Tierarztl. Wochenschr.*, 107, 151–155.
47. Wong, J.P., Yuang, H., Blasetti, K.L., Schnell, G., Conley, J., and Schofield, L.N. (2003), Liposome delivery of ciprofloxacin against Francisella tularensis infection, *J. Control. Rel.*, 92, 265–273.
48. Ellbogen, M.H., Olsen, K.M., Gentry-Nielsen, M.J., and Preheim, L.C. (2003), Efficacy of liposome-encapsulated ciprofloxacin compared with ciprofloxacin and ceftriaxone in a rat model of pneumococcal pneumonia, *J. Antimicrob. Chemother.*, 51, 83–91.
49. Cordeiro, C., Wiseman, D.L., Lutwyche, P., Uh, M., Evans, J.C., Finlay, B.B., and Webb, M.S. (2000), Antibacterial efficacy of gentamicin encapsulated in pH-sensitive liposomes against an *in vivo* Salmonella enterica serovar typhimurium intracellular infection model, *Antimicrob. Agents Chemother.*, 44, 533–539.

50. Semple, S.C., Chonn, A., and Cullis P.R. (1998), Interactions of liposomes and lipid-based carrier systems with blood proteins: Relation to clearance behavior *in vivo*, *Adv. Drug Deliv. Rev.*, 32, 3–17.
51. Senior, J., Trimble, K.R., and Maskiewicz, R. (1991), Interaction of positively-charged liposomes with blood: Implications for their application in vivo, *Biochim. Biophys. Acta*, 1070, 173–179.
52. Oku, N., Tokudome, Y., Namba, Y., Saito, N., Endo, M., Hasegawa, Y., Kawai, M., Tsukada, H., and Okda, S. (1996), Effect of serum protein binding on real-time trafficking of liposomes with different charges analyzed by positron emission tomography, *Biochim. Biophys. Acta*, 1280, 149–154.
53. Hernandez-Caselles, T., Villalain, J., and Gomez-Fernandez, J.C. (1993), Influence of liposome charge and composition on their interaction with human blood serum proteins, *Mol. Cell Biochem.*, 120, 119–126.
54. Chonn, A., Semple, S.C., and Cullis, P.R. (1994), Protein-membrane interactions in the biological milieu, in: Op den Kamp, J.A.F., Ed., *Biological Membranes: Structure, Biogenesis and Dynamics NATO ASI Series H: Cell Biology*, Springer-Verlag, Heidelberg, Germany, Vol. 82, pp. 101–109.
55. Oja, C.D., Semple, S.C., Chonn, A., and Cullis, P.R. (1996), Influence of dose on liposome clearance: critical role of blood proteins, *Biochim. Biophys. Acta*, 1281, 31–37.
56. Allen, T.M., Hancen, C., Martin, F., Redemann, C., and Yau-Young, A. (1991), Liposomes containing synthetic lipid derivatives of poly(ethylene glycol) show prolonged circulation half-lives in vivo, *Biochim. Biophys. Acta*, 1066, 29–36.
57. Klibanov, A.L., Maruyama, K., Torchilin, V.P., and Huang, L. (1990), Amphipathic polyethyleneglycols effectively prolong the circulation time of liposomes, *FEBS Lett.*, 268, 235–237.
58. Bakker-Woudenberg, I.A.J.M., Schiffelers, R.M., ten Kate, M.T., Storm, G., Guo, L., and Working, P. (2000), Targeting of antibiotics in bacterial infection using pegylated long-circulating liposomes, *J. Liposome. Res.*, 10, 513–521.
59. Bakker-Woudenberg, I.A.J.M., ten Kate, M.T., Guo, L., Working, P., and Mouton, J.W. (2001), Improved efficacy of ciprofloxacin administered in polyethylene glycol-coated liposomes for treatment of *Klebsiella pneumoniae* pneumonia in rats, *Antimicrob. Agents Chemother.*, 45, 1487–1492.
60. Bakker-Woudenberg, I.A.J.M., ten Kate, M.T., Guo, L., Working, P., and Mouton, J.W. (2002), Ciprofloxacin in Polyethylene Glycol-Coated Liposomes: Efficacy in Rat Models of Acute or Chronic *Pseudomonas aeruginosa* Infection, *Antimicrob. Agents Chemother.*, 46, 2575–2581.
61. Parsons, C.L., Boychuk, D., Jones, S., Hurst, R., and Callahan, H. (1990), Bladder surface glycosaminoglycans: an epithelial permeability barrier, *J. Urol.*, 143, 139–142.
62. Mysorekar, I.U. and Hultgren, S.J. (2006), Mechanisms of uropathogenic *Escherichia coli* persistence and eradication from the urinary tract, *Proc. Natl. Acad. Sci. USA*, 103, 14170–14175.
63. Wright, K.J., Seed, P.C., and Hultgren, S.J. (2005), Uropathogenic *Escherichia coli* flagella aid in efficient urinary tract colonization, *Infect. Immun.*, 73, 7657–7668.
64. Christensen, D., Korsholm, K.S., Rosenkrands, I., Lindenstrøm, T., Andersen, P., and Agger, E.M. (2007), Cationic liposomes as vaccine adjuvants, *Expert Rev. Vaccines*, 6, 785–796.

65. Sapra, P. and Allen, T.M. (2003), Ligand-targeted liposomal anticancer drugs, *Prog. Lipid Res.*, 42, 439–462.
66. Tauber, A.L., Borregaard, N., Simon, E., and Wright, J. (1963), Chronic granulomatous disease: A syndrome of phagocyte oxidase deficiencies, *Medicine*, 62, 286.
67. Winkelstein, J.A., Marino, M.C., Johnston, R.B. Jr., Boyle, J., Cumutte, J., Gallin, J.I., Malech, H.L., Holland, S.M., Ochs, H., Quie, P., Buckley, R.H., Foster, C.B., and Chanock, S.J. (2000), Chronic granulomatous disease. Report on a national registry of 368 patients, *Medicine*, 79, 155–169.
68. Gallin, J.I. (1983), Recent advances in chronic glaucomatous disease, *Ann. Intern. Med.*, 99, 657–674.
69. Sloan, J.M., Cameron, C.H., Maxwell, R.J., McCluskey, D.R., and Collins, J.S. (1996), Colitis complicating chronic granulomatous disease. A clinicopathological case report, *Gut*, 38, 619–622.
70. McDermott, R.P. (1994), Inflammatory bowel disease, *Curr. Opin. Gastroenterol.*, 10, 355–357.
71. Al-Mobaireek, K.F. (2001), Ulcerative colitis and chronic granulomatous disease: A case report and review of the literature, *Saudi J. Gastroenterol.*, 7, 119–121.
72. Mitchell, T.G. and Perfect, J.R. (1995), Cryptococcosis in the era of AIDS-100 years after the discovery of *Cryptococcus neoformans*, *Clin. Microbiol. Rev.*, 8, 515–548.
73. Vecchiarelli, A. (2000), Immunoregulation by capsular components of *Cryptococcus neoformans*, *Med. Mycol.*, 38, 407–417.
74. Casadevall, A. and Perfect, J.R. (1998), *Cryptococcus neoformans*, ASM Press, Washington, DC.
75. Martinez, L.R. and Casadevall, A. (2005), Specific antibody can prevent fungal biofilm formation and this effect correlates with protective efficacy, *Infect. Immun.*, 73, 6350–6362.
76. Martinez, L.R. and Casadevall, A. (2006), *Cryptococcus neoformans* cells in biofilms are less susceptible than planktonic cells to antimicrobial molecules produced by the innate immune system, *Infect. Immun.*, 74, 6118–6123.
77. Martinez, L.R. and Casadevall, A. (2006), Susceptibility of *Cryptococcus neoformans* biofilms to antifungal agents *in vitro*, *Antimicrob. Agents Chemother.*, 50, 1021–1033.
78. Walsh, T.J., Schlegel, R., Moody, M.M., Costerton, J.W., and Salcman, M. (1986), Ventriculoatrial shunt infection due to *Cryptococcus neoformans*: an ultrastructural And quantitative microbiological study, *Neurosurgery*, 18, 373–375.
79. Banerjee, U., Gupta, K., and Venugopal, P. (1997), A case of prosthetic valve endocarditis caused by *Cryptococcus neoformans* var. neoformans, *J. Med. Vet. Mycol.*, 35,139–141.
80. Braun, D.K., Janssen, D.A., Marcus, J.R., and Kauffman, C.A. (1994), Cryptococcal infection of a prosthetic dialysis fistula, *Am. J. Kidney Dis.*, 24, 864–867.
81. Penk, A. and Pittrow, L. (1999), Role of fluconazole in the long-term suppressive therapy of fungal infections in patients with artificial implants, *Mycoses*, 42, 91–96.
82. Rabea, E.I., Badawy, M.E., Stevens, C.V., Smagghe, G., and Steurbaut, W. (2003), Chitosan as antimicrobial agent: applications and mode of action, *Biomacromolecules*, 4, 1457–1465.

83. Martinez, L.R., Mihu, M.R., Han, G., Frases, S., Cordero, R.J., Casadevall, A., Friedman, A.J., Friedman, J.M., and Nosanchuk, J.D. (2010), The use of chitosan to damage *Cryptococcus neoformans* biofilms, *Biomaterials*, 31, 669–679.

84. Carlson, R.P., Taffs, R., Davison, W.M., and Stewart, P.S. (2008), Anti-biofilm properties of chitosan-coated surfaces, *J. Biomater. Sci. Polym. Ed.*, 19, 1035–1046.

85. Pasquantonio, G., Greco, C., Prenna, M., Ripa, C., Vitali, L.A., Petrelli, D., Di Luca, M.C., and Ripa, S. (2008), Antibacterial activity and anti-biofilm effect of chitosan against strains of *Streptococcus mutans* isolated in dental plaque, *Int. J. Immunopathol. Pharmacol.*, 21, 993–997.

86. Busscher, H.J., Engels, E., Dijkstra, R.J., and van der Mei, H.C. (2008), Influence of a chitosan on oral bacterial adhesion and growth *in vitro*, *Eur. J. Oral. Sci.*, 116, 493–495.

87. Meshulam, T., Levitz, S.M., Christin, L., and Diamond, R.D. (1995), A simplified new assay for assessment of fungal cell damage with the tetrazolium dye, (2,3)-bis-(2-methoxy-4-nitro-5-sulphenyl)-(2H)-tetrazolium-5-carboxanilide (XTT), *J. Infect. Dis.*, 172, 1153–1156.

88. Banks, I.R., Specht, C.A., Donlin, M.J., Gerik, K.J., Levitz, S.M., and Lodge, J.K. (2005), A chitin synthase and its regulator protein are critical for chitosan production and growth of the fungal pathogen *Cryptococcus neoformans*, *Eukaryot. Cell*, 4, 1902–1912.

89. Bachmann, S.P., VandeWalle, K., Ramage, G., Patterson, T.F., Wickes, B.L., Graybill, J.R., and López-Ribot, J.L. (2002), *In vitro* activity of caspofungin against *Candida albicans* biofilms, *Antimicrob. Agents Chemother.*, 46, 3591–3596.

90. Soustre, J., Rodier, M.H., Imbert-Bouyer, S., Daniault, G., and Imbert, C. (2004), Caspofungin modulates *in vitro* adherence of *Candida albicans* to plastic coated with extracellular matrix proteins, *J. Antimicrob. Chemother.*, 53, 522–525.

91. Trampuz, A. and Zimmerli, W. (2005), New strategies for the treatment of infections associated with prosthetic joints, *Curr. Opin. Investig. Drugs*, 6, 185–190.

92. Nosanchuk, J.D. and Casadevall, A. (1997), Cellular charge of *Cryptococcus neoformans*: contributions from the capsular polysaccharide, melanin, and monoclonal antibody binding, *Infect. Immun.*, 65, 1836–1841.

93. Miyake, Y., Tsunoda, T., Minagi, S., Akagawa, Y., Tsuru, H., and Suginaka, H. (1990), Antifungal drugs affect adherence of *Candida albicans* to acrylic surfaces by changing the zeta-potential of fungal cells, *FEMS Microbiol. Lett.*, 57, 211–214.

94. Savard, T., Beaulieu, C., Boucher, I., and Champagne, C.P. (2002), Antimicrobial action of hydrolyzed chitosan against spoilage yeasts and lactic acid bacteria of fermented vegetables, *J. Food Prot.*, 65, 828–833.

95. Sudarshan, N.R., Hoover, D.G., and Knorr, D. (1992), Antibacterial action of chitosan, *Food Biotechnol.*, 6, 257–272.

96. Jung, B., Kim, C., Choi, K., Lee, Y.M., and Kim, J. (1999), Preparation of amphiphilic chitosan and their antimicrobial activities, *J. Appl. Polym. Sci.*, 72, 1713–1719.

CHAPTER 11

# POLYMER-BASED ANTIMICROBIAL DELIVERY CARRIERS

## 11.1. BASIC CONSIDERATION FOR THE PREVENTION OF DEVICE-RELATED INFECTIONS THROUGH DEVELOPMENT OF NEW DEVICES

As microbial adherence is an essential step in the pathogenesis of foreign body-related infection (FBRI), inhibition of adherence appears to be a very attractive approach for prevention. All important steps in the pathogenesis (e.g., adhesion, accumulation, and biofilm formation), represent possible targets against which prevention strategies may be directed (Table 11.1). Although there is now a more detailed insight into the molecular pathogenesis of device-related infection, as outlined in Chapter 3, this has not yet led to strategies directed against specific adherence mechanisms, especially because it is still unknown if a specific adhesin (e.g., protein and polysaccharide) is genus- or species-specific or merely strain-specific. Therefore, most of the already developed strategies have focused on the modification of medical devices, especially of catheters.

Alteration of the material surface (e.g., of a polymeric catheter) leads to a change in specific and nonspecific interaction with microorganisms. Such a surface modification of polymeric medical devices may lead to a reduced microbial adherence via altered interactions with proteins and platelets.

The development of so-called antimicrobial polymers is aimed predominantly at the prevention of microbial colonization rather than microbial

*Biofilm Eradication and Prevention: A Pharmaceutical Approach to Medical Device Infections*,
By Tamilvanan Shunmugaperumal

**TABLE 11.1. Possible Strategies Directed Against Specific Factors in the Pathogenesis of Catheter-Related Infection**

| Steps in Pathogenesis | Possible Preventive Strategy |
|---|---|
| Adhesion | Antiadhesive surfaces by polymer surface modification<br>Inhibition of specific adherence mechanisms<br>Antimicrobial devices? |
| Accumulation | Inhibition of specific factors involved in accumulation (e.g., antibiotics against polysaccharide–adhesin, accumulation-associated protein)<br>Antimicrobial devices |
| Biofilm formation | Antimicrobial devices<br>Interference with quorum-sensing (QS)<br>Electrical current, ultrasound + antimicrobials |

adherence. Devices containing antibacterials, disinfectants, or metals have been evaluated experimentally. Some of them are in clinical trials or are commercially available in part because such devices are already used in clinical applications (e.g., intravascular catheters). Destruction of the biofilm embedding surface-adherent microorganisms by enzymes or ultrasound plus subsequent antibacterial therapy, as well as the electrical enhancement of antibacterial penetration through biofilms [1–4], are all therapeutic strategies rather than preventive measures. Therefore, it seems obvious that, because of the particular pathogenesis of FBRI, approaches that are directed against bacterial colonizaion of a device are very promising. Medical devices made out of a material that would be antiadhesive or at least colonization-resistant *in vivo* would be the most suitable candidates to avoid colonization and subsequent infection. In the last 15–20 years, there have been a large number of studies dealing with this problem, in part using different strategies. A general overview is given in Table 11.2 and most of the studies have been performed with intravascular catheters because of their widespread use. Thus, the main focus of this Chapter is on the discussion of modified catheter materials.

## 11.2. POLYMER-BASED DRUG DELIVERY CARRIERS

Many reviews have highlighted the use of biodegradable polyesters as effective drug carriers including nano- or microparticles, hydrogels, micelles, and fibrous scaffolds [5–10]. Inevitably, there are advantages and disadvantages associated with each delivery system. However, these experimental approaches have been investigated in a number of infections, including periodontitis and osteomyelitis, as well as intracellular infections (e.g., tuberculosis and brucellosis). Small biodegradable microspheres are useful alternatives to liposomes for targeting drugs to the monocyte–macrophage system. They tend not to

**TABLE 11.2. Prevention Strategies of Device-Related Infections by Material Modification**

| *Catheters and Devices Used in Modification Processes* |
|---|
| Intravascular catheters |
| Urinary catheters |
| Ventricular catheters |
| Continuous ambulatory peritoneal dialysis catheters |
| Catheter hubs |
| Cuffs |
| Dressings |
| Tubing systems |
| *Process of Modification* |
| Modification of basic polymers (antiadhesive polymers) |
| Incorporation of superficial bonding of antimicrobial substances (antimicrobial polymers) like antibacterials and antiseptics |
| Metals with antimicrobial activity |

suffer from the same difficulties of low encapsulation efficiency and stability on storage typically exhibited by liposomal formulations. Moreover, biodegradable microspheres prepared from poly(lactic acid) (PLA) and poly(lactic-*co*-glycolic) acid (PLGA) can release encapsulated drugs in a controlled way, depending on the method of microencapsulation and the physicochemical properties of the polymer and drug. Table 11.3 [11,12] presents most of the antibacterial drugs that have been used in controlled-release systems to date. The molecular weight of the drug, its water solubility, as well as its solubility in organic solvent, its melting temperature, and its antibacterial spectrum, must be known in order to design an antibiotic-eluting system. These properties are presented for each drug mentioned in Table 11.3.

## 11.3. IMPLANTABLE MATRICES, BEADS, STRUT, MICROPARTICLES, FIBROUS SCAFFOLDS, THERMOREVERSIBLE GELS, AND SO ON

Polymeric materials from both natural and synthetic origins are widely recognized as effective delivery carriers of antimicrobial agents to infections associated with implants. Now, there is an enormous amount of literature in this area and no single book chapter could give comprehensive coverage. Therefore, only the main areas of research with some selected examples were indicated. I apologize in advance to colleagues whose work has been omitted through lack of space. However, interested readers are encouraged to refer to a very recent review by Zilberman and Elsner [13]. This review describes approaches

**TABLE 11.3. Antibacterial Drugs and Their Properties**[a]

| Class–Drug | Molecular Weight ($g\,mol^{-1}$) | Water Solubility ($mg\,mL^{-1}$) | pH Induced in the Surrounding | Solubility in Organic[b] Solvents | Melting Temperature (°C) | Antibacterial Spectrum |
|---|---|---|---|---|---|---|
| *Aminoglyoosides* | | | | | | |
| Amikacin | 585.6 | Highly soluble (185 $mg\,mL^{-1}$) | Base | Insoluble | 220–230 (decomposes) | Broad spectrum, many Gram-positive and -negative bacteria |
| Gentamicin | 477.6 | Highly soluble (100 $mg\,mL^{-1}$) | | DMF, MeOH, EtOH, ether, $CHCl_3$, acetone. Low solubility: DMSO | 102–108 (Hydrochloride 194–209) | |
| Tobramycin | 467.5 | Highly soluble (538 $mg\,mL^{-1}$) | | Low: EtOH | 168 | |
| *Cefalosporins* | | | | | | |
| Cefazolin | 454.5 | Slightly soluble (0.487 $mg\,mL^{-1}$) | Weak acid | DMF, pyridine, acetone. Low: EtOH, methanol | 198–200 (decomposes) | Predominantly active against Gram-positive bacteria |
| Cefoperazone | 645.7 | Slightly soluble (0.286 $mg\,mL^{-1}$) | | MeOH | 169–171 | Gram-positive, with increased activity against Gram-negative bacteria |

| | | | | | | |
|---|---|---|---|---|---|---|
| *Glycopeptides* | | | | | | |
| Vancomycin | 1449.3 | Highly soluble ($>100\,mg\,mL^{-1}$) | Amphoteric | DMSO | 185–188 | Mainly Gram-positive bacteria. Mycobacteria |
| *Macrolides* | | | | | | |
| Erythromycin | 733.9 | Slightly soluble ($1.44\,mg\,mL^{-1}$) | Weak base | High: MeOH, EtOH, acetone, ACN, $CHCl_3$, EtOAc, ether, DMF, DMSO. Low: hexane, toluene | 191 | Gram-positive and fastidious Gram-negative bacteria |
| *Nitromidazoles* | | | | | | |
| Metronidazole | 171.2 | Soluble ($10\,mg\,mL^{-1}$) | Lipophilic, low ionization | High: EtOH. Low: ether, $ChCl_3$ | 158–160 | Most anaerobes |
| *Penicillins* | | | | | | |
| Ampicillin | 349.4 | Soluble ($10.1\,mg\,mL^{-1}$) | Acid | High: MeOH, DMF. EtOH, acetone, DMF. DMSO. Low: $ChCl_3$ | 199–202 (decomposes) | Gram-positive and some Gram-negative bacteria. Broader spectrum than most penicillins. |

**TABLE 11.3** ***Continued***

| Class–Drug | Molecular Weight ($g\,mol^{-1}$) | Water Solubility ($mg\,mL^{-1}$) | pH Induced in the Surrounding | Solubility in Organic[b] Solvents | Melting Temperature (°C) | Antibacterial Spectrum |
|---|---|---|---|---|---|---|
| *Polypeptides* | | | | | | |
| Colistin (polymyxin E) | 1155.4 | Highly soluble ($564\,mg\,mL^{-1}$) | Base | MeOH, DMF, DMSO, Low: Dioxane | 200–220 | Mainly Gram-negative bacteria |
| *Quinolones* | | | | | | |
| Ciprofloxacin | 331.4 | Insoluble ($0.001\,mg\,mL^{-1}$) | Amphoteric | High: MeOH, DMF, DMSO. Low: Dioxane | 255–257 (decomposes) | Broad spectrum. Active against both Gram-positive and -negative bacteria |
| Ofloxacin | 361.4 | Soluble ($28.3\,mg\,mL^{-1}$) | | $CHCl_3$. Low: EtOH, MeOH | 250–257 (decomposes) | More effective against Gram-negative than Gram-positive bacteria, but active against several important pathogens in both groups |

| | | | | | | |
|---|---|---|---|---|---|---|
| *Rifamycins* | | | | | | |
| Rifampin/ rifampicin | 823.0 | Slightly soluble ($1.4\,mg\,mL^{-1}$) | Lipophilic, low ionization | DMSO, $CHC_{13}$, ethyl acetate methanol, THE. Low: acetone | 183–188 | | Gram-positive and fastidious Gram-negative bacteria. Mycobacteria |
| *Tetracydines* | | | | | | |
| Doxycycline | 444.5 | Slightly soluble ($0.63\,mg\,mL^{-1}$) | Amphoteric | High: MeOH, Dioxane, DMF | 201 | | Broad spectrum, many Gram-positive and -negative bacteria, mycoplasma. Doxycycline and minocycline are more active against *Streptococcus aurus* and various streptococci than tetracycline. |
| Minocycline | 457.5 | Soluble ($52\,mg\,mL^{-1}$) | | Low: EtOH | N.A | | |
| Tetracycline | 444.5 | Slightly soluble ($0.23\,mg\,mL^{-1}$) | | High: Toluene, ether, EtOAc, acetone. Low: MeOH, EtOH, $CHCl_3$, DMF, Dioxane | 165 | | |

[a]See Refs. [11,12].

[b]*N*,*N*-Dimethylformamide = DMF, dimethyl sulfoxide = DMSO.

not available = NA

for local prevention of bacterial infections based on antibiotic-eluting medical devices. These devices include bone cements, fillers and coatings for orthopedic applications, wound dressings based on synthetic and natural polymers, intravascular devices, vascular grafts, and periodontal devices. In addition, part of this review is dedicated to novel composite drug-eluting fibers and structured drug-eluting films, which are designed to be used as basic elements of various devices.

Several resorbable materials (e.g., collagen) [14], gelatin [15], polymers in different chemistries (e.g., polylactides) [16–19], copolymers of lactide and glycolide [17,20–23], polyanhydrides [24], polycaprolactone [25–27], biodegradable bone cements [28,29], hydroxyapatite and glass ceramics [30–33], calcium sulfate [34], and fibrin sealant implants [35] have been investigated for use as drug-delivery systems of various antibiotics. Limited clinical reports are available from collagen-gentamicin sponge [14], and antibiotic impregnated calcium phosphate [36]. However, no significant number of these materials has been approved yet by the U.S. Food and Drug Administration (FDA) meant for antibiotic therapy.

## 11.4. ANTIBIOTIC-LOADED BONE CEMENTS AND FILLERS

Prevention and treatment of osteomyelitis, particularly associated with orthopedic implant surgery, have been the focus of many studies. Systems implanted at the same time as the prosthesis may be either nonbiodegradable or biodegradable. Few selected examples are discussed below in each category. The most extensively studied and earliest commercially available device for controlled release of antibiotics was developed in the 1970s according to Buchholz and Engelbrecht's [37] innovative idea of releasing antibiotics from the newly introduced nonbiodegradable poly(methyl methacrylate) (PMMA) bone cement. This device is still widely accepted as a means for reducing bone infection. Thus, antibiotic–PMMA cement and spacer beads constitute an effective system of local drug delivery of antibiotic agents in patients with bone and soft-tissue infections. Debridement followed by implantation of antibiotic–PMMA beads and systemic administration of antibiotic agents has achieved a 100% success rate in treating chronic osteomyelitis.

The nondegradable PMMA matrices are produced by a polymerization reaction between a solid and a liquid component that are mixed together. The former typically contains PMMA powder, an initiator and the drug and additives. The latter contains methyl methacrylate (MMA) monomers and other additives. Curing of the cement mixture occurs within minutes, thus trapping the drug within the dense glassy bulk. Incorporation of antibiotics in this type of system is limited to antibacterial drugs that are able to withstand the heat generated by polymerization. Recorded polymerization temperatures range between 70 and 120 °C [38]. Loaded drugs are released through mechanisms of water pore penetration, soluble matrix dissolution, and outward diffusion of solubilized drug via matrix imperfections (accessible pores and cracks). Poly(methyl methacrylate) typically displays a biphasic release pattern characterized by an initial burst release followed by a long tail of low, ineffective and largely incomplete release that continues for days or months. With evidence of subtherapeutic release of gentamicin 25 years after the primary operation [39], a number of studies have revealed that <10% of the trapped drug is eventually released from the cement [40–43]. Furthermore, it has several drawbacks: The PMMA enables only a small fraction of the loaded drug to diffuse through the polymer pores [40–43] and may possibly shelter resistant bacteria, thus causing treatment failure. Moreover, PMMA is not biodegradable, and when clinical failure occurs, secondary surgery is necessary to remove the PMMA before new bone can regenerate.

In 1969, Buchholz and co-workers [37,44,45] introduced the technique of combining antibiotics with bone cements. Palacos bone cement containing gentamicin powder was introduced as a commercial product in 1970. The CMW bone cement containing gentamicin was introduced in 1990. Moreover, commercial acrylic antibiotic-impregnated bone cement products have been sold in Europe for >20 years. The antibiotics are either premixed by the manufacturer or added by the surgeon in the operating room. The FDA has recently approved the use of the following low-dose premixed cements: Cobalt™ G-HV (Biomet); Palacos® G (Biomet); DePuy 1 (DePuy Orthpedics); Cemex® Genta (Exactech); VersaBond™ AB (Smith & Nephew), which contain gentamicin; and Simplex® P (Stryker Orthpedics), which contains tobramycin. Nevertheless, FDA approval of these low-dose antibiotic-loaded bone cements is restricted to cases of joint revision following the elimination of an active infection. These cements are therefore more appropriate as a preventative measure than for the treatment of an established infection that still requires hand mixing of higher dosages of various antibiotics into the cement by the physician [46]. Surgeons have been hand mixing commonly used cements: Palacos® (Smith & Nephew), Simplex® (Howmedica), CMW (DePuy), and Zimmer (Zimmer) with antibiotics (e.g., penicillin, erythromycin, colistin, cephalosporines, gentamicin, polymyxin, vancomycin, and tobramycin). This pattern of use has primarily been the result of antibiotic selection based on identification of the infecting organism [46].

Díez-Peña et al. [47] reported that hand mixing additional antibiotics into the low-dose (2.89% wt. gentamicin) commercial bone cement CMW-1® (DePuy) to values ~20% significantly improves the drug's release mechanism and enables almost complete release of the incorporated drug. It is thought that where "reservoirs" of gentamicin exist in close vicinity, water is able to create elution paths that enable more efficient release of gentamicin from the inner domains. Loading of additional drug may, however, lead to an undesirable weakening of the bone cement. This is a major compromise if the cement is used for implant fixation, where mechanical strength is imperative [48]. A similar weakening effect is observed following incorporation of various antibiotics into biodegradable osteo-conductive calcium phosphate bone cements (CPCs) [49–52], due to an interaction between the water-soluble drug and the setting reaction of the cement after adsorption of the drug molecules [51,52].

Incidently, bone cements produced by different manufacturers vary in their mechanical properties and antibiotic elution characteristics. Small changes in the formulation of a bone cement, which may not be apparent to surgeons, can also affect these properties. The supplier of Palacos bone cement with added gentamicin changed in 2005. Therefore, Bridgens et al. [53] carried out a study to examine the mechanical characteristics and antibiotic elution of Schering-Plough Palacos, Heraeus Palacos, and Depuy CMW Smartset bone cements. Both Heraeus Palacos and Smartset bone cements performed significantly better than Schering-Plough Palacos in terms of mechanical characteristics, with and without additional vancomycin ($p < 0.001$). All cements show a deterioration in flexural strength with increasing addition of vancomycin, albeit staying above International Organization for Standardization (ISO) minimum levels. Both Heraeus Palacos and Smartset elute significantly more gentamicin cumulatively than Schering-Plough Palacos. Smartset elutes significantly more vancomycin cumulatively than Heraeus Palacos. The improved antibiotic elution characteristics of Smartset and Heraeus Palacos are not associated with a deterioration in mechanical properties. Although marketed as the "original" Palacos, Heraeus Palacos has significantly altered mechanical and antibiotic elution characteristics compared with the most commonly used previous version.

Schnieders et al. [54] reported that microencapsulation of gentamicin in biodegradable PLGA microspheres prior to mixing of the cement can prevent the negative interaction of antibiotic and cement and may also offer better control over drug release. This first example of a drug-eluting composite cement was found to be capable of up to 30% drug loading without compromising mechanical properties and demonstrated both a low burst and linear release of gentamicin over a period of 100 days.

Eptacin™, a biodegradable polyanhydride implant in the form of linked beads containing gentamicin for local delivery of the antibiotic to infected bone, presents an alternative to the nonbiodegradable acrylic fillers described thus far. The implant's capability of achieving a high local drug concentration at the implantation site while limiting systemic exposure to the drug has been shown in a safety study conducted with patients [55]. The fabrication of this implant requires the melting of the polymer at 125 °C in order to produce a polymer-drug mixture. It is therefore limited to thermally stable drugs. Krasko et al. [56] recently developed an injectable biodegradable synthetic polymeric device made of poly(sebacic-*co*-ricinoleic-esteranhydride) for treatment of osteomyelitis, which overcomes this drawback. The pasty hydrophobic copolymer is incorporated with 10–20% gentamicin by mixing the drug powder into the paste at room temperature, and gels *in situ* when exposed to aqueous surroundings to form a hydrophobic protective environment for the entrapped drug. The polymer degrades mainly from its surface, releasing the entrapped drug. The safety and positive effect of the device were confirmed *in vivo* on established osteomyelitis induced in a rat model.

Antibiotic-impregnated cement is used frequently in revision procedures of infected total hip and knee arthroplasties. Local antibiotic treatment is as effective as the use of systemic antibiotics. The purpose of such treatment is to provide high tissue concentrations of antibiotics and minimize systemic toxicity, especially nephrotoxicity. Though antibiotic-impregnated cement is considered safe in terms of nephrotoxicity, two cases that have implicated aminoglycoside-impregnated cement in acute renal failure (ARF) after surgery for an infected total knee arthroplasty (TKA) have been reported [57,58]. Aminoglycoside (Tobramycin)-impregnated cement is typically fashioned into beads or block spacers, which are temporarily placed in infected joint spaces. The use of aminoglycoside-impregnated bone cement has allowed the local concentration to exceed the minimum inhibitory concentration breakpoint of susceptible organisms while serum concentrations after 48 h were usually not detected. Nephrotoxic complications are rarely encountered with this type of antibiotic delivery method. However, Curtis et al. [57] reported the case of an 85-year-old man with a history of renal insufficiency who experienced acute renal failure after undergoing revision treatment of an infected knee arthroplasty with the combined use of tobramycin–cefazolin bone cement and a block spacer. Clinicians should be aware of the potential for aminoglycoside-induced nephrotoxicity from the use of this combination. Similarly, although nephrotoxic side effects are uncommon, van Raaij et al. [58] reported a case of acute renal failure after two-stage revision treatment of an infected knee prosthesis with gentamicin-impregnated beads and block spacers. The combined use of beads and a cement block spacer, both gentamicin impregnated, may have induced this severe complication. Use of this procedure in elderly patients warrants careful follow-up of renal function.

More than 250,000 joint replacements are performed yearly in the United States. A common complication is infection, which occurs in 1–2% of primary replacements and 3–4% of revisions of previously infected prostheses. Antibiotic-laden cement is used for prosthesis placement to prevent or treat infection, while minimizing systemic drug exposure. Two more cases of postoperative ARF in conjunction with elevated serum tobramycin concentrations, after use of combined tobramycin- plus vancomycin-impregnated cement, this time in total hip arthroplasty (THA), have been reported recently [59]. Use of the Naranjo probability scale and consideration of possible contributing factors suggest a probable association of the antibiotic-laden cement and the development of ARF in these patients. Hence, antibiotic-laden cement with aminoglycosides and/or vancomycin has the potential for systemic toxicity. It should be used according to guidelines and with increased vigilance and prudent monitoring in patients at increased risk for nephrotoxicity. Recently, Dovas et al. [60] also report a case of ARF in a 61-year-old patient with a history of diabetes mellitus and hypertension after treatment of a febrile infection of a TKA with combined gentamicin- plus vancomycin-impregnated cement. The ARF could not sufficiently be attributed to other causes and though serum concentrations of antibiotics obtained from the 8th postoperative day and thereafter were far below the trough levels associated with nephrotoxicity, gentamicin, and vancomycin seem to have contributed significantly to ARF in this case.

## 11.5. ANTIBIOTIC-LOADED IMPLANT COATINGS

Antibiotic-loaded implant coatings present a straightforward approach for the prevention of implant-associated infections. They can provide an immediate response to the threat of implant contamination, but do not necessitate use of an additional carrier for the antibacterial agent other than the orthopedic implant itself. This is most relevant for "cementless" implantation procedures that have gained popularity due to better early- and intermediate-term results in young patients compared to cemented prostheses [61].

Unlike "passive" coating techniques that aim to reduce bacterial adhesion by altering the physiochemical properties of the substrate so that bacteria–substrate interactions are not favorable, "active" coatings are designed to temporarily release high fluxes of antibacterial agents immediately following the implantation [62]. High local doses of antibiotics against specific pathogens associated with implant infections can thus be administered without reaching systemic toxicity levels with enhanced efficacy and less probability for bacterial resistance. Recent studies have also raised the possibility of incorporating growth factors in order to promote tissue healing responses [63,64].

An antibiotic–PMMA strut was used by Chen and Lee [65] for treating spinal pyogenic spondylitis in a case report of a 57-year-old woman with C5–C6 pyogenic spondylitis, progressive kyphotic deformity, and neurological deficits. The patient underwent anterior C5 and C6 corpectomy and spinal reconstruction in which the antibiotic–PMMA strut was used. The strut was 14 mm in diameter and contained PMMA and vancomycin powder. The operation was technically successful, and no complication related to anesthesia or the surgical procedure occurred. At the 12-month follow-up examination, dynamic radiographs revealed cervical spine stabilization. The patient's neck pain subsided and she recovered neurologically with no residual infection. No antibiotic–PMMA strut dislodgment or failure was identified although a 9.8% subsidence of the strut into the vertebrae was observed [65].

The utilization of a bioactive ceramic coating containing hydroxyapatite (HA), calcium phosphate, and other osteoconductive materials as antibiotic carriers offers the added value of providing the physiochemical environment and structural scaffold required for bone-implant integration. *In vitro* release of antibiotics from hydroxyapetite-coated implants has been reported for chlorhexidine, vancomycin, gentamicin, tobramycin, and several other antibiotics [66–70] whose antibacterial efficacy was shown *in vitro* by the formation of inhibition zones in agar plate testing. The conventional plasma spraying technique for HA coating is associated with high processing temperatures, and therefore does not enable the incorporation of antibiotics in the process. Most reported work therefore focuses on soaking antibiotics onto plasma-sprayed HA. Stigter et al. [71] were the first to report the incorporation of tobramycin into HA coatings using a "biomimetic" coating technology at a mild temperature (37 °C). In short, a supersaturated solution of calcium phosphate containing ~3% w/w tobramycin was coprecipitated onto Ti alloy plates, forming an ~40-μm thick carbonated hydroxyapatite layer. In their later work [66], it was concluded that antibiotics containing carboxylic groups have a better interaction with Ca, resulting in improved binding and higher incorporation into the calcium phosphate coating. Alas, the longest antibacterial effect achieved still does not exceed 3 days [72]. To date, the only *in vivo* examination of a hydroxyapetite-coated implant in a rabbit infection model supports the concept by showing a significant decrease in infection rates. However, further substantiation of its biocompatibility and osseo-integration must be carried out [73].

The study of biodegradable polymeric coatings made from polylactic acid and its copolymers with glycolic acid is more established. Release profiles last from several hours to 12 days after exposure to an aqueous environment [74–77]. An additional advantage of such coatings is the relative ease with which the polymer can be applied to both alloys and plastics with polished, irregular, or porous surfaces using a simple dip-coating technique [75]. The implant can be dipped several times in a solution of polymer and antibiotics in an organic solvent to achieve a dense or thick polymer coating. The promising results displayed in an animal model for this type of coating [76] were taken a step further and its first use in humans was investigated for internal fixation of open tibial fractures using gentamicin poly(D,L-lactide) (PDLLA) coated tibial nails (UTN, Synthes, Bochum, Germany) [63,77]. Gentamicin was not detected in the serum and no adverse events were observed during a 1-year follow-up.

Alternative biodegradable coating materials that have been studied in recent years include natural rosin-based biopolymers and polyhydroxyalkanoates. Rosin is a natural polymer obtained from pine trees that is composed of a mixture of diterpene acids, known as resin acids, and a smaller amount of other acidic and neutral bodies. It demonstrates excellent film forming, coating, and microencapsulating properties. Its suitability has been confirmed by Fulzele et al. [78], who demonstrated permanent release of ciprofloxacin over a period of 90 days with 90% of the encapsulated drug released and good compatibility *in vivo*. Polyhydroxyalkanoates incorporated with Sulperazone® (cefoperazone) and Duocid® (ampicillin) in the form of rods have already shown promising results in treating implant-related osteomyletis in rabbits [79]. In addition to their biodegradability and biocompatibility, they also feature piezoelectricity, which is claimed to induce bone growth in load-bearing areas [79]. Rossi et al. [80] reported the coating of disks cut from a femoral hip implant with polyhydroxyalkanoates loaded with gentamicin. The coating was prepared by pouring dissolved polymer and gentamicin in chloroform onto metal specimens followed by drying of the mixture. The coating exhibited an initial burst release followed by continuous *in vitro* release of gentamicin over a period of 6 weeks, with bacterial eradication within 24–48 h, depending on the copolymer composition [80].

Antibiotics are given systemically prior to surgery in orthopedic and trauma surgery to prevent implant-related infection. However, due to the disturbed bony structure and the local vascularity of trauma patients, an appropriate local antibiotic level might not be achieved by circulating antibiotics. In addition, the dose required for systemic administration of antibiotics is relatively high in comparison to the dose required for local administration at the implant–bone interface. In most surgical procedures that include the incorporation of implants, the tissue–implant interface is especially prone to microbial contamination. Aiming for high protective tissue levels of the antibiotic agent at the interface by local application of prophylaxis appears to be a reasonable approach. Systemic side effects of the antibiotic can be avoided and higher local drug levels can be achieved without risking systemic toxicity. Impaired local blood supply due to surgical trauma, hematoma, and edema may affect the delivery of the antibiotic when administered systemically. Therefore, several strategies for local antibiotic prophylaxis have been attempted (e.g., antibiotic-loaded bone cements, antibiotic-impregnated collagen sponges, and PMMA beads) [81–83]. However, certain aspects need to be considered if local prophylaxis is to be performed: The technique of delivery must guarantee a rapid release of the antibiotic from the carrier and local drug levels well above the minimum inhibitory concentration (MIC) of current microorganisms need to be achieved. The drug release must be restricted to a limited period of time to prevent development of resistant bacterial strains. Bactericidal antibiotics should be favored over bacteriostatic. The use of self-dissolving (i.e., biodegradable) drug carriers is of advantage as secondary surgery for removal is not necessary. Considering these points, a local drug delivery system for gentamicin application was developed by Schmidmaier et al. [77]. Gentamicin was chosen as the antibiotic as it has been used successfully as a locally applied antibiotic in orthopedic surgery [84,85]. Its broad antimicrobial spectrum, covering most bacteria commonly involved in osteomyelitis, and its bactericidal effect, even on nonproliferating microorganisms [86], make it favorable for local application. In an animal experiment, the efficacy of local prophylaxis of gentamicin was compared to a systemic single shot of gentamicin and to a combination of both administrations [77]. The medullary cavities of rat tibiae were contaminated with *S. aureus* and titanium K-wires were implanted into the medullary canals.

For local antibiotic therapy, the implants were coated with biodegradable PDLLA loaded with gentamicin. All the animals not treated with local and systemic application of the antibiotic developed osteomyelitis and all cultures of the implants tested positive for *S. aureus*. Onset of infection was prevented in 80–90% of animals treated with gentamicin-coated K-wires, with and without systemic prophylaxis. For readers' interest, gentamicin-coated intramedullary tibial nails are CE (Conformité Européene, European Conformity) certified for Europe and Canada and several patients have already been treated for implant-related infection. Up to now, eight patients with open tibia fractures have been treated with an unreamed tibial nail (UTN) coated with PDLLA and gentamicin. In the 1-year follow up, none of the patients developed an infection. So far, the results suggest that a local application of gentamicin from PDLLA coated implants might support systemic antibiotic prophylaxis in preventing implant-associated osteomyelitis [77].

## 11.6. STRUCTURED BIORESORBABLE FILMS LOADED WITH ANTIMICROBIAL AGENT

As mentioned above, bacterial adhesion to biomaterials and the ability of many microorganisms to form biofilms on foreign bodies are well established as major contributors to the pathogenesis of implant-associated infections. Major problems in treating osteomyelitis include poor distribution of the antimicrobial agent at the site of infection due to limited blood circulation to infected skeletal tissue, and inability to directly address the biofilm pathogen scenario. Controlled antimicrobial release systems inside orthopedic devices thus represent alternatives to conventional systemic treatments [87]. In one of the recent studies, Aviv et al. [88] developed and studied gentamicin-loaded bioresorbable films that can be "bound" to orthopedic implants (by slightly dissolving their surface before attaching them to the implant surface) and prevent bacterial infections by a gentamicin-controlled release phase for at least 1 month. These systems provide desired drug delivery profiles and do not require an additional implant.

Poly(L-lactic acid) (PLLA) and poly(D,L-lactic-*co*-glycolic acid) (PDLGA) films containing gentamicin were prepared by solution processing accompanied by a postpreparation isothermal heat treatment. In the process of film preparation, the solvent evaporation rate determines the kinetics of drug and polymer solidification and thus the drug dispersion–location in the film. The resulting drug-eluting systems are therefore termed "structured films". In general, two types of polymer–gentamicin film structures were created and studied for all matrix polymer types:

1. A polymer film with drug particles located on its surface. This structure, which is derived from a dilute solution, was obtained using a slow solvent evaporation rate that enables prior drug nucleation and growth on the polymer solution surface. This skin formation is accompanied by a later polymer core formation–solidification. This structure was named the "A-type".
2. A polymer film with most of the drug particles distributed within the bulk. This structure, which is derived from a concentrated solution, was obtained using a fast solvent evaporation rate and resulted from drug

nucleation and segregation within a dense polymer solution. Solidification of drug and polymer occurred concomitantly. This structure was named the "B-type".

Gentamicin is a water-soluble drug that practically does not dissolve in chloroform. Some of its particles thus diffused out toward the surface during solvent evaporation. The drug concentration near the surface is therefore probably higher than in the center. Gentamicin's therapeutic level in serum is 4–8 μg mL$^{-1}$ and its toxic level is 12 μg mL$^{-1}$ (values are available at http://www.healthdigest.org/drugs/gentamicinsulfate.html, accessed on 15-08-2009). All studied films released gentamicin at levels higher than the MIC. As expected, lower molecular weight polymers exhibited higher burst effects and higher release rates, due to a higher quantity of hydroxylic and carboxylic edge groups, which make it more hydrophilic. Furthermore, a lower molecular weight results in a lower glass transition temperature, which facilitates faster drug release from the polymer. Processing conditions strongly affect the release profile through morphology. Thus, dilute solutions and slow evaporation rates resulted in A-type films with the drug located on the surface. These films exhibited a relatively high burst effect followed by a slow release rate. In contradistinction, concentrated solutions and fast evaporation rates resulted in B-type films, in which most of the drug is located in the polymeric film and some is located on the surface. These films exhibited a relatively low burst effect followed by a lower release rate. Thus, the gentamicin release profiles from the various systems is determined by the host polymer, its initial molecular weight, and the processing conditions, which affect the drug location–dispersion in the film. Drug loading has a minor effect on the release profile. In two separate recent studies, a mathematical model for predicting drug release profiles from structured bioresorbable films and microstructure of the structured bioresorbable films were reported [89,90]. The new mathematical model exhibits a potential for simulating the release profile of bioactive agents from structured films for a wide variety of biomedical applications.

Microbiological evaluation of the effect of gentamicin release on bacterial viability was also performed [88]. These experiments were carried out in order to monitor the effectiveness of various concentrations of the antibiotic released from the films in terms of the residual bacteria compared with the initial bacterial concentration. Bacteria present in phosphate-buffered saline (PBS) only served as the control. In all experiments, the bacteria were added at the beginning of the films' release, in order to simulate contamination at the time of implantation. The results are presented in Fig. 11.1. No bacteria were left after 1–3 days compared to the control, where all bacteria survived even after 7 days in the presence of a very high concentration of the starter ($1 \times 10^8$ mL$^{-1}$ CFU). All films exhibited marked gentamicin release, which was responsible for the dramatic decrease in bacterial survival ($10^3$ mL$^{-1}$ CFU after 1 day). Moreover, the polymer–gentamicin film preparation did not affect gentamicin's activity as an antimicrobial agent. This study enabled in-depth understanding of

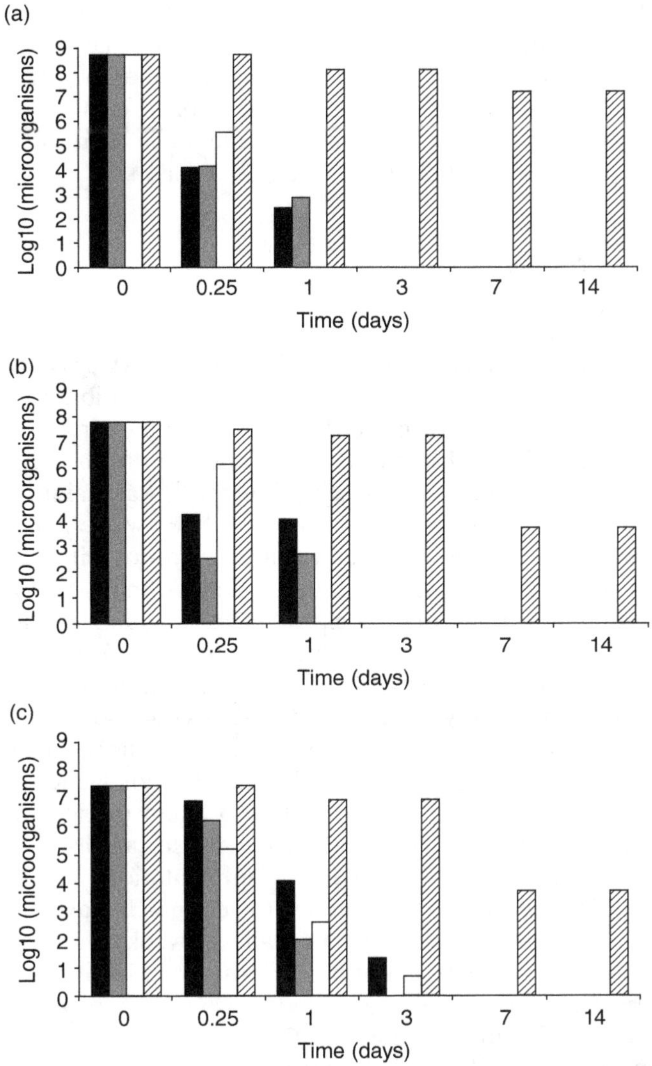


**Figure 11.1.** Number of colony forming units (CFU) versus time for microbiological experiment: (a) *Pseudomonas aeruginosa*, (b) *Staphylococcus epidermidis*, (c) *S. aureus*. The releasing films are ■ A-type PLLA film containing 30% w/w gentamicin; ▩ B-type PLLA film containing 30% w/w gentamicin; □ B-Type PDLGA film containing 10% w/w gentamicin; ▨ control-A-type PLLA film without gentamicin. (Reproduced with permission from Zilberman and Elsner *J. Control. Res.*, 130, 202–215, 2008 [13].)

gentamicin-loaded structured films. Consequently, the production of systems with the desired controlled gentamicin release profiles (i.e., with the desired burst effect and continuing release rate, within the therapeutic window) for several weeks. The developed systems can be applied on the surface of any metallic or polymeric fracture fixation device, and can therefore make a significant contribution to the field of orthopedic implants.

## 11.7. MISCELLANEOUS

Koort et al. [91] designed cylindrical composite pellets ($1.0 \times 0.9$ mm) from the bioabsorbable PDLLA matrix and ciprofloxacin (7.4 wt%). *In vitro* studies were carried out to delineate the release profile of the antibiotic and to verify its antimicrobial activity by means of MIC testing. A long-term study in rabbits was performed to validate the release of ciprofloxacin from the composite *in vivo*. A therapeutic level of ciprofloxacin (>2 $\mu g\,mL^{-1}$) was maintained between 60 and 300 days and the concentration remained below the potentially detrimental level of 20 $\mu g\,mL^{-1}$ *in vitro*. The released ciprofloxacin had retained its antimicrobial properties against common pathogens. In an exploratory long-term *in vivo* study with three rabbits, ciprofloxacin could not be detected from the serum after moderate filling (160 mg) of the tibia (follow-up 168 days), whereas after high dosing (a total dose of 1000 mg in both tibias) ciprofloxacin was found temporarily at low serum concentrations (14–34 $ng\,mL^{-1}$) during the follow-up of 300 days. The bone concentrations of ciprofloxacin could be measured in all samples at 168 and 300 days. The tested copolylactide matrix seems to be a promising option in selection of resorbable carriers for sustained release of antibiotics, but the composite needs modifications to promote ciprofloxacin release during the first 60 days of implantation.

Although PMMA beads impregnated with gentamicin have been available for ~20 years, they have to be removed usually ~4 weeks after insertion since they are nonbiodegradable [92]. A number of osteoconductive and biodegradable alternatives have been studied, including calcium phosphates (e.g., HA) whose chemical composition is similar to the bone mineral phase. Studies with ciprofloxacin incorporated into HA and PDLLA formulations implanted in the femur of rabbits, indicated that therapeutic bone levels were achieved over 6 weeks with release enhanced by erosion-disintegration and bone ingrowth into the implant [93]. Other studies using the glycopeptide antibiotic teicoplanin, effective against *S. aureus*, indicate that it too can be effective over several weeks when incorporated into microspheres prepared from PLGA (75:25) (mol.wt. 136,000) polymer [94,95]. Other materials studied against implant-related osteomyelitis include poly(3-hydroxybutyrate-*co*-3-hydroxyvalerate) (PHBV). The release behavior of sulbactam: cefoperazone from rods comprising 7, 14, or 22% (mol) 3-hydroxyvalerate were representative of typical monolithic devices where a rapid early release phase is followed by a slower and prolonged phase. With PHBV 22 rods, this extended phase lasted for up to 2 months, making it a promising controlled release vehicle, since treatment of device-related infections is typically up to 6 weeks [96]. Release studies of antibiotics adsorbed onto the surface of HA cylinders having a bimodal pore-size distribution also have a prolonged duration of release, attributed to the small pores, combined with favorable osteoconduction properties into the large pores [97]. A number of matrices have been formed using PLGA, including disks [98] and electrospun nanofibrous scaffolds [99]. In an effort to achieve the ideal drug release pattern of no lag time and zero-order release, lactide monomer or glycolide monomer have been incorporated into PLGA disks loaded with gentamicin. The idea here is that channels will form in the disk following dissolution of the monomer to aid the release of gentamicin. Disks containing 10% monomers showed nearly zero-order release kinetics for >1 month [98]. Evidence for the channel-forming properties of the monomers came from the water uptake by the disks. After 7 days, the amount of water absorbed by the control disks was 20%, compared with 60% in monomer-containing disks.

Fibrous scaffolds are currently receiving attention both as a means to prevent postsurgery adhesions and also to release drugs in a site-specific manner. While surgical implantation is required, they could be used where surgery is already indicated and the drug-release profile can be controlled by varying the scaffold's morphology, porosity, and composition. Cefoxitin was incorporated into PLGA scaffolds by Kim et al. [99]. Here, PLGA controls the rate of degradation while high molecular weight PLA confers mechanical strength to the scaffold. An amphiphilic diblock copolymer comprising PEG-*b*-poly(lactide) was added to the polymer solution to encapsulate the hydrophilic cefoxitin sodium, which otherwise has poor solubility in the PLGA solvent DMF. The drug–polymer solution was electrospun in a spinneret at 23–27 kV. As the concentration of cefoxitin increased, the scaffolds changed from a bead and string morphology, attributed to insufficient stretching of the polymer jet, to a fine fibrous structure of diameter $260 \pm 90$ nm. The spinning process did not affect the cefoxitin and following an initial burst, prolonged release was measured for up to 1 week.

Other applications of PLGA have included incorporation of antibiotic-loaded microparticles into an injectable collagen sponge, resulting in local drug delivery combined with the tissue regeneration properties of collagen [100]. Several other matrices have been used, including the aluminosilcate material halloysite. While chemically similar to kaolin, halloysite has a tubular structure that can be loaded with drug. Moreover, surface charge neutralization using cationic polymers can place an additional level of control on drug release.

The polyoxyethylene–polyoxypropylene copolymer (poloxamer 407) was used for its thermoreversible properties because it is a liquid <20 °C and forms a hydrogel at higher temperatures if it is in sufficient concentration. Furthermore, poloxamer 407 is biodegradable relatively nontoxic. The idea here is that the delivery system should be a liquid at room temperature, thus avoiding the need for refrigeration, and a device-retaining gel at the temperature within the gingival crevice. In this particular study, the transition temperature range was narrowed by the addition of PEGs, however, this caused the storage modulus (solidity) of the gel to drop as well [101]. This was overcome by the addition of 1% octyl cyanoacrylate (OCA), a powerful tissue adhesive that polymerizes at neutral pH. For storage, the pH of the system was held at 4 by addition of glacial acetic acid, and upon application to the gingival crevice the natural buffer capacity of the tissue slowly brought the pH of the formulation up to neutral, allowing the OCA to polymerize. Thermoreversible polymers have also been studied as delivery matrices in their own right [102–104]. For example, pluronic F-127 has been used to deliver vancomycin to treat methicillin-resistant *S. aureus* otitis media [105]. Based on the sol–gel phase transition measurements, a 25% w/v pluronic solution was loaded with vancomycin and injected through a 26 gauge needle. It changed to a gel at 36–37 °C. This gel consisted of large populations of the micelles and aqueous channels from which the vancomycin was released. Bioabsorption studies followed over 50 days indicated that the gel gradually disappeared, leaving open spaces that initially showed some inflammatory exudate, but by day 50 normal tissue was observed without any inflammation, fibrosis, or open spaces from the gel.

## 11.8. SURFACE MODIFIED POLYMERIC CATHETER MATERIALS

The most attractive approach to obviate FBRIs would be to develop a medical material that proves to be resistant against microbial adherence, even after insertion into the bloodstream and despite the ever-occurring interactions of the device surface with host factors (e.g., proteins and cells). There is evidence that the intrinsic properties of a material might be of advantage regarding resistance to infection. Thus, improvement of the surface texture, tailoring the protein adsorption characteristics, and improving the antithrombogenicity of a given material would be key factors in the development of innovative, infection-resistant materials. However, this goal has not yet been reached satisfactorily.

Several research groups have tried to develop polymers with new surface properties that would lead to a reduction of microbial adhesion onto medical devices, especially of catheters. Reported modified polymeric catheter materials include polystyrene modified with copolymers of poly(ethylene oxide) and poly(polypropylene oxide) [106,107], photochemical coating of polymers [108], polyvinylpyrrolidone-coated polyurethane catheter (called as Hydrocath®) [109], polyvinylpyrrolidone-coated polyurethane catheter with benzalkonium chloride [110], polyurethane surfaces modified by radiation grafting of 2-hydroxymethyl methacrylate [111], polyurethane coated with sulfonated poly(ethylene oxide), and polyurethane possessing a chain extension provided by glycerophosphorylcholine [112]. The last two approaches lead to increased water uptake and to lower bacterial adhesion. An overview mainly on experimental research on the surface modification of polymers and on binding macromolecules (e.g., albumin to surfaces) in order to prevent bacterial adherence can be found elsewhere [113,114]. So far, the only modified polymer used in clinical applications is a hydrophilic, polyvinylpyrrolidone-coated catheter (Hydrocath) based on polyurethane. Its relatively low thrombogenicity and low *in vitro* bacterial adherence should also be of benefit regarding infection resistance, although this has not yet been demonstrated in a clinical trial.

A major disadvantage of all of the previously described approaches, which aim primarily at the modification of the surface properties of basic materials (e.g., catheters or other devices) is the fact that, for thermodynamic reasons, the creation of surfaces that show a "zero" adhesion is probably not feasible. In an experimental study that investigated the relationship between bacterial adhesion and the free surface enthalpy of adhesion of a large number of differently modified polymers, Jansen and Kohnen [115] demonstrated that it is impossible to develop a polymer surface that shows an absolute bacterial "zero" adherence *in vitro*. In particular, adherence of *S. epidermidis* to a variety of polymers with different surface properties, generated by means of the glow discharge technique, was investigated [115]. The same authors found that adhesion of the bacteria to the modified materials decreased with increasing negative free enthalpy values. A certain minimum number of adherent *S. epidermidis* cells could be proved at positive free enthalpy values at which adhesion should be thermodynamically impeded. Hence, it seems impossible to design an absolute antiadhesive material that retains its properties even in the more complex *in vivo* situation, in which the native surface properties are masked by adsorption of bacterial and host components.

The biomaterials community over the past 25 years has attempted to produce anti-infective devices or implants by either (1) mechanical design alternatives (liquid–air breaks; skin cuffs; antibiotic fills; all for indwelling catheters), (2) tethered anti-infective agents, bound directly to the surface of the material (Ag coatings, tethered quaternary ammonium, synthetic antibiotics), or (3) the release of soluble toxic agents (chlorhexidine, antibiotics) into the adjacent surroundings. Mechanical design alternatives have had only marginal success and are only applicable for short-term indwelling catheters. Tethered anti-infective agents are only toxic to the initial wave of incoming bacteria and provide little residual effects once layers of dead cells accumulate, which are also inflammatory. Finally, regardless of the type of "drug-release" method used (passive vs sustained vs responsive), release of a toxic agent from a biomaterial of a soluble anti-infective agent will inevitably stop once the entrapped agent is depleted. Nevertheless, delivery of sublethal dosages of antibiotics can also lead to accelerated biofilm formation and induced virulence factor expression. In spite of the cited deficiencies, the usefulness of the polymeric catheter materials containing antimicrobial agents, anti-infective substances and antiseptic agents, as well as the catheter materials coated with metals, metal fluoride, and chelating agents in controlling biofilm-related nosocomial infections are discussed one by one in the following sections.

### 11.8.1. Antimicrobial-Containing Polymeric Catheter Materials

The loading of medical polymers with antimicrobial substances either for therapeutic or preventive purposes has a long tradition. The best known anti-infective, polymeric drug delivery systems are the PMMA–gentamicin bone cement and the PMMA–gentamicin beads (Septopal®) used for treatment of bone and soft-tissue infections [116,117]. Vascular prostheses made from Dacron® have been treated with various antibacterial agents to create infection-resistant grafts, but without routine clinical application to date [118–120].

Catheters or parts of the catheter system have been coated with an antimicrobial substance (e.g., an antibacterial, disinfectant or metal ion) is bound superficially to a catheter, either directly or by means of a carrier (as shown previously) or incorporated into the interior of the polymer. If such a device comes into contact with an aqueous environment, release of the drug into the near vicinity occurs. The amount of the antimicrobial substance released is influenced by the processing parameters, loading dose, applied technique, molecular size of the drug, and the physicochemical properties of the polymeric device. A high antimicrobial concentration is reached (at least initially) in the very near vicinity of the device surface, mostly exceeding the MIC and minimum bactericidal concentration of susceptible organisms. Most such materials exhibit a release pattern according to first-order kinetics, with an initially high drug release and subsequent exponential decrease of the released drug. However, more sophisticated drug-release systems with defined release kinetics have also been developed.

To date, it is unclear whether such a device is capable of inhibition of microbial adherence *per se*, or if its activity is more directed against colonization. However, at least an elimination of already adherent microorganisms should be achieved for the duration that the antimicrobial compound is released in the necessary concentration. Thus, such materials are especially suitable to prevent, for example, infection in short-term catheters, which originates from contamination during the insertion or from hub contamination.

There are a large number of studies on the bonding of antibacterials to biomaterials. Solovskij et al. [121] prepared polymers to which ampicillin and 6-aminopenicillanic acid were covalently bonded and that inhibited the *in vitro* growth of *S. aureus*. However, most studies have focused on the incorporation or superficial coating of antimicrobials rather than on covalent bonding by chemical reaction. Sherertz et al. [122] used a rabbit model to investigate intravascular catheters coated with several antimicrobial substances (dicloxacillin, clindamycin, fusidic acid, and chlorhexidine). The frequency of catheter infections was significantly reduced compared with the control group when the dicloxacillin-coated catheter was used. Similarly, the incorporation of flucloxacillin, clindamycin, and ciprofloxacin into polyurethane polymers demonstrated a considerable reduction of the *in vitro* adherence of *S. epidermidis* [111,123]. Another approach involves the loading glycopeptide teicoplanin into a commercially available central venous, hydrophilic-coated polyurethane catheter (Hydrocath) and shows the capability of this catheter to prevent colonization with *S. epidermidis* and *S. aureus* for a period of at least 48 h [124,125]. When teicoplanin and Ag were incorporated into the Hydrocath, the catheter showed a considerable activity against *S. epidermidis*, *E. coli*, and *Candida albicans* [115].

Kamal et al. [126] evaluated the efficacy of a cefazolin-containing catheter (in which cefazolin was bound to benzalkonium chloride) in a prospective, randomized trial. There was a significant decrease in catheter colonization (sevenfold) as determined by the semiquantitative tip culture method [127,128] and no catheter-related bloodstream infection (CRBSI) was observed in this study. A comparative study involving before and after the routine use of cefazolin catheters in the intensive care unit (ICU), the authors described a marked reduction in the rate of CRBSI from 11.5 to 5.1 infections/1000 catheter days [129].

Raad et al. [130,131] reported on the broad-spectrum activity against Gram-negative and -positive organisms and *C. albicans* of a minocycline–rifampicin catheter based on *in vitro* and animal data. This catheter has been marketed as the Cook Spectrum™ catheter (Cook Critical Care, Bloomington, IN) and is coated on the inner and outer surface with minocycline and rifampicin, which have a synergistic or additive action in combination. In a prospective, randomized clinical trial [132], the minocycline–rifampicin catheter was compared with an uncoated control catheter and demonstrated a statistically significant decrease in catheter colonization (8 vs 26% for the control catheter, $p < 0.001$) and in CRBSI (0 vs 5%, $p < 0.01$). In a large multicenter trial, the minocycline–rifampicin catheter was compared with another commercially available catheter containing chlorhexidine and silver sulfadiazine (the CHSS catheter), which is being described further in Section 11.8.2 [133]. It was found that the minocycline–rifampicin catheter was threefold less likely to be colonized (7.8 vs 22.6% for the CHSS catheter, $p < 0.0001$) and 12-fold less likely to lead to CRBSI (0.3 vs 3.4%, $p < 0.002$). This difference has been explained by the fact that minocycline–rifampicin catheters are coated internally and externally (in contrast with the first-generation CHSS catheter), the combination of minocycline and rifampicin showing superior surface activity than chlorhexidine and, finally, that the minocycline–rifampicin catheter retain surface antimicrobial activity longer *in situ* [134]. Although resistance against minocycline and rifampicin could not be detected in clinical trials, this remains of concern as *in vitro* development of resistance has been demonstrated [135]. Additional approaches on antibacterial-containing catheters include the adsorption of cefamandole nafate on functionalized urethane catheters that were then used to coat a commercial central venous catheters (CVC) [136] and the use of a combination of an antibacterial substance (rifampicin) in combination with an antifungal substance (miconazole) in a polyurethane catheter [137].

A disadvantage of these approaches might result from the risk for development of resistance against the antimicrobial agents, especially if antibacterials considered as first-line drugs in the therapy of infections are used as an active part of the modified catheters.

### 11.8.2. Antiseptics-Containing Polymeric Catheter Materials

Antimicrobial substances that differ from antibacterials (e.g., antiseptics) also have been used to develop new catheter materials. The disinfectant Irgasan® was incorporated into several polymer catheters, showing a reduction of infections in rabbits [138]. Jansen et al. [124] used the hydrophilic Hydrocath catheter to incorporate iodine, leading to a polyvinylpyrrolidone–iodine complex on the inner- and outer-catheter surface. *In vitro* adherence of various microorganisms (*Staphylococcus spp.*, *E. coli*, *Candida spp.*, *Pseudomonas spp.*) was completely inhibited for the time of iodine release. After iodine exhaustion, reloading of the catheter was possible. Tebbs and Elliott [110] incorporated benzalkonium chloride into triple-lumen Hydrocath catheters and demonstrated a long-lasting antimicrobial activity of the catheters against *staphylococci* and a somewhat lesser activity against Gram-negative bacteria and *C. albicans*.

The most promising development in this field was a catheter using a combination of an antiseptic CHSS catheter. This catheter became available ~12 years ago, is polyurethane-based, and impregnated with minute amounts of chlorhexidine and silver sulfadiazine (ArrowGard, Arrow International, Reading, PA). A synergistic effect of chlorhexidine and sulfadiazine has been shown *in vitro* [139]. This first-generation CHSS catheter is coated only on the exterior surface and exhibits antimicrobial properties for ~15 days. Since its introduction >8 million catheters have been sold worldwide and a considerable number of randomized clinical trials have been performed with this type of catheter [140–150]. In the study with the greatest patient numbers, which also used molecular methods for the confirmation of CRBSI, the CHSS catheter was associated with a twofold reduction in the incidence of catheter colonization and a fivefold reduction of CRBSI (RR 0.21, 95% CI 0.03, 0.95; $p = 0.03$) [147]. As the first-generation CHSS catheters are coated only externally, colonization of the inner lumen as a result of hub contamination might also be of greater relevance with longer duration of placement. For these reasons, a new second-generation CHSS catheter has been developed that is coated both internally and externally, and that exhibits enhanced chlorhexidine activity (ArrowGard Plus, Arrow International, Reading, PA). Clinical trials with this new type of catheter are also carried out and a significant reduction in catheter colonization was observed [151]. Development of resistance to chlorhexidine has been demonstrated *in vitro* [152]. However, *in vitro* resistance to either chlorhexidine or silver sulfadiazine associated with the use of the antimicrobial catheter has not yet been reported. Anaphylactoid reactions, probably due to chlorhexidine, have been reported from Japan and the United Kingdom, but have not been observed in the United States so far [153].

## 11.9. NEW ANTI-INFECTIVE COATINGS OF MEDICAL IMPLANTS

Infection of intravascular devices for vascular access and vascular prostheses for the replacement or bypass of damaged arteries is a rare but serious event. The infection of a vascular graft is a rare complication, with an estimated incidence of 0.5–2.5% of bypass procedures. However, the mortality and morbidity rates due to this complication are high (25–75%) [154], especially when the aorta is involved [155]. Once a prosthetic graft is infected, it almost always necessitates excision and replacement with a new prosthetic bypass. The development of infection-resistant vascular prostheses may therefore contribute to the prevention and treatment of this complication.

Surgical placement of medical implants [e.g., synthetic vascular grafts made up of polytetrafluoroethylene (PTFE) prostheses] are easily accessible to pathogens, mostly *S. aureus* and *S. epidermidis*. These pathogens colonize the implant by adhering to the patient's own proteins located on the surface of the graft and form a biofilm [156–160]. It is therefore of great importance to prevent bacterial adhesion on vascular grafts [161]. This can be achieved by antibiotic surface coatings. There have been several approaches to equipping vascular grafts with anti-infective agents to prevent bacterial colonization. Different antimicrobial agents have been used [162,163], as well as different ways to bind those drugs onto the surface of a PTFE prosthesis. A common method used to bind hydrophilic drugs onto the lipophilic surfaces of PTFE grafts is the use of surfactant-mediated agents (e.g., benzalkonium chloride [164,165] or tridodecylmethylammonium chloride) [166]. Another method of drug binding is the incorporation of drugs into biodegradable polymer carriers [75].

Polyethyleneterephtalate (PET, Dacron™) and ePTFE (expanded polytetrafluorethylene) vascular prostheses soaked in an antibiotic solution produce a wash-out release of antibiotics within minutes after placement [167,168]. Several approaches have been proposed for extending release over days and weeks. Antibiotics have been "bonded" by soaking collagen [118,169], albumin [170], and gelatin [171–173] sealed grafts to produce extended antibacterial activity. A comprehensive study on the effect of sealant matter and type of antibiotic used has been reported by Galdbart et al. [170]. The antibiotic release rate was found to vary with the type of antibiotic and protein support. Excess antibiotic unable to bind to the protein sealants was released immediately after soaking the graft in water, reaching up to 50%. Albumin- and gelatin-sealed grafts displayed relatively longer elution periods, especially for rifampicin, although none of the combinations displayed quantifiable amounts of antibiotics for periods exceeding 48 h. Succinylation of gelatin-sealed grafts has been used to improve matrix-drug bonding via ionic reactions between the drug and the matrix [173]. Overall, a prosthesis soaking in antibiotic (passive adsorption) provides immediate preventive protection of the graft as the drug reservoir is depleted within 4–7 days after implantation. Ginalska et al. recently reported an attempt to covalently immobilize gentamicin [174] and amikacin [175] to a gelatin-sealed PET graft via glutaraldehyde activation. They found that the antibiotic was bound in mixed-type way via three types of interactions, predominately strong covalent bonds, but also weak interactions: physical adsorption and ionic bonds. Only 3 and 15% of the total drug amount was released *in vitro* within 7 days for gentamicin and amikacin, respectively, and the remaining drug was bound to the biomaterial surface at high concentrations for at least 30 days. During this period, the prostheses exhibited growth inhibition of several bacterial strains at low inoculum concentrations. They may thus offer better protection against bacterial infection and biofilm formation than previously described [175]. The mode of action of very firmly bound antibiotics against bacteria remains unknown, but it is possible that they alter bacterial adherence to the prosthesis without being released as free molecules [155].

An alternative to modified gelatin binding is offered by Blanchemain et al. [168,176,177] who demonstrate the feasibility of coating cyclodextrins (CDs) on vascular Dacron grafts. Cyclodextrins are truncated torus-shaped cyclic oligosaccharides that have a hydrophobic internal cavity and a hydrophilic external wall. They are able to capture various active molecules and progressively release them unmodified. Dacron fibers are coated by a polycondensation reaction between CDs and citric acid as a cross-linking agent at 90 °C to form a polymer network of cross-linked CDs that physically adhere to the Dacron fibers. An *in vitro* drug release study of coated grafts demonstrated a linear release of vancomycin >50 days [168,177].

The work by Matl et al. [178] presents new lipid-based formulations to incorporate antibiotics for anti-infective action in grafts. Commercially available PTFE grafts with a diameter of 6 mm (Alpha Graft PTFE; Alpha Research Deutschland GmbH, Berlin, Germany) (Fig. 11.2) were coated with lipophilic agents (e.g., PDLLA) (Resomer R203H, Boehringer Ingelheim, Ingelheim, Germany), tocopherol acetate (Sigma-Aldrich AG, Deisenhofen, Germany), the diglyceride Softisan 649, and the triglyceride Dynasan 118 (Sasol Germany GmbH, Witten, Germany) as carriers for gentamicin and teicoplanin. The implants were coated with the carrier containing the drug by two dip-coating procedures. The dip-coating procedure was carried out in sterile sealable glass vials in the presence of a magnetic stir bar on a magnetic stirrer (RET basic IKAMAG; IKA) for 5 min, with a drying period of 5 min between the two coating procedures. All coating steps were carried out under aseptic conditions in a laminar airflow hood. The coatings developed with PDLLA, tocopherol acetate, or Dynasan 118, as the drug carrier completely inhibited the proliferation of *S. aureus* in pathologically relevant concentrations, while preserving biocompatible and hemocompatible characteristics. Because gentamicin and teicoplanin do not dissolve in the organic solvents used, samples were coated in drug-carrier suspensions. As a result, coatings consisted of antibiotic parti-

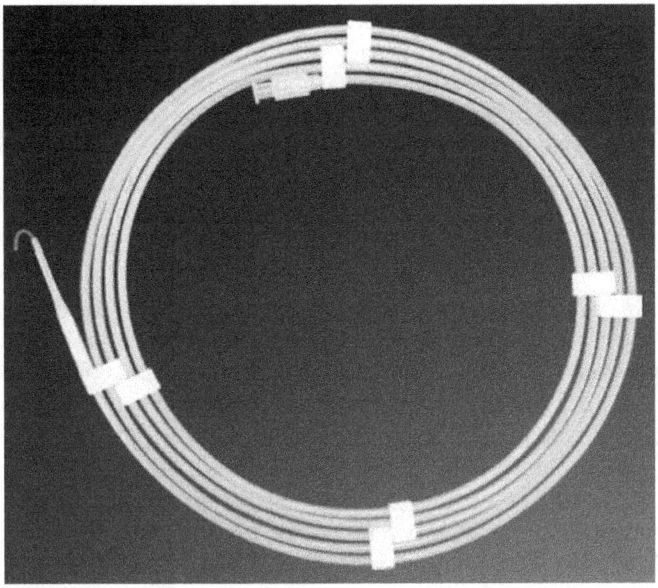

**Figure 11.2.** Commercially available PTFE graft with a diameter of 6 cm.

cles incorporated into the polymer. An initial drug burst in the first hour of elution is the consequence, since antibiotic particles from the surface of the coating dissolve rapidly after contact with elution buffer, *in vitro*. Particles located deeper inside the lipid-based polymer are released only after polymer degradation or diffusion through the polymer. The development of the biodegradable drug delivery systems described by Matl et al. [178] and *in vitro* studies of those systems highlight the most important requirements for effective as well as compatible anti-infective coatings of PTFE grafts. If these results can be confirmed *in vivo*, these drug delivery systems could be of great interest for vascular surgery.

### 11.9.1. Metals-Coated Polymeric Catheter Materials

Among metals with antimicrobial activity, Ag has raised the interest of many investigators because of its good antimicrobial action and low toxicity [179]. Silver also has extensively been used for the development of infection-resistant urinary catheters.

Sioshansi [180] used ion implantation to deposit Ag-based coatings on a Si rubber, which thereafter demonstrated antimicrobial activity. Silver–copper surface film, sputter-coated onto catheters materials, also showed antimicrobial activity *Pseudomonas aeruginosa* biofilm formation [181]. In a recent research, an ion beam technique applying low implantation energy has been used for the formation of Ag nanoparticles on the surface of polymers that exhibited and improved effect on bacterial adhesion [182]. Jansen and Kohnen

[115] developed an antimicrobial polymer by binding Ag ions to an acid modified, negatively charged polyurethane surface. Another approach is loading of a hydrophilic polyurethane catheter with silver nitrate [183]. In addition, surface-coated polyurethane catheters with a Ag surface thickness of 15–20 have been investigated with regards to their biocompatibility and antimicrobial efficacy. They showed markedly decreased adherence of Gram-positive and -negative microorganisms *in vitro* [184]. Further interest has been raised regarding devices in which Ag is distributed in the form of nanoparticles or in combination with other elements (e.g., C and Pt). The "Erlanger" Ag catheter used microdispersed Ag technology to increase the quantity of available ionized Ag [185]. The "Oligon" catheters are composed of polyurethane in which C, Ag, and Pt particles are incorporated, which leads to an electrochemically driven release of Ag ions in the outer and inner vicinity of the catheter surface. However, a peripherally implanted central catheter based on this technology (Olympicc™, Vygon, Cirencester, UK) has been withdrawn from the market, at least in Germany, because of mechanical problems associated with this type of catheter. A more recent development is the Oligon Vantex® catheter (Edwards Life Science, Irvine, CA) [186]. Other approaches are catheters with the "active iontophoresis" technology in which microorganisms are repelled by current generated from a carbon impregnated catheter [187] or where low amperage current is produced by two electrically charged parallel Ag wires helically wrapped around the proximal segment of Si catheters [188].

Several clinical studies have been performed with Ag containing intravascular catheters. In a randomized, prospective study in hemato–oncological patients, a silver sulfate–polyurethane catheter (Fresenius AG, Bad Homburg, Germany and UK) was associated with a significantly lower rate of CRBSI compared with the control group (10.2 vs 22.5%, $p = 0.01$) [189]. In three trials, the "Erlanger" Ag catheter in which the Ag is microdispersed was evaluated [185,190,191]. In the adult population, a reduction in catheter colonization and in "catheter associated sepsis" was observed.

However, the authors used criteria for determining CRBSI that differed from most other studies. Furthermore, recent clinical investigation failed to show a statistically significant difference in the colonization rate of the Ag catheter compared with a control catheter [191]. Ranucci et al. [186] compared the Oligon Vantex® catheter, composed of Ag, C, and Pt with a benzalkonium chloride treated catheter (Multi-Med, Edwards Life Sciences, Irvine, CA) in a prospective randomized trial. Use of the Oligon Ventax catheter decreased the rate of catheter colonization by 11%, while the rate of CRBSI did not differ significantly between the Oligon Vantex and control group.

### 11.9.2. Metal Oxide–Fluoride Nanoparticle Coated Sterile Surfaces to Inhibit Biofilm Formation

The inherent resistance of biofilms to killing and their pervasive involvement in implant-related infections has prompted the search for surfaces–coatings that inhibit bacterial colonization. One approach comes from recent progress in nanotechnology, which offers an opportunity for the discovery of compounds with antimicrobial activity, as well as the use of "nanofunctionalization" surface techniques. Recent examples include the direct antibacterial properties of colloidal ZnO nanoparticles toward a broad range of microorganisms [192,193] or the selective targeting of Au nanorods toward pathogenic bacteria and killing them by applying photothermal treatment [194]. Other examples include the functionalization of biomaterials with antibacterial properties by coating [195], impregnation [196–198], or embedding nanomaterials [199,200].

Fluorides are well known for their antimicrobial activity [201,202]. This activity is mediated via three major mechanisms: (1) the formation of metal-fluoride complexes, especially with Al and Be cations, which interact with F-ATPase and nitrogenase enzymes inhibiting their activities [203]; (2) the formation of hydrogen fluoride (HF), which disturbs the proton movement through the cell membrane [204]; and finally (3) $F^-$ or HF can directly bind and inhibit specific cellular enzymes. For example, enolase (an important enzyme in glycolysis) is known to be inhibited by a complex of $F^-$ and $Mg^{2+}$ at micromolar concentrations in low pH [205].

Recently, Lellouche et al. (206), utilized a simple and fast microwave-based synthesis method to synthesize $MgF_2$ nanoparticles ($MgF_2$.Nps), and characterized their activity against two common nosocomial biofilm-forming pathogens (i.e., *E. coli* and *S. aureus*). Scanning and transmission electron microscopic techniques indicated that the $MgF_2$.Nps attach and penetrate into the cells. Flow cytometry analysis revealed that the Nps caused a disruption in the membrane potential. The $MgF_2$.Nps also induced membrane lipid peroxidation and once internalized can interact with chromosomal DNA. Based on these findings, these authors further explored the possibility of using the $MgF_2$.Nps to coat surfaces and inhibit biofilm formation. A microwave synthesis and coating procedure was utilized to coat glass coupons. The $MgF_2$ coated surfaces effectively restricted biofilm formation of the tested bacteria. The effectiveness of $MgF_2$ coated surfaces to inhibit bacterial colonization as a function of time was examined [206]. As can be seen in Fig. 11.3 (*a* and *b*), the coated surfaces are able to restrict *S. aureus* and *E. coli* biofilm formation throughout the entire 3 days. Microscope evaluation of the surfaces clearly shows that the coated surfaces do not allow bacterial colonization and biofilm formation compared to the untreated controls. It is important to note that only on the third day do single cells begin to appear on the $MgF_2$ coated surfaces and many of those (~50%) are dead (i.e., stained red) based on a live–dead staining [Fig. 11.3(*a*)]. This data is also supported by viable counts obtained directly from the biofilm formed on the surfaces. Uncoated glass surfaces supported a massive biofilm formation ($12.6 \times 10^{11}$ and $11.6 \times 10^{11}$ CFU $cm^{-2}$ for *E. coli* and *S. aureus*, respectively, for the 3rd day) while $MgF_2$ coated surfaces dramatically restricted bacterial colonization (9.3 and 8.0 CFU $cm^{-2}$ for *E. coli* and *S. aureus*, respectively, in the last day). These results suggest that $MgF_2$

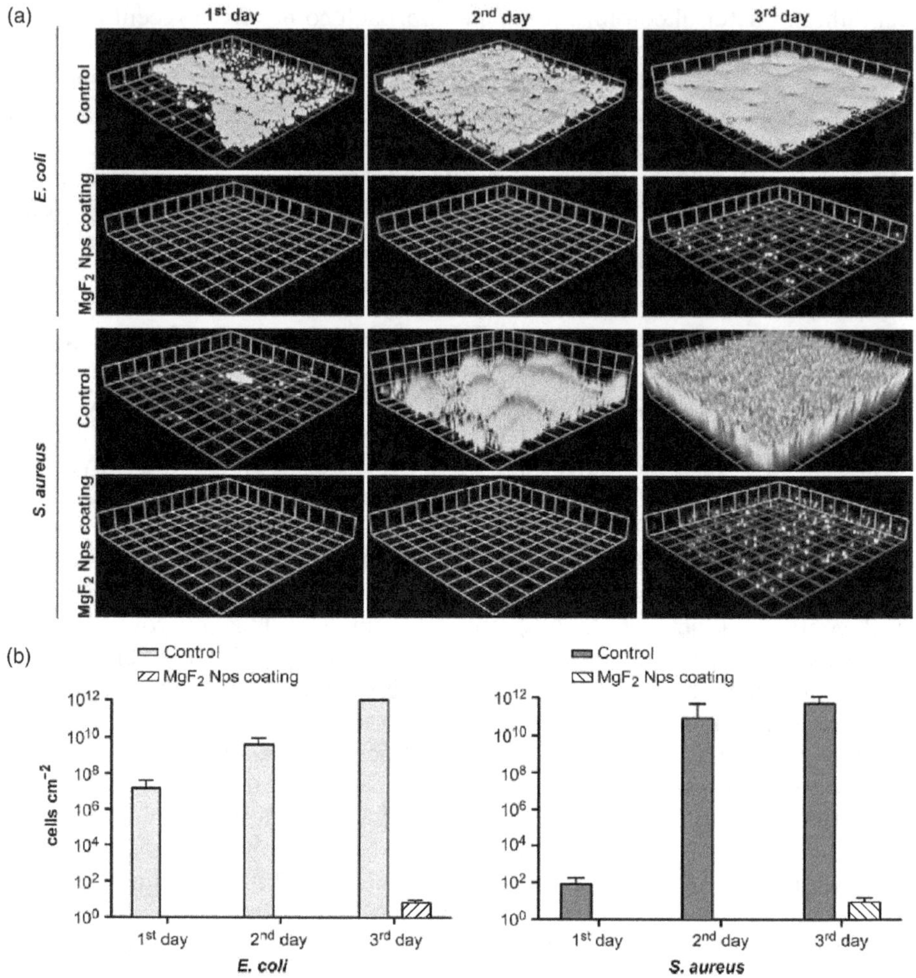


**Figure 11.3.** Extended antibiofilm activity of $MgF_2$.Nps coatings on glass surfaces. (a) Confocal laser scanning microscopy (CLSM) images of *E. coli* and *S. aureus* following biofilms formation over the course of three consecutive days on uncoated and $MgF_2$. Nps coated surfaces. Green and red staining represents, respectively, live and dead bacterial cells. In all images, 1 unit equals 13.8 mm. (b) Viable count of the biofilm cells. (control refers to the biofilm development on uncoated surface). (Reproduced with permission Lellouche et al. *Biomaterials*, 30, 5969–5978, 2009 [206].) (See color insert.)

nanoparticles are effective in restraining bacterial colonization of the surface. Furthermore, these results also highlight the potential of using $MgF_2$ nanoparticles for the design of sterile surface coatings that may be useful for various medical applications including the management of device-related nosocomial infections.

## 11.10. METAL CHELATORS TO PREVENT BIOFILM FORMATION AND CRBSI

Since metallic cations (e.g., Fe, Ca, and Mg are essential to microbial adherence, biofilm formation, and bacterial growth (Table 11.4), efforts have been directed toward utilizing high-affinity metal-binding agents that can chelate these ions thereby inhibiting bacterial growth by disturbing surface adherence and preventing biofilm production [217–241]. Figure 11.4 shows how chelators

**TABLE 11.4. Divalent Metal's Role in Cell Growth, Adherence, and Biofilm Formation**

| Divalent Metals | Organism | Function | References |
|---|---|---|---|
| Calcium | *C. albicans, P. aeruginosa* | Involved in morphogenesis, Increases and stabilizes extracellular matrix of biofilm | 207–209 |
| Magnesium | *S. epidermidis* | Increases adhesion and slime production | 207,210,211 |
| Iron | *P. aeruginosa, Actinobacillus actinomycetemcomitans* | Serves as a signal in biofilm production<br>Promotes biofilm formation | 207,210–216 |

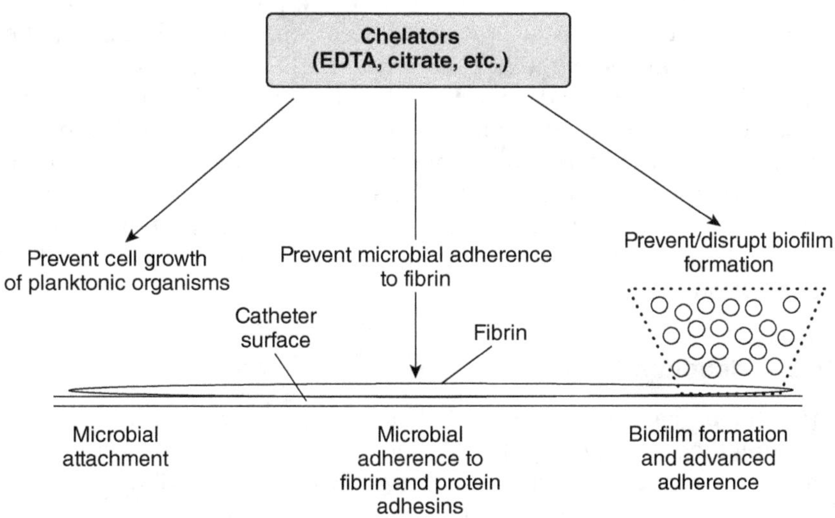


**Figure 11.4.** The role of chelators in disrupting surface adherence, preventing biofilm formation and inhibiting bacterial growth.

play a role in disrupting surface adherence, preventing biofilm formation, and inhibiting bacterial growth. Examples of metallic chelators that are demonstrated to prevent biofilm production include ethylenediaminetetraacetic acid (EDTA), sodium citrate, trisodium citrate, and so on.

### 11.10.1. Ethylenediaminetetraacetic Acid

Ethylenediaminetetraacetic acid is a metal chelator with established anticoagulant and inhibitory activity against methieillin-resistant *S. aureus* (MRSA), Gram-negative bacilli and Candida species as well as other organisms [219, 223–226,242]. It was reported that 40 mg $mL^{-1}$ (tetrasodium EDTA used in a lock solution for 21 h significantly reduced or potentially eradicated CVC associated biofilm growth of clinically relevant microorganisms [220]. Ethylenediaminetetraacetic acid, when used in combination with antibiotics [e.g., minocycline (M-EDTA)] in a lock solution, has been shown both *in vitro* and *in vivo* to significantly reduce the density of colonization by *S. epidermidis*, *S. aureus*, and *C. albicans* embedded in a biofilm [221,223]. It was demonstrated in rabbits that the M-EDTA catheter lock solution was highly efficacious in preventing catheter-related colonization, bacteremia, septic phlebitis, and endocarditis [223,229].

There have been four clinical studies conducted to determine the results of using M-EDTA as a lock solution in patients [243–246]. The M-EDTA was utilized as a lock solution in indwelling ports inserted in 14 children with cancer [245]. The authors found that no port infections, thrombotic events, or other adverse events were observed in the M-EDTA group, which was significant when compared with 10 port infections in heparin-flush group that consists of 48 control patients. While Raad et al. [243] discovered that M-EDTA was significantly efficacious at preventing recurrent CRBSI in three patients, Bleyer's [244] team compared heparin with M-EDTA as a flush solution and suggested that M-EDTA had a better 90-day catheter survival and significantly decreased the rate of catheter colonization. In addition, M-EDTA lock solution was shown to be effective in preventing catheter-related infections in patients receiving long-term parenteral nutrition [246]. The group concluded that compared to standard heparin flush, M-EDTA lock solution significantly decreases the incidence of CRBSI in the high-risk long-term parenteral nutrition population.

### 11.10.2. Trisodium Citrate

Several studies have demonstrated that TSC is synergistically effective in antimicrobial lock solutions [233,234]. Depending on the concentration used, 15 and 30% trisodium citrate (TSC) significantly reduced the number of CFUs $mL^{-1}$) of *S. aureus*, *S. epidermidis*, and *E. coli* over a period of 24 h. Moreover 30% TSC also reduced the number of *C. albicans* and *P. aeruginosa*. In addition to inhibiting bacterial growth, the researchers also suggested that

TSC, through its chelating activity on Ca and Mg, could potentially disrupt the biofilm formation [222,233,234].

To further investigate the minimum effective concentration of citrate against bacteria, Shanks et al. [228] demonstrated that sodium citrate at concentrations >0.5% efficiently inhibits biofilm formation and cell growth of *S. aureus* and *S. epidermidis*. They further showed that sodium citrate at 2% concentration or greater powerfully inhibits *in vitro* biofilm production by *S. aureus* and CoNS. Furthermore, a lower concentration of citrate (4%) was reported to be highly effective, when used in combination with taurolidine, in killing a diverse group of bacteria, including *S. aureus*, *S. epidermidis*, *P. aeruginosa*, and *Enterococcus faeclis* within biofilm [218]. These results suggested that citrate-based catheter lock solution is promising for reducing the risk of biofilm-associated infections in indwelling catheter. In addition, citrate, like EDTA, was found to increase the permeability of the outer membrane of microorganisms, thereby increasing their susceptibility to antimicrobial agents [235].

A randomized clinical trial comparing 30 TSC to high-dose heparin used in lock solutions demonstrated that TSC significantly reduced catheter-related infections and the number of major bleeding episodes associated with these lock solutions [234]. Moreover, taurolidine and 4% citrate as a catheter lock solution has been demonstrated to dramatically reduce the frequency of catheter-related bacteremia [231,232]. Recently, a randomized controlled study comparing gentamicin and 3.13% citrate to heparin alone as a lock solution in the prevention of catheter-related infections discovered that the infection-free duration of catheter use was significantly higher in the gentamicin and citrate group than in the heparin group [247]. In conclusion, gentamicin-citrate lock solution appears to be a highly effective strategy for the reduction of morbidity, and potential mortality and costs, associated with catheter-related infections.

### 11.10.3. Other Chelators

Transferrin (Tf) belongs to a family of Fe-binding monomeric glycoproteins and has been reported to possess a broad spectrum of antimicrobial properties attributed to its ability of chelating environmental Fe, thus making this essential nutrient inaccessible to an invading microorganism [236]. Lactoferrin (Lf), another Fe binding protein, has been shown by Singh et al. [217] to have the capacity of blocking biofilm development by *P. aeruginosa*. Several other chelators, including ethylene glycol bis(β-aminoethyl ether)-*N.N.N'N'*-tetraacetic acid (EGTA), deferoxamine, bismuth dimercaprol, 2,3-dimercaptosuccinic acid (DMSA), diethylene-triamine-pentaacetic acid (DTPA), and *N,N'*-ethylenebis[2-(2-hydroxyphenyl)-glycine] (EHPG), have also been demonstrated *in vitro* for their ability to disrupt biofilm formation and inhibit bacterial growth [225,226,237–239]. Recently, Ibrahim et al. [240,241] showed that Fe chelators deferiprone and deferasirox synergistically improved survival and

**TABLE 11.5. Nonexhaustive List of Iron Chelators**

| Iron Chelators |
|---|
| 1, 10-Phenanthroline |
| 3-Hydroxy-2-methyl-4-pyrone |
| Acetohydroxamic acid |
| Deferiprone |
| Deferoxamine |
| Kojic acid |
| Picolinic acid |
| Ferric ammonium citrate |
| Diethylenetriaminepentaacetic acid |

decreased fungal burden when combined with liposomal amphotericin B. Table 11.5 shows the nonexhaustive list of Fe chelators and chemical structures of selected Fe chelator's molecules are depicted in Fig. 11.5.

## 11.11. ETHANOL IN ANTIBIOTICS-CHELATOR LOCK SOLUTION

Although the combination of antibiotics with chelators is synergistically active in eradicating organisms embedded in biofilm on catheter surfaces, they (like all other antibiotic catheter locks) require a prolonged dwell time of at least 16–24 h in order to demonstrate significant activity against a high inoculum of organisms embedded in biofilm [223,230]. This prolonged dwell time might not be feasible or achievable in most clinical situations. Ethanol, on the contrary, has been shown to have broad-spectrum antimicrobial activity against microbial organisms embedded with side effects [248], lower and safer concentration of 50% ethanol alone has limited activity against staphylococcal organisms embedded in biofilm as tested in an animal model [249]. However, ethanol activity can be significantly enhanced when combined with antibiotics and chelators in lock solutions. Recently, it was demonstrated that a triple combination of 3 mg $mL^{-1}$ of minocycline and 30 mg $mL^{-1}$ EDTA (M-EDTA) in 25% ethanol used as a catheter lock solution is rapidly and synergistically active in eradicating staphylococcal and Candida organisms embedded in biofilm, within a dwell time ranging from 15 to 60 min [230].

Further studies by Chandra et al. [250,251] have shown that a triple combination of trimethoprim (TMP), EDTA, and ethanol is superior to any of these components alone in prevention and treatment of bacterial and fungal biofilms. The authors tested the combination against *C. albicans*, MRSA, and *P. aeruginosa* biofilms and demonstrated that it was able to prevent biofilm

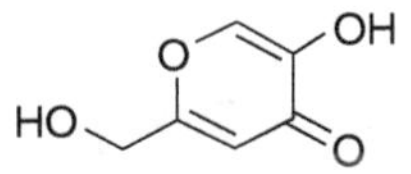

Kojic acid (log $K_a$ = 27)

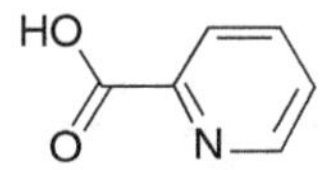

picolinic acid

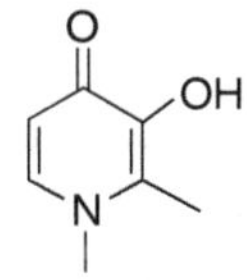

Deferiprone (log $K_a$ = 37.2)

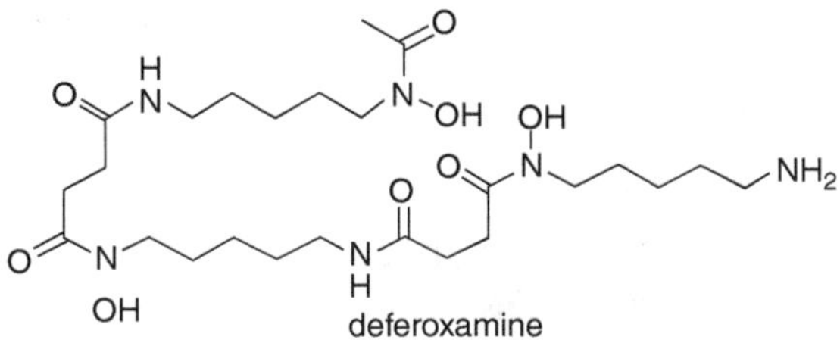

deferoxamine
log$K_a$ = 30.6

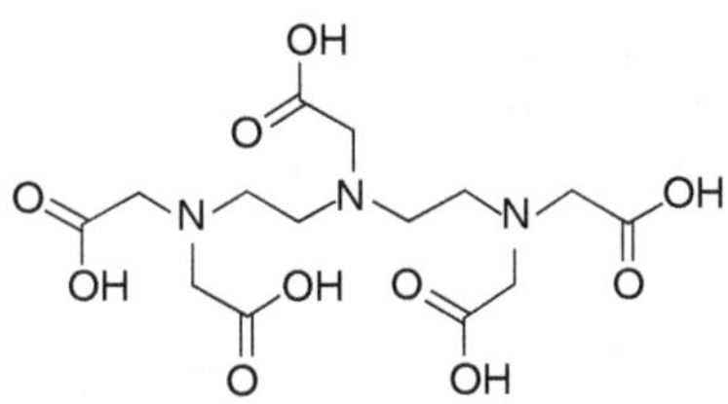

Diethylenetriaminepentaacetic acid

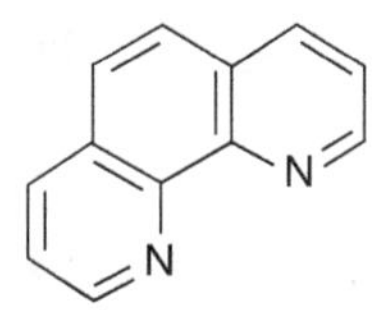

1,10-Phenanthroline (log $K_a$ = 14.1) (DTPA) (log $K_a$ = 28.6)

**Figure 11.5.** Chemical structures of selected iron chelator's molecules.

formation even after short-term exposure (15 min) and eradicated mature-phase biofilm after long-term exposure (2–4 h). On the basis of these studies, the triple combination of an antibiotic, a chelator, and a low concentration of ethanol (25%) provides an optimal antimicrobial catheter lock solution that is rapidly active within a short time (within 2 h) [247,250,251].

## 11.12. CHELATORS AS ALTERNATIVE TO HEPARIN

Today's standard of care for maintaining catheter patency is heparin or isotonic saline lock solutions. Although the use of heparin in an antibiotic lock solution is comparable and well tolerated, it has not been proven to enhance or complement antimicrobials in preventing or treating CVC related infections [252]. This finding might be related in part to the fact that heparin alone does not have any antimicrobial activity [253]. Furthermore, the use of heparin in lock solutions is associated with some risk factors (e.g., heparin-induced thrombocytopenia and thrombosis) [254]. Moreover, several investigators recently reported that heparin enhances the formation of *S. aureus* biofilm on catheter surfaces [228,249] possibly mediated by *S. aureus'* ability of producing a heparin-binding protein [223,249,255,256]. Therefore, as heparin alternatives, chelators (e.g., EDTA and citrate) have significant advantages, would include providing three necessary and important functions: an anticoagulant effect, an antibiofilm effect, and a synergistic antimicrobial effect, enhancing the antimicrobial activity of the antibiotic or antiseptic used and in turn, help prevent CRBSI and eradicate established infections [257].

## 11.13. NOVEL SMALL MOLECULE CONTROL OF BACTERIAL BIOFILM FORMATION

*Staphylococcus aureus* is one of the most frequent causes of bacterial keratitis. Infection can be severe, leading to corneal ulceration and perforation if not treated effectively. The ability of *S. aureus* to adhere to the epithelial cell glycocalyx is thought to be one of the first steps in the colonization and infection of wet mucosal surfaces. Recent evidence has shown that cell surface-associated mucins, major components of apical membranes in wet-surfaced epithelia, are critical elements of the mucosal barrier to infection [258–260]. Although mucin carbohydrates, or O-glycans, constitute up to 80% of the mucin mass of cell surface mucins [261], little is known about their contribution to a host's defense against bacterial adhesion and infection.

Membrane-anchored, cell surface-associated mucins are defined by the presence of long extracellular amino terminal domains containing hundreds of clustered O-linked glycans. The human ocular surface epithelia express at least three membrane-associated mucins, mucin1, -4, and -16 [262,263]. Structurally, they are defined by the presence of central tandem repeats of amino acids rich in serine, threonine, and proline residues, and by their extensive O-glycosylation terminal domains. These domains can extend 200–500 nm above the cell membrane, well beyond other glycoproteins on the glycocalyx, and therefore, constitute the initial site of interaction between the cell and the extracellular milieu [264]. The biosynthesis of O-glycans is enzymatically initiated by transfer of *N*-acetylgalactosamine to the side chain of a serine or threonine within the peptide core of the mucin molecule [265]. Further elongation of this structure leads to various linear and branched extensions, which may bear different terminal carbohydrate structures. Some O-glycans are known to be targeted as ligands for carbohydrate binding by adhesins on the bacterial cell surface, thus facilitating attachment [266,267]. Nevertheless, the contribution of O-glycans on cell surface-associated mucins to *S. aureus* keratitis has not been elucidated.

Benzyl-*N*-acetyl-α-D-galactosaminide (benzyl-α-GalNAc) is a chemical primer commonly used to suppress the elongation of cell-surface mucin-type

*O*-linked glycans [268]. The primer competes for elongation of the core GalNAc residue (GalNAc-*O*-Ser/Thr) found in mucin-type O-linked glycans. Benzyl-α-GalNAc has been used extensively to study physiological consequences of mucin-type O-glycans [269,270] and does not interfere with N-glycosylation [271] or with the quantity or expression pattern of glycolipids [272]. In a recent study, Ricciuto et al. [273] used benzyl-α-GalNAc to determine the role of mucin-type O-glycans in preventing *S. aureus* adhesion to differentiated human corneal epithelial cells. They have concluded that further characterization of the glycosylation changes on the epithelial cell surface in patients with higher risk of infection could, therefore, prove relevant to the development of pharmacological drugs aimed at restoring the normal composition of the glycocalyx and to the prophylactic inhibition of bacterial adhesion and invasion.

The exploitation of stresses already imposed on microorganisms by the *in vivo* environment or host defense systems represents an intriguing new approach to combating infections [274]. Furthermore, since it has been demonstrated that medically important antibiotics, including aminoglycosides, fluoroquinolones, and tetracycline, among others, work poorly in chronic infections and, in contrast, even act as intermicrobial signaling agents that stimulate bacterial biofilm formation at subinhibitory concentrations [275,276], new antimicrobial agents are needed to combat chronic infections. With this in mind, Hancock and co-workers [277] analyzed the interaction between cationic host defense (antimicrobial) peptides and *P. aeruginosa*. These peptides represent a promising class of antimicrobials and are ubiquitous in nature as components of innate immune defense systems [278–280]. They are found at mucosal surfaces or in phagocytic granules. They are characterized as having 12–50 amino acids, including 2–9 basic (Arg or Lys) residues and ~50% hydrophobic amino acids [280]. Certain peptides possess direct antimicrobial activity against Gram-positive and -negative bacteria, fungi, and protozoa. Synthetic peptides, in particular, can demonstrate minimum inhibitory concentrations (MICs) as low as 0.25–4 $\mu g\,mL^{-1}$ [281]. These peptides often have a broad spectrum of abilities to modulate immunity as part of the innate immune response. They demonstrate promise as a new approach to antimicrobial therapy [278–280]. The major human cationic host defense peptide, LL-37, is found at mucosal surfaces, in the granules of phagocytes, and in most bodily fluids at concentrations of ~2–5 $\mu g\,mL^{-1}$. It is found at much larger concentrations at sites of chronic inflammation (e.g., 30 $\mu g\,mL^{-1}$ in the cystic fibrosis (CF) lung). Although this peptide often is designated a cationic antimicrobial peptide, Hancock and co-workers [278,279] argued that its antimicrobial activity is strongly antagonized under physiological salt concentrations (e.g., its MIC for many common pathogens is 32–96 $\mu g\,mL^{-1}$ in the growth medium that usually is utilized for the assessment of antibiotic MICs). Thus its most important antimicrobial property *in vivo* relates to its potent anti-inflammatory (antiendotoxic) activity and selective ability to modulate favorable immune responses. In addition to its key role in modulating the innate immune response and weak antimicrobial activity, LL-37 potently inhibited the formation of bacterial biofilms *in vitro*. This occurred at the very low and physiologically meaningful concentration of $0.5\,\mu g\,mL^{-1}$, far below that required to kill or inhibit growth (MIC = $64\,\mu g\,mL^{-1}$). The LL-37 also affected existing, pregrown *P. aeruginosa* biofilms in polypropylene 96-well microtiter plates after 20h of incubation at 37°C. Similar results were obtained using the bovine neutrophil peptide indolicidin, but no inhibitory effect on biofilm formation was detected using subinhibitory concentrations of the mouse peptide CRAMP, which shares 67% identity with LL-37, polymyxin B, or the bovine bactenecin homologue Bac2A. By using microarrays and follow-up studies, Hancock and coworkers [277] were able to demonstrate that LL-37 affected biofilm formation by decreasing the attachment of bacterial cells, stimulating twitching motility, and influencing two major quorum-sensing (QS) systems (Las and Rhl), leading to the downregulation of genes essential for biofilm development. Results similar to this finding were reported by other research groups for the inhibition of *P. aeruginosa* biofilms by lactoferrin [217]. This cationic human glycoprotein (lactoferrin), which is present in external secretions, especially milk [282], was found to inhibit bacterial biofilm formation due to its Fe chelating properties, which also resulted in increased twitching motility.

## 11.14. CONCLUSION

The pathogenesis of a wide variety of human infections, including device-related infections, as well as infections not associated with devices, is now recognized to relate to the presence of microorganisms (bacteria, fungus, and yeast) in biofilms.

As our population ages, there will be an increase in the number of people experiencing hospitalization and receiving short- or long-term biomedical implants. As engineered biomaterials and tissue regenerative medicine advance, an increasing portion of the population will receive one or multiple biomedical devices, ranging from disposable contact lenses, dental implants, orthopedic implants, and vascular grafts to tissue engineered livers, small diameter vascular grafts that promote stem cells differentiation into endothelial cells, and polymer transfection systems that deliver micro-RNA knockout therapy to control chronic inflammation. The current healthcare approach to clean and sterilize has done little to prevent an epidemic in nosocomial infections. Biomaterials technologies employing disinfectant rinses, tethered, or release antibiotics have also done little to reduce this epidemic and may have contributed to the raise of antibiotic resistant bacteria.

The use of surgically implanted devices is increasing as a means to improve quality of life, and in some cases, to survival rates. However, these foreign bodies, once implanted, are sites of competition between host cell integration and bacterial adhesion. If bacteria are able to adhere successfully, they will undergo biofilm formation, which alters their properties and renders them resistant to commonly used antibiotics. In light of the emergence of multiresistant microorganisms (e.g., meticillin-resistant staphylococci) with reduced susceptibility or with resistance toward glycopeptides, prudent use of antimicrobials is advised. Innovative and multidisciplinary approaches for prophylaxis and management should result in novel, more effective control strategies.

Some of the new developments (e.g., antimicrobial catheters), have already been adopted, but more, good quality clinical studies are needed to better define their impact on reducing FRBIs, patient morbidity and mortality, and their cost effectiveness. These studies are needed before recommending a broader use of these devices. Although antimicrobial materials obviously have the potential to decrease infections, there is a major point of criticism–concern: that of development of antimicrobial resistance against the agents used. Still, this should be carefully monitored when using such devices and should be an important issue in forthcoming clinical studies. The most important challenge will be to implement all of the current knowledge in daily practice. In addition, translating the recent data on the mechanisms of biofilm formation and bacterial interference into applicable strategies and innovative materials may avoid the unnecessary and expensive removal of, in particular, highly needed and/or difficult to replace medical devices.

On the basis of combating biofilm antibiotic resistance by enhanced or more efficient delivery of antimicrobial agents, much research has been focused on engineering better materials and methods for treatment of biofilms [283,284]. For example, electrical, ultrasound, and photodynamic stimulation can disrupt biofilms and enhance the efficiency of certain antimicrobial agents. Aerosolization of antibiotics has been shown to be quite effective for direct application of these drugs to the respiratory system. In particular, aerosolized tobramycin, and more recently nebulized hypertonic saline, have achieved clinical efficacy in treating *P. aeruginosa* lung infection in patients with CF [285–287]. In this manner, higher concentrations of drug can be delivered directly to the site of infection. Treatment strategies for biofilms are constantly evolving. The synergy between natural compounds and traditional antibiotics seems quite promising for future clinical applications. Coupled with improved delivery mechanisms, these molecules may prove to be a boon to the medical field. Indeed, much progress has already been achieved, as seen with aerosolized delivery of tobramycin. While much research is still needed, novel treatments and biofilm inhibitory molecules are constantly being identified. These potential therapies offer much hope for the future of combating biofilm infections. Similarly, in the future, the development of medical devices based on modified anti-infective materials will lead to a further reduction of the incidence of FRBIs. However, even the best technology will fail if standard hygienic procedures, with their often easy-to-perform preventive techniques based on recommendations of the respective national guidelines, are not implemented.

Increasing scientific research over the past 10 years in biofilm formation has provided a wealth of possible targets with which to prevent or eradicate biofilm infections. Advances in the understanding of biofilm formation, coupled with emerging engineered biomaterials, provide many potential platforms and strategies to prevent or significantly reduce biofilm infections in susceptible populations.

Nevertheless, considering the additional medical expenses required for the removal of already implanted but infected medical devices, it becomes necessary to look for alternative (pharmaceutical) ways of eradicating the device-related nosocomial infections. Native and stealth (pegylated) liposomes were in fact investigated extensively for improving antiadhesive property of the implant material, for concentrating the encapsulated antimicrobial agents in an adequate amount at the infected surfaces of the medical devices, and for targeting the antimicrobial agents to biofilm-associated intracellular infections. On the other hand, biodegradable and nonbiodegradable polymer-based matrices, beads, microspheres, strut, gels, fibrous scaffolds, and so on, and surface (properties) modified polymeric catheter materials (e.g., antimicrobial, antiseptic, or metallic substances-coated polymeric materials) were also developed in an attempt to eradicate biofilm-associated infections, especially in implant and the periodontal cavity by the local delivery of the entrapped antibiotic substances. The advantages of these novel drug delivery carriers are mirrored by a high number of high-quality scientific papers, published in conventional and open-access journals. However, the potential of lipid- and polymer-based drug delivery carriers in eradicating biofilm consortia in device-related nosocomial infections is not achieved fully in terms of further clinical application and subsequent approval from healthcare authorities. Recent developments in microscopy imaging and surface-analytical techniques allowed the quantitative *in situ* investigation of cell–surface interactions at the submicron scale, providing information on the strength of microbial cell attachment to solid substrata and the properties of macromolecules involved in this process. (See details in a review by Beech et al. [288].) Gaining deeper insight into the fundamental mechanisms of biofilm-mediated deleterious interfacial processes together with understanding of the physiology of biofilm bacteria at the genomic and proteomic levels will, undoubtedly, result in the development of practices that will aid in their control especially through lipid- and polymer-based drug delivery carriers.

## REFERENCES

1. Rediske, A.M., Roeder, B.L., Brown, M.K., Nelson, J.L., Robison, R.L., Draper, D.O., Schaalje, G.B., Robison, R.A., and Pitt, W.G. (1999), Ultrasonic enhancement of antibiotic action on *Escherichia coli* biofilms: An *in vivo* model, *Antimicrob. Agents Chemother.*, 43, 1211–1214.
2. Rediske, A.M., Roeder, B.L., Nelson, J.L., Robison, R.L., Schaalje, G.B., Robison, R.A., and Pitt, W.G. (2000), Pulsed ultrasound enhances the killing of *Escherichia coli* biofilms by aminoglycoside antibiotics *in vivo*, *Antimicrob. Agents Chemother.*, 44, 771–772.

3. Khoury, A.E., Lam, K., Ellis, B., and Costerton, J.W. (1992), Prevention and control of bacterial infections associated with medical devices, *ASAIO J.*, 38, M174–M178.
4. Ascher, D.P., Shoupe, B.A., Maybee, D., and Fischer, G.W. (1993), Persistent catheter related bacteraemia: clearance with antibiotics and urokinase, *J. Pediatr. Surg.*, 28, 627–629.
5. Freiberg, S. and Zhu, X.X. (2004), Polymer microspheres for controlled drug release, *Int. J. Pharm.*, 282, 1–18.
6. Varde, N.K. and Pack, D.W. (2004), Microspheres for controlled drug delivery, *Expert. Opin. Biol. Ther.*, 4, 35–51.
7. Rabinow, B.E. (2004), Nanosuspensions in drug delivery, *Nat. Rev. Drug Discov.*, 3, 785–796.
8. Moses, M.A., Brem, H., and Langer, R. (2003), Advancing the field of drug delivery: taking aim at cancer, *Cancer Cell*, 4, 337–341.
9. Sinha, V.R. and Trehan, A. (2003), Biodegradable microspheres for protein delivery, *J. Control. Rel.*, 90, 261–280.
10. Zhao, Z., Wang, J., Mao, H.Q., and Leong, K.W. (2003), Polyphosphoesters in drug and gene delivery, *Adv. Drug Deliv. Rev.*, 55, 483–499.
11. Bryskier, A. (2005), *Antimicrobial agents: antibacterials and antifungals*, ASM Press, Washington, DC (various pagings).
12. Budavari, S. (1989) (1v.), *The Merck index: an encyclopedia of chemicals, drugs, and biologicals*, 11th ed., Merck & Co., Rahway, NJ, (various pagings).
13. Zilberman, M. and Elsner, J.J. (2008), Antibiotic-eluting medical devices for various applications, *J. Control. Rel.*, 130, 202–215.
14. Stemberger, A., Grimm, H., Bader, F., Rahn, H.D., and Ascherl, R. (1997), Local treatment of bone and soft tissue infections with the collagen-gentamicin sponge, *Eur. J. Surg.*, 578, 17–26.
15. Keogh, B.S., Triplett, R.G., Aufdemorte, T.B., and Boyan, B.D. (1989), The effect of local antibiotics in treating chronic osseous *Staphylococcus aureus* infection, *J. Oral Maxillofac. Surg.*, 47, 940–945.
16. Kanellakopoulou, K., Kolia, M., Anastassiadis, A., Korakis, T., Giamarellous-Bourboulis, K.J., Andreopoulos, A., Dounis, E., and Giamarellou, H. (1999), Lactic acid polymers as biodegradable carriers of fluoroquinolones: An *in vitro* study, *Antimicrob. Agents Chemother.*, 43, 714–716.
17. Mader, J.T., Calhoun, J., and Cobos, J. (1997), In vitro evaluation of antibiotic diffusion from antibiotic-impregnated biodegradable beads and polymethylmetacrylate beads, *Antimicrob. Agents Chemother.*, 41, 415–418.
18. Zhang, X., Wyss, U.P., Pichora, D., and Goosen, M.F. (1994), Biodegradable controlled antibiotic release devices for osteomyelitis: Optimization of release properties, *J. Pharm. Pharmacol.*, 46, 718–724.
19. Wei, G., Kotoura, Y., Oka, M., Yamamuro, T., Wada, R., Hyon S-H., and Ikada, Y. (1991), A bioabsorbable delivery system for antibiotic treatment of osteomyelitis, *J. Bone Joint Surg.*, 73, 246–252.
20. Ambrose, C.G., Gogola, G.R., Clyburn, T.A., Raymond, A.K., Peng, A.S., and Mikos, A.G. (2003), Antibiotic microspheres: Preliminary testing for potential treatment of osteomyelitis, *Clin. Orthop.*, 415, 279–285.

21. Overbeck, J.P., Winckler, S.T., Meffert, R., Törmälä, P., Spiegel, H.U., and Brug, E. (1995), Penetration of ciprofloxacin into bone: A new bioabsorbable implant, *J. Invest. Surg.*, 8, 155–162.

22. Nie, L., Nicolau, D.P., Nightingale, C.H., Browner, B.D., and Quintiliani, R. (1995), *In vitro* elution of ofloxacin from a bioabsorbable polymer, *Acta. Orthop. Scand.*, 66, 365–368.

23. Garvin, K.L., Miyano, J.A., Robinson, D., Giger, D., Novak, J., and Radio, S. (1994), Polylactide/polyglycolide antibiotic implants in the treatment of osteomyelitis, *J. Bone Joint Surg.*, 76, 1500–1506.

24. Jacob, E., Setterstrom, J.A., Bach, D.E., Heath III, J.R., McNiesh, L.M., and Cierny III., G. (1991), Evaluation of biodegradable ampicillin anhydrate microcapsules for local treatment of experimental staphylococcal osteomyelitis, *Clin. Orthop.*, 267, 237–244.

25. Le Rey, A.M., Chiffoleau, S., Iooss, P., Grimandi, G., Gouyette, A., Daculsi, A., and Merle, C. (2003), Vancomycin encapsulation in biodegradable poly(ε-caprolactone) microparticles for bone implantation. Influence of the formulation process on size, drug loading, *in vitro* release and cytocompatibility, *Biomaterials*, 24, 443–449.

26. Burd, T.A., Anglen, J.O., Lowry, K.J., Hendricks, K.J., and Day, D. (2001), *In vitro* elution characteristics of tobramycin from bioabsorbable polycaprolactone bead, *J. Orthop. Trauma*, 15, 424–428.

27. Hendricks, K.J., Lane, D.L., Burd, T.A., Lowry, K.J., Day, D., Phaup, J.G., and Anglen, J.O. (2001), Elution characteristics of tobramycin from polycaprolactone in a rabbit model, *Clin. Orthop.*, 392, 418–426.

28. Solberg, B.D., Gutow, A.P., and Baumgaertner, M.R. (1999), Efficacy of gentamycin-impregnated resorbable hydroxyapatite cement in treating osteomyelitis in a rat model, *J. Orthop. Trauma*, 13, 102–106.

29. Gerhart, T.N., Roux, R.F., Horowitz, G., Miller, R.L., Hanff, P., and Hayes, W.C. (1988), Antibiotic release from an experimental biodegradable bone cement, *J. Orthop. Res.*, 6, 585–592.

30. Mäkinen, T.J., Veiranto, M., Lankinen, P., Moritz, N., Jalava, J., Törmälä, P., and Aro, H.T. (2005), *In vitro* and *in vivo* release of ciprofloxacin from osteoconductive bone defect filler, *J. Antimicrob. Chemother.*, 56, 1063–1068.

31. Shirtliff, M.E., Calhoun, J.H., and Mader, J.T. (2002), Experimental osteomyelitis treatment with antibiotic-impregnated hydroxyapatite, *Clin. Orthop.*, 401, 239–247.

32. Saito, T., Takeuchi, R., Hirakawa, K., Nagata, N., Yoshida, T., Koshino, T., Okuda, K., Takema, M., and Hori, T. (2002), Slow-releasing potential of vancomycin-loaded porous hydroxyapatite blocks implanted into MRSA osteomyelitis, *J. Biomed. Mater. Res.*, 63, 245–251.

33. Nolan, P.C., Wilson, D.J., and Mollan, R.A. (1993), Calcium hydroxyapatite ceramic delivery system, *J. Bone Joint Surg.*, 75, 600–604.

34. Nelson, C.L., McLaren, S.G., Skinner, R.A., Smeltzer, M.S., Thomas, J.R., and Olsen, K.M. (2002), The treatment of experimental osteomyelitis by surgical debridement and the implantation of calcium sulfate tobramycin pellets, *J. Orthop. Res.*, 20, 643–647.

35. Mader, J.T., Stevens, C.M., Stevens, J.H., Roble, R., Lathrop, J.T., and Calhoun, J.H. (2002), Treatment of experimental osteomyelitis with a fibrin sealant antibiotic implant, *Clin. Orthop.*, 403, 58–72.

36. McKee, M.D., Wild, L.M., Schemitsch, E.H., and Waddell, J.P. (2002), The use of an antibiotic-impregnated, osteoconductive, bioabsorbable bone substitute in the treatment of infected long bone defects: Early results of a prospective trial, *J. Orthop. Trauma*, 16, 622–627.

37. Buchholz, H.W. and Engelbrecht, H. (1970), Uber die Depotwirkung einiger Antibiotica bei Vermischung mit dem Kunstharz Palacos [Depot effects of various antibiotics mixed with Palacos resins], *Chirurg*, 41, 511–515.

38. DiPisa, J.A., Sih, G.S., and Berman, A.T. (1976), The temperature problem at the bone-acrylic cement interface of the total hip replacement, *Clin. Orthop. Relat. Res.*, 121, 95–98.

39. Webb, J.C.J., Spencer, R.F., Lovering, A.M., and Learmonth, I.D. (2005), Very late release of gentamicin from bone cement in total hip arthroplasty (THA), *J. Bone Jt. Surg. Br.*, 87-B, (Suppl. 1), 52.

40. Levin, P.D. (1975), The effectiveness of various antibiotics in methyl methacrylate, *J. Bone Jt. Surg. Br.*, 57, 234–237.

41. Bunetel, L., Segui, A., Cormier, M., Percheron, E., and Langlais, F. (1989), Release of gentamicin from acrylic bone cement, *Clin. Pharmacol.*, 17, 291–297.

42. DiMaio, F.R., O'Halloran, J.J., and Quale, J.M. (1994), *In vitro* elution of ciprofloxacin from polymethylmethacrylate cement beads, *J. Orthop. Res.*, 12, 79–82.

43. Miclau, T., Dahners, L.E., and Lindsey, R.W. (1993), *In vitro* pharmacokinetics of antibiotic release from locally implantable materials, *J. Orthop. Res.*, 11, 627–632.

44. Wahlig, H. and Buchholz, H.W. (1972), Experimental and clinical studies on the release of gentamicin from bone cement, *Chirurg*, 43, 441–445.

45. Buchholz, H.W. and Gartmann, H.D. (1972), Infection prevention and surgical management of deep insidious infection in total endoprothesis, *Chirurg*, 43, 446–453.

46. Hanssen, A.D. (2004), Prophylactic use of antibiotic bone cement: an emerging standard-in opposition.[comment], *J. Arthroplast.*. 19 (4 Suppl. 1), 73–77.

47. Díez-Peňa, E., Frutos, G., Frutos, P., and Barrales-Rienda, J.M. (2002), Gentamicin sulphate release from a modified commercial acrylic surgical radiopaque bone cement. I. Influence of the gentamicin concentration on the release process mechanism, *Chem. Pharm. Bull.*, 50, 1201–1208.

48. Joseph, T.N., Chen, A.L., and Di Cesare, P.E. (2003), Use of antibiotic-impregnated cement in total joint arthroplasty, *J. Am. Acad. Orthop. Surg.*, 11, 38–47.

49. Joosten, U., Joist, A., Frebel, T., Brandt, B., Diederichs, S., and von Eiff, C. (2004), Evaluation of an *in situ* setting injectable calcium phosphate as a new carrier material for gentamicin in the treatment of chronic osteomyelitis: studies *in vitro* and *in vivo*, *Biomaterials*, 25, 4287–4295.

50. Armstrong, M.S., Spencer, R.F., Cunningham, J.L., Gheduzzi, S., Miles, A.W., and Learmonth, I.D. (2002), Mechanical characteristics of antibiotic-laden bone cement, *Acta Orthop. Scand.*, 73, 688–690.

51. Takechi, M., Miyamoto, Y., Ishikawa, K., Nagayama, M., Kon, M., Asaoka, K., and Suzuki, K. (1998), Effects of added antibiotics on the basic properties of anti-washout-type fast-setting calcium phosphate cement, *J. Biomed. Mater. Res.*, 39, 308–316.

52. Gbureck, U., Probst, J., and Thull, R. (2002), Surface properties of calcium phosphate particles for self setting bone cements, *Biomol. Eng.*, 19, 51–55.

53. Bridgens, J., Davies, S., Tilley, L., Norman, P., and Stockley, I. (2008), Orthopaedic bone cement: do we know what we are using? *J. Bone Joint Surg. Br.*, 90, 643–647.

54. Schnieders, J., Gbureck, U., Thull, R., and Kissel, T. (2006), Controlled release of gentamicin from calcium phosphate-poly(lactic acid-*co*-glycolic acid) composite bone cement, *Biomaterials*, 27, 4239–4249.

55. Li, L.C., Deng, J., and Stephens, D. (2002), Polyanhydride implant for antibiotic delivery-from the bench to the clinic, *Adv. Drug Deliv. Rev.*, 54, 963–986.

56. Krasko, M.Y., Golenser, J., Nyska, A., Nyska, M., Brin, Y.S., and Domb, A.J. (2007), Gentamicin extended release from an injectable polymeric implant, *J. control. Rel.*, 117, 90–96.

57. Curtis, J.M., Sternhagen, V., and Batts, D. (2005), Acute renal failure after placement of tobramycin-impregnated bone cement in an infected total knee arthroplasty, *Pharmacotherapy*, 25, 876–880.

58. van Raaij, T.M., Visser, L.E., Vulto, A.G., and Verhaar, J.A. (2002), Acute renal failure after local gentamicin treatment in an infected total knee arthroplasty, *J. Arthroplasty*, 17, 948–950.

59. Patrick, B.N., Rivey, M.P., and Allington, D.R. (2006), Acute renal failure associated with vancomycin- and tobramycin-laden cement in total hip arthroplasty, *Ann. Pharmacother.*, 40, 2037–2042.

60. Dovas, S., Liakopoulos, V., Papatheodorou, L., Chronopoulou, I., Papavasiliou, V., Atmatzidis, E., Giannopoulou, M., Eleftheriadis, T., Simopoulou, T., Karachalios, T., and Stefanidis, I. (2008), Acute renal failure after antibiotic-impregnated bone cement treatment of an infected total knee arthroplasty, *Clin. Nephrol.*, 69, 207–212.

61. Berger, R.A., Jacobs, J.J., Quigley, L.R., Rosenberg, A.G., and Galante, J.O. (1997), Primary cementless acetabular reconstruction in patients younger than 50 years old. 7- to 11-year results, *Clin. Orthop. Relat. Res.*, 344, 216–226.

62. Hetrick, E.M. and Schoenfisch, M.H. (2006), Reducing implant-related infections: active release strategies, *Chem. Soc. Rev.*, 35, 780–789.

63. Raschke, M.J. and Schmidmaier, G. (2004), [Biological coating of implants in trauma and orthopedic surgery], *Unfallchirurg*, 107, 653–663.

64. Wildemann, B., Sander, A., Schwabe, P., Lucke, M., Stöckle, U., Raschke, M., Haas, N.P., and Schmidmaier, G. (2005), Short term *in vivo* biocompatibility testing of biodegradable poly(D,L-lactide)-growth factor coating for orthopaedic implants, *Biomaterials*, 26, 4035–4040.

65. Chen, J.F. and Lee, S.T. (2006), Antibiotic-polymethylmethacrylate strut: an option for treating cervical pyogenic spondylitis, Case report, *J. Neurosurg. Spine*, 5, 90–95.

66. Stigter, M., Bezemer, J., de Groot, K., and Layrolle, P. (2004), Incorporation of different antibiotics into carbonated hydroxyapatite coatings on titanium implants, release and antibiotic efficacy, *J. Control. Rel.*, 99, 127–137.

67. DeJong, E.S., DeBerardino, T.M., Brooks, D.E., Nelson, B.J., Campbell, A.A., Bottoni, C.R., Pusateri, A.E., Walton, R.S., Guymon, C.H., and McManus, A.T. (2001), Antimicrobial efficacy of external fixator pins coated with a lipid stabilized hydroxyapatite/chlorhexidine complex to prevent pin tract infection in a goat model, *J. Trauma*, 50, 1008–1014.

68. Campbell, A.A., Song, L., Li, X.S., Nelson, B.J., Bottoni, C., Brooks, D.E., and DeJong, E.S. (2000), Development, characterization, and anti-microbial efficacy of hydroxyapatite-chlorhexidine coatings produced by surface-induced mineralization, *J. Biomed. Mater. Res.*, 53, 400–407.

69. Rauschmann, M.A., Wichelhaus, T.A., Stirnal, V., Dingeldein, E., Zichner, L., Schnettler, R., and Alt, V. (2005), Nanocrystalline hydroxyapatite and calcium sulphate as biodegradable composite carrier material for local delivery of antibiotics in bone infections, *Biomaterials*, 26, 2677–2684.

70. Teller, M., Gopp, U., Neumann, H.G., and Kühn, K.D. (2007), Release of gentamicin from bone regenerative materials: an *in vitro* study, *J. Biomed. Mater. Res. B. Appl. Biomater.*, 81, 23–29.

71. Stigter, M., de Groot, K., and Layrolle, P. (2002), Incorporation of tobramycin into biomimetic hydroxyapatite coating on titanium, *Biomaterials*, 23, 4143–4153.

72. Chai, F., Hornez, J.C., Blanchemain, N., Neut, C., Descamps, M., and Hildebrand, H.F. (2007), Antibacterial activation of hydroxyapatite (HA) with controlled porosity by different antibiotics, *Biomol. Eng.*, 24, 510–514.

73. Alt, V., Bitschnau, A., Osterling, J., Sewing, A., Meyer, C., Kraus, R., Meissner, S.A., Wenisch, S., Domann, E., and Schnettler, R. (2006), The effects of combined gentamicin-hydroxyapatite coating for cementless joint prostheses on the reduction of infection rates in a rabbit infection prophylaxis model, *Biomaterials*, 27, 4627–4634.

74. Price, J.S., Tencer, A.F., Arm, D.M., and Bohach, G.A. (1996), Controlled release of antibiotics from coated orthopedic implants, *J. Biomed. Mater. Res.*, 30, 281–286.

75. Gollwitzer, H., Ibrahim, K., Meyer, H., Mittelmeier, W., Busch, R., and Stemberger, A. (2003), Antibacterial poly(D,L-lactic acid) coating of medical implants using a biodegradable drug delivery technology, *J. Antimicrob. Chemother.*, 51, 585–591.

76. Lucke, M., Schmidmaier, G., Sadoni, S., Wildemann, B., Schiller, R., Haas, N.P., and Raschke, M. (2003), Gentamicin coating of metallic implants reduces implant-related osteomyelitis in rats, *Bone*, 32, 521–531.

77. Schmidmaier, G., Lucke, M., Wildemann, B., Haas, N.P., and Raschke, M. (2006), Prophylaxis and treatment of implant-related infections by antibiotic-coated implants: a review, *Injury*, 37 (Suppl. 2), S105–S112.

78. Fulzele, S.V., Satturwar, P.M., and Dorle, A.K. (2007), Novel biopolymers as implant matrix for the delivery of ciprofloxacin: biocompatibility, degradation, and *in vitro* antibiotic release, *J. Pharm. Sci.*, 96, 132–144.

79. Gürsel, I., Korkusuz, F., Türesin, F., Alaeddinoglu, N.G., and Hasirci, V. (2001), *In vivo* application of biodegradable controlled antibiotic release systems for the treatment of implant-related osteomyelitis, *Biomaterials*, 22, 73–80.

80. Rossi, S., Azghani, A.O., and Omri, A. (2004), Antimicrobial efficacy of a new antibiotic-loaded poly(hydroxybutyric-co-hydroxyvaleric acid) controlled release system, *J. Antimicrob. Chemother.*, 54, 1013–1018.

81. Nelson, C.L., Hickmon, S.G., and Harrison, B.H. (1994), Elution characteristics of gentamicin-PMMA beads after implantation in humans, *Orthopedics*, 17, 415–416.

82. Hettfleisch, J. and Schottle, H. (1993), Local preventive antibiotic treatment in intramedullary nailing with gentamycin impregnated biomaterials, *Aktuelle Traumatol.*, 23, 68–71.

83. Letsch, R., Rosenthal, E., and Joka, T. (1993), Local antibiotic administration in osteomyelitis treatment-a comparative study with two different carrier substances, *Aktuelle Traumatol.*, 23, 324–329.

84. Bauer, T.W. and Schils, J. (1999), The pathology of total joint arthroplasty. I. Mechanisms of implant fixation, *Skeletal Radiol.*, 28, 423–432.

85. Taylor, G.J., Bannister, G.C., and Calder, S. (1990), Perioperative wound infection in elective orthopaedic surgery [published erratum appears in J. Hosp. Infect. 1991 Feb; 17(2):155], *J. Hosp. Infect.*, 16, 241–247.

86. Eron, L.J. (1985), Prevention of infection following orthopedic surgery, *Antibiot. Chemother.*, 33, 140–164.

87. Baro, M., Sánchez, E., Delgado, A., Perera, A., and Evora, C. (2002), *In vitro-in vivo* characterization of gentamicin bone implants, *J. Control. Rel.*, 83, 353–364.

88. Aviv, M., Berdicevsky, I., and Zilberman, M. (2007), Gentamicin-loaded bioresorbable films for prevention of bacterial infections associated with orthopedic implants, *J. Biomed. Mater. Res. A*, 83, 10–19.

89. Zilberman, M., Shifrovitch, Y., Aviv, M., and Hershkovitz, M. (2009), Structured drug-eluting bioresorbable films: microstructure and release profile, *J. Biomater. Appl.*, 23, 385–406.

90. Zilberman, M. and Malka, A. (2009), Drug controlled release from structured bioresorbable films used in medical devices—a mathematical model, *J. Biomed. Mater. Res. B Appl. Biomater.*, 89, 155–164.

91. Koort, J.K., Suokas, E., Veiranto, M., Mäkinen, T.J., Jalava, J., Törmälä, P., and Aro, H.T. (2006), *In vitro* and *in vivo* testing of bioabsorbable antibiotic containing bone filler for osteomyelitis treatment, *J. Biomed. Mater Res.*, 78A, 532–540.

92. Kanellalopoulou, K. and Giamarellos-Bourboulis, E.J. (2000), Carrier systems for the local delivery of antibiotics in bone infections, *Drugs*, 59, 1223–1232.

93. Castro, C., Sánchez, E., Delgado, A., Soriano, I., Núñez, P., Baro, M., Perera, A., and Évora, C. (2003), Ciprofloxacin implants for bone infection. *In vitro-in vivo* characterization, *J. Control. Rel.*, 93, 341–354.

94. Yenice, I., Çalis, S., Atilla, B., Kas, H.S., Ozalp, M., Ekizoglu, M., Bilgili, H., and Hincal, A.A. (2003), *In vitro/in vivo* evaluation of the efficiency of teicoplanin-loaded biodegradable microparticles formulated for implantation to infected bone defects, *J. Microencapsul.*, 20, 705–717.

95. Yenice, I., Çalis, S., Kas, H.S., Ozalp, M., Ekizoglu, M., and Hincal, A.A. (2002), Biodegradable implantable teicoplanin beads for the treatment of bone Infections, *Int. J. Pharm.*, 242, 271–275.

96. Gürsel, I., Yagmurlu, F., Korkusuz, F., and Hasirci, V. (2002), *In vitro* antibiotic release from poly(3hydroxybutyrate-*co*-3-hydroxyvalerate) rods, *J. Microencapsul.*, 19, 153–164.
97. Hasegawa, M., Sudo, A., Komlev, V.S., Barinov, S.M., and Uchida, A. (2004), High release of antibiotic from a novel hydroxyapatite with bimodal pore size distribution, *J. Biomed. Mater Res.*, 70B, 332–339.
98. Yoo, J.Y., Kim, J.M., Khang, G., Kim, M.S., Cho, S.H., Lee, H.B., and Kim, Y.S. (2004), Effect of lactide/glycolide monomers on release behaviours of gentamicin sulfate-loaded PLGA discs, *Int. J. Pharm.*, 276, 1–9.
99. Kim, K., Luu, Y.K., Chang, C., Fang, D., Hsiao, B.S., Chu, B., and Hadjiargyrou, M. (2004), Incorporation and controlled release of a hydrophilic antibiotic using poly(lactide-*co*-glycolide)-based electrospun nanofibrous scaffolds, *J. Control. Rel.*, 98, 47–56.
100. Schlapp, M. and Friess, W. (2003), Collagen/PLGA microparticle controlled composites for local delivery of gentamicin, *J. Pharm. Sci.*, 92, 2145–2151.
101. Kelly, H.M., Deasy, P.B., Ziaka, E., and Claffey, N. (2004), Formulation and preliminary *in vivo* dog studies of a novel drug delivery system for the treatment of periodontitis, *Int. J. Pharm.*, 274, 167–183.
102. Veyries, M.L., Couarraze, G., Geiger, S., Agnely, F., Massias, L., and Kunzli, B. (1999), Controlled release of vancomycin from poloxamer 407 gels, *Int. J. Pharm.*, 192, 183–193.
103. Esposito, E., Carotta, V., Scabbia, A., Trombelli, L.D., Antona, P., and Menegatti, E. (1996), Comparative analysis of tetracycline-containing dental gels: poloxamer- and monoglyceride-based formulations, *Int. J. Pharm.*, 142, 9–23.
104. Miyazaki, S., Tobiyama, T., Takada, M., and Attwood, D. (1995), Percutaneous absorption of indomethacin from pluronic F 127 gels in rats, *J. Pharm. Pharmacol.*, 47, 455–457.
105. Lee, S.H., Lee, J.E., Baek, W.Y., and Lim, J.O. (2004), Regional delivery of vancomycin using pluronic F-127 to inhibit methicillin resistance *Staphylococcus aureus* (MRSA) growth in chronic otitis media *in vitro* and *in vivo*, *J. Control Rel.*, 96, 1–7.
106. Bridgett, M.J., Davies, M.C., and Denyer, S.P. (1992), Control of staphylococcal adhesion to polystyrene surfaces by polymer surface modification with surfactants, *Biomaterials*, 13, 411–416.
107. Desai, N.P., Hossainy, S.F., and Hubbell, J.A. (1992), Surface-immobilized polyethylene oxide for bacterial repellence, *Biomaterials*, 13, 417–420.
108. Dunkirk, S.G., Gregg, S.L., Duran, L.W., Monfils, J.D., Haapala, J.E., Marcy, J.A., Clapper, D.L., Amos, R.A., and Gurie, P.E. (1991), Photochemical coatings for the prevention of bacterial colonization, *J. Biomater. Appl.*, 6, 131–156.
109. Tebbs, S.E., Sawyer, A., and Elliott, T.S. (1994), Influence of surface morphology on *in vitro* bacterial adherence to central venous catheters, *Br. J. Anaesth.*, 72, 587–591.
110. Tebbs, S.E. and Elliott, T.S. (1994), Modification of central venous catheter polymers to prevent *in vitro* microbial colonization, *Eur. J. Clin. Microbiol. Infect. Dis.*, 13, 111–117.
111. Jansen, B., Schareina, S., and Steinhauser, H. (1987), Development of polymers with antiinfective properties, *Polym. Mater. Sci. Eng.*, 57, 43–46.

112. Baumgartner, J.N., Yang, C.Z., and Cooper, S.L. (1997), Physical property of analysis and bacterial adhesion on a series of phosphonated polyurethanes, *Biomaterials*, 18, 831–837.

113. Kohnen, W. and Jansen, B. (2000), Changing material surface chemistry for preventing bacterial adhesion, in: An, Y.H. and Friedman, R.J. Eds., *Handbook of Bacterial Adhesion: Principles, Methods and Applications*, Humana Press, Totowa, NJ, pp. 581–589.

114. An, Y.H., Blair, B.K., Martin, K.L., and Friedman, R.J. (2000), Macromolecule surface coating for preventing bacterial adhesion, in: An, Y.H. and Friedman, R.J., Eds., *Handbook of Bacterial Adhesion: Principles, Methods and Applications*, Humana Press, Totowa, NJ, pp. 609–625.

115. Jansen, B. and Kohnen, W. (1995), Prevention of biofilm formation by polymer modification, *J. Ind. Microbiol.*, 15, 391–396.

116. Marcinko, D.E. (1985), Gentamicin-impregnated PMMA beads: an introduction and review, *J. Foot. Surg.*, 24, 116–121.

117. Welch, A. (1978), Antibiotics in acrylic bone cement: *in vitro* studies, *J. Biomed. Mater. Res.*, 12, 679–700.

118. Moore, W.S., Chvapil, M., Seiffert, G., and Keown K. (1981), Development of an infection-resistant vascular prosthesis, *Arch. Surg.*, 116, 1403–1407.

119. Powell, T.W., Burnham, S.J., and Johnson, G., Jr. (1983), A passive system using rifampin to create an infection-resistant vascular prosthesis, *Surgery*, 94, 765–769.

120. McDougal, E.G., Burnham, S.J., and Johnson, G., Jr. (1986), Rifampin protection against experimental graft sepsis, *J. Vasc. Surg.*, 4, 5–7.

121. Solovskij, M.V., Ulbrich, K., and Kopecek, J. (1983), Synthesis of N-(2-hydroxypropyl) methacrylamide copolymers with antimicrobial activity, *Biomaterials*, 4, 44–48.

122. Sherertz, R.J., Carruth, W.A., Hampton, A.A., Byron, M.P., and Solomon, D.D. (1993), Efficacy of antibiotic-coated catheters in preventing subcutaneous *Staphylococcus aureus* in rabbits, *J. Infect. Dis.*, 167, 98–106.

123. Jansen, B. and Peters, G. (1991), Modern strategies in the prevention of polymer-associated infections, *J. Hosp. Infect.*, 19, 83–88.

124. Jansen, B., Jansen, S., Peters, G., and Pulverer, G. (1992), *In vitro* efficacy of a central venous catheter ('Hydrocath') loaded with teicoplanin to prevent bacterial colonization, *J. Hosp. Infect.*, 22, 93–107.

125. Romano, G., Berti, M., Goldstein, B.P., and Borghi, A. (1993), Efficacy of a central venous catheter (Hydrocath) loaded with teicoplanin in preventing subcutaneous staphylococcal infection in the mouse, *Zentralbl. Bakteriol.*, 279, 426–433.

126. Kamal, G.D., Pfaller, M.A., Rempe, L.E., and Jebson, P.J. (1991), Reduced intravascular catheter infection by antibiotic bonding: a prospective, randomized, controlled trial, *JAMA*, 265, 2364–2368.

127. Maki, D.G., Jarrett, F., and Sarafin, H.W., (1977), A semiquantitative culture method for identification of catheter-related infection in the burn patient, *J. Surg. Res.*, 22, 513–520.

128. Maki, D.G., Weise, C.E., and Sarafin, H.W. (1977), A semiquantitative culture method for identifying intravenous-catheter-related infection, *New Engl. J. Med.*, 296, 1305–1309.

129. Kamal, G.D., Divishek, D., Kumar, G.C., Porter, B.R., Tatman, D.J., and Adams, J.R. (1998), Reduced intravascular catheter-related infection by routine use of antibiotic-bonded catheters in a surgical intensive care unit, *Diagn. Microbiol. Infect. Dis.*, 30, 145–152.
130. Raad, I., Darouiche, R., Hachem, R., Sacilowski, M., and Bodey, G.P. (1995), Antibiotics and prevention of microbial colonization of catheters, *Antimicrob. Agents Chemother.*, 39, 2397–2400.
131. Raad, I., Darouiche, R., Hachem, R., Mansouri, M., and Bodey, G.P. (1996), The broad-spectrum activity and efficacy of catheters coated with minocycline and rifampin, *J. Infect. Dis.*, 173, 418–424.
132. Raad, I., Darouiche, R., Dupuis, J., Abi-Said, D., Gabrielli, A., Hachem, R., Wall, M., Harris, R., Jones, J., Buzaid, A., Robertson, C., Shenaq, S., Curling, P., Burke, T., and Ericsson, C. (1997), Central venous catheters coated with minocycline and rifampin for the prevention of catheter-related colonization and bloodstream infections: a randomized, double-blind trial, The Texas Medical Center Catheter Study Group, *Ann. Intern. Med.*, 127, 267–274.
133. Johnson, J.R., Delavari, P., and Azar, M. (1999), Activities of a nitrofurazone containing urinary catheter and a silver hydrogel catheter against multidrug resistant bacteria characteristic of catheter-associated urinary tract infection, *Antimicrob. Agents Chemother.*, 43, 2990–2995.
134. Crnich, C.J. and Maki, D.G. (2002), The promise of novel technology for the prevention of intravascular device-related bloodstream infection, I: pathogenesis and short-term devices, *Clin. Infect. Dis.*, 34, 1232–1242.
135. Tambe, S.M., Sampath, L., and Modak, S.M. (2001), *In vitro* evaluation of the risk of developing bacterial resistance to antiseptics and antibiotics used in medical devices, *J. Antimicrob. Chemother.*, 47, 589–598.
136. Bach, A., Eberhardt, H., Frick, A., Schmidt, H., Bottinger, B.W., and Martin, E. (1999), Efficacy of silver-coating central venous catheters in reducing bacterial colonization, *Crit. Care Med.*, 27, 515–520.
137. Schierholz, J.M., Fleck, C., Beuth, J., and Pulverer, G. (2000), The antimicrobial efficacy of a new central venous catheter with long-term broad-spectrum activity, *J. Antimicrob. Chemother.*, 46, 45–50.
138. Kingston, D., Seal, D.V., and Hill, I.D. (1986), Self-disinfecting plastics for intravenous catheters and prosthetic inserts, *J. Hyg., (London)* 96 (2), 185–198.
139. Quesnel, L.B., Al Najjar, A.R., and Buddhavudhikrai, P. (1978), Synergism between chlorhexidine and sulphadiazine, *J. Appl. Bacteriol.*, 45, 397–405.
140. Heard, S.O., Wagle, M., Vijayakumar, E., McLean, S., Brueggemann, A., Napolitano, L.M., Edwards, L.P., O'Connell, F.M., Puyana, J.C., and Doern, G.V. (1998), Influence of triple-lumen central venous catheters coated with chlorhexidine and silver sulfadiazine on the incidence of catheter-related bacteraemia, *Arch. Intern. Med.*, 158, 81–87.
141. Bach, A., Schmidt, H., Bottiger, B., Schreiber, B., Bohrer, H., Motsch, J., Martin, E., and Sonntag, H.G. (1996), Retention of antibacterial activity and bacterial colonization of antiseptic-bonded central venous catheters, *J. Antimicrob. Chemother.*, 37, 315–322.
142. Hannan, M., Juste, R.N., Umasanker, S., Glendenning, A., Nightingale, C., Azadian, B., and Soni, N. (1999), Antiseptic-bonded central venous catheters and bacterial colonisation, *Anaesthesia*, 54, 868–872.

143. Ciresi, D.L., Albrecht, R.M., Volkers, P.A., and Scholten, D.J. (1996), Failure of antiseptic bonding to prevent central venous catheter-related infection and sepsis, *Am. Surg.*, 62, 641–646.

144. Pemberton, L.B., Ross, V., Cuddy, P., Kremer, H., Fessler, T., and McGurk, E. (1996), No difference in catheter sepsis between standard and antiseptic central venous catheters: a prospective randomized trial, *Arch. Surg.*, 131, 986–989.

145. George, S.J., Vuddamalay, P., and Boscoe, M.J. (1997), Antiseptic-impregnated central venous catheters reduce the incidence of bacterial colonization and associated infection in immunocompromised transplant patients, *Eur. J. Anaesthesiol.*, 14 (4), 428–431.

146. Logghe, C., Van Ossel, C., D'Hoore, W., Ezzedine, H., Wauters, G., and Haxhe, J.J. (1997), Evaluation of chlorhexidine and silver-sulfadiazine impregnated central venous catheters for the prevention of bloodstream infection in leukaemic patients: a randomized controlled trial, *J. Hosp. Infect.*, 37, 145–156.

147. Maki, D.G., Stolz, S.M., Wheeler, S., and Mermel, L.A. (1997), Prevention of central venous catheter-related bloodstream infection by use of an antiseptic-impregnated catheter: a randomized, controlled trial, *Ann. Intern. Med.*, 127, 257–266.

148. Tennenberg, S., Lieser, M., McCurdy, B., Boomer, G., Howington, E., Newman, C., and Wolf, I. (1997), A prospective randomized trial of an antibiotic- and antiseptic-coated central venous catheter in the prevention of catheter-related infections, *Arch. Surg.*, 132, 1348–1351.

149. Collin, G.R. (1999), Decreasing catheter colonization through the use of an antiseptic-impregnated catheter: a continuous quality improvement project, *Chest*, 115, 1632–1640.

150. Sheng, W.H., Ko, W.J., Wang, J.T., Chang, S.C., Hsueh, P.R., and Luh, K.T. (2000), Evaluation of antiseptic-impregnated central venous catheters for prevention of catheter-related infection in intensive care unit patients, *Diagn. Microbiol. Infect. Dis.*, 38, 1–5.

151. Brun-Buisson, C., Doyon, F., Sollet, J.P., Cochard, J.F., Cohen, Y., and Nitenberg, G. (2004), Prevention of intravascular catheter-related infection with newer chlorhexidine-silver-sulfadiazine-catheters, *Intensive Care Med.*, 30, 837–843.

152. Tattawasart, U., Maillard, J.Y., Furr, J.R., and Russell, A.D. (1999), Development of resistance to chlorhexidine diacetate and cetylpyridinium chloride in *Pseudomonas stutzeri* and changes in antibiotic susceptibility, *J. Hosp. Infect.*, 42, 219–229.

153. Oda, T., Hamasaki, J., Kanda, N., and Mikami, K. (1997), Anaphylactic shock induced by an antiseptic-coated central venous [correction of nervous] catheter [published erratum appears in Anesthesiology 1998 Feb., 88 (2), 560], *Anesthesiology*, 87 (5), 1242–1244.

154. Soetevent, C., Klemm, P.L., Stalenhoef, A.F., and Bredie, S.J. (2004), Vascular graft infection in aortoiliac and aortofemoral bypass surgery: clinical presentation, diagnostic strategies and results of surgical treatment, *Neth. J. Med.*, 62, 446–452.

155. Malassiney, P., Goëau-Brissonniére, O., Coggia, M., and Pechére, J.C. (1996), Rifampicin loading of vascular grafts, *J. Antimicrob. Chemother.*, 37, 1121–1129.

156. Khardori, N. and Yassien, M. (1995), Biofilms in device related infections, *J. Ind. Microbiol.*, 15, 141–147.

157. Barton, A.J., Sagers, R.D., and Pitt, W.G. (1996), Bacterial adhesion to orthopedic implant polymers, *J. Biomed. Mater. Res.*, 30, 403–410.

158. Gotz, F. (2002), *Staphylococcus* and biofilms, *Mol. Microbiol.*, 43, 1367–1378.

159. Gristina, A.G. (1994), Implant failure and the immuno-incompetent fibroinflammatory zone, *Clin. Orthop. Relat. Res.*, 106–118.

160. Schmitt, D.D., Bandyk, D.F., Pequet, A.J., and Towne, J.B. (1986), Bacterial adherence to vascular prostheses. A determinant of graft infectivity, *J. Vasc. Surg.*, 3, 732–740.

161. Darouiche, R.O. (2001), Device-associated infections: a macroproblem that starts with microadherence, *Clin. Infect. Dis.*, 33, 1567–1572.

162. Bergamini, T.M., McCurry, T.M., Bernard, J.D., Hoeg, K.L., Corpus, R.A., James, B.E., Peyton, J.C., Brittian, K.R., and Cheadle, W.G. (1996), Antibiotic efficacy against *Staphylococcus epidermidis* adherent to vascular grafts, *J. Surg. Res.*, 60, 3–6.

163. Camiade, C., Goldschmidt, P., Koskas, F., Ricco, J.B., Jarraya, M., Gerota, J., and Kieffer, E. (2001), Optimization of the resistance of arterial allografts to infection: comparative study with synthetic prostheses, *Ann. Vasc. Surg.*, 15, 186–196.

164. Greco, R.S., Harvey, R.A., Smilow, P.C., and Tesoriero, J.V. (1982), Prevention of vascular prosthetic infection by a benzalkonium-oxacillin bonded polytetrafluoroethylene graft, *Surg. Gynecol. Obstet.*, 155, 28–32.

165. Harvey, R.A. and Greco, R.S. (1981), The noncovalent bonding of antibiotics to a polytetrafluoroethylene-benzalkonium graft, *Ann. Surg.*, 194, 642–647.

166. Harvey, R.A., Alcid, D.V., and Greco, R.S. (1982), Antibiotic bonding to polytetrafluoroethylene with tridodecylmethylammonium chloride, *Surgery*, 92, 504–512.

167. Richardson, R.L., Jr., Pate, J.W., Wolf, R.Y., Ledes, C., and Hopson, W.B., Jr. (1970), The outcome of antibiotic-soaked arterial grafts in guinea pig wounds contaminated with *E. coli* or *S. aureus*, *J. Thorac. Cardiovasc. Surg.*, 59, 635–637.

168. Blanchemain, N., Haulon, S., Martel, B., Traisnel, M., Morcellet, M., and Hildebrand, H.F. (2005), Vascular PET prostheses surface modification with cyclodextrin coating: development of a new drug delivery system, *Eur. J. Vasc. Endovasc. Surg.*, 29, 628–632.

169. Chervu, A., Moore, W.S., Gelabert, H.A., Colburn, M.D., and Chvapil, M. (1991), Prevention of graft infection by use of prostheses bonded with a rifampin/collagen release system, *J. Vasc. Surg.*, 14, 521–524 discussion 524–525.

170. Galdbart, J.O., Branger, C., Andreassian, B., Lambert-Zechovsky, N., and Kitzis, M. (1996), Elution of six antibiotics bonded to polyethylene vascular grafts sealed with three proteins, *J. Surg. Res.*, 66, 174–178.

171. Goeau-Brissonniere, O., Leport, C., Bacourt, F., Lebrault, C., Comte, R., and Pechère, J.C. (1991), Prevention of vascular graft infection by rifampin bonding to a gelatin-sealed Dacron graft, *Ann. Vasc. Surg.*, 5, 408–412.

172. Avramovic, J.R. and Fletcher, J.P. (1991), Rifampicin impregnation of a protein-sealed Dacron graft: an infection-resistant prosthetic vascular graft, *Aust. N.Z. J. Surg.*, 61, 436–440.

173. Strachan, C.J., Newsom, S.W., and Ashton, T.R. (1991), The clinical use of an antibiotic-bonded graft, *Eur. J. Vasc. Surg.*, 5, 627–632.

174. Ginalska, G., Osinska, M., Uryniak, A., Urbanik-Sypniewska, T., Belcarz, A., Rzeski, W., and Wolski, A. (2005), Antibacterial activity of gentamicin-bonded gelatin-sealed polyethylene terephthalate vascular prostheses, *Eur. J. Vasc. Endovasc. Surg.*, 29, 419–424.

175. Ginalska, G., Kowalczuk, D., and Osinska, M. (2007), Amikacin-loaded vascular prosthesis as an effective drug carrier, *Int. J. Pharm.*, 339, 39–46.

176. Blanchemain, N., Haulon, S., Boschin, F., Marcon-Bachari, E., Traisnel, M., Morcellet, M., Hildebrand, H.F., and Martel, B. (2007), Vascular prostheses with controlled release of antibiotics Part1: Surface modification with cyclodextrins of PET prostheses, *Biomol. Eng.*, 24, 149–153.

177. Blanchemain, N., Haulon, S., Boschin, F., Traisnel, M., Morcellet, M., Martel, B., and Hildebrand, H.F. (2007), Vascular prostheses with controlled release of antibiotics Part 2. *In vitro* biological evaluation of vascular prostheses treated by cyclodextrins, *Biomol. Eng.*, 24, 143–148.

178. Matl, F.D., Obermeier, A., Repmann, S., Friess, W., Stemberger, A., and Kuehn, K.D. (2008), New anti-infective coatings of medical implants, *Antimicrob. Agents Chemother.*, 52, 1957–1963.

179. Gosheger, G., Hardes, J., Ahrens, H., Streitburger, A., Buerger, H., Erren, M., Günsel, A., Kemper, F.H., Winkelmann, W., and Von Eiff, C. (2004), Silver-coated megaendoprostheses in a rabbit model: an analysis of the infection rate and toxicological side effect, *Biomaterials*, 25, 5547–5556.

180. Sioshansi, P. (1994), New processes for surface treatment of catheters, *Artif. Organs*, 18, 266–271.

181. Mclean, R.J., Hussain, A.A., Sayer, M., Vincent, P.J., Hughes, D.J., and Smith, T.J. (1993), Antibacterial activity of multilayer silver–copper surface film on catheter material, *Can. J. Microbiol.*, 39, 895–899.

182. Davenas, J., Thevenard, P., Philippe, F., and Arnaud, M.N. (2002), Surface implantation treatments to prevent infection complications in short term devices, *Biomol. Eng.*, 19, 263–268.

183. Gatter, N., Kohnen, W., and Jansen, B. (1998), *In vitro* efficacy of hydrophilic central venous catheter loaded with silver to prevent microbial colonization, *Zentralbl Bakteriol.*, 287, 157–169.

184. Jansen, B., Rinck, M., Wolbring, P., Strohmeier, A., and Jahns, T. (1994), *In vitro* evaluation of the antimicrobial efficacy and biocompatibility of a silver-coated central venous catheter, *J. Biomater. Appl.*, 9, 55–70.

185. Böswald, M., Lugauer, S., Regenfus, A., Braun, G.G., Martus, P., Geis, C., Scharf, J., Bechert, T., Greil, J., and Guggenbichler, J.P. (1999), Reduced rates of catheter-associated infection by use of a new silver-impregnated central venous catheter, *Infection*, 27 Suppl. 1, S56–S60.

186. Ranucci, M., Isgro, G., Giomarelli, P.P., Pavesi, M., Luzzani, A., Cattabriga, I., Carli, M., Giomi, P., Compostella, A., Digito, A., Mangani, V., Silvestri, V., and Mondelli, E. (2003), Catheter Related Infection Trial (CRIT) Group, Impact of Oligon central venous catheters on catheter colonization and catheter-related bloodstream infection, *Crit. Care Med.*, 31, 52–59.

187. Liu, W.K., Tebbs, S.E., Byrne, P.O., and Elliott, T.S. (1993), The effect of electric current on bacteria colonising intravenous catheters, *J. Infect.*, 27, 261–269.

188. Raad, I., Hachem, R., Zermeno, A., Dumo, M., and Bodey, G.P. (1996), *In vitro* antimicrobial efficacy of silver iontophoretic catheter, *Biomaterials*, 17, 1055–1059.

189. Goldschmidt, H., Hahn, U., Salwender, H.J., Haas, R., Jansen, B., Wolbring, P., Rinck, M., and Hunstein, W. (1995), Prevention of catheter-related infection by silver coated central venous catheter in oncological patients, *Zentralbl Bakteriol.*, 283, 215–23.

190. Carbon, R.T., Lugauer, S., Geitner, U., Regenfus, A., Böswald, M., Greil, J., Bechert, T., Simon, S.I., Hümmer, H.P., and Guggenbichler, J.P. (1999), Reducing catheter associated infections with silver-impregnated catheters in long-terms therapy of children, *Infection*, 27 Suppl. 1, S69–S73.

191. Stoiser, B., Kofler, J., Staudinger, T., Georgopoulos, A., Lugauer, S., Guggenbichler, J.P., Burgmann, H., and Frass, M. (2002), Contamination of central venous catheters in immunocompromised patients; a comparison between two different types of central venous catheters, *J. Hosp. Infect.*, 50, 202–206.

192. Brayner, R., Ferrari-Iliou, R., Brivois, N., Djediat, S., Benedetti, M.F., and Fievet, F. (2006), Toxicological impact studies based on Escherichia coli bacteria in ultrafine ZnO nanoparticles colloidal medium, *Nano Lett.*, 6, 866–870.

193. Jones, N., Ray, B., Ranjit, K.T., and Manna, A.C. (2008), Antibacterial activity of ZnO nanoparticle suspensions on a broad spectrum of microorganisms, *FEMS Microbiol. Lett.*, 279, 71–76.

194. Norman, R.S., Stone, J.W., Gole, A., Murphy, C.J., and Sabo-Attwood, T.L. (2008), Targeted photothermal lysis of the pathogenic bacteria, *Pseudomonas aeruginosa*, with gold nanorods, *Nano Lett.*, 8, 302–306.

195. Roe, D., Karandikar, B., Bonn-Savage, N., Gibbins, B., and Roullet, J.B. (2008), Antimicrobial surface functionalization of plastic catheters by silver nanoparticles, *J. Antimicrob. Chemother.*, 61, 869–876.

196. Flemming, R.G., Capelli, C.C., Cooper, S.L., and Proctor, R.A. (2000), Bacterial colonization of functionalized polyurethanes, *Biomaterials*, 21, 273–281.

197. Raad, I., Reitzel, R., Jiang, Y., Chemaly, R.F., Dvorak, T., and Hachem, R. (2008), Anti-adherence activity and antimicrobial durability of anti-infective-coated catheters against multidrug-resistant bacteria, *J. Antimicrob. Chemother.*, 62, 746–750.

198. Shi, Z.L., Neoh, K.G., Kang, E.T., and Wang, W. (2006), Antibacterial and mechanical properties of bone cement impregnated with chitosan nanoparticles, *Biomaterials*, 27, 2440–2449.

199. Chang, Q.Y., He, H., and Ma, Z.C. (2008), Efficient disinfection of *Escherichia coli* in water by silver loaded alumina, *J. Inorg. Biochem.*, 102, 1736–1742.

200. Beyth, N., Houri-Haddad, Y., Baraness-Hadar, L., Yudovin-Farber, I., Domb, A.J., and Weiss, E.I. (2008), Surface antimicrobial activity and biocompatibility of incorporated polyethylenimine nanoparticles, *Biomaterials*, 29, 4157–4163.

201. Marquis, R.E. (1995), Antimicrobial actions of fluoride for oral bacteria, *Can. J. Microbiol.*, 41, 955–964.

202. Marquis, R.E., Clock, S.A., and Mota-Meira, M. (2003), Fluoride and organic weak acids as modulators of microbial physiology, *FEMS Microbiol. Rev.*, 26, 493–510.

203. Sturr, M.G. and Marquis, R.E. (1990), Inhibition of proton-translocating ATPases of *Streptococcus mutans* and *Lactobacillus casei* by fluoride and aluminum, *Arch. Microbiol.*, 155, 22–27.

204. Guha-Chowdhury, N., Iwami, Y., and Yamada, T. (1997), Effect of low levels of fluoride on proton excretion and intracellular pH in glycolysing streptococcal cells under strictly anaerobic conditions, *Caries Res.*, 31, 373–378.

205. Guha-Chowdhury, N., Clark, A.G., and Sissons, C.H. (1997), Inhibition of purified enolases from oral bacteria by fluoride, *Oral Microbiol. Immunol.*, 12, 91–97.

206. Lellouche, J., Kahana, E., Elias, S., Gedanken, A., and Banin, E. (2009), Antibiofilm activity of nanosized magnesium fluoride, *Biomaterials*, 30, 5969–5978.

207. Banin, E., Brady, K.M., and Greenberg, E.P. (2006), Chelator-induced dispersal and killing of *Pseudomonas aeruginosa* cells in a biofilm, *Appl. Environ. Microbiol.*, 72, 2064–2069.

208. Roy, B. and Datta, A.A. (1987), Calmodulin inhibitor blocks morphogenesis in *Candida albicans*, *FEMS Microbiol Lett.*, 41, 327–329.

209. Sarkisova, S., Patrauchan, M., Berglund, D., Nivens, D.E., and Franklin, M.J. (2005), Calcium-induced virulence factors associated with extracellular matrix of mucoid *Pseudomonas aeruginosa* biofilms, *J. Bacteriol.*, 187, 4327–4337.

210. Dunne, W. and Burd, E. (1992), The effects of magnesium, calcium, EDTA, and pH on the in vitro adhesion of *Staphylococcus epidermidis* to plastic, *Microbiol Immunol.*, 36, 1016–1027.

211. Ozerdem Akpolat, N., Elci, S., Atmaca, S., Akbayin, H., and Gül, K. (2003), The effects of magnesium, calcium and EDTA on slime production by *Staphylococcus epidermidis* strains, *Folia Microbiol. (Praha)*, 48, 649–653.

212. Brubaker, R. (1985), Mechanisms of bacterial virulence, *Annu. Rev. Microbiol.*, 39, 21–50.

213. Ratledge, C. and Dover, L. (2000), Iron metabolism in pathogenic bacteria, *Annu. Rev. Microbiol.*, 54, 881–941.

214. Weinberg, E.D. (2004), Suppression of bacterial biofilm formation by iron limitation, *Med. Hypotheses*, 63, 863–865.

215. Banin, E., Vasil, M., and Greenberg, E. (2005), Iron and *Pseudomonas aeruginosa* biofilm formation, *Proc. Natl. Acad. Sci. USA.*, 102, 11076–11081.

216. Rhodes, E., Shoemaker, C., Menke, S.M., Edelmann, R.E., and Actis, L.A. (2007), Evaluation of different iron sources and their influence in biofilm formation by the dental pathogen *Actinobacillus actinomycetemcomitans*, *J. Med. Microbiol.*, 56(Pt 1), 119–128.

217. Singh, P.K., Parsek, M.R., Greenberg, E.P., and Welsh, M.J. (2002), A component of innate immunity prevents bacterial biofilm development, *Nature (London)*, 417, 552–555.

218. Shah, C.B., Mittelman, M.W., Costerton, J.W., Parenteau, S., Pelak, M., Arsenault, R., and Mermel, L.A. (2002), Antimicrobial activity of a novel catheter lock solution, *Antimicrob. Agents Chemother.*, 46, 1674–1679.

219. Root, J.L., McIntyre, O.R., Jacobs, N.J., and Daghlian, C.P. (1988), Inhibitory effect of disodium EDTA upon the growth of Staphylococcus epidermidis *in vitro*:

relation to infection prophylaxis of Hickman catheters, *Antimicrob. Agents Chemother.*, 32, 1627–1631.

220. Percival, S.L., Kite, P., Eastwood, K., Murga, R., Carr, J., Arduino, M.J., and Donlan, R.M. (2005), Tetrasodium EDTA as a novel central venous catheter lock solution against biofilm, *Infect. Control Hosp. Epidemiol.*, 26, 515–519.
221. Raad, I., Chatzinikolaou, I., Chaiban, G., Hanna, H., Hachem, R., Dvorak, T., Cook, G., and Costerton, W. (2003), *In vitro* and *ex vivo* activities of minocycline and EDTA against microorganisms embedded in biofilm on catheter surfaces, *Antimicrob. Agents Chemother.*, 47, 3580–3585.
222. Chen, X. and Stewart, P.S. (2000), Biofilm removal caused by chemical treatments, *Water Res.*, 34, 4229–4233.
223. Raad, I., Hachem, R., Tcholakian, R.K., and Sherertz, R. (2002), Efficacy of minocycline and EDTA lock solution in preventing catheter-related bacteremia, septic phlebitis, and endocarditis in rabbits, *Antimicrob. Agents Chemother.*, 46, 327–332.
224. Gil, M.L., Casanova, M., and Martínez, J.P. (1994), Changes in the cell wall glycoprotein composition of Candida albicans associated to the inhibition of germ tube formation by EDTA, *Arch. Microbiol.*, 161, 489–494.
225. Bergan, T., Klaveness, J., and Aasen, A.J. (2001), Chelating agents, *Chemotherapy*, 47, 10–14.
226. Vineeta, N., Singh, V., and Makkar, S. (2001), Anti microbial activity of dimercaptosuccinic acid (DMSA): a new chelating agent, *J. Indian Soc. Pedod. Prev. Dent.*, 19, 160–163.
227. Rose, R. (2000), The role of calcium in oral streptococcal aggregation and the implications for biofilm formation and retention, *Biochim. Biophys. Acta.*, 1475, 76–82.
228. Shanks, R.M., Sargent, J.L., Martinez, R.M., Graber, M.L., and O'Toole, G.A. (2006), Catheter lock solutions influence staphylococcal biofilm formation on abiotic surfaces, *Nephrol. Dial. Transplant.*, 21, 2247–2255.
229. Sherertz, R.J., Boger, M.S., Collins, C.A., Mason, L., and Raad, I.I. (2006), Comparative *in vitro* efficacies of various catheter lock solutions, *Antimicrob. Agents Chemother.*, 50, 1865–1868.
230. Raad, I., Hanna, H., Dvorak, T., Chaiban, G., and Hachem, R. (2007), Optimal antimicrobial catheter lock solution, using different combinations of minocycline, EDTA, and 25-percent ethanol, rapidly eradicates organisms embedded in biofilm, *Antimicrob. Agents Chemother.*, 51, 78–83.
231. Allon, M. (2003), Prophylaxis against dialysis catheter-related bacteremia with a novel antimicrobial lock solution, [Comments in: Clin Infect. Dis. 2004 June 1:38(11):1641; author reply 1641–1642] *Clin. Infect. Dis.*, 36, 1539–1544.
232. Betjes, M.G. and van Agteren, M. (2004), Prevention of dialysis catheter-related sepsis with a citrate-taurolidine-containing lock solution, *Nephrol. Dial. Transplant.*, 19, 1546–1551.
233. Weijmer, M.C., Debets-Ossenkopp, Y.J., Van De Vondervoort, F.J., and ter Wee, P.M. (2002), Superior antimicrobial activity of trisodium citrate over heparin for catheter locking, *Nephrol. Dial. Transplant.*, 17, 2189–2195.
234. Weijmer, M.C., van den Dorpel, M.A., Van de Ven, P.J., ter Wee, P.M., van Geelen, J.A., Groeneveld, J.O., van Jaarsveld, B.C., Koopmans, M.G., le Poole, C.Y.,

Schrander-Van der Meer, A.M., Siegert, C.E., and Stas, K.J. CITRATE Study Group, (2005), Randomized, clinical trial comparison of trisodium citrate 30% and heparin as catheter-locking solution in hemodialysis patients, *J. Am. Soc. Nephrol.*, 16, 2769–2777.

235. Guha, A., Choudhury, A., Unni, B.G., and Roy, M.K. (2002), Effect of outer-membrane permeabilizers on the activity of antibiotics and plant extracts against *Pseudomonas aeruginosa*, *Folia Microbiol.* (Praha), 47, 379–384.

236. Ardehali, R., Shi, L., Janatova, J., Mohammad, S.F., and Burns, G.L. (2002), The effect of apo-transferrin on bacterial adhesion to biomaterials, *Artif. Organs*, 26, 512–520.

237. Arakawa, Y., Saito, T., Saito, H., Kakegawa, T., and Kobayashi, H. (2000), Ethylene glycol bis (beta-aminoethyl ether)-*N,N,N′,N′*-tetraacetic acid (EGTA) inhibits the growth of *Escherichia coli* at alkaline pH, *J. Gen. Appl. Microbiol.*, 46, 127–131.

238. van Asbeck, B.S., Marcelis, J.H., Marx, J.J., Struyvenberg, A., van Kats, J.H., and Verhoef, J. (1983), Inhibition of bacterial multiplication by the iron chelator deferoxamine: potentiating effect of ascorbic acid, *Eur. J. Clin. Microbiol.*, 2, 426–431.

239. Zhang H., Tang, J., Meng X., Tsang, J., and Tsang, T.K. (2005), Inhibition of bacterial adherence on the surface of stents and bacterial growth in bile by bismuth dimercaprol, *Dig. Dis. Sci.*, 50, 1046–1051.

240. Ibrahim, A., Gebermariam, T., Fu, Y., Lin, L., Husseiny, M.I., French, S.W., Schwartz, J., Skory, C.D., Edwards, J.E., Jr., and Spellberg, B.J. (2007), The iron chelator deferasirox protects mice from mucormycosis through iron starvation, *J. Clin. Invest.*, 117, 2649–2657.

241. Ibrahim, A., Edwards, J., Fu, J., and Spellberg, B. (2006), Deferiprone iron chelator as a novel therapy for experimental mucormycosis, *J. Antimicrob. Chemother.*, 58, 1070–1073.

242. Flemming, C.A., Palmer, R.J., Jr., Arrage, A.A., van der Mei, H.C., and White, D.C. (1998), Cell surface physiochemistry alters biofilm development of *Pseudomonas aeruginosa* lipopolysaccharide mutants, *Biofouling*, 13, 213–231.

243. Raad, I., Buzaid, A., Rhyne, J., Hachem, R., Darouiche, R., Safar, H., Albitar, M., and Sherertz, R.J. (1997), Minocycline and ethylenediaminetetraacetate for the prevention of recurrent vascular catheter infections, *Clin. Infect. Dis.*, 25, 149–151.

244. Bleyer, A.J., Mason, L., Russell, G., Raad, I.I., and Sherertz, R.J. (2005), A randomized, controlled trial of a new vascular catheter flush solution (minocycline-EDTA) in temporary hemodialysis access, [Infect. Control Hosp. Epidemiol., 2005 June, 26(6):511–514], *Infect. Control Hosp. Epidemiol.*, 26, 520–524.

245. Chatzinikolaou, I., Zipf, T.F., Hanna, H., Umphrey, J., Roberts, W.M., Sherertz, R., Hachem, R., and Raad, I. (2003), Minocycline-ethylenediaminetetraacetate lock solution for the prevention of implantable port infections in children with cancer, *Clin. Infect. Dis.*, 36, 116–119.

246. McGovern, P., Comher, C., Evans-Stoner, N., and Raad, I., Minocycline-disodium EDTA lock solution in a long-term parenteral nutrition population [abstract], In: The 16th Annual Scientific Meeting of the Society for Healthcare and Epidemiology of America, Chicago, IL, 18–21 March 2006.

247. Dogra, G.K., Herson, H., Hutchison, B., Irish, A.B., Heath, C.H., Golledge, C., Luxton, G., and Moody, H. (2002), Prevention of tunneled hemodialysis catheter-related infections using catheter-restricted filling with gentamicin and citrate: a randomized controlled study, *J. Am. Soc. Nephrol.*, 13, 2133–2139.

248. Dannenberg, C., Bierbach, U., Rothe, A., Beer, J., and Körholz, D. (2003), Ethanol-lock technique in the treatment of bloodstream infections in pediatric oncology patients with broviac catheter, *J. Pediatr. Hematol. Oncol.*, 25, 616–621.

249. Shanks, R.M., Donegan, N.P., Graber, M.L., Buckingham, S.E., Zegans, M.E., Cheung, A.L., and O'Toole, G.A. (2005), Heparin stimulates *Staphylococcus aureus* biofilm formation, *Infect. Immun.*, 73, 4596–4606.

250. Chandra, J., Mukherjee, P., and Ghannoum, M. Optimization of antibiofilm lock solution for activity against fungal and bacterial biofilms [abstract], in: The 47th Annual Interscience Conference on Antimicrobial Agents and Chemotherapy (ICAAC), Chicago, IL, 17–20 Sept. 2007.

251. Chandra, J., Mukherjee, P., and Ghannoum, M. B-Lock is superior to citrate-based lock therapy in prevention and treatment of bacterial and fungal biofilms [abstract], In: The 47th Annual Interscience Conference on Antimicrobial Agents and Chemotherapy (ICAAC), Chicago, IL, 17–20 Sept. 2007.

252. Bestul M. and Vandenbussche, H. (2005), Antibiotic lock technique: review of the literature, *Pharmacotherapy*, 25, 211–227.

253. Capdevila, J.A., Gavalda, J., Fortea, J., López, P., Martin, M.T., Gomis, X., and Pahissa, A. (2001), Lack of antimicrobial activity of sodium heparin for treating experimental catheter related infection due to *Staphylococcus aureus* using the antibiotic-lock technique, *Clin. Microbiol. Infect.*, 7, 206–212.

254. Arepally, G. and Cines, D. (1998), Heparin-induced thrombocytopenia and thrombosis, *Clin. Rev. Allergy Immunol.*, 16, 237–247.

255. Fallgren, C., Utt, M., and Ljungh, A. (2001), Isolation and characterization of a 17-kDa staphylococcal heparin-binding protein with broad specificity, *J. Med. Microbiol.*, 50, 547–557.

256. Liang, O., Ascencio, F., Fransson, L., and Wadström, T. (1992), Binding of heparin sulfate to *Staphylococcus aureus*, *Infect. Immun.*, 60, 899–906.

257. Raad, I.I., Fang, X., Keutgen, X.M., Jiang, Y., Sherertz, R., and Hachem, R. (2008), The role of chelators in preventing biofilm formation and catheter-related bloodstream infections, *Curr. Opin. Infect. Dis.*, 21, 385–392.

258. Blalock, T.D., Spurr-Michaud, S.J., Tisdale, A.S., Heimer, S.R., Gilmore, M.S., Ramesh, V., and Gipson, I.K. (2007), Functions of MUC16 in corneal epithelial cells, *Investig. Ophthalmol. Vis. Sci.*, 48, 4509–4518.

259. McAuley, J.L., Linden, S.K., Png, C.W., King, R.M., Pennington, H.L., Gendler, S.J., Florin, T.H., Hill, G.R., Korolik, V., and McGuckin, M.A. (2007), MUC1 cell surface mucin is a critical element of the mucosal barrier to infection, *J. Clin. Investig.*, 117, 2313–2324.

260. McGuckin, M.A., Every, A.L., Skene, C.D., Linden, S.K., Chionh, Y.T., Swierczak, A., McAuley, J., Harbour, S., Kaparakis, M., Ferrero, R., and Sutton, P. (2007), MUC1 mucin limits both *Helicobacter pylori* colonization of the murine gastric mucosa and associated gastritis, *Gastroenterology*, 133, 1210–1218.

261. Gendler, S.J. and Spicer, A.P. (1995), Epithelial mucin genes, *Annu. Rev. Physiol.*, 57, 607–634.

262. Sumiyoshi, M., Ricciuto, J., Tisdale, A., Gipson, I.K., Mantelli, F., and Argüeso, P. (2008), Antiadhesive character of mucin O-glycans at the apical surface of corneal epithelial cells, *Invest. Ophthalmol. Vis. Sci.*, 49, 197–203.

263. Gipson, I.K., Hori, Y., and Argueso, P. (2004), Character of ocular surface mucins and their alteration in dry eye disease, *Ocul. Surf.*, 2, 131–148.

264. Hilkens, J., Ligtenberg, M.J., Vos, H.L., and Litvinov, S.V. (1992), Cell membrane-associated mucins and their adhesion-modulating property, *Trends Biochem. Sci.*, 17, 359–363.

265. Van den Steen, P., Rudd, P.M., Dwek, R.A., and Opdenakker, G. (1998), Concepts and principles of O-linked glycosylation, *Crit. Rev. Biochem. Mol. Biol.*, 33, 151–208.

266. Costa, N.R., Mendes, N., Marcos, N.T., Reis, C.A., Caffrey, T., Hollingsworth, M.A., and Santos-Silva, F. (2008), Relevance of Muc1 mucin variable number of tandem repeats polymorphism in *H. pylori* adhesion to gastric epithelial cells, *World J. Gastroenterol.*, 14, 1411–1414.

267. Grange, P.A., Erickson, A.K., Anderson, T.J., and Francis, D.H. (1998), Characterization of the carbohydrate moiety of intestinal mucin-type sialoglycoprotein receptors for the K88ac fimbrial adhesin of *Escherichia coli*, *Infect. Immun.*, 66, 1613–1621.

268. Prescher, J.A. and Bertozzi, C.R. (2006), Chemical technologies for probing Glycans, *Cell*, 126, 851–854.

269. Huet, G., Hennebicq-Reig, S., de Bolos, C., Ulloa, F., Lesuffleur, T., Barbat, A., Carriere, V., Kim, I., Real, F.X., Delannoy, P., and Zweibaum, A. (1998), GalNAc-α-*O*-benzyl inhibits NeuAc α 2–3 glycosylation and blocks the intracellular transport of apical glycoproteins and mucus in differentiated HT-29 cells, *J. Cell Biol.*, 141, 1311–1322.

270. Tsuiji, H., Takasaki, S., Sakamoto, M., Irimura, T., and Hirohashi, S. (2003), Aberrant *O*-glycosylation inhibits stable expression of dysadherin, a carcinoma-associated antigen, and facilitates cell-cell adhesion, *Glycobiology*, 13, 521–527.

271. Bell, S.L., Khatri, I.A., Xu, G., and Forstner, J.F. (1998), Evidence that a peptide corresponding to the rat Muc2 C-terminus undergoes disulphidemediated dimerization, *Eur. J. Biochem.*, 253, 123–131.

272. Vyas, A.A., Blixt, O., Paulson, J.C., and Schnaar, R.L. (2005), Potent glycan inhibitors of myelin-associated glycoprotein enhance axon outgrowth *in vitro*, *J. Biol. Chem.*, 280, 16305–16310.

273. Ricciuto, J., Heimer, S.R., Gilmore, M.S., and Argüeso, P. (2008), Cell surface *O*-glycans limit *Staphylococcus aureus* adherence to corneal epithelial cells, *Infect. Immun.*, 76, 5215–5220.

274. Kaneko, Y., Thoendel, M., Olakanmi, O., Britigan, B.E., and Sink, P.K. (2007), The transition metal gallium disrupts *Pseudomonas aeruginosa* iron metabolism and has antimicrobial and antibiofilm activity, *J. Clin. Invest.*, 117, 877–888.

275. Hoffman, L.R., D'Argenio, D.A., MacCoss, M.J., Zhang, Z., Jones, R.A., and Miller, S.I. (2005), Aminoglycoside antibiotics induce bacterial biofilm formation, *Nature*, 436, 1171–1175.

276. Linares, J.F., Gustafsson, I., Baquero, F., and Martinez, J.L. (2006), Antibiotics as intermicrobial signaling agents instead of weapons, *Proc. Natl. Acad. Sci. USA*, 103, 19484–19489.

277. Overhage, J., Campisano, A., Bains, M., Torfs, E.C., Rehm, B.H., and Hancock, R.E. (2008), Human host defense peptide LL-37 prevents bacterial biofilm formation, *Infect. Immun.*, 76, 4176–4182.

278. Bowdish, D.M.E., Davidson, D.J., Scott, M.G., and Hancock, R.E.W. (2005), Immunomodulatory activities of small host defense peptides, *Antimicrob. Agents Chemother.*, 49, 1727–1732.

279. Bowdish, D.M.E., Davidson, D.J., Lau, Y.E., Lee, K., Scott, M.G., and Hancock, R.E.W. (2005), Impact of LL-37 on anti-infective immunity, *J. Leukoc. Biol.*, 77, 451–459.

280. Brown, K.L. and Hancock, R.E.W. (2006), Cationic host defense (antimicrobial) peptides, *Curr. Opin. Immunol.*, 18, 24–30.

281. Powers, J.P. and Hancock, R.E.W. (2003), The relationship between peptide structure and antibacterial activity, *Peptides*, 24, 1681–1691.

282. Andersen, J.H., Jenssen, H., and Gutteberg, T.J. (2003), Lactoferrin and lactoferricin inhibit herpes simplex 1 and 2 infection and exhibit synergy when combined with acyclovir, *Antiviral Res.*, 58, 209–215.

283. Tamilvanan, S., Venkateshan, N., and Ludwig, A., (2008), The potential of lipid- and polymer-based drug delivery carriers for eradicating biofilm consortia on device-related nosocomial infections, *J. Controlled Rel.*, 128, 2–22.

284. Smith, A.W. (2005), Biofilms and antibiotic therapy: Is there a role for combating bacterial resistance by the use of novel drug delivery systems? *Adv. Drug Del. Rev.*, 57, 1539–1550.

285. Donaldson, S.H., Bennett, W.D., Zeman, K.L., Knowles, M.R., Tarran, R., and Boucher, R.C. (2006), Mucus clearance and lung function in cystic fibrosis with hypertonic saline, *N. Engl. J. Med.*, 354, 241–250.

286. Elkins, M.R., Robinson, M., Rose, B.R., Harbour, C., Moriarty, C.P., Marks, G.B., Belousova, E.G., Xuan, W., and Bye, P.T. (2006), A controlled trial of long-term inhaled hypertonic saline in patients with cystic fibrosis, *N. Engl. J. Med.*, 354, 229–240.

287. Gibson, R.L., Burns, J.L., and Ramsey, B.W. (2003), Pathophysiology and management of pulmonary infections in cystic fibrosis, *Am. J. Respir. Crit. Care Med.*, 168, 918–951.

288. Beech, I.B., Sunner, J.A., and Hiraoka, K. (2005), Microbe-surface interactions in biofouling and biocorrosion processes, *Inter. Microbiol.*, 8, 157–168.

# INDEX



*Biofilm Eradication and Prevention: A Pharmaceutical Approach to Medical Device Infections*, By Tamilvanan Shunmugaperumal

www.ingramcontent.com/pod-product-compliance
Lightning Source LLC
LaVergne TN
LVHW011903200926
848484LV00002B/514
*9780470479964*